National Center for Construction Education and Research

Electrical Level Three

Annotated Instructor's Guide

Prentice Hall

Upper Saddle River, New Jersey
Columbus, Ohio

contren™
Learning Series

National Center for Construction Education and Research

President: Dan Bennet
Vice President of Training Operations and Program Development: Don Whyte
Director of Curriculum Revision and Development: Daniele Dixon
Electrical Project Manager: Daniele Dixon
Production Manager: Debie Ness
Editor: Tara Cohen
Desktop Publisher: Jessica Martin

The NCCER would like to acknowledge the contract service provider for this curriculum:
Topaz Publications, Liverpool, New York.

Pearson Education, Inc.

Editor in Chief: Stephen Helba
Product Manager: Lori Cowen
Production Editor: Stephen C. Robb
Design Coordinator: Karrie M. Converse-Jones
Text Design Concept: Rebecca Bobb
Cover Designer: Linda Smith
Copy Editor: Sheryl Rose
Scanning Coordinator: Karen L. Bretz
Scanning Technician: Janet Portisch
Production Manager: Pat Tonneman

This book was set in Palatino and Helvetica by Carlisle Communications, Ltd. It was printed and bound by Document Technology Resources. The cover was printed by Phoenix Color Corp.

Pearson Education Ltd.
Pearson Education Australia Pty. Limited
Pearson Education Singapore Pte. Ltd.
Pearson Education North Asia Ltd.
Pearson Education Canada, Ltd.
Pearson Educación de Mexico, S.A. de C.V.
Pearson Education—Japan
Pearson Education Malaysia Pte. Ltd.
Pearson Education, *Upper Saddle River, New Jersey*

This information is general in nature and intended for training purposes only. Actual performance of activities described in this manual requires compliance with all applicable operating, service, maintenance, and safety procedures under the direction of qualified personnel. References in this manual to patented or proprietary devices do not constitute a recommendation of their use.

---

10 9 8 7 6 5 4 3 2
ISBN 0-13-047225-5

Preface

This volume was developed by the National Center for Construction Education and Research (NCCER) in response to the training needs of the construction, maintenance, and pipeline industries. It is one of many in NCCER's *Contren™ Learning Series.* The program, covering training for close to 40 construction and maintenance areas, and including skills assessments, safety training, and management education, was developed over a period of years by industry and education specialists.

NCCER also maintains a National Registry that provides transcripts, certificates, and wallet cards to individuals who have successfully completed modules of NCCER's *Contren™ Learning Series,* when the training program is delivered by an NCCER Accredited Training Sponsor.

The NCCER is a not-for-profit 501(c)(3) education foundation established in 1995 by the world's largest and most progressive construction companies and national construction associations. It was founded to address the severe workforce shortage facing the industry and to develop a standardized training process and curricula. Today, NCCER is supported by hundreds of leading construction and maintenance companies, manufacturers, and national associations, including the following partnering organizations:

PARTNERING ASSOCIATIONS

- American Fire Sprinkler Association
- American Petroleum Institute
- American Society for Training & Development
- American Welding Society
- Associated Builders & Contractors, Inc.
- Association for Career and Technical Education
- Associated General Contractors of America
- Carolinas AGC, Inc.
- Carolinas Electrical Contractors Association
- Citizens Democracy Corps
- Construction Industry Institute
- Construction Users Roundtable
- Design-Build Institute of America
- Merit Contractors Association of Canada
- Metal Building Manufacturers Association
- National Association of Minority Contractors
- National Association of State Supervisors for Trade and Industrial Education
- National Association of Women in Construction
- National Insulation Association
- National Ready Mixed Concrete Association
- National Utility Contractors Association
- National Vocational Technical Honor Society
- North American Crane Bureau
- Painting & Decorating Contractors of America
- Portland Cement Association
- SkillsUSA–VICA
- Steel Erectors Association of America
- Texas Gulf Coast Chapter ABC
- U.S. Army Corps of Engineers
- University of Florida
- Women Construction Owners & Executives, USA

Some features of NCCER's *Contren™ Learning Series* are:

- An industry-proven record of success
- Curricula developed by the industry for the industry
- National standardization providing portability of learned job skills and educational credits
- Credentials for individuals through NCCER's National Registry
- Compliance with Apprenticeship, Training, Employer, and Labor Services (ATELS) requirements for related classroom training (CFR 29:29)
- Well-illustrated, up-to-date, and practical information

FEATURES OF THIS BOOK

Capitalizing on a well-received campaign to redesign our textbooks, NCCER and Prentice Hall are continuing to publish select textbooks in color. *Electrical Level Three* incorporates the new design and layout, along with color photos and illustrations, to present the material in an easy-to-use format. This edition has been fully updated to meet the 2002 *National Electrical Code*®.

Special pedagogical features augment the technical material to maintain the trainees' interest and foster a deeper appreciation of the trade.

Inside Track provides a head start for those entering the field by presenting tricks of the trade from master electricians in a variety of disciplines.

Think About It uses "What If?" questions to help trainees apply theory to real-world experiences and put ideas into action.

What's Wrong with This Picture? includes photos of actual code violations for identification and encourages trainees to approach each installation with a critical eye.

Case Histories emphasize the importance of safety by citing examples of the costly (and often devastating) consequences of ignoring NEC or OSHA regulations.

We're excited to be able to offer you these improvements and hope they lead to a more rewarding learning experience. As always, your feedback is welcome. Please let us know how we are doing by visiting NCCER at www.nccer.org or e-mail us at info@nccer.org.

ANNOTATED INSTRUCTOR'S GUIDES—DESIGNED TO MAKE YOUR JOB EASIER!

We understand that the course material you choose may have a profound effect on your success. The best way we can help to ensure your success is to provide you with the finest tools we have available. Since 2000, NCCER and Prentice Hall have released select titles as Annotated Instructor's Guides. Each Annotated Instructor's Guide is actually the Trainee Module enhanced with specific directions to the instructor, space for the instructor's notes, suggestions for session break-outs, a comprehensive Materials and Equipment list, and teaching tips to coincide with the performance examinations, laboratories, and demonstrations.

The *Electrical Level Three Annotated Instructor's Guide* is packaged with an accompanying test booklet that includes the performance tests, written module exams, and answer keys. Also included is the Instructor's version of the Trainee Workbook. This supplemental workbook has practice exercises that can be assigned as homework or used to complement classroom instruction. Please note that the Transparency Masters package and Test Scrambling Software are sold separately for an additional charge.

For more information on the Annotated Instructor's Guides or to find out which titles are planned for annotations in the future, please contact your Prentice Hall Sales Specialist or NCCER Customer Service at info@nccer.org.

NOTES

The designations "National Electrical Code," "NE Code," and "NEC," where used in this document, refer to the *National Electrical Code*®, which is a registered trademark of the National Fire Protection Association, Quincy, MA. All National Electrical Code (NEC) references in this module refer to the 2002 edition of the NEC.

If you feel that additional math instruction would be helpful, Prentice Hall offers a basic math textbook entitled *Fundamentals of Electrical and Mechanical Mathematics*. It covers the basic math requirements for electrical trainees and may be ordered by contacting Prentice Hall Customer Service at 1-800-922-0579.

Acknowledgments

This curriculum was revised as a result of the farsightedness and leadership of the following organizations:

Beacon Electric Company
Circle Electric, Inc.
Cypress Electric, an IES Company
Duck Creek Engineering
Encompass Electrical Technologies –Florida, LLC
Leading Edge Electrical Services
Pelican Chapter ABC
San Diego ABC
TIC, The Industrial Company
Washington CITC
Washington Rust Constructors, Inc.
Wisconsin ABC
Zachry Construction

This curriculum would not exist were it not for the dedication and unselfish energy of those volunteers who served on the Authoring Team. A sincere thanks is extended to:

Clarence "Ed" Cockrell
Randy Cole
Gary Edgington
Tim Ely
E.L. Jarrell
Dan Lamphear
Leonard R. "Skip" Layne
L.J. LeBlanc
Daniel M. LePage
Neil Matthes
Jim Mitchem
Robert Mueller
Christine Porter
Mike Powers

Contents

Load Calculations—Branch and Feeder Circuits

26301-02

Load Calculations—Branch and Feeder Circuits
Annotated Instructor's Guide

Module 26301-02

MODULE OVERVIEW

This course introduces the Electrical trainee to the load calculations and NEC requirements for branch and feeder circuits.

PREREQUISITES

Please refer to the Course Map in the Trainee Module. Prior to training with this module, it is recommended that the trainee shall have successfully completed the following:

Core Curriculum; Electrical Level One; Electrical Level Two.

OBJECTIVES

Upon completion of this module, the trainee will be able to:

1. Calculate loads for single-phase and three-phase branch circuits.
2. Size branch circuit overcurrent protection devices (circuit breakers and fuses) for noncontinuous duty and continuous duty circuits.
3. Apply derating factors to size branch circuits.
4. Calculate ampacity for single-phase and three-phase loads.
5. Use load calculations to determine branch circuit conductor sizes.
6. Use *NEC Table 220.19* to calculate residential cooking equipment loads.
7. Select branch circuit conductors and overcurrent protection devices for electric heat, air conditioning equipment, motors, and welders.

PERFORMANCE TASKS

This is a knowledge-based module; there is no performance examination.

NCCER STANDARDIZED CRAFT TRAINING PROGRAMS

The National Center for Construction Education and Research (NCCER) provides a standardized national program of accredited craft training. Key features of the program include instructor certification, competency-based training, and performance testing. The program provides trainees, instructors, and companies with a standard form of recognition through a National Craft Training Registry. The program is described in full in the *Guidelines for Accreditation,* published by the NCCER. For more information on standardized craft training, contact the NCCER by writing us at P.O. Box 141104, Gainesville, FL 32614-1104, calling 352-334-0911, or e-mailing info@nccer.org. More information may be found at our Web site, www.nccer.org.

HOW TO USE THIS ANNOTATED INSTRUCTOR'S GUIDE

Each page presents two sections of information. The larger section displays each page exactly as it appears in the Trainee Module. The narrow column ties suggested trainee and instructor actions to each page and provides icons to call your attention to materials, safety, audiovisual, or testing requirements. The bottom of each page includes space for your notes.

If you see the Teaching Tip icon, that means there is a teaching tip associated with this section. Also refer to the suggested teaching tips at the end of the module.

SAFETY CONSIDERATIONS

Ensure that the trainees are equipped with appropriate personal protective equipment.

PREPARATION

Before teaching this module, you should review the Module Outline, Objectives, and the Materials and Equipment List. Be sure to allow ample time to prepare your own training or lesson plan and gather all required equipment and materials.

MATERIALS AND EQUIPMENT LIST

Materials:
Transparencies
Markers/chalk
Copy of the latest edition of the *National Electrical Code®*
Module Examinations*

Equipment:
Overhead projector and screen
Whiteboard/chalkboard
Appropriate personal protective equipment
Calculator

*Located in the Test Booklet packaged with this Annotated Instructor's Guide.

ADDITIONAL RESOURCES

This module is intended to present thorough resources for task training. The following reference work is suggested for both instructors and motivated trainees interested in further study. This is optional material for continued education rather than for task training.

National Electrical Code Handbook. Quincy, MA: National Fire Protection Association, 2002.

TEACHING TIME FOR THIS MODULE

An outline for use in developing your lesson plan is presented below. Note that each Roman numeral in the outline equates to one session of instruction. Each session has a suggested time period of 2½ hours. This includes 10 minutes at the beginning of each session for administrative tasks and one 10-minute break during the session. Approximately 12½ hours are suggested to cover *Load Calculations—Branch and Feeder Circuits*. You will need to adjust the time required for hands-on activity and testing based on your class size and resources.

Topic	Planned Time
Session I. Introduction	
A. Introduction	________
1. Branch Circuit Ratings	________
2. Derating	________
a. Temperature Derating	________
b. Voltage Drop Derating for Single-Phase Circuits	________
c. Voltage Drop Derating for Three-Phase Circuits	________
3. Calculating Branch Circuit Ampacity	________
Session II. Lighting Loads; Receptacle Loads; Multi-Outlet Assemblies; Show Window Loads	
A. Lighting Loads	________
1. Recessed Lighting	________
2. Heavy-Duty Lamp Holder Outlets	________
B. Receptacle Loads	________
C. Multi-Outlet Assemblies	________
D. Show Window Loads	________

Session III. Sign Loads; Residential Branch Circuits

A. Sign Loads ____________

B. Residential Branch Circuits ____________

1. Small Appliance Load ____________
2. Laundry Circuit ____________
3. Dryers ____________
4. Cooking Appliances ____________

Session IV. Commercial Kitchen Equipment; Water Heaters; Electric Heating Loads; Air Conditioning Loads; Motor Loads; Welders

A. Commercial Kitchen Equipment ____________

B. Water Heaters ____________

C. Electric Heating Loads ____________

D. Air Conditioning Loads ____________

E. Motor Loads ____________

F. Welders ____________

Session V. Review; Module Examination

A. Review ____________

B. Module Examination ____________

1. Trainees must score 70% or higher to receive recognition from the NCCER.
2. Record the testing results on Craft Training Report Form 200 and submit the results to the Training Program Sponsor.

Module 26301-02

Load Calculations—Branch and Feeder Circuits

Course Map

This course map shows all of the modules in the third level of the Electrical curriculum. The suggested training order begins at the bottom and proceeds up. Skill levels increase as you advance on the course map. The local Training Program Sponsor may adjust the training order.

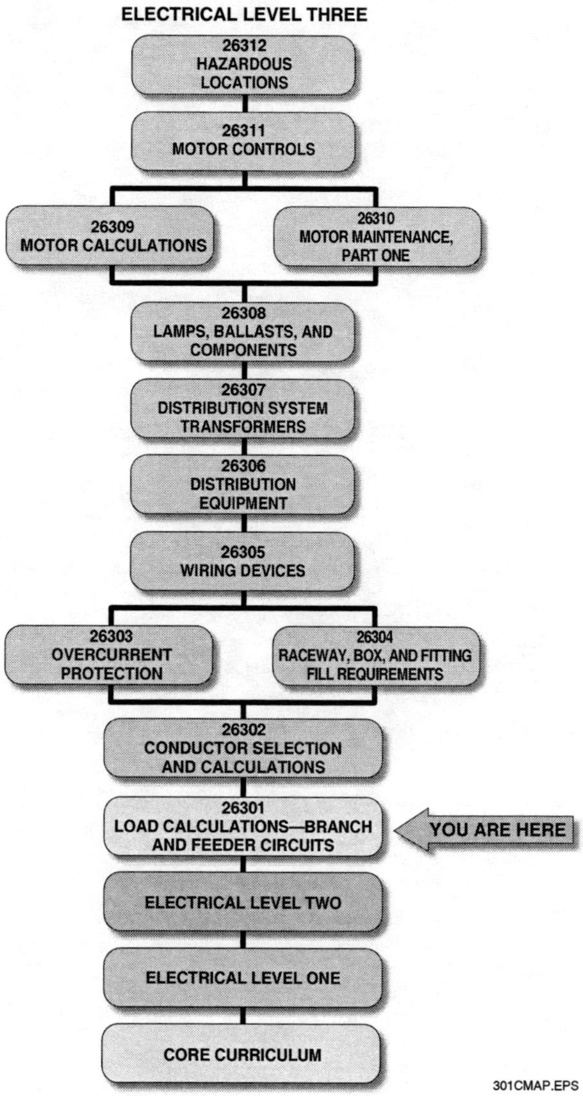

301CMAP.EPS

Homework

Assign reading of Module 26301.

MODULE 26301 CONTENTS

Figure

MODULE 26301

Load Calculations—Branch and Feeder Circuits

Ensure that you have everything required to teach the course. Check the Materials and Equipment List at the front of this Instructor's Guide.

Objectives

When you have completed this module, you will be able to do the following:

1. Calculate loads for single-phase and three-phase branch circuits.
2. Size branch circuit overcurrent protection devices (circuit breakers and fuses) for noncontinuous duty and continuous duty circuits.
3. Apply derating factors to size branch circuits.
4. Calculate ampacity for single-phase and three-phase loads.
5. Use load calculations to determine branch circuit conductor sizes.
6. Use *NEC Table 220.19* to calculate residential cooking equipment loads.
7. Select branch circuit conductors and overcurrent protection devices for electric heat, air conditioning equipment, motors, and welders.

Prerequisites

Before you begin this module, it is recommended that you successfully complete the following: Core Curriculum; Electrical Level One; Electrical Level Two.

Required Trainee Materials

1. Paper and pencil
2. Appropriate personal protective equipment
3. Copy of the latest edition of the *National Electrical Code*

1.0.0 ◆ INTRODUCTION

The purpose of **branch circuit** load calculations is to determine the size of branch circuit **overcurrent** protection **devices** and branch circuit conductors using *National Electrical Code* (NEC) requirements. Once the branch circuit load is accurately calculated, branch circuit components may be sized to serve the load safely. Branch circuits supply **utilization equipment**. Utilization equipment is defined by the NEC as equipment that utilizes electric energy.

NEC Article 210 covers branch circuits (except for branch circuits that supply only motor loads). ***NEC Section 210.2*** provides a listing of other code articles for specific-purpose branch circuits. Per ***NEC Section 210.3***, branch circuits are rated by the maximum rating or setting of the overcurrent device. Except for circuits serving individual utilization equipment (dedicated circuits), branch circuits shall be rated 15A, 20A, 30A, 40A, and 50A. Branch circuits designed to serve individual loads can supply any size load with no restrictions to the ampere rating of the circuit.

Per ***NEC Section 210.19***, branch circuit conductors are required to be sized with an **ampacity** rating that is no less than the maximum load to be served. Branch circuit overcurrent protection is required to have a rating or setting not exceeding the rating specified in ***NEC Section 240.4*** for conductors, ***NEC Section 240.3*** for equipment, and ***NEC Section 210.21*** for **outlet** devices including lamp holders and **receptacles.** ***NEC Article 430*** applies to branch circuits supplying only motor loads and ***NEC Article 440*** applies to branch circuits supplying only air conditioning equipment, refrigerating equipment, or both.

Review the goals of the course.

Explain that terms shown in bold (blue) are defined in the Glossary at the back of this module.

Show Transparency 1, Course Objectives.

Explain the purpose of branch circuit calculations. Point out the applicable sections of the NEC.

Note: The designations "National Electrical Code," "NE Code," and "NEC," where used in this document, refer to the *National Electrical Code®*, which is a registered trademark of the National Fire Protection Association, Quincy, MA. *All National Electrical Code (NEC) references in this module refer to the 2002 edition of the NEC.*

Teaching Tip

Discuss the "Think About It" features.

Classroom

Emphasize that the NEC provides only the minimum requirements for branch circuit calculations. Note that local codes often contain additional requirements.

Demonstrate how to calculate the maximum load on a branch circuit.

Explain the difference between continuous and noncontinuous loads. Point out the applicable sections of the NEC.

THINK ABOUT IT

Load Calculations

What factors need to be taken into consideration when adding a large load to an existing service?

THINK ABOUT IT

Unit Load per Square Foot

Referring to ***NEC Table 220.3(A)***, what is the unit load per square foot for an office building where the actual number of general-purpose receptacle outlets is unknown?

Branch circuit conductors must have an ampacity rating equal to, or greater than, the noncontinuous load plus 125% of the **continuous load** before the application of any adjustment or correction factors per *NEC Section 210.19(A)*.

NEC Sections 210.23(A) through (D) define permissible loads for branch circuits. This is important information because it lists the types of loads that may be served according to the size of the branch circuit. *NEC Section 210.24* and *NEC Table 210.24* summarize the branch circuit requirements.

NEC Article 220 includes the requirements used to determine the number of branch circuits required and the requirements used to compute branch circuit, feeder, and service loads. *NEC Section 210.20(A)* states that the rating of a branch circuit overcurrent protection device shall not be less than the noncontinuous load plus 125% of the continuous load. *NEC Table 220.3(A)* gives general lighting loads listed by types of occupancies. These general lighting loads are expressed as a unit load per square foot in volt-amperes (VA).

For example, the unit lighting load for a barber shop is 3VA/sq. ft., a store is also 3VA/sq. ft., and a storage warehouse is ¼VA/sq. ft. *NEC Sections 220.3(B)(1) through (11)* list minimum loads for **outlets** used in all occupancies—these outlets include general-use receptacles and outlets not used for general illumination.

Note

Local codes may require different values than the minimum NEC values. For instance, local codes may limit the number of outlets on a branch circuit to less than the calculated value, or they may require dedicated circuits other than those that are listed in the NEC. Always check local codes before beginning any installation.

1.1.0 Branch Circuit Ratings

The maximum load that a single-phase branch circuit may serve is determined by multiplying the rating or setting of the overcurrent protection device (circuit breaker or fuse) by the circuit voltage. For example, the maximum load that may be supplied by a 20A, two-wire, 120V circuit is calculated by multiplying 20A by 120V.

$$20A \times 120V = 2{,}400VA$$

The maximum load that can be supplied by a 20A, three-wire, 120/240V circuit is:

$$20A \times 240V = 4{,}800VA$$

The maximum load that can be supplied by a 20A, 208V, three-phase circuit is determined by multiplying 20A by 208V by $\sqrt{3}$.

Note

The square root of 3 (written as $\sqrt{3}$) is approximately 1.732. You may wish to make a note of this for use in future calculations.

$$20A \times 208V \times 1.732 = 7{,}205.12VA$$

Branch circuits may supply noncontinuous loads, continuous loads, or a combination of the two. Per *NEC Section 210.20(A)*, the branch circuit rating shall not be less than the noncontinuous load plus 125% of the continuous load. A continuous load is defined by *NEC Article 100* as a load whose maximum current is expected to continue for three hours or more. Continuous loads are calculated at 125% of the maximum current rating of the load. Both *NEC Sections 210.19(A) Exception and 210.20(A) Exception* state that, except where the assembly is listed for operation at 100% of its rating (exception), overcurrent devices located in

Instructor's Notes:

panelboards shall not exceed 80% of their rating where the overcurrent device supplies a continuous duty load.

For 15A and 20A branch circuits, *NEC Section 210.23(A)(2)* allows fastened-in-place utilization equipment to be connected in the same circuit with lighting units, cord- and plug-connected utilization equipment not fastened in place, or both. Under this condition, the fastened-in-place utilization equipment shall not exceed 50% of the branch circuit ampere rating.

Example 1:
A store has fluorescent lighting fixtures consisting of nine fluorescent ballasts rated at 1.5A at 120V. The fixtures will operate continuously during normal business hours from 9:00 A.M. until 9:00 P.M. daily. What is the minimum size circuit breaker required for a branch circuit to serve this load?

Determine the branch circuit load:

$$9 \times 1.5A = 13.5A$$

Determine the continuous duty load (this is a continuous duty load because the lighting fixtures will stay on for more than three hours):

$$13.5A \times 125\% = 16.88A$$

The minimum size circuit breaker required is 20A.

THINK ABOUT IT

Service Size

Why couldn't this 30A service be used in new residential construction?

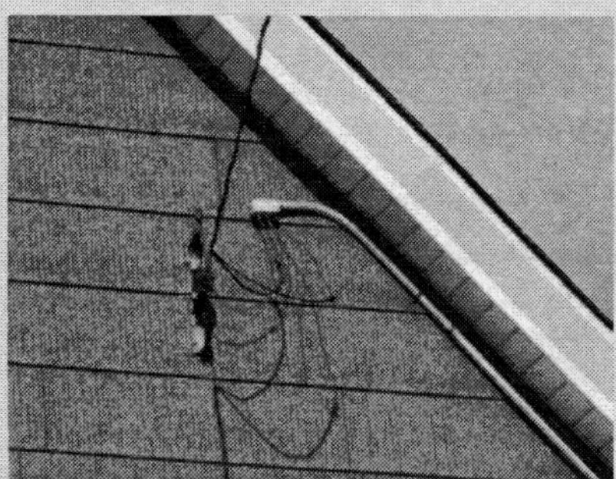

301P0101.EPS

Example 2:
What is the maximum continuous load that may be connected to a 30A, 120V fuse?

Per *NEC Section 210.20(A)*:

$$30A \times 80\% = 24A$$

The maximum continuous load cannot exceed 24A.

Example 3:
How many **receptacle outlets** in other-than-residential occupancies can be connected to a 20A, two-wire, 120V circuit? (The receptacle outlets serve noncontinuous duty loads.)

Determine branch circuit capacity:

$$20A \times 120V = 2,400VA$$

Per *NEC Section 220.3(B)(9)*, each outlet is assigned a load of 180VA.

$$2,400VA \div 180VA = 13.33$$

Thirteen receptacle outlets can be connected to this circuit.

Example 4:
An office manager has purchased a new state-of-the-art copy machine. The nameplate rating on the copy machine is 17A, 120V.

What is the minimum size branch circuit required to serve this equipment?

This is not a continuous load; however, per *NEC Sections 210.23(A)(1) and (B)*, the rating of any one cord-connected utilization equipment shall not exceed 80% of the branch circuit ampere rating for 15A, 20A, and 30A circuits. Since the load is greater than 15A, determine the current-carrying capacity of a 20A (the smallest logical size) circuit to serve this equipment:

$$20A \times 80\% = 16A$$

A 20A circuit is not sufficient; determine the current-carrying capacity of a 30A circuit:

$$30A \times 80\% = 24A$$

The minimum size branch circuit required is 30A.

Note that this solution assumes that only multi-receptacle 20A and 30A branch circuits exist in the office. An alternate solution would be to install a 20A **individual branch circuit** exclusively for the copy machine.

Example 5:
What is the maximum lighting load that may be connected to a 20A branch circuit supplying a piece of fixed equipment that has a rating of 8.5A, 120V?

Classroom

Review the example calculations provided in the Trainee Module.

Teaching Tip

Discuss the "Think About It."

Classroom

List the situations in which the NEC requires ampacity adjustments.

Discuss the relationship between temperature ratings and ampacity ratings. Work through the examples provided in the Trainee Module.

The equipment rating is smaller than 50% of the rating of the 20A branch circuit. Therefore, 11.5A of noncontinuous lighting may be added.

$$20A - 8.5A = 11.5A$$

Example 6:
A restaurant dishwasher has a nameplate rating of 14.7A, 208V, 3Ø. During busy times in the restaurant, it is anticipated that the dishwasher will be turned on and operated for more than three hours at a time. What is the minimum size branch circuit required to supply this equipment?

This equipment is considered a continuous load (operated for more than three hours). Therefore, the load is to be multiplied by 125% to determine the branch circuit size.

$$14.7A \times 125\% = 18.38A$$

The minimum size branch circuit required is 20A.

1.2.0 Derating

The current-carrying capacity (ampacity) of branch circuit conductors must be adjusted when any of the following circumstances apply:

- There are more than three current-carrying conductors in a raceway per ***NEC Section 310.15(B)(2)(a).***
- The ambient temperature that the conductors will pass through exceeds the temperature ratings for conductors listed in ***NEC Table 310.16.***
- A voltage drop exists that exceeds the recommended 3% for branch circuits or 5% for the combination of the feeder and branch circuits that is caused when the distance of branch circuit conductors becomes excessive, per ***NEC Section 210.19(A), FPN No. 4.***
- You bundle any cable assemblies more than 24" in length per ***NEC Section 310.15(B)(2)(a).***

NEC Table 310.16 lists allowable ampacities for insulated conductors rated 0 to 2,000V. The ampacities listed apply when no more than three current-carrying conductors are installed in a raceway, are part of a cable assembly, or are for direct-buried conductors. The ampacities are based upon an ambient temperature of 30°C (86°F). If the number of current-carrying conductors in the conduit or cable exceeds three, the ampacity of the conductors must be derated using ***NEC Section 310.15(B)(2)(a).*** If the ambient temperature exceeds 30°C (86°F), the ampacity of the conductors must be derated using the temperature correction factors listed at the bottom of ***NEC Table 310.16.*** Note that the multipliers listed at the bottom of ***NEC Table 310.16*** are called *correction factors.* NEC references to adjustment and correction factors mean ambient temperatures other than 30°C and more than three current-carrying conductors in a raceway or cable.

1.2.1 Temperature Derating

NEC Section 110.14(C) states that the lowest temperature rating of any component in a branch circuit (circuit breaker, fuse, receptacle, conductors, etc.) must be used to determine the ampacity rating of the branch circuit conductors. Conductors that have a higher temperature rating per ***NEC Table 310.16*** can be used for ampacity adjustment, correction, or both. For example, if the termination rating of a circuit breaker is 60°C , the ampacity rating of the branch circuit conductor rating cannot exceed the value given in the 60°C column of ***NEC Table 310.16*** for overcurrent protection. However, if the branch circuit conductor has a higher temperature rating (i.e., 90°C THHN), the higher ampacity rating can be used for ampacity adjustment, correction, or both.

Example 1:
It is determined that by combining branch circuits we can eliminate two runs of conduit. The branch circuits are to be pulled in a single conduit. All of the conductors in the conduit will be current-carrying conductors of the same size and insulation type.

What is the derated ampacity for each of eight No. 12 THHN copper conductors when pulled in a single conduit?

Per ***NEC Table 310.16,*** the ampacity of No. 12 THHN is 30A. Per ***NEC Table 310.15(B)(2)(a),*** the ampacity shown in ***NEC Table 310.16*** must be adjusted to 70% of its value (seven to nine current-carrying conductors).

$$30A \times 70\% = 21A$$

Example 2:
What is the maximum load that may be connected on each of six No. 2 current-carrying THWN copper conductors in a single conduit?

Per ***NEC Table 310.16,*** the ampacity of No. 2 THWN is 115A. Per ***NEC Table 310.15(B)(2)(a),*** the ampacity shown in ***NEC Table 310.16*** must be adjusted to 80% of its value (four to six current-carrying conductors).

$$115A \times 80\% = 92A$$

The maximum load cannot exceed 92A.

Example 3:
A branch circuit is required to supply a noncontinuous equipment load with a nameplate rating

Instructor's Notes:

of 55A. The equipment is located in a room with plastic extrusion equipment. The ambient temperature in the room is 92°F.

What is the minimum size THHN copper conductors required to supply the equipment?

Per *NEC Table 310.16,* No. 8 THHN copper conductors have an ampacity rating of 55A. Per correction factors to *NEC Table 310.16,* the ampacity of the conductors must be multiplied by a factor of 0.96.

$$55A \times 0.96 = 52.8A$$

The ampacity of No. 8 THHN copper conductors in an ambient temperature of 92°F will not meet the requirements of *NEC Table 310.16* for a load of 55A. A No. 6 THHN copper conductor has a rating of 75A.

$$75A \times 0.96 = 72A$$

No. 6 THHN copper conductors will be required.

Example 4:
A branch circuit supplies 16A worth of continuous loads. The conductors to be used are Type THHN copper. The conductors will be installed in a raceway with a total of 12 current-carrying conductors and run in an ambient temperature of 105°F. What is the minimum wire size and overcurrent protective device required for this circuit?

Per *NEC Section 210.19(A)(1)* for conductors and *NEC Section 210.20(A)* for overcurrent, the minimum current rating for the conductor and the overcurrent protective device is calculated as follows:

$$16A \times 125\% = 20A \text{ (for a continuous load)}$$

Per *NEC Table 310.16* using the 60° column, a No. 12 THHN conductor is required since *NEC Section 240.4(D)* prohibits a No. 14 from having an overcurrent protective device rated higher than 15A.

The correction factor for 12 current-carrying conductors is 50%. The correction factor for 105°F is 0.87. The allowable ampacity for a No. 12 THHN would be:

$$30 \times 50\% \times 0.87 = 13.05A$$

The allowable ampacity for a No. 10 THHN would be:

$$40 \times 50\% \times 0.87 = 17.4A$$

The No. 10 would be required as the minimum safe conductor and must be protected at 20A.

1.2.2 Voltage Drop Derating for Single-Phase Circuits

NEC Section 210.19(A), FPN No. 4 states that reasonable efficiency will be provided by branch circuits in which the voltage drop for branch circuit conductors does not exceed 3% to the farthest outlet and 5% for the combination of feeder and branch circuit distance to the farthest outlet. There are two formulas used to calculate voltage drop for single-phase circuits.

Formula 1:

$$VD = \frac{2 \times L \times R \times I}{1,000}$$

Where:

L = load center or total length in feet

R = conductor resistance per ***NEC Chapter 9, Table 8***

I = current

The number 1,000 represents 1,000' or less of conductor.

Formula 2:

$$VD = \frac{2 \times L \times K \times I}{CM}$$

Where:

L = load center or total length in feet

K = constant (12.9 for copper conductors; 21.2 for aluminum conductors)

I = current

CM = area of the conductor in circular mils (from ***NEC Chapter 9, Table 8***)

The result of these formulas is in volts. The result is divided by circuit voltage, and the second result is multiplied by 100 to determine the percent voltage drop. If the branch circuit voltage

Classroom

Discuss the relationship between voltage drop and circuit efficiency. Point out the NEC recommendations for voltage drop.

Demonstrate how to use the two voltage drop formulas.

THINK ABOUT IT

Voltage Drop Formula Factors

What do you think the number 2 in voltage drop Formula 1 represents?

Teaching Tip

Discuss the "Think About It."

Classroom

Review the single-phase voltage drop calculations provided in the Trainee Module.

Audiovisual

Show Transparency 2 (Figure 1).

Classroom

Explain how to calculate the total load and load center length of a typical circuit.

drop exceeds 3%, the conductor size should be increased to compensate. For a 120V circuit, the maximum voltage drop (in volts) is 120V × 3% = 3.6V. For a 240V single-phase circuit, the maximum voltage drop (in volts) is 240V × 3% = 7.2V.

In commercial fixed load applications, if the branch circuit load is not concentrated at the end of the branch circuit but is spread out along the circuit due to multiple outlets, the load center length of the circuit should be calculated and used in the formulas just listed. This is because the total current does not flow the complete length of the circuit. If the full length is used in computing the voltage drop, the drop determined would be greater than what would actually occur. The load center length of a circuit is that point in the circuit where, if the load were concentrated at that point, the voltage drop would be the same as the voltage drop to the farthest load in the actual circuit. To determine the load center length of a branch circuit with multiple outlets, as shown in *Figure 1,* multiply each outlet load by its actual physical routing distance from the supply end of the circuit. Add these products for all loads fed from the circuit and divide this sum by the sum of the individual loads. The resulting distance is the load center length (L) for the total load (I) of the branch circuit.

Example 1:
The length of a 120V, two-wire branch circuit is 95'. The noncontinuous load is 14.5A.

If No. 12 THHN solid copper conductors are used, will the voltage drop for this branch circuit exceed 3%?

Use the first voltage drop formula to determine voltage drop for this branch circuit. Look up the resistance for No. 12 solid copper in *NEC Chapter 9, Table 8.* It is 1.93Ω.

$$VD = \frac{2 \times L \times R \times I}{1{,}000}$$

$$VD = \frac{2 \times 95' \times 1.93\Omega \times 14.5A}{1{,}000}$$

$$VD = 5.32V$$

$$5.32V \div 120V = 0.0443 \times 100 = 4.43\%$$

Since this percentage exceeds the 3% allowable voltage drop, the answer to the question is yes, and larger conductors would be required for this circuit. Note that this solution uses the first formula for single-phase (1∅) voltage drop.

Example 2:
What size THHN solid copper branch circuit conductors would be recommended for a noncontinuous branch circuit load of 23A, 240V, 1∅? The length of the circuit is 130'.

Use the second voltage drop formula to determine the voltage drop for this branch circuit. Since No. 10 THHN copper would be the smallest size permitted for a branch circuit load of 23A, this size will be evaluated first. Look up the area in CM for No. 10 solid conductor in *NEC Chapter 9, Table 8.*

$$VD = \frac{2 \times L \times K \times I}{CM}$$

$$VD = \frac{2 \times 130' \times 12.9 \times 23A}{10{,}380}$$

$$VD = 7.43$$

$$7.43V \div 240V = 0.031 \times 100 = 3.10\%$$

This percentage exceeds 3%; therefore, No. 10 THHN copper conductors would not meet NEC

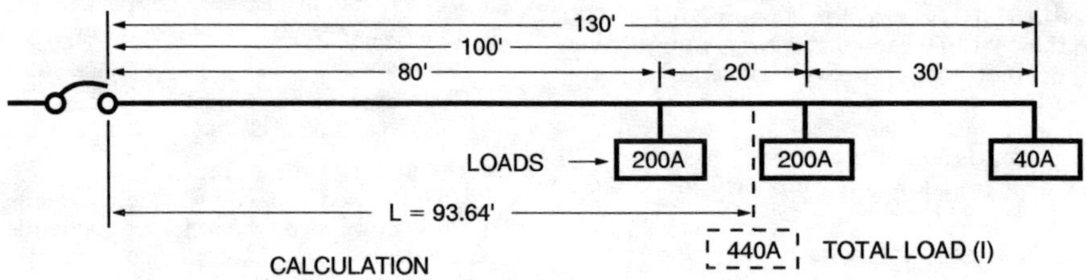

CALCULATION

LOAD 1 =	80'	×	200A	=	16,000
LOAD 2 =	100'	×	200A	=	20,000
LOAD 3 =	130'	×	40A	=	5,200
TOTALS			440A		)41,200 = 93.64' (LOAD CENTER LENGTH)

TOTAL LOAD (I) = 440A
LOAD CENTER LENGTH (L) = 93.64'

301F01.EPS

Figure 1 ◆ Calculating a load center length and total load for multiple fixed loads on a circuit.

Instructor's Notes:

Finding the Minimum Conductor Size

INSIDE TRACK

A quicker way to find the minimum conductor size for the situation described in *Example 2* is to use the following formula:

$$CM = \frac{2 \times L \times K \times I}{VD}$$

Using this with the values from *Example 2:*

$$CM = \frac{2 \times 130' \times 12.9 \times 23A}{240V \times .03}$$

$$CM = \frac{77{,}142}{7.2}$$

$$CM = 10{,}714$$

This is the minimum size conductor that is acceptable. Using *NEC Chapter 9, Table 8*, the circular mil area of No. 10 is 10,380CM. Since that is less than 10,714CM, we must use a No. 8 that has an area of 16,510CM.

recommendations, and the next larger size would be required, which in this case is No. 8 THHN.

Example 3:
What size THHN copper conductors would be recommended for a 240V, single-phase, three-wire branch circuit with multiple power outlets consisting of three fixed 240V loads of 30A at 60', 30A at 80', and 20A at 100'?

Determine the load center length of the circuit by multiplying the outlet loads by their distance from the circuit source and then by dividing the sum of the three products by the sum of the three loads as follows:

Outlet 1 = 60' × 30A = 1,800
Outlet 2 = 80' × 30A = 2,400
Outlet 3 = 100' × 20A = 2,000

Sum the products:

1,800 + 2,400 + 2,000 = 6,200

Divide by the sum of the loads:

30A + 30A + 20A = 80A

6,200 ÷ 80A = 77.5' for the load center

For a load of 80A (per ***NEC Table 310.16***), No. 4 copper THWN/THHN conductors at 75°C will be selected. Use the second formula and substitute values to check voltage drop. In this example, all terminals are rated for 75°C.

$$VD = \frac{2 \times L \times K \times I}{CM}$$

$$VD = \frac{2 \times 77.5' \times 12.9 \times 80A}{41{,}740}$$

$$VD = 3.8V$$

Since the permissible voltage drop is 240V × 3% or 7.2V, the No. 4 THWN/THHN conductors are satisfactory for this application.

1.2.3 Voltage Drop Derating for Three-Phase Circuits

The voltage drop for balanced three-phase circuits (with negligible reactance and a power factor of 1) can be calculated using the two formulas above by substituting $\sqrt{3}$ (1.732) for the value of 2 in the formulas:

Formula 1:

$$VD = \frac{\sqrt{3} \times L \times R \times I}{1{,}000}$$

Formula 2:

$$VD = \frac{\sqrt{3} \times L \times K \times I}{CM}$$

Example 1:
What is the voltage drop of a 208V, 3Ø branch circuit with a load of 32A, a distance from the circuit breaker to the load of 115', and using No. 8 stranded copper conductors?

Use the first three-phase voltage drop formula to determine voltage drop for this 3Ø circuit. Look up the resistance for No. 8 stranded copper conductor in ***NEC Chapter 9, Table 8***. It is .778Ω.

$$VD = \frac{\sqrt{3} \times L \times R \times I}{1{,}000}$$

$$VD = \frac{1.732 \times 115' \times .778\Omega \times 32A}{1{,}000}$$

$$VD = 4.96V$$

Classroom

Explain how to calculate the voltage drop of a three-phase circuit. Work through the examples provided in the Trainee Module.

Classroom

Explain how to calculate the branch circuit ampacity for a single-phase circuit. Work through the examples provided in the Trainee Module.

Point out that lighting calculations are based on three factors:

- **Type of lighting**
- **Branch circuit voltage**
- **Whether the load is continuous or noncontinuous**

Demonstrate how to calculate the lighting load for a typical branch circuit.

The maximum voltage drop permitted for a 208V circuit is:

$$208V \times 3\% = 6.24V$$

Since, in this problem, 4.96V is less than the maximum allowable voltage drop, this circuit will not have a voltage drop problem.

Example 2:
What size THHN copper conductors would be recommended for a 208V, 3Ø, four-wire feeder with a length of 150' from the source to a fixed continuous load of 160A?

Determine the minimum ampacity required by multiplying 160 by 125% to obtain 200A. Using *NEC Table 310.16*, determine that 3/0 THHN copper conductors at 75°C are satisfactory for the load of 200A. To determine if the 3/0 conductors are adequately sized to prevent unacceptable voltage drop, use the second 3Ø voltage drop formula and substitute values. Note that the current used will be 160A and not the computed value of 200A, which is only used to size branch circuit or feeder conductor sizes and minimum overcurrent protective device sizes.

$$VD = \frac{\sqrt{3} \times L \times K \times I}{CM}$$

$$VD = \frac{1.732 \times 150' \times 12.9 \times 160A}{167{,}800}$$

$$VD = 3.2V \text{ (rounded)}$$

Since the maximum recommended voltage drop is 208 × 3% or 6.24V, the 3/0 conductors are satisfactory.

1.3.0 Calculating Branch Circuit Ampacity

The branch circuit ampacity for single-phase circuits is calculated by dividing VA by the circuit voltage. For example, the ampacity of a 120V, 1Ø load rated at 1,600VA is determined by dividing 1,600VA by 120V.

$$1{,}600VA \div 120V = 13.33A$$

The ampacity of a 3,450VA load at 277V, 1Ø is 12.45A.

$$3{,}450VA \div 277V = 12.45A$$

The branch circuit ampacity for three-phase circuits is calculated by dividing VA by the circuit voltage times $\sqrt{3}$ (1.732). For example, the ampacity (I) of a 208V, three-phase load rated at 5,200VA is determined as follows:

$$I = \frac{VA}{V \times \sqrt{3}}$$

$$I = \frac{5{,}200VA}{208V \times 1.732}$$

$$I = 14.43A$$

Using the same equation, the ampacity of a 14,000VA three-phase load at 480V is:

$$\frac{14{,}000VA}{480V \times 1.732} = 16.84A$$

Example 1:
What is the ampacity of a single-phase load with a nameplate rating of 5.5kW, 240V?

Multiply 5.5kW by 1,000 to determine VA (watts = volt-amperes) and then divide the result by 240V.

$$5.5kW \times 1{,}000 = 5{,}500VA \div 240V = 22.92A$$

The ampacity of this load is 22.92A.

Example 2:
What is the ampacity of a three-phase electric water heater with a nameplate rating of 20kW, 208V?

Multiply 20kW by 1,000 and divide the result by (208V × 1.732).

$$20kW \times 1{,}000 = 20{,}000kW$$
$$\div (208V \times 1.732) = 55.52A$$

The ampacity of this load is 55.52A.

2.0.0 ◆ LIGHTING LOADS

Lighting load branch circuit calculations are based upon the type of lighting (incandescent or electric discharge), the branch circuit voltage, and whether or not the lighting is to be used for more than three hours without an off period (continuous duty).

Branch circuit loads for incandescent lighting are determined by adding the total incandescent load (watts = VA) and dividing the total by the circuit voltage. For example, three 500W quartz lamps connected on a 120V circuit would equal 12.5A, as shown here.

$$3 \times 500VA = 1{,}500VA \div 120V = 12.5A$$

Branch circuit loads for electric discharge lighting (lighting units having ballasts, transformers,

Instructor's Notes:

or autotransformers) are determined by multiplying the number of fixtures times the ampacity rating of the ballast, assuming that all of the ballasts are identical. Per *NEC Section 220.4(B)*, the calculated load shall be based upon the total ampere rating of the fixture (ballast or ballasts) and not the total watts of the lamps. For example, if fifteen 150W, high-pressure sodium fixtures connected on a 277V circuit each have a ballast amperage of .79A, the total load is 11.85A, as shown in the following equation.

$$15 \times 0.79A = 11.85A$$

Example 1:
Incandescent lighting utilizing 150W medium base lamps is required for temporary lighting on a construction site. This lighting will remain on all day and all night.

How many 150W lamps can be connected on a 20A, 120V circuit?

Since the lighting will remain on for longer than three hours, this is a continuous load. For continuous duty, multiply 150VA (watts) × 125% = 187.5VA. Determine the total capacity (in VA) for the 20A circuit:

$$20A \times 120V = 2{,}400VA$$

Divide the circuit ampacity by the lamp demand load:

$$2{,}400VA \div 187.5VA = 12.8 \text{ lamps}$$

Twelve 150W lamps may be safely connected to a 20A, 120V circuit.

Example 2:
Two 400W metal halide high bay fixtures are required to provide additional lighting for a production machine. There is an existing 20A, 277V lighting circuit available near the equipment and, after investigation, it is determined that the circuit has six of the same type of fixtures connected. The nameplate ampere rating for these fixtures is 2.0A at 277V. During normal operation, the fixtures are turned on for 12 hours every day.

Can two 400W fixtures be safely added to the existing circuit?

This is a continuous lighting load. The maximum continuous load that may be connected to the 20A circuit is 16A.

$$20A \times 80\% = 16A$$

The load for eight fixtures (six existing and two new fixtures) is 16A.

$$8 \text{ fixtures} \times 2.0A = 16A$$

Per NEC requirements, two new fixtures may safely be added to this circuit.

2.1.0 Recessed Lighting

The load for recessed lighting fixtures (excluding the residential general lighting load that is calculated at 3VA per square foot) is calculated per *NEC Section 220.3(B)(4)* by using the maximum VA rating of the equipment (fixture) and lamps. For example, a load of 250VA would be used to calculate the load for a recessed incandescent fixture rated at 250W. To calculate the load for a recessed fixture that uses compact fluorescent lamps, we would need to know the voltage and ampacity rating of the fluorescent ballast.

Example 1:
What is the load in amps for seven incandescent recessed cans that have a nameplate indicating a maximum lamp size of 150W, and the fixtures are to be connected to a 120V circuit?

The number of fixtures is multiplied by the VA rating of each fixture to obtain the total VA load.

$$7 \times 150VA \text{ (watts)} = 1{,}050VA$$

To determine ampacity, the total VA rating is divided by the circuit voltage.

$$1{,}050VA \div 120V = 8.75A$$

Example 2:
What is the total load in amps for seven fluorescent recessed fixtures operating continuously? Each fixture has a ballast ampacity of 0.20A at 277V, and each fixture takes two compact fluorescent lamps rated at 26 watts each.

Because these fixtures are of the electric discharge type, we need to know the ampacity of the fixture and the number of fixtures in order to calculate the total ampacity.

$$7 \times 0.20A = 1.4A \times 125\%$$
$$\text{(for a continuous load)} = 1.75A$$

2.2.0 Heavy-Duty Lamp Holder Outlets

Per *NEC Section 220.3(B)(5)*, outlets for heavy-duty lamp holders are calculated at a minimum of 600VA each. For example, a noncontinuous load consisting of four 120V outlets for heavy-duty lamp holders could be connected to a 20A branch circuit:

$$2{,}400VA \div 600VA = 4 \text{ outlets}$$

3.0.0 ◆ RECEPTACLE LOADS

When the exact VA rating of a load that is to be cord- and plug-connected to a receptacle outlet is not known, a VA rating of not less than 180VA per outlet is used per *NEC Section 220.3(B)(9)*. For

Classroom

Work through the lighting load examples provided in the Trainee Module.

Discuss the requirements for recessed lighting load calculations and review the examples provided in the Trainee Module.

Point out that heavy-duty lamp holders are calculated at a minimum of 600VA each.

Classroom

Explain how to calculate the number of receptacles that can be served by a branch circuit.

Discuss the NEC requirements for multi-outlet assemblies. Work through the examples provided in the Trainee Module.

noncontinuous receptacle loads, the total noncontinuous load is calculated at 100%. For continuous loads, the total continuous load is calculated at 125%. This NEC reference does not apply to residential receptacles. Residential receptacles are included in the general illumination load (**general-purpose branch circuits**) or in specific residential loads such as receptacles required for small **appliance** and laundry loads.

For example, a 15A, 120V circuit is to be added to supply general-purpose receptacles. The receptacles are to be located above workbenches that are to be positioned against a wall. The owner states that the workers using the benches will plug small tools into the outlets, but the tools will only be occasionally used for short periods of time.

How many general-purpose duplex receptacles can be connected to a 15A, 120V circuit when the outlets are to be rated for noncontinuous duty?

Per *NEC Section 220.3(B)(9)*, each receptacle is assigned a value of 180VA. The capacity of a 15A, 120V circuit is 1,800VA.

$$15A \times 120V = 1{,}800VA$$

The total circuit capacity is then divided by the rating per receptacle to determine the total number that may be connected.

$$1{,}800VA \div 180VA = 10 \text{ receptacles}$$

4.0.0 ◆ MULTI-OUTLET ASSEMBLIES

A **multi-outlet assembly** is defined by *NEC Article 100* as a type of surface or flush raceway designed to hold conductors and receptacles, assembled in the field or at the factory. Multi-outlet assemblies may consist of single outlets wired to one or more circuits and typically spaced equally apart at distances of 6", 12", 18", etc. The NEC rules for calculating loads for multi-outlet assemblies do not apply to dwelling units.

Per *NEC Section 220.3(B)(8)*, each 5' of multi-outlet assembly is considered as one outlet of at least 180VA capacity. In locations where many appliances are likely to be used at one time, each 1' of multi-outlet assembly is considered as one outlet of at least 180VA capacity.

Example 1:
A total of 40' of Plugmold® includes one single receptacle per foot of the assembly. The multi-outlet assembly is to be connected to a single 120V circuit.

What is the minimum size 120V circuit that would safely supply this light-duty multi-outlet assembly?

Since this is a light-duty application, the load is to be calculated at 180VA per 5' of multi-outlet assembly. Divide 40' of multi-outlet assembly by 5' and multiply the result by 180VA to obtain the total VA load. Divide the total VA load by 120V to determine the circuit ampacity and minimum circuit size.

$$40' \div 5' = 8 \times 180VA = 1{,}440VA$$

$$1{,}440VA \div 120V = 12A$$

The minimum 120V circuit size would be a 15A circuit.

Teaching Tip

Discuss the "Think About It."

THINK ABOUT IT

Multi-Outlet Assemblies

This assembly is 5' in length and will be used in an other-than-residential application in which the appliances are *not* likely to be used simultaneously. According to ***NEC Section 220.3(B)(8)***, what is the load for this device?

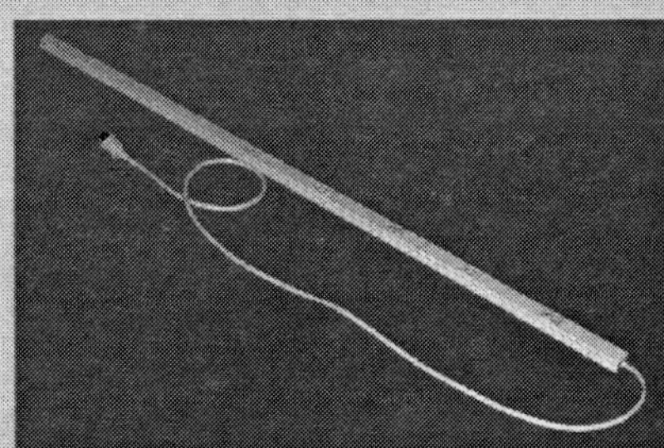

301P0102.EPS

Instructor's Notes:

Example 2:
The conditions listed for the previous example have changed, and it is necessary to feed this Plugmold® with three circuits. The reason for the change is that now more workers are using equipment plugged into the assembly. This change means that the assembly will now be rated to supply several loads simultaneously.

How many 20A, 120V circuits would be required to safely supply this heavy-duty multi-outlet assembly?

Since this is a heavy-duty application, the load is to be calculated at 180VA per foot of multi-outlet assembly. First, determine the VA capacity for a 20A, 120V circuit.

$$20A \times 120V = 2{,}400VA$$

Multiply 40' of multi-outlet assembly by 180VA to obtain the total VA load.

$$40' \times 180VA = 7{,}200VA \text{ (total VA)}$$

Divide the total VA by 2,400VA (20A circuit capacity) to determine the number of circuits required.

$$7{,}200VA \div 2{,}400VA = 3$$

Three 20A, 120V circuits would be required to supply this assembly.

5.0.0 ◆ SHOW WINDOW LOADS

The NEC addresses three areas of requirements with regard to show window lighting: receptacle requirements, branch circuit load calculations, and feeder or service load calculations.

NEC Section 210.62 requires that at least one receptacle outlet be installed directly above a show window for each 12 linear feet of window, or major fraction thereof.

NEC Section 220.3(B)(7) provides two options for calculating branch circuit loads for show window lighting, with option (a) using the computed receptacle load of 180VA per receptacle, and option (b) computing the load at 200VA per linear foot of show window.

Note
Since show window lighting is likely to be on for three hours or more, it must be considered as a continuous load when computing the branch circuit load.

NEC Section 220.12(A) addresses the computed load of show windows when calculating feeder or service loads and requires that the total load be calculated at not less than 200VA for each linear foot of show window.

Example 1:
According to ***NEC Section 210.62***, how many receptacles would be required for a show window area that measures 67' in length?

Divide the length of show window by 12 to determine the number of receptacles required.

$$67 \div 12 = 5.58$$

Since ***NEC Section 210.62*** requires one receptacle for each 12 linear feet or major fraction thereof, six receptacles would be required.

THINK ABOUT IT

Show Window Load Calculations

These individual show windows are each 5' in length. Assuming that we are using option (b) of the requirements given in ***NEC Section 220.3(B)(7)***, what is the branch circuit load for each window?

301P0103.EPS

Classroom

Discuss the NEC requirements for show windows. Work through the examples provided in the Trainee Module.

Teaching Tip

Discuss the "Think About It."

Classroom

Refer the trainees to *NEC Article 600* and discuss the requirements for sign loads. Point out the special requirements for neon tubing. Work through the examples provided in the Trainee Module.

Emphasize that the NEC contains specific requirements for residential branch circuit calculations.

Note that the NEC requires a minimum of two 20A small appliance branch circuits to serve the kitchen and other listed areas.

Teaching Tip

Discuss the "Think About It."

Example 2:
Using option (a) of *NEC Section 220.3(B)(7)*, what is the total load for the receptacles required in the previous example?

Multiply the number of receptacles by 180VA per receptacle and multiply the result by 125% to determine the load.

$$6 \times 180VA = 1,080VA$$

$$1,080 \times 125\% = 1,350VA$$

The load for these show window receptacles is 1,350VA.

Example 3:
Using option (b) of *NEC Section 220.3(B)(7)*, what is the load for a 30'-long show window?

Multiply the total length of show window area by 200VA and multiply the result by 125% to determine the load.

$$30' \times 200VA = 6,000VA \times 125\% = 7,500VA$$

6.0.0 ◆ SIGN LOAD

NEC Article 600 covers requirements for signs and outline lighting. *NEC Sections 600.5(B) (1) and (2)* specify that sign circuits that supply incandescent and fluorescent lighting shall not exceed 20A, and sign circuits that supply neon tubing shall not exceed 30A. *NEC Section 600.5(A)* requires at least one 20A sign circuit that supplies no other load. This sign circuit must be provided for each commercial building and each commercial occupancy accessible to pedestrians. *NEC Section 220.3(B)(6)* requires the load for the sign circuit to be computed at a minimum of 1,200VA. Since signs for commercial occupancies are expected to operate for more than three hours at a time, the sign circuit is typically considered as a continuous load. Under this condition, the branch circuit rating must be calculated at 125% of 1,200VA, which is equal to 1500VA. The actual sign load cannot exceed 80% of the branch circuit rating.

For example, what is the maximum continuous load in VA that may be connected to a 20A, 120V sign circuit?

Multiply 20A by 120V to determine the total VA load for a 20A circuit.

$$20A \times 120V = 2,400VA$$

Multiply the result by 80% to determine the maximum continuous VA load.

$$2,400VA \times 80\% = 1,920VA$$

7.0.0 ◆ RESIDENTIAL BRANCH CIRCUITS

There are a number of branch circuits that are required for dwelling units. The NEC lists the requirements for these circuits including the requirements for calculating branch circuit loads for specific dwelling unit (residential) loads. These loads, which are unique to dwelling units, will be covered here.

7.1.0 Small Appliance Load

NEC Sections 210.11(C)(1) and 210.52(B) require at least two 20A small **appliance branch circuits** to be installed to supply receptacles installed in the kitchen, pantry, breakfast room, and dining room. *NEC Section 220.16(A)* requires that the feeder load be computed at 1,500VA for each required small appliance branch circuit. No small appliance branch circuit may serve more than one kitchen.

> **Note**
> Most jurisdictions interpret this to include only the two circuits required. Others require a unit load of 1,500VA for each small appliance branch circuit required.

For example, what is the total feeder load for two small appliance branch circuits rated at 20A, 120V?

Multiply the number of small appliance branch circuits by 1,500VA to determine the total load.

$$2 \text{ circuits} \times 1,500VA = 3,000VA$$

> THINK ABOUT IT
>
> ***Adding a Kitchen***
>
> You are adding a new kitchen to the basement of a residence. How does this affect the load calculation for this residence?

Instructor's Notes:

THINK ABOUT IT

Residential Laundry Room Circuit

Does the NEC specifically require a dedicated branch circuit for a washing machine?

301P0104.EPS

7.2.0 Laundry Circuit

NEC Sections 210.11(C)(2) and 210.52(F) require the installation of at least one receptacle to supply laundry equipment. ***NEC Section 220.16(B)*** requires that the feeder load be computed at 1,500VA for each laundry circuit.

7.3.0 Dryers

As specified in ***NEC Section 220.18,*** the load for household electric dryers is calculated at 5,000VA or the nameplate rating of the dryer, whichever is larger. ***NEC Table 220.18*** lists **demand factors** for dryers. Note that this table applies to residential dryers used in single-family dwelling units or in multi-family dwelling units. The demand factors listed do not apply to commercial dryers used in commercial facilities.

Example 1:
What is the demand load for one household electric dryer with a nameplate rating of 5,500 watts?

Per ***NEC Table 220.18,*** the demand factor for one dryer is 100%. Therefore, the load would be calculated as 5,500VA.

Example 2:
A new apartment building is planned. It consists of 10 dwelling units. Each unit has an electric dryer with a nameplate rating of 5,500 watts. There will be one utility transformer and one service entrance serving the building.

What is the demand load, in VA, for these 10 dryers?

Per ***NEC Table 220.18,*** the demand factor for 10 dryers is 50%. Add the total VA for the 10 dryers and multiply the result by 50%.

$$10 \text{ dryers} \times 5{,}500\text{W} = 55{,}000\text{VA}$$
$$55{,}000\text{VA} \times 50\% = 27{,}500\text{VA}$$

7.4.0 Cooking Appliances

Loads for ranges, wall-mounted ovens, counter-mounted cooking units, and other household cooking appliances are calculated using the demand factors listed in ***NEC Table 220.19*** and the notes to ***NEC Table 220.19.*** The most important single piece of information required to size the circuit for residential cooking equipment is the nameplate rating of the equipment. Although ***NEC Table 220.19*** lists loads in kW, ***NEC Section 220.19*** states that kVA shall be considered equivalent to kW for loads calculated using ***NEC Table 220.19.***

Demand factors are calculated using ***NEC Table 220.19*** based upon the number of appliances, the maximum demand listed in Column C, and a demand factor percentage that is found using Columns A and B. Column C is used when the nameplate rating is over 8¾kW but not over 12kW. Column C is also used to calculate larger range loads per Notes 1 and 2. Column A is used when the nameplate rating is less than 3½kW. Column B is used when the nameplate rating is from 3½kW to 8¾kW. Note 3 provides a method to calculate the demand for multiple ranges, each of which has a nameplate rating of more than 1¾kW but less than 8¾kW. Note 4 provides a method to determine the size for a single branch circuit supplying one counter-mounted unit and no more than two wall-mounted ovens, provided the equipment is located in the same room.

Example 1:
What is the demand load for one range with a nameplate rating of 11.3kW?

Since the nameplate rating is greater than 8¾kW and not over 12kW, Column C is used. The demand rating for one range, not over 12kW, is listed as 8kW.

Example 2:
What is the demand load, in amps, for one range with a nameplate rating of 15.7kW, 1∅, 240V?

Since the nameplate rating exceeds 12kW, ***NEC Table 220.19, Note 1*** is used. Subtract 12kW from

Teaching Tip

Discuss the "Think About It."

Classroom

Discuss the NEC requirements for laundry loads. Point out the demand factors that apply to multi-family dwelling units. Work through the examples provided in the Trainee Module.

Explain how to calculate the loads for household cooking appliances. Refer the trainees to *NEC Table 220.19.* Note the demand factors applied to multiple appliances and work through the examples provided in the Trainee Module.

Classroom

Continue to work through the appliance examples provided in the Trainee Module.

Teaching Tip

Discuss the "Think About It."

the nameplate rating of the range to determine the number of kW exceeding 12.

15.7kW − 12kW = 3.7kW

Since 0.7 is a major fraction (≥.5), the result is rounded off to 4kW. Multiply 5% by 4 to obtain the demand increase percentage.

4kW × 5% = 20%

The maximum demand listed in Column C for one range (8kW) is then multiplied by 120% to determine the demand load for this range.

8kW × 120% = 9.6kW

To determine the ampacity of this load, first convert to VA (kW × 1,000), then divide VA by the circuit voltage:

9.6kW × 1,000 = 9,600VA

9,600VA ÷ 240V = 40A

THINK ABOUT IT

One Cooktop and One Oven

Which note under ***NEC Table 220.19*** refers to this installation?

301P0105.EPS

Example 3:
One wall-mounted oven and one counter-mounted cooking unit are to be installed in the kitchen. To reduce the cost of the electrical installation, both pieces of cooking equipment are to be connected to the same 240V, 1Ø circuit. The oven has a nameplate rating of 12.5kW and the counter-mounted unit has a nameplate rating of 8.0kW.

What is the demand load, in amps, for a single circuit to supply this cooking equipment?

NEC Table 220.19, Note 4 may be used for this calculation. First, obtain the total load by adding the nameplate ratings of the individual pieces of cooking equipment.

12.5kW + 8.0kW = 20.5kW

The equivalent of one range is 20.5kW. Using ***NEC Table 220.19, Note 1,*** subtract 12kW from 20.5kW to determine the number of kW exceeding 12.

20.5kW − 12kW = 8.5kW

Since .5 is a major fraction, the result is rounded off to 9kW. Multiply 5% by 9:

9 × 5% = 45%

The maximum demand in Column C for one range (8kW) is then multiplied by 145% to determine the demand load.

8kW × 145% = 11.6kW

Divide total VA by circuit voltage to determine branch circuit ampacity.

11,600VA ÷ 240V = 48.33A

Example 4:
Using the resulting ampacity for the branch circuit in the previous example, what is the minimum size conductor that may be used for this branch circuit, and what size circuit breaker is required?

According to ***NEC Section 110.14(C)(1),*** branch circuits of 100A or less must be terminated using the 60°C column of ***NEC Table 310.16.*** Therefore, we will select a No. 6 copper conductor rated at 55A. A 50A circuit breaker would protect this circuit.

INSIDE TRACK

Calculating Range Loads

NEC Table 220.19, Note 1 states that when calculating demand loads for individual ranges over 12kW but not more than 27kW, the maximum demand in Column C of the table must be increased by 5% for each additional kilowatt of rating (or major fraction thereof) over 12kW.

Instructor's Notes:

Demand Loads

THINK ABOUT IT

You are adding a hot tub with a nameplate rating of 4,000VA and have three other fixed appliances with a total rating of 13,000VA. Assuming that this feeder does not also serve an electric range, clothes dryer, space heating units, or air conditioning equipment, how will this addition affect the total demand load?

8.0.0 ◆ COMMERCIAL KITCHEN EQUIPMENT

Loads for commercial kitchen equipment are calculated based upon the nameplate rating of the equipment. *NEC Table 220.20* lists demand factors for commercial kitchen equipment. *NEC Section 220.20* applies to commercial kitchen equipment, including cooking equipment, dishwasher booster heaters, water heaters, and other kitchen equipment.

Example 1:
What is the load, in amps, for one dishwasher booster heater? The booster heater has a nameplate rating of 15kW, 208V, 3Ø.

Under this condition, kW = kVA. Divide total kVA by voltage × $\sqrt{3}$ (1.732).

$$15\text{kVA} \times 1{,}000 = 15{,}000\text{VA}$$

$$\frac{15{,}000\text{VA}}{208\text{V} \times 1.732} = 41.64\text{A}$$

Example 2:
What is the demand load for six pieces of commercial kitchen equipment with a total VA rating of 47,000VA?

Per *NEC Table 220.20*, the demand factor for six units is 65%. Multiply the total VA rating by 65%.

$$47{,}000\text{VA} \times 65\% = 30{,}550\text{VA}$$

The demand load is 30,550VA.

9.0.0 ◆ WATER HEATERS

As specified in *NEC Section 422.13*, fixed storage water heaters having a storage capacity of 120 gallons or less shall have the branch circuit rated at no less than 125% of the nameplate rating of the water heater.

For example, what are the minimum size circuit breaker and branch circuit conductors using Type NM cable for a water heater with a nameplate rating of 9,000VA at 240V?

Divide total VA by voltage to determine ampacity; then multiply the result by 125%.

$$9{,}000\text{VA} \div 240\text{V} = 37.5\text{A}$$

$$37.5\text{A} \times 125\% = 46.88\text{A}$$

Per *NEC Table 310.16*, 60°C column, No. 6 copper conductor is rated (for NM cable) at 55A. The circuit breaker size would be 50A.

Water Heater Load Calculation

THINK ABOUT IT

This photo shows a water heater nameplate. What is the required conductor size and branch circuit rating for this appliance if the supply voltage is 240V?

301P0106.EPS

Teaching Tip

Discuss the "Think About It."

Classroom

Discuss the NEC requirements for commercial kitchen equipment. Demonstrate how to calculate a typical load and apply the appropriate demand factor for multiple appliances.

Note that water heaters are typically calculated at 125% of the heater nameplate rating.

Teaching Tip

Discuss the "Think About It."

Classroom

Refer the trainees to *NEC Article 424* and discuss the requirements for electric space heating equipment. Work through the examples provided in the Trainee Module.

Refer the trainees to *NEC Article 440* and discuss the requirements for air conditioning loads.

Demonstration

Demonstrate how to calculate the branch circuit size and overcurrent protection for air conditioning loads.

10.0.0 ◆ ELECTRIC HEATING LOADS

NEC Article 424 covers fixed electric space heating equipment, including heating cable, unit heaters, boilers, central systems, or other approved heating equipment. This article does not apply to process heating or room air conditioning. Per *NEC Section 424.3(A),* branch circuits are permitted to supply any size fixed electrical space heating equipment. If two or more outlets for heating equipment are supplied, the branch circuit size is limited to 15A, 20A, or 30A. When space heating equipment utilizes electric resistance heating elements, protection for the resistance elements shall not exceed 60A (*NEC Section 424.22[B]*). If the equipment is rated at more than 48A, the heating elements must be subdivided and each subdivided load must not exceed 48A.

For space heating equipment consisting of resistance heating elements with or without a blower motor, the ampacity of the branch circuit shall not be less than 125% of the total load per *NEC Section 424.3(B).*

Example 1:
What is the minimum conductor ampacity for the following baseboard electric heaters rated 240V, 1Ø: one 1,500W unit, one 1,000W unit, and two 500W units?

First, determine the connected load by adding the load of each baseboard unit and then divide the total load in VA by the circuit voltage.

1,500W + 1,000W + 500W + 500W = 3,500W (VA)

3,500VA ÷ 240V = 14.58A

Next, multiply the load by 125% to determine the minimum conductor ampacity.

14.58A × 125% = 18.23A

Example 2:
What size NM conductors and circuit breaker would be required to supply the load for the previous question?

Per *NEC Table 310.16,* 60°C column, No. 12 copper conductor (for Type NM) is rated at 25A and may be connected to a 20A circuit breaker. The load, 18.23A, would require a 20A circuit breaker.

Example 3:
What is the minimum conductor ampacity for an electric forced air furnace with heating elements totaling 25kW and a blower motor with a nameplate full-load current of 4.9A? The resistance heat and the motor are rated at 240V, 1Ø.

Divide the total of the resistance heat, in VA, by the circuit voltage to determine the ampacity for the resistance heat. Add the motor ampacity to the result and multiply the total by 125% to determine total circuit ampacity.

25,000VA ÷ 240V = 104.17A

104.17A + 4.9A = 109.07A

The total circuit ampacity is 109.07A.

109.07A × 125% = 136.34A

Note that since this equipment is rated at more than 48A, the heating elements must be subdivided, and each subdivided load cannot exceed 48A.

11.0.0 ◆ AIR CONDITIONING LOADS

Branch circuit conductors supplying a single motor compressor are rated at least 125% of the air conditioning compressor full-load current or the branch circuit selection current, whichever is greater, per *NEC Section 440.32.* Branch circuit selection current is determined by the equipment manufacturer, and this current is required to be greater than or equal to the compressor full-load current. If the branch circuit selection current is higher, its rating shall be used instead of the compressor full-load current. Per *NEC Section 440.22(A),* overcurrent protection devices shall not exceed 175% of the compressor full-load current or the branch circuit selection current. If this breaker is not sufficient to hold the starting current of the motor, the rating is permitted to be increased but cannot exceed 225% of the motor-rated load current or the branch circuit selection current, whichever is greater.

Example 1:
What size THHN copper conductors are required for a branch circuit supplying an air conditioner with a nameplate rating of 45A, 480V, 3Ø?

Multiply the nameplate ampacity by 125% and, using the result, select the proper THHN conductor size from *NEC Table 310.16.*

45A × 125% = 56.25A

NEC Table 310.16 gives an ampacity of 70A for No. 4 THHN copper. (Use the 60°C column—see *NEC Section 110.14(C)*). Therefore, the conductor size would be No. 4 THHN copper.

Example 2:
What size Type NM cable and circuit breaker would be required to supply an air conditioner with a nameplate rating of 16.5A, 240V, 1Ø?

Multiply the nameplate ampacity by 125% and, using the result, select the proper NM conductor size from the 60°C column of *NEC Table 310.16.*

16.5A × 125% = 20.63A

Instructor's Notes:

INSIDE TRACK

Air Conditioning Nameplates

The nameplate provides information about the compressor's operating frequency, voltage, and current. An important piece of information given on the nameplate is the RLA or FLA value. RLA stands for *rated-load amps* and FLA stands for *full-load amps.* The two terms are used interchangeably. Nameplates for hermetic compressors built after 1972 are marked with the RLA, as shown here.

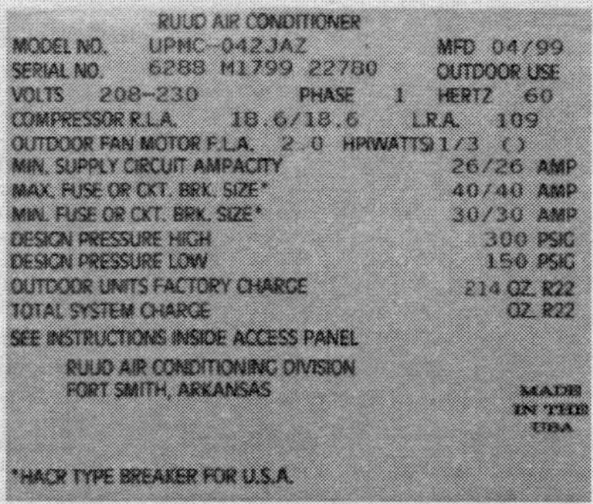

301P0107.EPS

The 60°C column gives a rating of 25A for No. 12 copper conductors (for Type NM). Multiply the nameplate ampacity by 175% to determine the circuit breaker size.

16.5A × 175% = 28.88A

Since *NEC Section 440.22(A)* states that the rating of the overcurrent device cannot exceed 175%, a 25A circuit breaker would be required.

12.0.0 ◆ MOTOR LOADS

Branch circuit conductors and overcurrent protection devices for motors are determined by using *NEC Table 430.148* for single-phase motors and *NEC Table 430.150* for three-phase motors. To use the NEC motor tables, it is necessary to know the motor horsepower, voltage, phase, and design letter. Motor information is obtained from the motor nameplate. *NEC Section 430.6(A)(1)* requires that the values given in *NEC Tables 430.147 through 430.150* be used to determine the ampacity of branch circuit conductors and the ampere rating of switches, branch circuit short circuit and ground fault protection, etc. The motor nameplate current data is not permitted to be used in sizing these components unless otherwise specified.

To determine the branch circuit conductor size for motor circuits, the motor full-load current taken from the appropriate table is multiplied by 125% and the resulting ampacity is used to select branch circuit conductors. To determine motor full-load current using the tables, it is necessary to know the motor size in horsepower, motor voltage, and motor phase. If the motor nameplate includes a value for amps but not horsepower, *NEC Section 430.6(A)* states that the horsepower rating shall be assumed to be the value given in the tables. To assume a horsepower value, the horsepower value that most closely corresponds to the ampere value in the table is selected.

Example 1:
What is the motor full-load current for a 2hp, 230V, 1∅ motor?

THINK ABOUT IT

Add-On Air Conditioning

This homeowner wants to add a 240V central air conditioner. Will the existing service be adequate to support this load?

301P0108.EPS

Teaching Tip

Discuss the "Think About It."

Classroom

Refer the trainees to *NEC Article 430* and discuss the requirements for motor loads.

Demonstrate how to calculate the branch circuit size and overcurrent protection for motor loads.

Classroom

Discuss the NEC requirements for welding equipment. Refer the trainees to *NEC Article 630.*

Demonstrate how to calculate the branch circuit size and overcurrent protection for welding loads.

NEC Table 430.148 is used to determine full-load current for single-phase AC motors. The left column in the table is used to find the correct motor hp and the ampacity is then determined based upon the motor voltage. A 2hp, 230V motor has a full-load current of 12A.

Example 2:
What is the motor full-load current for a 10hp, 460V, 3∅ motor?

NEC Table 430.150 is used to determine full-load current for three-phase AC motors. The left column is used to locate the motor hp and the ampacity is then obtained under the voltage column. A 10hp, 460V, 3∅ motor has a full-load current of 14A.

Example 3:
What size copper THWN branch circuit conductors would be required to supply a 25hp, 3∅, 460V motor?

Using *NEC Table 430.150,* the full-load current for this motor is 34A. Multiply the motor full-load current by 125% to determine the minimum ampacity for branch circuit conductors.

$$34A \times 125\% = 42.5A$$

Branch circuit conductor size is then determined using *NEC Table 310.16.* No. 6 THWN copper is rated at 55A using the 60°C column. The branch circuit conductors required to supply this 25hp motor would be No. 6 THWN copper.

Motor short circuit and ground fault protection is sized using *NEC Table 430.52.* To use this table, it is helpful to know the type of motor, horsepower, phase, motor ampacity (full-load current), and the design letter, if any.

Example 4:
What size dual-element time-delay fuse would be required for a squirrel cage motor with the following rating: 50hp, 3∅, 460V, Design Letter D?

The motor full-load current is first determined using *NEC Table 430.150.* Then, using *NEC Table 430.52,* the full-load current is multiplied by the proper percentage listed for the characteristics of the motor. From *NEC Table 430.150,* the full-load current for a 50hp, 3∅, 460V motor is 65A. From *NEC Table 430.52,* the full-load current is multiplied by 175% to select a time-delay fuse for a polyphase motor (3∅) with a design letter other than E.

$$65A \times 175\% = 113.75A$$

Per *NEC Section 430.52(C)(1),* the next lower standard fuse size is 110A. A fuse with a 110A rating would be used. If the 110A fuse proves to be inadequate to carry the load, *NEC Section 430.52(C)(1), Exception 2* allows the next higher size fuse to be used. In no case, however, can the rating exceed 225% of the full-load current.

Example 5:
What size inverse-time circuit breaker would be required for a squirrel cage motor with the following rating: 10hp, 3∅, 208V, Design Letter E?

The motor full-load current is 30.8A per *NEC Table 430.150.* The full-load current is multiplied by 250% per *NEC Table 430.52* to size an inverse-time circuit breaker for a 3∅ motor with Design Letter E.

$$30.8A \times 250\% = 77A$$

The next lower size circuit breaker is 70A, and *NEC Section 430.52(C)(1), Exception 2* would permit an 80A circuit breaker if the 70A breaker is inadequate for the load.

13.0.0 ◆ WELDERS

Branch circuit conductors and overcurrent protection devices for welders are sized using the primary current and duty cycle for the welder. This information is obtained from the nameplate on the welder. Using the relevant NEC section, the multiplier (demand factor) is determined by using the nameplate duty cycle of the welder. The primary current is multiplied by the NEC multiplier to determine the ampacity that is then used to size branch circuit conductors and overcurrent protection. Typical welders include transformer arc welders, motor-generator welders, and resistance welders.

NEC Article 630 covers electric welders. Part B covers arc welders with or without motor-generators; Part C covers resistance welders. Each NEC section lists duty cycle multipliers for individual welders based upon the type of welder. *NEC Section 630.11(A)* provides multipliers for arc welders and *NEC Section 630.31(A)(2)* provides multipliers for resistance welders.

Per *NEC Section 630.12(A),* the overcurrent protection device for an individual welder branch circuit shall have a rating or setting not exceeding 200% of the primary current rating of the welder. This NEC section applies to arc welders with or without motor-generators. For resistance welders, *NEC Section 630.32(A)* provides that the rating or setting for the overcurrent protection device shall not exceed 300% of the primary current rating of the welder.

Instructor's Notes:

> **Note**
>
> It is advisable to check with the local inspection authority or project specifications regarding the setting or rating for welder branch circuit overcurrent protection devices. Many authorities will not permit the overcurrent device rating to exceed the rating of the branch circuit conductors.

Example 1:
What size THWN copper branch circuit conductors would be required to supply an individual arc welder with a nonmotor-generator having a nameplate primary current of 70A and a duty cycle of 80%?

The nameplate primary current is multiplied by 0.89. This multiplier is obtained from ***NEC Section 630.11(A)*** for an arc welder with a nonmotor-generator and a duty cycle of 80%.

$$70A \times 0.89 = 62.3A$$

The copper THWN conductor size is selected per ***NEC Table 310.16.*** No. 4 THWN copper conductors, with a rating of 70A, would be required using the 60°C column.

Example 2:
What size THWN copper branch circuit conductors and what size circuit breaker would be required to supply a resistance welder with a nameplate primary current of 125A and a duty cycle of 40%?

The nameplate primary current is multiplied by 0.63. This multiplier is obtained from ***NEC Section 630.31(A)(2)*** for an arc welder with a duty cycle of 40%.

$$125A \times 0.63 = 78.75A$$

Per ***NEC Table 310.16,*** No. 3 THWN conductors (85A using the 60°C column) would be used. Per ***NEC Section 630.32(A),*** the primary current is multiplied by 300% to size the overcurrent protection device. The nameplate rating for this resistance welder is 125A.

$$125A \times 300\% = 375A$$

Since this value exceeds the rating of a standard 300A overcurrent protection device, ***NEC Section 630.32*** permits the next higher standard size listed in ***NEC Section 240.6(A)*** (400A) to be used.

Summary

In order to determine the size of branch circuit overcurrent protection devices and branch circuit conductors, it is important to accurately calculate branch circuit loads using NEC requirements. When the branch circuit load is calculated, branch circuit components can be sized to safely serve the load.

NEC articles cover branch circuits ranging from lighting loads to welders. As a trainee in the electrical field, you must understand branch circuit requirements.

Classroom

Have the trainees complete the Review Questions and go over the answers prior to administering the Module Examination.

Examination

Administer the Module Examination. Record the results on Craft Training Report Form 200 and submit the results to the Training Program Sponsor.

Review Questions

1. The general lighting load for a store is ____ VA/sq. ft.
   a. 180
   b. 4½
   c. 3½
   d. 3

2. The general lighting load (expressed as a unit load per square foot) for a warehouse is ____ VA/sq. ft.
   a. ½
   b. ¼
   c. 1
   d. 1½

3. A continuous load is defined by the NEC as a load whose maximum current is expected to continue for at least ____.
   a. 30 minutes
   b. 1 hour
   c. 8 hours
   d. 3 hours

4. What is the capacity, in amperes, for a 20A circuit breaker supplying a continuous load?
   a. 20
   b. 12
   c. 16
   d. 18

5. Branch circuit conductors supplying continuous duty loads are calculated at ____% of the rated load.
   a. 100
   b. 125
   c. 80
   d. 115

6. What is the current-carrying ampacity for each of 10 No. 10 THHN copper conductors installed in a single conduit?
   a. 20A
   b. 28A
   c. 32A
   d. 40A

7. The voltage drop should not exceed ____% to the farthest outlet in a branch circuit.
   a. 2
   b. 3
   c. 5
   d. 10

8. The voltage drop for a 208V, 1Ø circuit with a load of 26.5A, using No. 8 stranded copper conductors at a circuit length of 145' is ____.
   a. 5.87V
   b. 5.98V
   c. 5.08V
   d. 5.18V

9. The voltage drop for a 480V, 3Ø circuit with a load of 26.5A, using No. 8 stranded copper conductors at a circuit length of 145' is ____.
   a. 5.87V
   b. 5.98V
   c. 5.08V
   d. 5.18V

10. Using the resistance formula, the voltage drop for an 18A, 208V, 3Ø load with a total circuit length of 105' using No. 10 solid copper conductors is ____.
    a. 3.96V
    b. 4.57V
    c. 4.06V
    d. 4.69V

11. Load calculations for circuits supplying lighting units with ballasts are based upon ____.
    a. the total wattage of all lamps in the fixtures
    b. the circuit voltage times lamp wattage
    c. the ampere ratings of the ballasts
    d. 180VA per ballast

12. What is the calculated load, in amps, for a 120V circuit supplying four 150W recessed incandescent fixtures and six recessed fluorescent fixtures operating continuously? (Each fluorescent ballast is rated at .65A.)
    a. 3.9A
    b. 6.07A
    c. 12.5A
    d. 11.125A

13. What is the maximum number of general-purpose noncontinuous duty duplex receptacles that can be connected to a 20A, 120V circuit in a commercial building? What is the NEC required load rating for each receptacle?
    a. 10 receptacles; 1.5A
    b. 10 receptacles; 1.875A
    c. 16 receptacles; 150VA
    d. 13 receptacles; 180VA

14. The load for 6' of multi-outlet assembly used simultaneously is ____.
    a. 180VA per outlet
    b. 180VA per outlet times 125%
    c. 1,080VA
    d. 216VA

Instructor's Notes:

15. How many receptacles are required for 90 linear feet of show window area?
   a. 4
   b. 7
   c. 9
   d. 8

16. What is the maximum computed load for 55 linear feet of show window?
   a. 9,900VA
   b. 13,750VA
   c. 10,000VA
   d. 6,600VA

17. What is the total load, in VA, for two small appliance branch circuits in a single-family dwelling?
   a. 2,400VA
   b. 1,800VA
   c. 3,000VA
   d. 1,500VA

18. What is the demand load, in amps, for one household electric range with a nameplate rating of 16.75kW, 1Ø, 240V?
   a. 69.79A
   b. 41.67A
   c. 50A
   d. 33.33A

19. What is the demand load, in amps, for a single circuit supplying one wall-mounted oven rated 11.75kW, 1Ø, 240V, and a countertop cooking unit rated 9.6kW, 1Ø, 240V?
   a. 66.67A
   b. 48.33A
   c. 50A
   d. 88.96A

20. What is the demand load, in amps, for seven pieces of commercial cooking equipment with a total nameplate rating of 44.5kVA with thermostatic control operating at continuous duty? Each piece of equipment has a nameplate rating of 208V, 3Ø.
   a. 80.29A
   b. 108.08A
   c. 100.36A
   d. 123.52A

21. What is the minimum conductor ampacity for four 1,000W, 208V, 1Ø electric baseboard heaters?
   a. 24.04A
   b. 33.33A
   c. 20.83A
   d. 19.23A

22. What size THWN copper branch circuit conductors are required to supply an air conditioning unit with a nameplate rating of 33.5A, 208V, 3Ø?
   a. 10
   b. 8
   c. 6
   d. 4

23. What is the maximum size time-delay fuse that can be used for an AC motor rated at 15hp, 208V, 3Ø, with a Design Letter E?
   a. 60A
   b. 70A
   c. 90A
   d. 110A

24. What is the motor full-load current for a single-phase motor rated at 5hp, 230V, and what is the minimum ampacity for branch circuit conductors?
   a. 28A; 35A
   b. 15.2A; 28A
   c. 16.7A; 16.7A
   d. 30.8A ; 28A

25. An AC nonmotor-generator arc welder on an individual circuit has a nameplate primary current of 100A and a duty cycle of 70%. The branch circuit conductors must be a minimum size ____ AWG THHN copper.
   a. No. 3
   b. No. 6
   c. No. 2
   d. No. 4

GLOSSARY

Trade Terms Introduced in This Module

Ampacity: The current in amperes that a conductor can carry continuously under the conditions of use without exceeding its temperature rating.

Appliance: Utilization equipment, generally other than industrial, normally built in standardized sizes or types, that is installed or connected as a unit to perform one or more functions such as clothes washing, air conditioning, food mixing, deep frying, etc.

Appliance branch circuit: A branch circuit supplying energy to one or more outlets to which appliances are to be connected. Such circuits are to have no permanently connected lighting fixtures that are not part of an appliance.

Branch circuit: The circuit conductors between the final overcurrent device protecting the circuit and the outlet(s).

Continuous load: A load in which the maximum current is expected to continue for three hours or more.

Demand factors: The ratio of the maximum demands of a system, or part of a system, to the total connected load of a system or the part of the system under consideration.

Device: A unit of an electrical system that is intended to carry but not utilize electric energy.

General-purpose branch circuit: A branch circuit that supplies a number of outlets for lighting and appliances.

Individual branch circuit: A branch circuit that supplies only one piece of utilization equipment.

Multi-outlet assembly: A type of surface or flush raceway designed to hold conductors and receptacles, assembled in the field or at the factory.

Outlet: A point on the wiring system at which current is taken to supply utilization equipment.

Overcurrent: Any current in excess of the rated current of equipment or the ampacity of a conductor. It may result from overload, short circuit, or ground fault.

Receptacle: A contact device installed at an outlet for connection as a single contact device. A single receptacle is a single contact device with no other contact device on the same yoke. A multiple receptacle is a single device containing two (duplex) or more receptacles.

Receptacle outlet: An outlet where one or more receptacles are installed.

Utilization equipment: Equipment that utilizes electric energy for electronic, chemical, heating, lighting, electromechanical, or similar purposes.

Instructor's Notes:

Additional Resources

This module is intended to present thorough resources for task training. The following reference work is suggested for further study. This is an optional material for continued education rather than for task training.

National Electrical Code Handbook. Quincy, MA: National Fire Protection Association, 2002.

ANSWER KEY

Answers to Review Questions

Answer	Section
1. d	1.0.0
2. b	1.0.0
3. d	1.1.0
4. c	1.1.0
5. b	1.1.0
6. a	1.2.0
7. b	1.2.2
8. b	1.2.2
9. d	1.2.3
10. a	1.2.3
11. c	2.0.0
12. d	2.1.0
13. d	3.0.0
14. c	4.0.0
15. d	5.0.0
16. b	5.0.0
17. c	7.1.0
18. b	7.4.0
19. b	7.4.0
20. a	8.0.0
21. a	10.0.0
22. c	11.0.0
23. c	12.0.0
24. a	12.0.0
25. a	13.0.0

TEACHING TIPS—MODULE 26301-02

Section 1.0.0 *Think About It—Load Calculations*

1. Cost
2. Size of the existing service
3. Whether or not there is space in the panel
4. Installation considerations such as clearance requirements and unit disconnect requirements
5. Grounding and bonding requirements

Section 1.0.0 *Think About It—Unit Load per Square Foot*

An office building has a unit load of 4½VA square foot. ***NEC Table 220.3(A)*** provides a value of 3½ VA per square foot and an additional 1VA per square foot is added per ***Note B.***

Section 1.1.0 *Think About It—Service Size*

NEC Section 230.79(C) requires a minimum of 100A.

Section 1.2.1 *Think About It—Voltage Drop Formula Factors*

The number 2 in this formula represents the two current-carrying conductors in a single-phase circuit.

Section 4.0.0 *Think About It—Multi-Outlet Assemblies*

The load is 180VA.

Section 5.0.0 *Think About It—Show Window Load Calculations*

5 ft × 200VA/ft = 1,000VA × 125% = 1,250VA

Section 7.1.0 *Think About It—Adding a Kitchen*

A new kitchen adds 3,000VA to the load calculation because the NEC requires that each kitchen be served by at least two 20A small appliance branch circuits plus the individual loads of any fixed appliance, such as dishwashers, trash compactors, etc.

Section 7.2.0 *Think About It—Residential Laundry Room Circuit*

The NEC does not specifically require a separate circuit for the washing machine, but it does require one small appliance circuit in the laundry room, which is usually used for the washing machine.

Section 7.4.0 *Think About It—One Cooktop and One Oven*

NEC Table 220.19, Note 4 covers an installation with one cooktop and one oven.

Section 7.4.0 *Think About It—Demand Loads*

The total demand load is now 12,750VA, which is actually lower than the existing load. This is because *NEC Section 220.17* allows a demand factor of 75% for four or more appliances serving certain types of equipment in dwelling units.

Section 9.0.0 *Think About It—Water Heater Calculation*

4,500VA ÷ 240V = 18.75A

18.75A × 125% = 23.44A

This installation requires No. 10 AWG copper with a 30A circuit breaker.

Section 11.0.0 *Think About It—Add-On Air Conditioning*

No, a two-wire service can only supply 120V loads.

CONTREN™ LEARNING SERIES — USER UPDATES

The NCCER makes every effort to keep these textbooks up-to-date and free of technical errors. We appreciate your help in this process. If you have an idea for improving this textbook, or if you find an error, a typographical mistake, or an inaccuracy in NCCER's Contren™ textbooks, please write us, using this form or a photocopy. Be sure to include the exact module number, page number, a detailed description, and the correction, if applicable. Your input will be brought to the attention of the Technical Review Committee. Thank you for your assistance.

Instructors – If you found that additional materials were necessary in order to teach this module effectively, please let us know so that we may include them in the Equipment/Materials list in the Instructor's Guide.

Write: Curriculum Revision and Development Department
National Center for Construction Education and Research
P.O. Box 141104, Gainesville, FL 32614-1104

Fax: 352-334-0932

E-mail: curriculum@nccer.org

Craft ____________________ Module Name ____________________

Copyright Date ____________ Module Number ____________ Page Number(s) ____________

Description ____________________

____

____

____

(Optional) Correction ____________________

____

____

(Optional) Your Name and Address ____________________

____

____

Conductor Selection and Calculations

26302-02

Conductor Selection and Calculations
Annotated Instructor's Guide

Module 26302-02

MODULE OVERVIEW

This course introduces the Electrical trainee to conductor selection and calculations using various tables in the NEC.

PREREQUISITES

Please refer to the Course Map in the Trainee Module. Prior to training with this module, it is recommended that the trainee shall have successfully completed the following:

Core Curriculum; Electrical Level One; Electrical Level Two; Electrical Level Three, Module 26301.

OBJECTIVES

Upon completion of this module, the trainee will be able to:

1. Select electrical conductors for specific applications.
2. Calculate voltage drop in both single-phase and three-phase applications.
3. Apply NEC regulations governing conductors to a specific application.
4. Calculate and apply NEC tap rules to a specific application.
5. Size conductors for the load.
6. Derate conductors for fill, temperature, and voltage drop.
7. Select conductors for various temperature ranges and atmospheres.

PERFORMANCE TASKS

This is a knowledge-based module; there is no performance examination.

NCCER STANDARDIZED CRAFT TRAINING PROGRAMS

The National Center for Construction Education and Research (NCCER) provides a standardized national program of accredited craft training. Key features of the program include instructor certification, competency-based training, and performance testing. The program provides trainees, instructors, and companies with a standard form of recognition through a National Craft Training Registry. The program is described in full in the *Guidelines for Accreditation*, published by the NCCER. For more information on standardized craft training, contact the NCCER by writing us at P.O. Box 141104, Gainesville, FL 32614-1104, calling 352-334-0911, or e-mailing info@nccer.org. More information may be found at our Web site, www.nccer.org.

HOW TO USE THIS ANNOTATED INSTRUCTOR'S GUIDE

Each page presents two sections of information. The larger section displays each page exactly as it appears in the Trainee Module. The narrow column ties suggested trainee and instructor actions to each page and provides icons to call your attention to material, safety, audiovisual, or testing requirements. The bottom of each page includes space for your notes.

If you see the Teaching Tip icon, that means there is a teaching tip associated with this section. Also refer to the suggested teaching tips at the end of the module.

SAFETY CONSIDERATIONS

Ensure that the trainees are equipped with appropriate personal protective equipment.

PREPARATION

Before teaching this module, you should review the Module Outline, Objectives, and the Materials and Equipment List. Be sure to allow ample time to prepare your own training or lesson plan and gather all required equipment and materials.

MATERIALS AND EQUIPMENT LIST

Materials:
Transparencies
Markers/chalk
Copy of the latest edition of the *National Electrical Code®*
One length each of various solid and stranded conductors
Module Examinations*

Equipment:
Overhead projector and screen
Whiteboard/chalkboard
Appropriate personal protective equipment

*Located in the Test Booklet packaged with this Annotated Instructor's Guide.

ADDITIONAL RESOURCES

This module is intended to present thorough resources for task training. The following reference works are suggested for both instructors and motivated trainees interested in further study. These are optional materials for continued education rather than for task training.

American Electrician's Handbook, Terrell Croft and Wilfred I. Summers. New York, NY: McGraw-Hill, 1996.

National Electrical Code Handbook. Quincy, MA: National Fire Protection Association, 2002.

TEACHING TIME FOR THIS MODULE

An outline for use in developing your lesson plan is presented below. Note that each Roman numeral in the outline equates to one session of instruction. Each session has a suggested time period of 2½ hours. This includes 10 minutes at the beginning of each session for administrative tasks and one 10-minute break during the session. Approximately 15 hours are suggested to cover *Conductor Selection and Calculations.* You will need to adjust the time required for hands-on activity and testing based on your class size and resources.

Topic	Planned Time
Session I. Introduction	
A. Introduction	________
1. Compact Conductors	________
Session II. Conductor Applications	
A. Conductor Applications	________
1. Branch Circuits	________
2. Conductor Protection	________
a. Location of Overcurrent Protection in Circuits	________
Session III. Properties of Conductors	
A. Properties of Conductors	________
1. Identifying Conductors	________
a. Color-Coding	________
b. Changing Colors	________

Session IV. Voltage Drop

A. Voltage Drop ____________
   1. Wire Sizes Based on Resistance ____________
      a. Circular Mil—Unit of Conductor Area ____________
      b. Conversion of Square Mils to Circular Mils ____________
   2. Resistance of Conductors ____________
   3. Resistance of Copper per Mil Foot ____________

Session V. Voltage Drop Equations

A. Voltage Drop Equations ____________
   1. Equations for Voltage Drop Using Conductor Area or Conductor Resistance ____________
   2. Use of Voltage Drop Equations ____________
      a. Miscellaneous Voltage Drop Equations ____________

Session VI. Review; Module Examination

A. Review ____________

B. Module Examination ____________
   1. Trainees must score 70% or higher to receive recognition from the NCCER.
   2. Record the testing results on Craft Training Report Form 200 and submit the results to the Training Program Sponsor.

Module 26302-02

Conductor Selection and Calculations

Course Map

This course map shows all of the modules in the third level of the Electrical curriculum. The suggested training order begins at the bottom and proceeds up. Skill levels increase as you advance on the course map. The local Training Program Sponsor may adjust the training order.

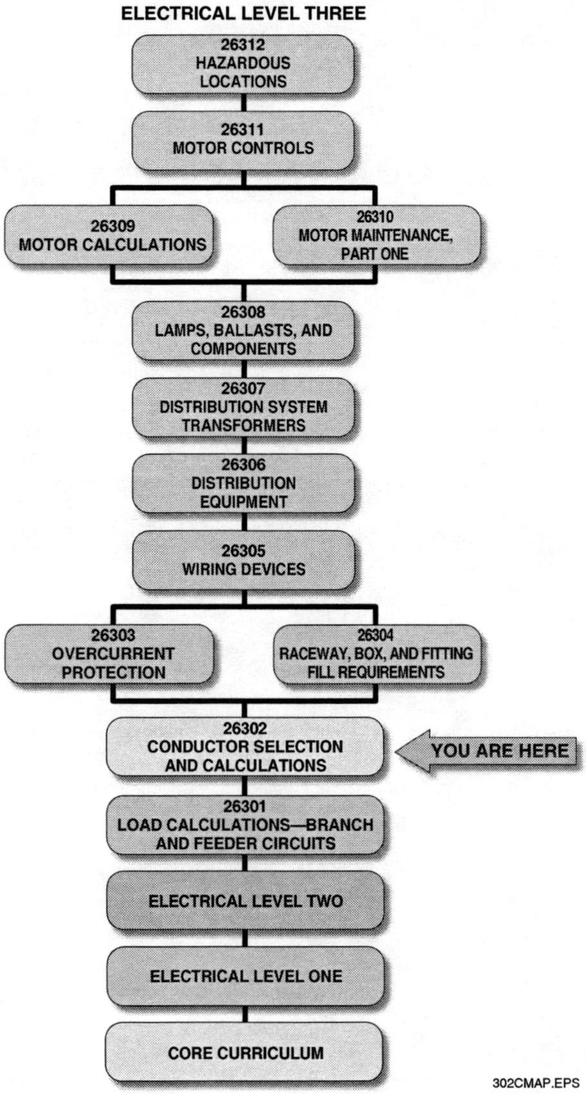

302CMAP.EPS

Homework

Assign reading of Module 26302.

MODULE 26302 CONTENTS

Figures

Tables

Instructor's Notes:

MODULE 26302

Conductor Selection and Calculations

Objectives

When you have completed this module, you will be able to do the following:

1. Select electrical conductors for specific applications.
2. Calculate voltage drop in both single-phase and three-phase applications.
3. Apply NEC regulations governing conductors to a specific application.
4. Calculate and apply NEC tap rules to a specific application.
5. Size conductors for the load.
6. Derate conductors for fill, temperature, and voltage drop.
7. Select conductors for various temperature ranges and atmospheres.

Prerequisites

Before you begin this module, it is recommended that you successfully complete the following: Core Curriculum; Electrical Level One; Electrical Level Two; Electrical Level Three, Module 26301.

Required Trainee Materials

1. Pencil and paper
2. Appropriate personal protective equipment
3. Copy of the latest edition of the *National Electrical Code*

1.0.0 ◆ INTRODUCTION

A variety of materials may be used to transmit electrical energy, but copper, due to its excellent cost-to-conductivity ratio, still remains the most ideal conductor. Electrolytic copper, the type used in most electrical conductors, has the following general characteristics:

- *Method of stranding*—Stranding refers to the relative flexibility of the conductor and may consist of only one strand or many strands, depending on the rigidity or flexibility required for a specific need. ***NEC Section 310.3*** requires all conductors size No. 8 and larger to be stranded (more than one strand) when installed in a raceway. For example, a small-gauge wire that is to be used in a fixed installation is normally solid (one strand), whereas a wire that will be constantly flexed requires a high degree of flexibility and would contain many strands.
  - –Solid wire is the least flexible form of a conductor and is merely one strand of copper.
  - –Stranded refers to more than one strand in a given conductor and may vary widely, depending on size. See *Figure 1* and *NEC Chapter 9, Table 8.*

> INSIDE TRACK
>
> ***Solid Conductors***
>
> According to ***NEC Chapter 9, Table 8,*** a solid conductor is considered a one-stranded conductor.

Note: The designations "National Electrical Code," "NE Code," and "NEC," where used in this document, refer to the *National Electrical Code*®, which is a registered trademark of the National Fire Protection Association, Quincy, MA. *All National Electrical Code (NEC) references in this module refer to the 2002 edition of the NEC.*

Materials

Ensure that you have everything required to teach the course. Check the Materials and Equipment List at the front of this Instructor's Guide.

Classroom

Review the goals of the course.

Explain that terms shown in bold (blue) are defined in the Glossary at the back of this module.

Audiovisual

Show Transparency 1, Course Objectives.

Classroom

Discuss methods of stranding and provide examples for the trainees to examine. Point out that the NEC requires all conductors size No. 8 and larger to be stranded when installed in a raceway.

Audiovisual

Show Transparency 2 (Figure 1).

Classroom

List the categories used to indicate the relative hardness of a conductor.

Discuss the various coatings used with bare copper.

Describe the American Wire Gauge (AWG) system of wire sizing.

Audiovisual

Show Transparency 3 (Figure 2).

Teaching Tip

See the Teaching Tip at the end of this module.

- *Degree of hardness (temper)*–Temper refers to the relative hardness of the conductor and is noted as soft drawn (SD), medium hard drawn (MHD), and hard drawn (HD). Again, the specific need of an installation will determine the required temper. Where greater tensile strength is indicated, MHD would be used over SD, and so on.
- *Bare, tinned, or coated*–Untinned copper is plain, bare copper that is available in either solid or stranded types and in the various tempers just described. In this form, it is often referred to as red copper. Bare copper is also available with a coating of tin, silver, or nickel to facilitate soldering, impede corrosion, and prevent adhesion of the copper conductor to rubber or other types of conductor insulation. The various coatings will also affect the electrical characteristics of copper.

The **American Wire Gauge (AWG)** is used in the United States to identify the sizes of wire and **cable** up to and including No. 4/0 (0000), which is commonly pronounced in the electrical trade as four-aught or four-naught. These numbers run in reverse order as to size; that is, No. 14 AWG is smaller than No. 12 AWG and so on up to size No. 1 AWG. Up to this size (No. 1 AWG), the larger the gauge number, the smaller the size of the conductor. However, the next larger size after No. 1 AWG is No. 1/0 AWG, then 2/0 AWG, 3/0 AWG, and 4/0 AWG. At this point, the AWG designations end and the larger sizes of conductors are identified by circular mils (CM or cmil). From this point, the larger the size of wire, the larger the number of circular mils. For example, 300,000 cmil is larger than 250,000 cmil. In writing these sizes in circular mils, the thousand decimal is replaced by the letter *k*, and instead of writing, say, 500,000 cmil, it is usually written 500 kcmil—pronounced five-hundred kay-cee-mil. See *Figure 2* for a comparison of the different wire sizes.

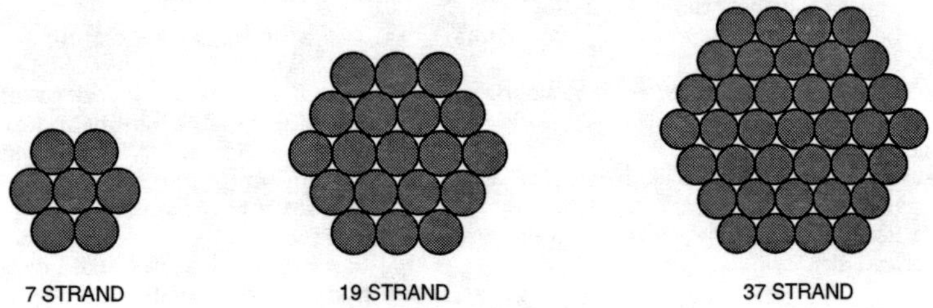

Figure 1 ◆ Common strand configurations.

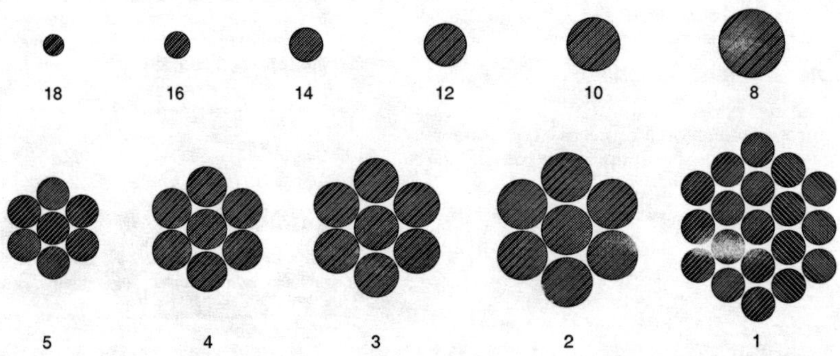

Figure 2 ◆ Comparison of various wire sizes.

Instructor's Notes:

1.1.0 Compact Conductors

Compact aluminum conductors are those conductors that have been compressed so as to reduce the air space between the strands. *Figure 3* shows a cross section of a 37-strand compact conductor.

The purpose of compact conductors is to reduce the overall diameter of the cable so that it may be installed in a conduit that is smaller than that required for standard stranded conductors of the same wire size. Compact conductors are especially useful when increasing the ampacity of an existing service or feeder circuits.

For example, an existing service is rated at 250A and is fed with four 350 kcmil THW conductors in 3" conduit. Should it become necessary to increase the ampacity of the service to 300A, 500 kcmil THW compact conductors may replace the 350 kcmil conductors without increasing the size of the conduit. Compact conductors are only available in aluminum, and their properties are listed in *NEC Chapter 9, Table 5A*.

Both standard and compressed conductors are covered in this module, including various types of conductor insulation and the practical applications of each type.

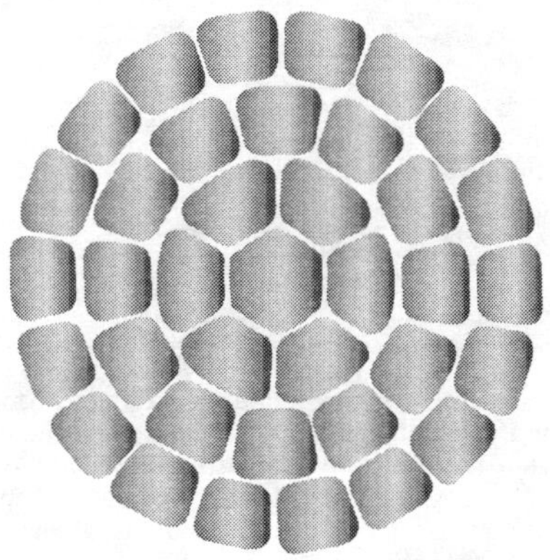

302F03.EPS

Figure 3 ◆ Cross section of a 37-strand compact conductor.

2.0.0 ◆ CONDUCTOR APPLICATIONS

NEC Article 100 defines feeders as the circuit conductors between the service equipment (or other power supply source) and the final branch circuit overcurrent device. A branch circuit is defined as the circuit conductors between the final overcurrent device protecting the circuit and the outlet(s). The power riser diagram in *Figure 4* shows examples of both feeders and branch circuits.

When current-carrying conductors are used in an electrical system, the NEC requires that each ungrounded conductor be protected from damage by an overcurrent protective device such as a fuse or circuit breaker. The conductors must also

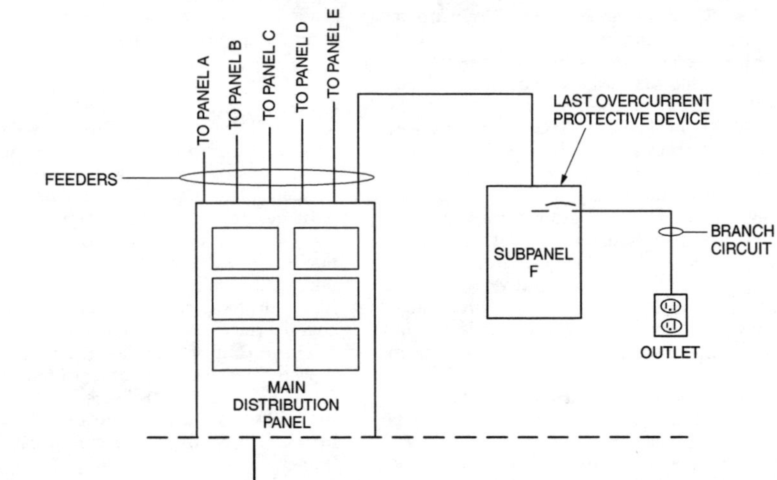

302F04.EPS

Figure 4 ◆ Power riser diagram showing feeders and branch circuits.

Classroom

Discuss compact aluminum conductors and their applications. Provide an example for the trainees to examine.

Audiovisual

Show Transparency 4 (Figure 3).

Classroom

Review the components of a typical power riser diagram.

Audiovisual

Show Transparency 5 (Figure 4).

Classroom

Explain that conductors must be protected from excess current caused by overloads, short circuits, or ground faults. Note that in most cases overcurrent devices are located at the point where the protected conductor receives its supply.

Audiovisual

Show Transparencies 6 and 7 (Figures 5 and 6).

INSIDE TRACK

Roughed-In Residential Feeders

In the rough-in installation shown here, the large blue cable will feed a subpanel on the inside of the house, while the other three insulated cables will probably feed electrical heating units, an air conditioning unit, and other relatively high loads. The small wire is a bare grounding conductor that will connect the grounding electrode to the inside panel grounding bus.

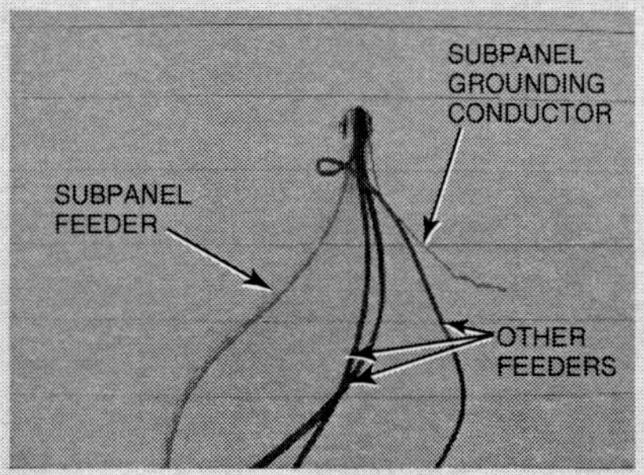

302P0201.EPS

be identified so that the ungrounded conductors may be distinguished from the grounded and grounding conductors. Minimum sizes or required ampacity for any of these conductors used in a circuit are selected based on NEC rules and tables. The NEC also provides tables that list the physical and electrical properties of conductors to allow selection of the proper conductor for any application.

NEC Section 240.4 covers the requirements for protecting conductors from excess current caused by overloads, short circuits, or ground faults. The setting or sizes of the protective device are based on the ampacity of the conductors as listed in *NEC Tables 310.16 through 310.19.* Under certain conditions, the overcurrent device setting may be larger than the ampacity rating of the conductors as listed in the exceptions to the basic rules. For convenience, a standard rating of a fuse or circuit breaker may be used even if this rating exceeds the ampacity of the conductor as long as the next higher standard rating does not exceed 800A. For example, a branch circuit with a load of 56A may be protected with a 60A overcurrent device, because 56A is not a standard overcurrent device size.

In most cases, an overcurrent device must be connected at the point where the conductor to be protected receives its supply (*NEC Section 240.21*). The most common situations are shown in *Figure 5,* which illustrates the basic rule and several exceptions, including the 10' tap rule. *Figure 6* illustrates the 25' tap rule.

Wiring systems require a grounded conductor in most installations. A grounded conductor, such as a neutral, or a grounding conductor must be identified either by the color of its insulation, by markings at the terminals, or by other suitable means (*NEC Section 200.6*). In general, a grounded conductor must have a white or gray finish. When this is not practical for conductors larger than No. 6 AWG, marking the terminations white is an acceptable method of identifying the conductor. Tagging is also acceptable.

2.1.0 Branch Circuits

Now that an overview of NEC overcurrent protection has been presented, we will examine the NEC installation requirements for conductors.

In general, the ampacity (current-carrying capacity) of a conductor must not be less than the maximum load served. However, there are exceptions to this rule, namely, a branch circuit supplying a motor. Motors and motor circuits are covered in *NEC Article 430.*

The rating of the branch circuit overcurrent device determines the rating of the branch circuit (*NEC Section 210.3*). For example, if a No. 10 AWG, 30A conductor is protected by a 20A circuit breaker, then the circuit is considered a 20A branch circuit.

Furthermore, the current-carrying capacity of branch circuit conductors must not be less than the maximum load to be served. Where the branch

Instructor's Notes:

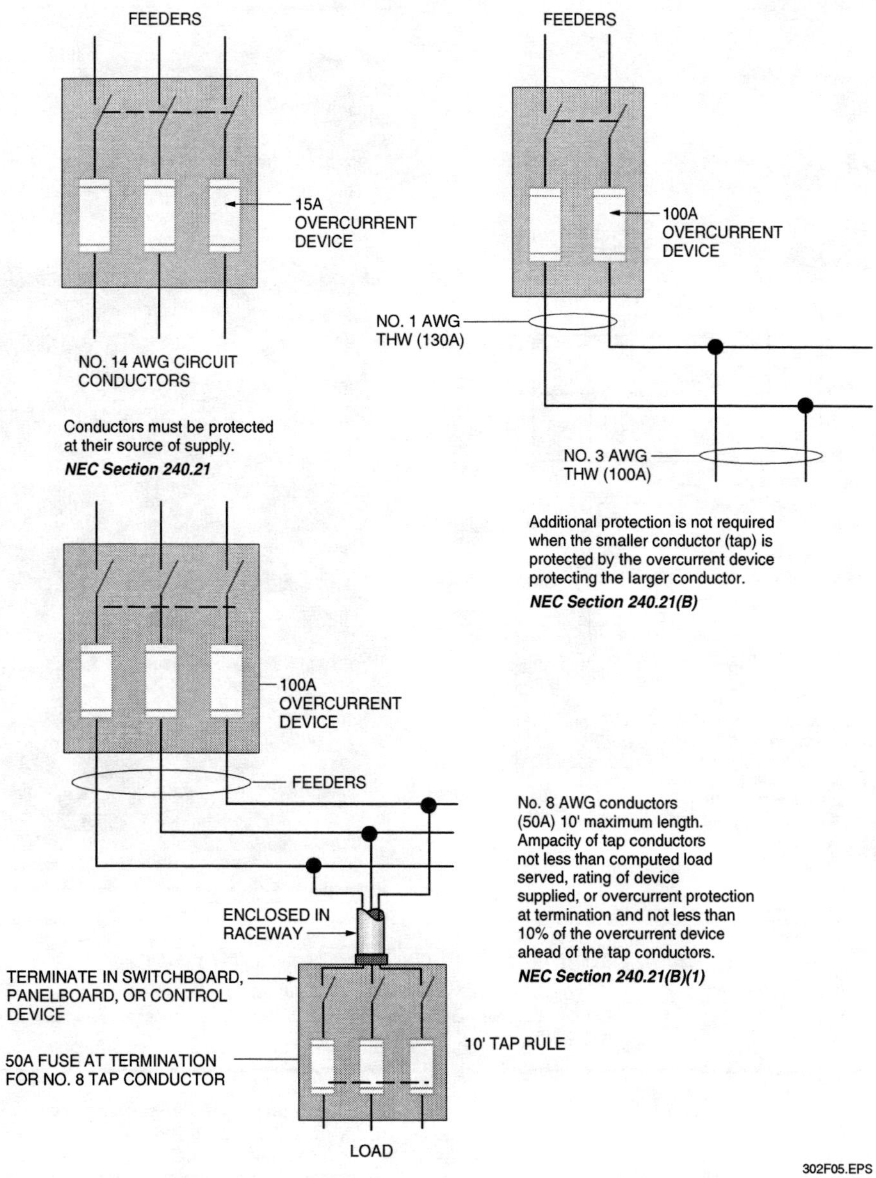

Figure 5 ◈ Location of overcurrent protection in circuits.

Classroom

Point out the requirements of the 10' tap rule.

Classroom

Point out the requirements of the 25' tap rule.

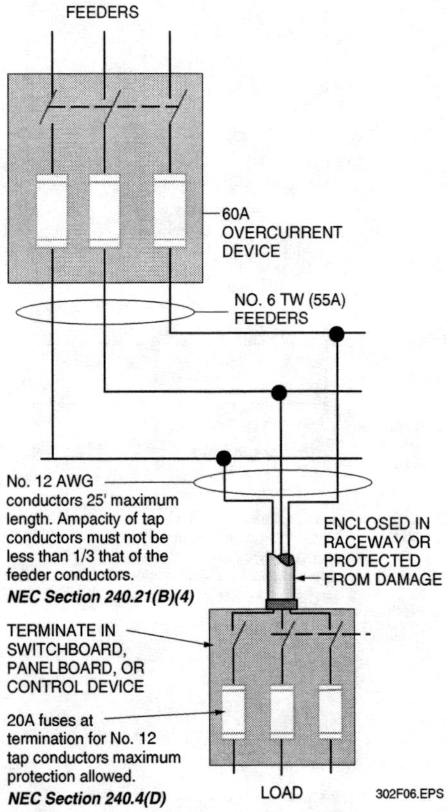

Figure 6 ◆ NEC 25' tap rule.

THINK ABOUT IT

Conductor Identification

Can you identify the ungrounded feeder conductors, the grounded feeder conductor, and the grounding feeder conductor in this picture?

302P0202.EPS

Teaching Tip

Discuss the "Think About It."

INSIDE TRACK

Minimum Branch Circuit Conductor Size

The smallest size conductor permitted by ***NEC Table 310.5*** for branch circuits in residential, commercial, and industrial locations is No. 14 copper. However, some local codes may require No. 12 or larger.

circuit supplies receptacle outlets for use with cord- and plug-connected appliances and other utilization equipment, the conductor's ampacity must not be less than the rating of the branch circuit overcurrent device.

As mentioned previously, when the ampacity of the conductor does not match up with a standard rating of fuses or circuit breakers, the next higher standard-size overcurrent device may be used, provided that the overcurrent device does not exceed 800A (*NEC Section 240.4[B]*). This exception is not permitted, however, when the branch circuit supplies receptacles where cord- and plug-connected appliances and similar electrical equipment could be used, because too many loads plugged into the circuit could result in an

Instructor's Notes:

overload condition. The next standard size fuse or circuit breaker may be used only when the circuit supplies a fixed load.

The allowable ampacity of conductors used on most electrical systems is found in *NEC Tables 310.16 through 310.19.* However, the ampacities are subject to correction factors that must be applied where high ambient temperatures are encountered, that is, when the ambient temperature for the conductor location exceeds 30°C (86°F). This reduction is required even if the reduction for more than three conductors in a raceway is also applied. For example, if six No. 10 AWG, TW current-carrying conductors are installed in a single raceway where the ambient temperature is 40°C, the ampacity of 30A must be derated or reduced to 80% of its value because of conduit fill (*NEC Table 310.15[B][2][a]*) and then reduced again by a correction factor of 0.82 (*NEC Table 310.16*) because of the ambient temperature. Therefore, when more than three current-carrying conductors are installed in a single raceway or cable in an ambient temperature of 40°C, the allowable ampacity for this condition is calculated as follows:

$$30A \times 0.80 \times 0.82 = 19.68A$$

In this situation, the listed ampacities must be reduced because of the heating effect of many current-carrying conductors in proximity. Grounding conductors are not counted as current-carrying conductors.

The rating of the branch circuit overcurrent device serving continuous loads must be not less than the noncontinuous load plus 125% of the continuous load (*NEC Section 210.20[A]*).

> **Note**
>
> ***NEC Article 100*** defines a continuous load as a load where the maximum current is expected to continue for three hours or more.

2.2.0 Conductor Protection

According to the NEC, conductors must be installed and protected from damage (both physically and electrically). Additional requirements specify the use of boxes or fittings for certain connections, specify how connections are made to terminals, and restrict the use of parallel conductors. When conductors are installed in enclosures or raceways, additional rules apply. Finally, if conductors are installed underground, the burial depth and other installation requirements are specified by the NEC. All conductors must be protected against overcurrent in accordance with their ampacities as set forth in the NEC. They must also be protected against overloads and ground fault/short circuit current damage.

According to *NEC Section 240.6*, standard overcurrent device sizes are 15A, 20A, 25A, 30A, 35A, 40A, 45A, 50A, 60A, 70A, 80A, 90A, 100A, 110A, 125A, 150A, 175A, 200A, 225A, 250A, 300A, 350A, 400A, 450A, 500A, 600A, 700A, 800A, 1,000A, 1,200A, 1,600A, 2,000A, 2,500A, 3,000A, 4,000A, 5,000A, and 6,000A. Additional standard ratings for fuses are 1A, 3A, 6A, 10A, and 601A.

> **Note**
>
> The small fuse ratings of 1A, 3A, 6A, and 10A were added to the NEC to provide more effective overcurrent protection for small loads.

Protection of conductors under short circuit conditions is accomplished by obtaining the maximum short circuit current available at the supply end of the conductor, the short circuit withstand rating of the conductor, and the short circuit let-through characteristics of the overcurrent device.

When a noncurrent-limiting device is used for short circuit protection, the conductor's short circuit withstand rating must be properly selected based on the overcurrent protective device's ability to protect the circuit. See *Figure 7*.

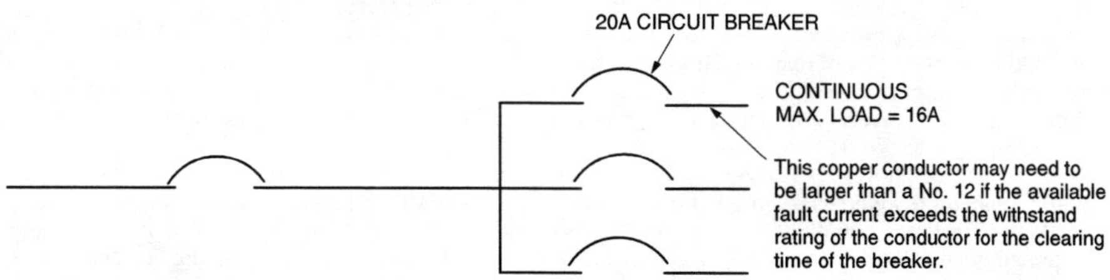

Figure 7 ◆ Noncurrent-limiting device.

302F07.EPS

Classroom

Explain how to adjust the ampacities listed in *NEC Tables 310.16 through 310.19* to account for high ambient temperatures and/or three or more conductors in a raceway.

Provide the trainees with several example problems and have them apply the correction factors. Point out that the branch circuit overcurrent device must be rated at 125% of the continuous load.

Discuss the NEC requirements for protecting conductors from damage.

List the standard overcurrent device sizes. Refer the trainees to *NEC Section 240.6*.

Define *noncurrent-limiting device*.

Audiovisual

Show Transparency 8 (Figure 7).

Classroom

Define *current-limiting device*. Discuss the advantages of this type of overcurrent device.

Audiovisual

Show Transparency 9 (Figure 8).

Classroom

Review the NEC requirements for the location of overcurrent protection in a circuit. Refer the trainees to *NEC Section 240.21*.

INSIDE TRACK

Circuit Breakers

Although a circuit breaker may appear to be a simple on/off switch, it is actually a complex mechanical device. This picture shows the internal components of a typical circuit breaker.

302P0203.EPS

It is necessary to check the energy let-through of the overcurrent device under short circuit conditions. Select a wire size of sufficient short circuit withstand ability.

In contrast, the use of a current-limiting device permits a device to be selected that limits short circuit current to a level substantially less than obtainable in the same circuit if the overcurrent protection device were replaced with a solid conductor having comparable impedance (resistance)—doing away with the need for oversized ampacity conductors. See *Figure 8.*

In many applications, it is desirable to use the convenience of a circuit breaker for a disconnecting means and general overcurrent protection, supplemented by current-limiting devices at strategic points in the circuits.

Per ***NEC Section 240.5***, flexible cords, including tinsel cords and extension cords, must be protected against overcurrent in accordance with their ampacities as listed in ***NEC Table 400.5(A)***. Supplementary overcurrent protection is acceptable. For example, with No. 18 AWG fixture wire that is 50' in length or more, a 6A overcurrent device would provide the necessary protection. For No. 16 AWG fixture wire of 100' or more, an 8A overcurrent device would provide the necessary protection. For No. 18 AWG extension cords, a 10A

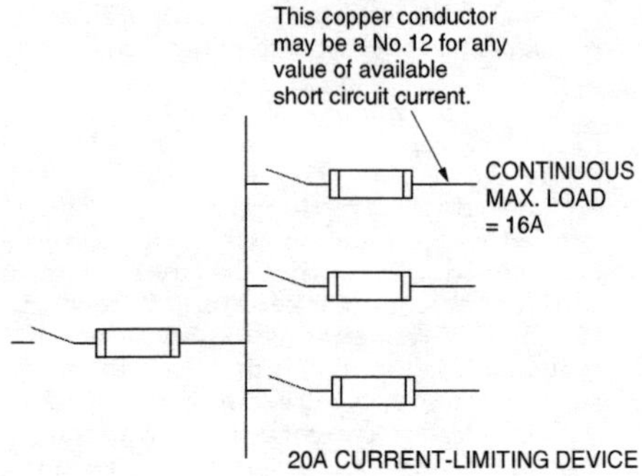

302F08.EPS

Figure 8 ◆ Current-limiting device.

overcurrent device would provide the necessary protection for a cord where only two conductors are carrying current; a 7A overcurrent device would provide the necessary protection for a cord where three conductors are carrying current.

2.2.1 Location of Overcurrent Protection in Circuits

In general, overcurrent protection must be installed at points where the conductors receive their supply, that is, at the beginning or line side of a branch circuit or feeder. See ***NEC Section 240.21.*** Exceptions to this rule follow:

- Per ***NEC Section 240.21(B)(1)***, overcurrent protection is not required at the conductor supply if a feed tap conductor is not over 10' long; is enclosed in raceway; does not extend beyond the switchboard, panelboard, disconnecting means, or control device that it supplies; and has an ampacity not less than the combined computed loads supplied and not less than the rating of the device supplied by the tap conductors or not less than the rating of the overcurrent device at the tap conductor termination. For field-installed taps, the ampacity of the overcurrent device on the line side of the tap conductor cannot exceed 10 times the ampacity of the tap conductor.
- Per ***NEC Section 240.21(B)(2)***, overcurrent protection is not required at the conductor supply if a feeder tap conductor is not over 25' long; is suitably protected from physical damage; has an ampacity not less than ⅓ that of the overcurrent device protecting the feeder conductors; and terminates in a single overcurrent device.

Instructor's Notes:

- Per *NEC Section 240.21(B)(3),* overcurrent protection is not required at the conductor supply if a transformer feeder tap has primary conductors at least $\frac{1}{3}$ ampacity and/or secondary conductors at least $\frac{1}{3}$ ampacity when multiplied by the approximate transformer turns ratio of the overcurrent device protecting the feeder conductors; the total length of one primary plus one secondary conductor (excluding any portion of the primary conductor that is protected at its ampacity) is not over 25' in length; the secondary conductors terminate in a single overcurrent device rated at the ampacity of the tap conductors; and if the primary and secondary conductors are suitably protected from physical damage.
- Per *NEC Section 240.21(B)(4),* overcurrent protection is not required at the conductor supply in high bay manufacturing buildings over 35' high at walls when only qualified persons will service such a system; if the tap conductors are not over 25' long horizontally and not over 100' long total length; the ampacity of the tap conductors is not less than $\frac{1}{3}$ of the rating of the overcurrent device protecting the feeder conductors; terminate in a single overcurrent device; are suitably protected from physical damage or are enclosed in a raceway; are at least No. 6 AWG copper or No. 4 AWG aluminum; are continuous from end-to-end and contain no splices; do not penetrate walls, floors, or ceilings; and are made no less than 30' from the floor.

WARNING!
Smaller conductors tapped to larger conductors can be a serious hazard. If not protected against short circuit conditions, these unprotected conductors can vaporize or incur severe insulation damage.

- Per *NEC Section 240.21(B)(5),* overcurrent protection is not required at the terminals of transformer secondaries located outdoors when the conductors are suitably protected from physical damage, and they terminate in a disconnecting means that is either immediately outside or inside the building with an overcurrent device that limits the load to the ampacity of the secondary conductors. In this instance, the length of the conductors outside of the buildings is not limited.
- Per *NEC Section 240.21(C)(2),* overcurrent protection is not required at a transformer secondary if the secondary conductor does not exceed 10', and the ampacity of the conductor is not less than the combined computed loads on the circuits supplied by the secondary conductors and not less than the rating of the device supplied by the secondary conductors or not less than the rating of the overcurrent protective device at the termination of the secondary conductors. In addition, the secondary conductors cannot extend beyond the switchboard, panelboard, disconnecting means, or control devices they supply. The secondary conductors must also be enclosed in a raceway that extends from the transformer to the enclosure of an enclosed switchboard, panelboard, or control devices or to the back of an open switchboard.
- Per *NEC Section 240.21(C)(6),* overcurrent protection is not required at a transformer secondary if the secondary conductor does not exceed 25', and the secondary conductors have an ampacity that, when multiplied by the ratio of the secondary-to-primary voltage, is at least $\frac{1}{3}$ of the rating of the overcurrent device protecting the primary of the transformer. In addition, the secondary conductors must be suitably protected from physical damage and must terminate in a single circuit breaker or set of fuses that limit the load current to not more than the conductor ampacity permitted by *NEC Section 310.15.*
- Other tap provisions for single-phase, two-wire, and three-phase delta transformer secondaries are found in *NEC Section 240.21(C)(1),* while *NEC Section 240.21(C)(3)* covers secondary conductors in industrial installations.

Note
Switchboard and panelboard protection, along with transformer protection, must still be observed. See ***NEC Sections 408.16 and 450.3.***

Safety

Emphasize the hazards of tapping smaller conductors to larger conductors.

Classroom

Review the rules for transformer taps and secondary conductors.

Teaching Tip

Discuss the "Think About It."

Classroom

Point out the conductor properties listed in *NEC Table 310.13 and NEC Chapter 9, Table 8.*

Demonstrate how to find selected conductor properties using *NEC Table 310.13.* Provide the trainees with a list of conductors and have them find the trade name, maximum operating temperature, applications, insulation thickness, and jacket type.

THINK ABOUT IT

Location of Overcurrent Protection

Is overcurrent protection required for the secondary conductors at the transformer terminals shown here?

302P0204.EPS

3.0.0 ◆ PROPERTIES OF CONDUCTORS

Various NEC tables define the physical and electrical properties of conductors. Electricians use these tables to select the type of conductor and the size of conduit, or other raceway, to enclose the conductors in specific applications. NEC tables list the properties of conductors as follows:

- Name
- Operating temperature
- Application
- Insulation
- Physical properties
- Electrical resistance
- AC resistance and reactance

NEC Table 310.13 lists the name, maximum operating temperature, applications, and insulation of various types of conductors. ***NEC Chapter 9, Table 8*** gives the physical properties and electrical resistance.

To gain an understanding of these tables and how they are used in practical applications, we will take a 4/0 THHN copper conductor and see what properties may be determined from the NEC tables.

Step 1 Turn to ***NEC Table 310.13*** and scan down the second column from the left *(Type Letter)* to find THHN. Scan to the left in this row to see that the trade name of this conductor is heat-resistant thermoplastic.

Step 2 Scanning to the right in this row, note that the maximum operating temperature for this wire type is 90°C (194°F). Continuing to the right in this row, under the column headed *Application Provisions,* we find that this wire type is suitable for use in dry and damp locations. The next column reveals that the insulation is flame-retardant, heat-resistant thermoplastic.

Step 3 Continuing to the right in this row, the next column lists insulation thickness for various AWG or kcmil wire sizes. The insulation thicknesses for Type THHN wire, from No. 14 AWG to 1,000 kcmil, are as follows:

- 14–12 15 mils
- 10 20 mils
- 8–6 30 mils
- 4–2 40 mils
- 1–4/0 50 mils
- 250–500 60 mils
- 501–1,000 70 mils

Consequently, the insulation thickness for 4/0 THHN is 50 mils.

Step 4 Looking in the rightmost column (*Outer Covering*), we see that this wire type has a nylon jacket or equivalent.

If you want to find the maximum current-carrying capacity of this conductor when used in

Instructor's Notes:

THINK ABOUT IT

Conductor Derating

If six current-carrying conductors are bundled tightly together with no spacing, but are installed in an open tray instead of an enclosed conduit, do they have to be derated?

a raceway, turn to *NEC Table 310.16* and proceed as follows:

Step 1 Scan down the left-hand column until the wire size is found.

Step 2 Scan to the right in this row until the 90°C column is found. This column covers the insulation types that are rated for 90°C maximum operating temperature and includes Type THHN conductor insulation.

Step 3 Note that the current-carrying rating for this size and type of conductor is 260A. This rating, however, is for not more than three conductors in a raceway. If we have more than three conductors, such as four conductors in a three-phase, four-wire feeder to a subpanel, this figure (260A) must be derated.

Step 4 Refer to *NEC Section 310.15(B)(2)(a)*, which states that where the number of current-carrying conductors in a raceway or cable exceeds three, the allowable ampacities shall be reduced as shown in *Table 1*.

Since we want to know the allowable maximum current-carrying capacity of four 4/0 THHN conductors in one raceway, it is necessary to multiply the previous amperage (260A) by 80% or 0.80.

$$260 \times .80 = 208\text{A}$$

These amperage tables are also based on the conductors being installed in areas where the ambient air temperature is 30°C (86°F). If the conductors are installed in areas with different ambient temperatures, a further deduction is required. For example, if this same set of four 4/0 THHN conductors were installed in an industrial area where the ambient temperature averaged 35°C, we would look in the correction factor tables at the bottom of *NEC Tables 310.16 through 310.19.* In doing so, we would find that the correction or derating factor for our situation is 0.96. Consequently, our present current-carrying capacity of 208A must be multiplied by 0.96 to obtain the actual current-carrying capacity of the four conductors:

$$208 \times 0.96 = 199.68\text{A}$$

Other sizes and types of conductors are handled in a similar manner; that is, find the appropriate table, determine the listed ampacity, and then multiply this ampacity by the appropriate factors in the correction factor tables.

It sometimes becomes necessary to know additional properties of conductors for some conductor calculations, especially for voltage drop calculations, which will appear in a later section in this module. There are many useful tables in *NEC Chapter 9*. Examples of their practical use are presented later in this module.

Table 1 Allowable Ampacities for Current-Carrying Conductors

Number of Current-Carrying Conductors	Percent of Values in Tables as Adjusted for Ambient Temperature, If Necessary
4 through 6	80
7 through 9	70
10 through 20	50
21 through 30	45
31 through 40	40
41 and above	35

3.1.0 Identifying Conductors

The NEC specifies certain methods of identifying conductors used in wiring systems of all types. For example, the high leg of a 120/240V, grounded three-phase, four-wire delta system

INSIDE TRACK

High Leg Connection

NEC Section 408.3(E) requires that panelboards supplied by a three-phase, four-wire delta service have the high leg connected to center phase B.

Teaching Tip

Discuss the "Think About It."

Classroom

Demonstrate how to find the maximum current-carrying capacity of any number of conductors of a given size in a raceway. Provide the trainees with example problems.

Explain that the NEC requires that certain conductors be identified by color-coding.

Classroom

Point out that the colors white and/or gray are reserved for the grounded circuit conductor, while green is reserved for the grounding conductor.

Note the NEC requirements for conductor reidentification.

Review the NEC requirements for voltage drop.

must be marked with an orange color for identification; a grounded conductor must be identified either by the color of its insulation, by markings at the terminals, or by other suitable means. Unless allowed by NEC exceptions, a grounded conductor must have a white or gray finish. When this is not practical for conductors larger than No. 6 AWG, marking the terminals white is an acceptable method of identifying the conductors.

3.1.1 Color Coding

Conductors contained in cables are color-coded for easy identification at each access point. When conductors are installed in raceway systems, any color insulation is permitted for the ungrounded phase conductors except the following:

- White or gray, which is reserved for use as the grounded circuit conductor
- Green, which is reserved for use as a grounding conductor only

3.1.2 Changing Colors

Should it become necessary to change the actual color of a conductor to meet NEC requirements or to facilitate maintenance on circuits and equipment, the conductors may be re-identified with nonconductive colored tape or paint.

For example, assume that a two-wire cable containing a black and white conductor is used to feed a 240V, two-wire, single-phase motor. Since the white-colored conductor is supposed to be reserved for the grounded conductor, and none is required in this circuit, the white conductor may be marked with a piece of black tape at each end of the circuit so that everyone will know that this wire is not a grounded conductor.

4.0.0 ◆ VOLTAGE DROP

In all electrical systems, the conductors should be sized so that the voltage drop does not exceed 3% for power, heating, and lighting loads, or combinations of these. Furthermore, the maximum total voltage drop for conductors, feeders, and branch circuits combined should not exceed 5%. These percentages are recommended by the NEC but are not requirements. However, it is considered to be good practice to incorporate these percentages into every electrical installation.

In some applications, such as for circuits feeding hospital X-ray equipment, the voltage drop is even more critical—requiring a maximum of 2% voltage drop throughout. With the higher ratings on the newer types of insulation, it is extremely important to keep volt loss in mind; otherwise, some very unsatisfactory problems are likely to be encountered.

For example, the resistance and voltage drop on long conductor runs may be great enough to seriously interfere with the efficient operation of the connected equipment. Resistance elements, such as those used in incandescent lamps and electric heating units, are particularly critical in this respect; a drop of just a few volts greatly reduces their efficiency.

Electric motors are not affected by small voltage variations to the same degree as pure resistance loads, but motors will not operate at their rated horsepower if the voltage is below that at which they are rated. When loaded motors are operated at reduced voltage, the current flow actually increases, as it requires more amperes to produce a given wattage and horsepower at low voltage than at normal voltage. This current increase is also caused by the fact that the opposition of the motor windings to current flow decreases as the motor speed decreases.

From the foregoing, we can see that it is very important to have all conductors of the proper size to avoid excessive heating and voltage drop, and that, in the case of long runs, it is necessary to determine the wire size by consideration of resistance and voltage drop, rather than by the heating effect or *NEC Tables 310.16 through 310.19* alone.

To solve the ordinary problems of voltage drop requires only a knowledge of a few simple facts about the areas and resistances of conductors and the application of a few mathematical equations.

Note
The actual conductor used must also meet the other sizing requirements such as full-load current, ambient temperature, number in a raceway, etc.

4.1.0 Wire Sizes Based on Resistance

Earlier modules covered wire sizes and how conductors are normally specified in kcmil or AWG sizes. This numbering system was originated by the Brown & Sharpe Company and was originally called the B & S gauge. However, the B & S gauge quickly evolved into the American Wire Gauge (AWG) and is now standard in the United States for indicating sizes of round wires and conductors.

Instructor's Notes:

AWG numbers are arranged according to the resistance of the wires, with the larger numbers representing the wires of greatest resistance and smallest area. A handy rule to remember is that decreasing the gauge by three numbers gives a wire of approximately twice the area and half the resistance. Conversely, increasing the gauge by three numbers gives a wire of approximately half the area and twice the resistance.

For example, if we increase the wire gauge from No. 3 AWG, which has a resistance of 0.245 ohm per 1,000', to a No. 6 AWG, we find it has a resistance of 0.491 ohm per 1,000', which is about double. See ***NEC Chapter 9, Table 8.***

Although ***NEC Chapter 9, Table 8*** only lists sizes down to No. 18 AWG, the American Wire Gauge numbers range from 0000 (4/0) down in size to No. 40. The 4/0 conductor is more than ½" in overall diameter and the No. 40 is as fine as a thin hair.

The most common sizes used for light and power installations range from 4/0 to No. 14 AWG. Lighting fixture wires are frequently size No. 16 or No. 18 AWG, and low-voltage control wiring sometimes drops down to size No. 22 AWG.

4.1.1 Circular Mil—Unit of Conductor Area

In addition to AWG gauge numbers, a unit called the mil is also used for measuring the diameter and area of conductors. The mil is equal to 1/1000 of an inch, so it is small enough to measure and express these sizes very accurately. For example, instead of saying a wire has a diameter of 0.055", or fifty-five thousandths of an inch, we can simply call it 55 mils. Consequently, a wire of 250 mils in diameter is also 0.250", or ¼" in diameter.

Since the resistance and current-carrying capacity of conductors both depend on the conductor's cross-sectional area, a unit is necessary to express this area. Electrical conductors are commonly made in several different shapes *(Figure 9)*. For

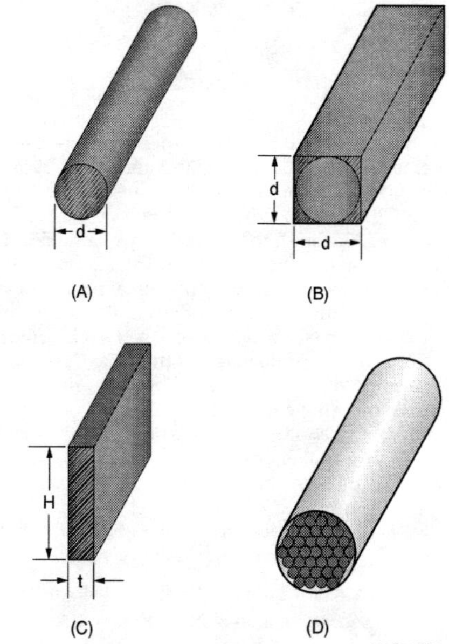

Figure 9 ◆ Electrical conductors in various shapes.

Classroom

Discuss the wire sizes commonly used in residential and light commercial applications.

Define *circular mil.*

Audiovisual

Show Transparency 10 (Figure 9).

INSIDE TRACK

Busbars

Busbars are sized by square mils. They are often used in industrial installations, such as the one shown here.

302P0205.EPS

Classroom

Demonstrate how to convert from square mils to circular mils and vice versa.

Demonstrate how to calculate the resistance of various circuits using *NEC Chapter 9, Table 8.*

square conductors, such as busbars, the square mil is used, which is a square ¹⁄₁₀₀₀" on each side. For round conductors, the circular mil unit is used, which is the area of a circle with a diameter of ¹⁄₁₀₀₀".

These units greatly simplify conductor calculations. For example, to determine the area of a square conductor, as shown in *Figure 9(B)*, multiply one side by the other, measuring the sides in either mils or thousandths of an inch.

4.1.2 Conversion of Square Mils to Circular Mils

When comparing round and square conductors, remember that the square mil and the circular mil are not quite the same units of area. For example, see *Figure 9(B)*, which shows a circle within a square. Although the circle has the same diameter as the square, the corners of the square make it larger in area. From this, we can say that the area of one circular mil is less than that of one square mil. The actual ratio between the two is 0.7854, or the circle has only 78.54% of the area of a square of the same diameter. Consequently, to find the circular mil area from the number of square mils, divide the square mils by 0.7854. If the reverse were true, that is, finding the square mil area from circular mils, multiply the circular mils by 0.7854.

For example, the conductor in *Figure 9(A)* is a No. 4/0 conductor with an area of 211,600 circular mils. What is its area in square mils?

$$\text{Circular mil area} = 211{,}600 \text{ circular mils}$$
$$\text{Square mil area} = 211{,}600 \times 0.7854 = 166{,}191 \text{ square mils}$$

If the busbar in *Figure 9(C)* is 1½" high and ¼" thick, what is its area in square mils, and what size of round conductor would be necessary to carry the same current as this busbar? First, the dimensions of a ¼" × 1½" busbar, stated in mils, are 250 mils × 1,500 mils. Therefore, the area in square mils may be determined by the following equation:

$$250 \times 1{,}500 = 375{,}000 \text{ square mils}$$

To find this area in circular mils, divide the square mil area (375,000) by 0.7854; the results are as follows:

$$375{,}000 \div 0.7854 = 477{,}464 \text{ circular mils}$$

The nearest standard size to this is a round conductor 500,000 circular mils in size (500 kcmil).

Busbars of the shape shown in *Figure 9(C)* are commonly used in panelboards and switchgear. These bars normally range in thickness from 0.250" to 0.375" or more, and in heights from 1" to 12".

Stranded conductors such as the one shown in *Figure 9(D)* are used on all conductor sizes No. 6 AWG and larger. Since these conductors are not solid throughout, their area cannot be determined accurately. This diameter also varies somewhat with the twist or lay of the strands.

To determine the cross-sectional area of such conductors, first determine the area of each strand, either from a wire table or by calculation from its diameter, and then multiply this by the number of strands to get the total area of the cable in circular mils. *NEC Chapter 9, Table 8* can also be used to find the area of most conductors in circular mils. It also provides information necessary for other wiring calculations.

4.2.0 Resistance of Conductors

It is often necessary to determine the exact resistance of a conductor of a certain length in order to calculate the voltage drop under a certain current load.

The resistance per 1,000' of various conductors can be obtained from *NEC Chapter 9, Table 8.* These conductor specifications are necessary to accurately determine the voltage drop for various sizes of conductors.

For example, to find the total resistance of a 120V, two-wire circuit consisting of two No. 10 AWG solid copper conductors, each 150' long, proceed as follows:

Step 1 The length of one conductor (150') must first be multiplied by 2 to obtain the entire length of both conductors:

$$2 \times 150' = 300'$$

Step 2 Refer to *NEC Chapter 9, Table 8* to find the resistance of No. 10 solid copper wire. The table gives a resistance of 1.21 ohms per 1,000' for uncoated copper wire at 75°C.

Step 3 Since our circuit is less than 1,000', we must determine the resistance of 300'. This is accomplished by dividing the actual footage (300' in this case) by 1,000'.

$$\frac{300'}{1{,}000'} = 0.30$$

Step 4 Multiply this result (0.30) times the resistance of 1.21Ω:

$$0.30 \times 1.21\Omega = 0.363\Omega$$

In another situation, we want to install an outside 120V, two-wire line between two buildings a

Instructor's Notes:

distance of 1,650' using No. 1 AWG copper wire. What would be the total resistance of this circuit?

Step 1 Determine the total length of both conductors by multiplying the length (one way) by 2:

$$1{,}650' \times 2 = 3{,}300'$$

Step 2 Referring again to ***NEC Chapter 9, Table 8***, we see that No. 1 AWG uncoated copper wire has a resistance of 0.154 ohm per 1,000'.

Step 3 The total length of 3,300' must be divided by 1,000'.

$$\frac{3{,}300'}{1{,}000'} = 3.3$$

Step 4 Multiply this result (3.3) times the resistance found in ***NEC Chapter 9, Table 8*** (0.154Ω).

$$3.3 \times 0.154\Omega = 0.5082\Omega$$

Now we will see what happens if we apply the full current to these conductors allowed by ***NEC Table 310.16***. Assuming that Type THHN conductors are used, ***NEC Table 310.16*** allows a maximum load on these conductors of 150A at 90°C. (Always use the 90°C column in this table when derating conductors.) If this much current flowed through this circuit for a distance of 3,300' (both ways), the voltage drop in the circuit would be:

$$\text{Voltage drop} = I \times R$$
$$= 150A \times 0.5082\Omega$$
$$= 76.2V$$

If the initial voltage is only 120V, this means that the voltage at the far end of the circuit will be only 120V − 76.2V or 43.8V and few, if any, 120V loads will operate at this low voltage. Consequently, the load will have to be reduced to make this circuit of any use.

Our goal is to keep the voltage drop within 3% of the original voltage. Therefore, since the original voltage is 120V, the allowable voltage drop may be found by using the following equation:

$$120V \times 0.03 = 3.6V$$

Now, what size load may be applied to this circuit to stay within this 3% range? The equation for finding the maximum current on this circuit to keep the voltage drop within 3% or 3.6V is as follows:

$$\text{Maximum current (I)} \times \text{total resistance } (\Omega)$$
$$= \text{original voltage (V)} \times \text{allowable voltage drop\%}$$

Substituting our known values in this equation, we have:

$$I \times 0.5082\Omega = 3.6V$$

To solve for I, we divide both sides of the equation by 0.5082, which results in the following:

$$I = \frac{3.6V}{0.5082\Omega} = 7.084A$$

Therefore, to keep the voltage drop within 3% for this length of circuit with only 120V applied, the amperage must be held to 7.084A or below.

If a larger load is connected to this circuit, and we need to hold the voltage drop to within 3%, then either a larger conductor will have to be used, or else the voltage will have to be increased. For example, assume that this circuit feeds a 120/240V dual-voltage pump motor. If the motor connections were rewired to accept a 240V branch circuit, then the allowable voltage drop (at 3%) would be:

$$0.03 \times 240V = 7.2V$$

Continuing as before:

$$I \times \frac{7.2V}{0.5082} = 14.17A$$

If 480V single phase were applied to the circuit (even though it is not feasible with this two-wire arrangement), the amperage would again double. These examples should show why long electric transmission lines utilize extremely high voltages—up to 250,000V or more—to keep the current, resulting voltage drop, and conductor size to the bare minimum.

4.3.0 Resistance of Copper per Mil Foot

In many cases, it may be necessary to calculate the resistance of a certain length of wire or a busbar of a given size.

This can be done very easily if the unit resistance of copper is known. A unit called the mil foot may be used. A mil foot represents a piece of round wire that is 1 mil in diameter and 1' in length and is a small enough unit to be very accurate for all practical calculations. A round wire of 1 mil in diameter has an area of 1 circular mil, as the diameter multiplied by itself, or squared, is:

$$1 \times 1 = 1 \text{ circular mil}$$

The resistance of ordinary copper is 12.9Ω per mil foot. This constant is important and should be remembered. It is normally represented by the letter *K*.

Classroom

Provide the trainees with example circuit values. For each circuit, have them calculate the total resistance and then determine the voltage drop.

Point out that the resistance of copper is 12.9 ohms per mil foot. Demonstrate how to calculate the resistance of various lengths of wire using this constant.

Classroom

Have the trainees calculate the resistance of various lengths of wire. Point out that the values listed in *NEC Tables 310.16* through *310.19* do not account for the length of the conductors or the voltage drop.

List the factors that influence the selection of a specific wire size for a given application.

Demonstrate how to calculate the resistance and voltage drop of various types of wire using the equations in the Trainee Module.

Suppose we want to determine the resistance of a piece of No. 12 copper wire that is 50' long. We know that the resistance of any conductor increases as its length increases and decreases as its area increases. So, for a wire that is 50' long, we first multiply and get 50' × 12.9Ω = 645Ω/mil ft.^{2}, which would be the resistance of a wire that is 1 circular mil in area and 50' long. Then we find in *NEC Chapter 9, Table 8* that the area of a No. 12 wire is 6,530 circular mils, which will reduce the resistance in proportion. So we now divide:

$$\frac{645}{6{,}530} = 0.0988\Omega/\text{mil ft.}^2$$

In another case, we wish to find the resistance of a coil containing 3,000' of No. 18 stranded wire. We would first multiply and get 3,000' × 12.9Ω = 38,700Ω/mil ft.^{2} Since the area of No. 18 wire is 1,620 circular mils, the total resistance may be found using the following equation:

$$\frac{38{,}700}{1{,}620} = 23.89\Omega$$

Checking this with *NEC Chapter 9, Table 8,* we find a resistance of 7.95Ω per 1,000' for No. 18 stranded AWG wire. For 3,000', we use the following equation:

$$3 \times 7.95\Omega = 23.85\Omega$$

The small difference between this answer and the one obtained by the first calculation is caused by using approximate figures instead of lengthy decimal places.

The mil foot unit and its resistance of 12.9Ω for copper may also be used to calculate the resistance of square busbars by simply using the figure 0.7854 to change from square mils to circular mils.

Suppose we wish to find the resistance of a square busbar that is ¼" × 2" and 100' long. The dimensions in mils will be 250 × 2,000 = 500,000 square mils. To find the circular mil area, we divide 500,000 by 0.7854 and get 636,618.3 circular mils. Then we multiply and get 100' × 12.9Ω = 1,290Ω, or the resistance of 100' of copper that is 1 mil in area. Because the area of this bar is 636,618.3 circular mils, we divide:

$$\frac{1{,}290\Omega}{636{,}618.3} = 0.002026\Omega \text{ total resistance}$$

NEC Tables 310.16 through 310.19 list the allowable current-carrying capacities of conductors with various types of insulation. These tables, however, do not take into consideration the length of the conductors or the voltage drop. Consequently, it is often necessary to use a larger size conductor than is indicated in the tables.

4.4.0 Equations for Voltage Drop Using Conductor Area or Conductor Resistance

The size of the conductor required to connect an electrical load to the source of supply is determined by several factors, including:

- Load current in amperes
- Permissible voltage drop between source and load
- Total length of conductor
- Type of wire (i.e., copper, aluminum, copper-clad aluminum, etc., and its permissible load-carrying capability based on *NEC Tables 310.16 through 310.19*)

To perform voltage drop calculations, it is also necessary to recall the following:

- The resistance of a wire varies directly with its length:

$$R = \text{resistance per foot (K)} \times \text{length in feet (L)}$$

- The resistance varies inversely with its cross-sectional area:

$$R = \frac{1}{A}$$

Combining both statements, we obtain the following equation:

$$R = \frac{L \times K}{A}$$

Where:

A = wire size (first from *NEC Tables 310.16 through 310.19* for a given load, and then, if necessary, from *NEC Chapter 9, Table 8* for circular mils)

L = length of wire in feet

R = total resistance of wire

K = resistance per mil foot

K is a constant whose value depends upon the units chosen and the type of wire (12.9 for copper and 21.2 for aluminum). Using the foot as the unit of length and the circular mil as the unit of area, the values for K represent the resistance in ohms per mil foot.

The length (L) in the equation just given is the total length of a single current-carrying conductor. For a two-conductor, 120V circuit or a 240V, balanced three-wire circuit (neutral current equals zero), the length in the equation must be multiplied by 2. The equation now becomes:

$$R = \frac{2 \times L \times K}{A} \text{ for single-phase circuits}$$

Instructor's Notes:

For a three-phase, four-wire balanced circuit with a power factor near the value of 1, the equation must be multiplied by $\sqrt{3} \div 2$. (The value of $\sqrt{3}$ is approximately 1.732; you may wish to make a note of this for use in future calculations.) The equation now becomes:

$$R = \frac{\sqrt{3}}{2} \times \frac{2 \times L \times K}{A}$$

The 2s cancel each other out and the equation reduces to:

$$R = \frac{\sqrt{3} \times L \times K}{A} \text{ for three-phase balanced circuits}$$

Since $R = E \div I$ (where E = voltage drop [VD]), substitute $VD \div I$ for R in the previous equations:

$$\frac{VD}{I} = \frac{2 \times L \times K}{A} \text{ for single-phase circuits}$$

and

$$\frac{VD}{I} = \frac{\sqrt{3} \times L \times K}{A} \text{ for three-phase balanced circuits}$$

Multiply each side of both equations by I:

$$VD = \frac{2 \times L \times K \times I}{A} \text{ for single-phase circuits}$$

and

$$VD = \frac{\sqrt{3} \times L \times K \times I}{A}$$

for three-phase balanced circuits

Substitute circular mils (CM) for A where CM is initially determined from the wire size ampacities listed in *NEC Tables 310.16 through 310.19* for the desired load and where *NEC Chapter 9, Table 8* is used to convert the wire size so determined to circular mils, if required. The equations become:

$$VD = \frac{2 \times L \times K \times I}{CM} \text{ for single-phase circuits}$$

and

$$VD = \frac{\sqrt{3} \times L \times K \times I}{CM}$$

for three-phase balanced circuits

You should recognize these equations as the same ones used to calculate voltage drop for branch circuits in the previous module. Using an exercise similar to the one just listed, the following equations (also used in the previous module) can be derived for a given wire size resistance and load:

$$VD = \frac{2 \times L \times R \times I}{1{,}000} \text{ for single-phase circuits}$$

and

$$VD = \frac{\sqrt{3} \times L \times R \times I}{1{,}000}$$

for three-phase balanced circuits

4.5.0 Use of Voltage Drop Equations

If we want to determine the voltage drop on either an existing 120V, two-wire installation or a proposed project, we can find the voltage drop of the

THINK ABOUT IT

Voltage Drops

This parking lot fixture has three 1,000W lamps, is 300' from the panel, uses No. 12 wire, and is supplied by a 20A, 208V circuit. The fixture manufacturer specifies a maximum voltage drop of 3%. Does this circuit adequately support this load?

302P0206.EPS

Classroom

Discuss the applications of the voltage drop equations.

Teaching Tip

Discuss the "Think About It."

Classroom

Explain how to rearrange the voltage drop equations to solve for the maximum circuit length at a given voltage drop, load, and wire size.

Explain how to rearrange the voltage drop equations to solve for the wire size in circular mils for a given voltage drop, wire length, and load. Point out the two rules associated with this alternate equation.

circuit by using one of the single-phase formulas previously given:

$$VD = \frac{2 \times L \times K \times I}{CM}$$

For example, assume we have a 120V, two-wire circuit feeding a total load of 30A and the branch circuit length is 120'.

NEC Table 310.16 allows a No. 10 AWG conductor at 60°C to carry this load. (This does not consider voltage drop or derating factors.) Referring to ***NEC Chapter 9, Table 8,*** we find that the area (A) of a No. 10 AWG conductor is 10,380 circular mils. Then, substituting these values and a value of 12.9 (copper) for K in the equation, we have:

$$VD = \frac{2 \times 120' \times 12.9 \times 30A}{10,380}$$

$$VD = 8.95V$$

The voltage at the load is 120V − 8.95V = 111.05V, which is usually not acceptable on most installations since it exceeds the 3% recommended maximum voltage drop; that is, 3% of 120 = 3.6V. In fact, it is more than double. There are three ways to correct this situation:

- Reduce the load
- Increase the conductor size
- Increase the voltage

4.5.1 *Miscellaneous Voltage Drop Equations*

The final voltage drop (VD) equations discussed earlier can be solved for length (L), yielding the following equations:

$$L = \frac{VD \times CM}{2 \times K \times I} \text{ or } L = \frac{1,000 \times VD}{2 \times R \times I}$$

for single-phase circuits

and

$$L = \frac{VD \times CM}{\sqrt{3} \times K \times I} \text{ or } L = \frac{1,000 \times VD}{\sqrt{3} \times R \times I}$$

for three-phase balanced circuits

These equations can be used to determine the maximum circuit length for a specified voltage drop at a given load using a wire size selected from ***NEC Tables 310.16 through 310.19*** for the ampacity of the load. ***NEC Chapter 9, Table 8*** may be used to convert the selected wire size to circular mils, if required.

For example, determine the maximum circuit length for a feeder serving a 208V balanced three-phase load of 400A at a voltage drop not to exceed 3%. From ***NEC Table 310.16,*** determine that a 750 kcmil conductor at 60°C is rated to carry 400A. Using the three-phase length equation above containing the CM term, substitute values and calculate length:

$$L = \frac{VD \times CM}{\sqrt{3} \times K \times I}$$

$$L = \frac{208V \times 3\% \times 750,000\ CM}{1.732 \times 12.9 \times 400A}$$

$$L = 523.6'$$

Therefore, in this example the 400A load may be positioned up to 523.6' from the source and the voltage drop will be 3% or less using 750 kcmil conductors at 60°C.

The two final voltage drop equations given earlier may also be solved for wire size in circular mils (CM) to directly determine, under certain conditions, the wire size for a given voltage drop, wire length, and load. Solving for CM yields:

$$CM = \frac{2 \times L \times K \times I}{VD} \text{ for single-phase circuits}$$

and

$$CM = \frac{\sqrt{3} \times L \times K \times I}{VD} \text{ for three-phase circuits}$$

The results from these two CM equations *must* be used in accordance with the following two rules:

- *Rule 1*–If the wire size calculated using the CM equation results in a wire size with less ampacity than the given load as determined from ***NEC Tables 310.16 through 310.19,*** discard the calculated result and use the wire size with the rated ampacity for the given load.
- *Rule 2*–If the wire size calculated using the CM equation results in a wire size equal to or greater than the wire size determined from ***NEC Tables 310.16 through 310.19*** for the given load, use the calculated wire size or the next larger standard wire size.

The reason for these rules is that the equation results are for non-derated bare wire. The wire sizes determined from the NEC tables are derated for the number of wires, temperature, and other factors. Calculated sizes that are smaller than those sizes rated at the desired load ampacity cannot be used. However, calculated wire sizes that are larger may be used when excessive voltage drop occurs that is a function of only the size of wire (that has already been derated) and that is caused by additional length beyond the maxi-

Instructor's Notes:

mum length for a specified voltage drop and load ([L] equations previously shown.)

To illustrate, we will solve for CM using the terms of the example given for determining length (L) (i.e., a 400A, three-phase load at 523.6'):

$$CM = \frac{\sqrt{3} \times L \times K \times I}{VD}$$

Substituting values, we have:

$$CM = \frac{1.732 \times 523.6' \times 12.9 \times 400A}{208V \times 3\%}$$

$$CM = 749{,}916 \text{ or } 750 \text{ kcmil}$$

In this instance, of course, the length and load correlated exactly with the 3% voltage drop and a 750 kcmil conductor was obtained from the calculation. In accordance with *Rule 2,* the size obtained equaled the size of a 60°C conductor with a rated ampacity of 400A. Note that longer lengths for this same load will result in larger conductor sizes, each of which will be proven correct if cross-checked using the VD equations.

Now take the same problem but change the 523.6' to 100' and solve for CM:

$$CM = \frac{\sqrt{3} \times L \times K \times I}{VD}$$

$$CM = \frac{1.732 \times 100' \times 12.9 \times 400A}{208V \times 3\%}$$

$$CM = 143{,}223$$

From ***NEC Chapter 9, Table 8,*** the next largest standard conductor is 167,800 CM or a No. 3/0 AWG conductor. However, when ***NEC Table 310.16*** is checked, the maximum permissible ampacity of a No. 3/0 AWG conductor is found to be only 165A—far below the load of 400A.

Therefore, in accordance with *Rule 1,* the 3/0 AWG conductor solution would be discarded and 750 kcmil 60°C conductors, determined from ***NEC Table 310.16*** for the 400A load, would be used for the 100' run.

Even when sizing wire for low-voltage (24V) control circuits, the voltage drop should be limited to 3% because excessive voltage drop causes:

- Failure of control coil to activate
- Control contact chatter
- Erratic operation of controls
- Control coil burnout
- Contact burnout

The voltage drop calculations described previously may also be used for low-voltage wiring, but tables are quite common and can save much calculation time. One example is shown in *Table 2.*

To use *Table 2,* assume a load of 35VA with a 50' run for a 24V control circuit. Referring to the table, scan the 50' column. Note that No. 18 AWG wire will carry 29VA and No. 16 wire will carry 43VA while still maintaining a maximum of 3% voltage drop. In this case, No. 16 wire would be the size to use.

The 3% voltage drop limitation is imposed to assure proper operation when the power supply is below the rated voltage. For example, if the rated 240V primary supply is 10% low (216V), the transformer secondary side does not produce 24V but rather 21.6V. When normal voltage drop is taken from this 21.6V, it approaches the lower operating limit of most controls. If it is assured that the primary voltage to the transformer will

Table 2 Table for Calculating Voltage Drop in Low-Voltage (24V) Wiring

	Length of Circuit (One Way in Feet)											
AWG Wire Size	**25**	**50**	**75**	**100**	**125**	**150**	**175**	**200**	**225**	**250**	**275**	**300**
20	29	14	10	7.2	5.8	4.8	4.1	3.8	3.2	2.9	2.8	2.4
18	58	29	19	14	11	9.6	8.2	7.2	6.4	5.8	5.2	4.8
16	86	43	29	22	17	14	12	11	9.6	8.7	7.8	7.2
14	133	67	44	33	27	22	19	17	15	13	12	11

THINK ABOUT IT

Voltage Drops in Low-Voltage Devices

You are adding a 12V/12W head 50' away from an emergency pack. Use the voltage drop calculations to determine what size wire you should run.

Classroom

Provide example problems for the trainees to solve.

Teaching Tip

Discuss the "Think About It."

Classroom

Have the trainees complete the Review Questions and go over the answers prior to administering the Module Examination.

always be at rated value or above, the control circuit will operate satisfactorily with more than 3% voltage drop.

In most installations, several lines connect the transformer to the control circuit. One line usually carries the full load of the control circuit from the secondary side of the transformer to one control, with the return perhaps coming through several lines of the various other controls. Therefore, the line from the secondary side of the transformer is the most critical regarding voltage drop and VA capacity and must be properly sized.

When low-voltage lines are installed, it is suggested that one extra line be run for emergency purposes. This can be substituted for any one of the existing lines that may be defective.

Summary

A great deal of valuable information about conductors, such as dimensions, resistance, and current-carrying capacity, can be obtained from convenient NEC tables. Use these tables whenever possible; they are great time-saving devices.

Sometimes appropriate tables are not available, or do not give the exact information needed for a certain project or situation. This is when a knowledge of simple conductor calculations pays off.

For example, *NEC Tables 310.16 through 310.19* give the allowable current-carrying capacities of conductors with various types of insulation based on the heating of the conductors, but these tables do not account for voltage drop due to resistance of long circuit runs. This is very important and should always be kept in mind when planning or installing any electrical installation.

Instructor's Notes:

Review Questions

1. All things considered, the most ideal conductor is made of ____.
   a. bronze
   b. brass
   c. aluminum
   d. copper

2. Each of the following is a standard configuration for stranded wire *except* ____.
   a. 5-strand
   b. 7-strand
   c. 19-strand
   d. 37-strand

3. Of the following, the largest standard AWG wire size is ____.
   a. No. 40 AWG
   b. No. 1 AWG
   c. No. 1/0 AWG
   d. No. 4/0 AWG

4. A branch circuit is best described as ____.
   a. the circuit conductors between the service equipment and the final branch circuit overcurrent device
   b. the circuit conductors between the final overcurrent device protecting the circuit and the outlet(s)
   c. a conductor tapped onto another conductor
   d. two conductors run in parallel

5. The rating of a branch circuit is determined by the ____.
   a. type and size of the branch circuit conductors
   b. rating of the branch circuit overcurrent device
   c. total connected load
   d. characteristics of the supply circuit

6. A continuous load is best described as one in which the maximum current is expected to continue for ____ hour(s) or more.
   a. one
   b. two
   c. three
   d. four

7. All of the following are standard ratings for fuses *except* ____.
   a. 600A
   b. 601A
   c. 650A
   d. 700A

8. You would find information on conductor applications and descriptions of insulation types in ____.
   a. *NEC Table 310.15(B)(2)*
   b. *NEC Table 310.13*
   c. *NEC Table 310.16*
   d. *NEC Chapter 9, Table 8*

9. Use *NEC Table 310.16* to find the maximum ampacity for a No. 2 AWG copper wire at 90°C (assume THHN insulation).
   a. 95A
   b. 110A
   c. 130A
   d. 150A

10. The constant (K) for the resistance of ordinary copper is ____.
    a. 12.9Ω per mil foot
    b. 21.2Ω per mil foot
    c. 1.732Ω per mil foot
    d. 3.141Ω per mil foot

Examination

Administer the Module Examination. Record the results on Craft Training Report Form 200 and submit the results to the Training Program Sponsor.

GLOSSARY

Trade Terms Introduced in This Module

American Wire Gauge (AWG): The U.S. standard for measuring wires.

Cable: An assembly of two or more insulated or bare wires.

Instructor's Notes:

Additional Resources

This module is intended to present thorough resources for task training. The following reference works are suggested for further study. These are optional materials for continued education rather than for task training.

American Electrician's Handbook. Terrell Croft and Wilfred I. Summers. New York, NY: McGraw-Hill, 1996.

National Electrical Code Handbook. Quincy, MA: National Fire Protection Association, 2002.

Answers to Review Questions

Answer	Section
1. d	1.0.0
2. a	1.0.0
3. d	1.0.0
4. b	2.0.0
5. b	2.1.0
6. c	2.1.0
7. c	2.2.0
8. b	3.0.0
9. c	4.2.0
10. a	4.3.0

Section 1.0.0 *Introduction*

Wire sizing is not particularly logical; the numbers run in reverse order, then switch at 1/0 and run in numerical order, then jump to kcmil. It is very difficult for trainees to visualize that No. 14 wire is smaller than No. 8, but No. 1/0 is larger than both of the others and yet smaller than No. 4/0, and so on. This exercise provides a hands-on opportunity to understand this somewhat complicated system. For this exercise, you will need two or more sets of index cards written out with the various wire sizes, along with examples of each wire size. Allow 30 minutes for this exercise.

1. Shuffle each set of index cards.
2. Divide the trainees into groups and provide one set to each group.
3. Have the trainees put the index cards in order of wire size from smallest to largest.
4. Have the trainees match their cards to the lengths of wire. Any errors they have made will be immediately apparent.

Section 2.1.0 *Think About It—Conductor Identification*

The two black conductors that connect to the compression lugs are the ungrounded feeder conductors, the wire connecting to the upper right side bus terminal is the grounded (neutral) feeder conductor, and the silver-colored conductor connecting to the left-side bus terminal is the grounding feeder conductor.

Section 2.2.1 *Think About It—Location of Overcurrent Protection*

Overcurrent protection is not required per *NEC Section 240.21(C)(3).*

Section 3.0.0 *Think About It—Conductor Derating*

Yes, derating is required per *NEC Chapter 9, Table 5A.*

Section 4.5.0 *Think About It—Voltage Drops*

No. The voltage drop in this circuit is 16.5V, which exceeds the 3% guideline of the manufacturer (7.2V). Remember, *NEC Section 110.3(B)* requires adherence to the manufacturer's instructions for the installed device. At this voltage drop, the fixture is only receiving 191.5V. This will result in reduced lumen output and flickering, and may even prevent the device from starting.

Section 4.5.1 *Think About It—Voltage Drops in Low-Voltage Devices*

No. 14 wire should be used. It is important to recognize that a small applied voltage doesn't necessarily dictate the use of smaller wire.

CONTREN™ LEARNING SERIES — USER UPDATES

The NCCER makes every effort to keep these textbooks up-to-date and free of technical errors. We appreciate your help in this process. If you have an idea for improving this textbook, or if you find an error, a typographical mistake, or an inaccuracy in NCCER's Contren™ textbooks, please write us, using this form or a photocopy. Be sure to include the exact module number, page number, a detailed description, and the correction, if applicable. Your input will be brought to the attention of the Technical Review Committee. Thank you for your assistance.

Instructors – If you found that additional materials were necessary in order to teach this module effectively, please let us know so that we may include them in the Equipment/Materials list in the Instructor's Guide.

Write: Curriculum Revision and Development Department
National Center for Construction Education and Research
P.O. Box 141104, Gainesville, FL 32614-1104

Fax: 352-334-0932

E-mail: curriculum@nccer.org

Craft ______ Module Name ______

Copyright Date ______ Module Number ______ Page Number(s) ______

Description ______

______

______

______

(Optional) Correction ______

______

______

(Optional) Your Name and Address ______

______

______

Overcurrent Protection

26303-02

Overcurrent Protection
Annotated Instructor's Guide

Module 26303-02

MODULE OVERVIEW

This module covers the procedures used when sizing and selecting overcurrent protection, along with the applicable NEC requirements.

PREREQUISITES

Please refer to the Course Map in the Trainee Module. Prior to training with this module, it is recommended that the trainee shall have successfully completed the following:

Core Curriculum; Electrical Level One; Electrical Level Two; Electrical Level Three, Modules 26301 and 26302.

OBJECTIVES

Upon completion of this module, the trainee will be able to:

1. Apply the key NEC requirements regarding overcurrent protection.
2. Check specific applications for conformance to NEC sections that cover short circuit current, fault currents, interrupting ratings, and other sections relating to overcurrent protection.
3. Determine let-through current values (peak and rms) when current-limiting overcurrent devices are used.
4. Select and size overcurrent protection for specific applications.

PERFORMANCE TASKS

This is a knowledge-based module; there is no performance examination.

NCCER STANDARDIZED CRAFT TRAINING PROGRAMS

The National Center for Construction Education and Research (NCCER) provides a standardized national program of accredited craft training. Key features of the program include instructor certification, competency-based training, and performance testing. The program provides trainees, instructors, and companies with a standard form of recognition through a National Craft Training Registry. The program is described in full in the *Guidelines for Accreditation,* published by the NCCER. For more information on standardized craft training, contact the NCCER by writing us at P.O. Box 141104, Gainesville, FL 32614-1104, calling 352-334-0911, or e-mailing info@nccer.org. More information may be found at our Web site, www.nccer.org.

HOW TO USE THIS ANNOTATED INSTRUCTOR'S GUIDE

Each page presents two sections of information. The larger section displays each page exactly as it appears in the Trainee Module. The narrow column ties suggested trainee and instructor actions to each page and provides icons to call your attention to material, safety, audiovisual, or testing requirements. The bottom of each page includes space for your notes.

If you see the Teaching Tip icon, that means there is a teaching tip associated with this section. Also refer to the suggested teaching tips at the end of the module.

SAFETY CONSIDERATIONS

Ensure that the trainees are equipped with appropriate personal protective equipment.

PREPARATION

Before teaching this module, you should review the Module Outline, Objectives, and the Materials and Equipment List. Be sure to allow ample time to prepare your own training or lesson plan and gather all required equipment and materials.

MATERIALS AND EQUIPMENT LIST

Materials:
Transparencies
Markers/chalk
Copy of the latest edition of the *National Electrical Code*®
Various types of circuit breakers
Various types of fuses, including electronic fuses
Sample fuse blocks and holders with nonrejection bases
Sample fuse blocks and holders with rejection clips that accept only Class R fuses
Module Examinations*

Equipment:
Overhead projector and screen
Whiteboard/chalkboard
Appropriate personal protective equipment

*Located in the Test Booklet packaged with this Annotated Instructor's Guide.

ADDITIONAL RESOURCES

This module is intended to present thorough resources for task training. The following reference works are suggested for both instructors and motivated trainees interested in further study. These are optional materials for continued education rather than for task training.

American Electrician's Handbook, Terrell Croft and Wilfred I. Summers. New York, NY: McGraw-Hill, 1996.

National Electrical Code Handbook. Quincy, MA: National Fire Protection Association, 2002.

TEACHING TIME FOR THIS MODULE

An outline for use in developing your lesson plan is presented below. Note that each Roman numeral in the outline equates to one session of instruction. Each session has a suggested time period of 2½ hours. This includes 10 minutes at the beginning of each session for administrative tasks and one 10-minute break during the session. Approximately 12½ hours are suggested to cover *Overcurrent Protection.* You will need to adjust the time required for hands-on activity and testing based on your class size and resources.

Topic	Planned Time
Session I. Introduction	
A. Introduction	______
1. Overcurrents	______
a. Overloads	______
b. Short Circuits	______
c. Ground Faults	______
Session II. Fuses; Operating Principles of Fuses	
A. Fuses	______
1. Types of Fuses	______
2. Voltage Rating	______
3. Ampere Rating	______
4. Interrupting Rating	______
5. Selective Coordination	______
6. Current Limitation	______
B. Operating Principles of Fuses	______
1. Nontime-Delay Fuses	______
2. Dual-Element, Time-Delay Fuses	______

Session III. UL Fuse Classes

A. UL Fuse Classes
   1. Branch Circuit Listed Fuses
   2. Medium-Voltage Fuses
      a. Application of Medium-Voltage Fuses
   3. Current-Limiting Fuses
   4. Fuses for Selective Coordination
   5. Fuse Time-Current Curves
      a. Peak Let-Through Charts

Session IV. Motor Overload and Short Circuit Protection; Circuit Breakers; Circuit Protection

A. Motor Overload and Short Circuit Protection

B. Circuit Breakers
   1. Interrupting Capacity Rating

C. Circuit Protection
   1. Lighting/Appliance Branch Circuits
      a. Feeder Circuits
      b. Service Equipment
      c. Transformer Secondary Circuits
      d. Motor Circuit Protection

Session V. Review; Module Examination

A. Review

B. Module Examination
   1. Trainees must score 70% or higher to receive recognition from the NCCER.
   2. Record the testing results on Craft Training Report Form 200 and submit the results to the Training Program Sponsor.

Module 26303-02

Overcurrent Protection

Course Map

This course map shows all of the modules in the third level of the Electrical curriculum. The suggested training order begins at the bottom and proceeds up. Skill levels increase as you advance on the course map. The local Training Program Sponsor may adjust the training order.

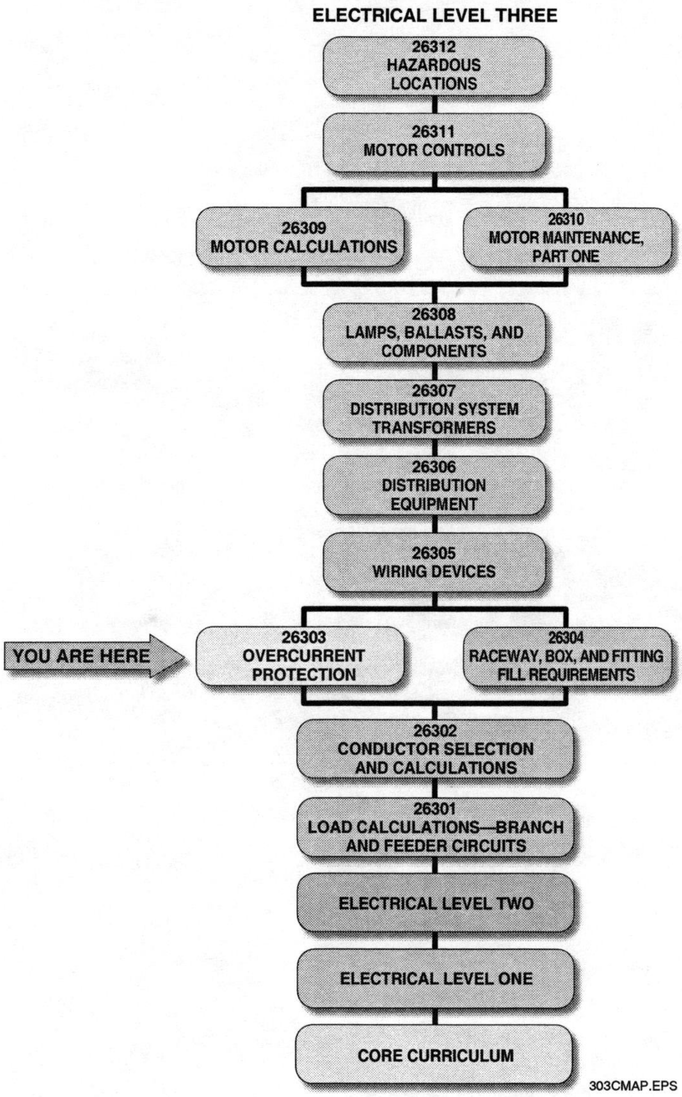

303CMAP.EPS

Assign reading of Module 26303.

MODULE 26303 CONTENTS

Figures

Tables

Instructor's Notes:

MODULE 26303

Overcurrent Protection

Objectives

When you have completed this module, you will be able to do the following:

1. Apply the key NEC requirements regarding overcurrent protection.
2. Check specific applications for conformance to NEC sections that cover short circuit current, fault currents, interrupting ratings, and other sections relating to overcurrent protection.
3. Determine let-through current values (peak and rms) when current-limiting overcurrent devices are used.
4. Select and size overcurrent protection for specific applications.

Prerequisites

Before you begin this module, it is recommended that you successfully complete the following: Core Curriculum; Electrical Level One; Electrical Level Two; Electrical Level Three, Modules 26301 and 26302.

Required Trainee Materials

1. Pencil and paper
2. Appropriate personal protective equipment
3. Copy of the latest edition of the *National Electrical Code*

1.0.0 ◆ INTRODUCTION

Electrical distribution systems are often quite complicated. They cannot be absolutely fail-safe. Circuits are subject to destructive overcurrents. Factors that contribute to the occurrence of such overcurrents are harsh environments, general deterioration, accidental damage or damage from natural causes, excessive expansion, or electrical distribution system overload. Reliable protective devices prevent or minimize costly damage to transformers, conductors, motors, and many other components and loads that make up the complete distribution system. Reliable circuit protection is essential to avoid the severe monetary losses that can result from power blackouts and prolonged downtime of facilities. It is the need for reliable protection, safety, and freedom from fire hazards that has made overcurrent protective devices absolutely necessary in all electrical systems, both large and small.

Overcurrent protection of electrical circuits is so important that the NEC devotes an entire article to this subject. *NEC Article 240* provides the general requirements for overcurrent protection and overcurrent protective devices. *NEC Article 240, Parts I through VIII* cover systems 600V (nominal) and under; *NEC Article 240, Part IX* covers overcurrent protection over 600V (nominal). This entire article will be covered in this module, along with practical examples.

All conductors must be protected against overcurrents in accordance with their ampacities as set forth in *NEC Section 240.3.* They must also be protected against **short circuit current** damage as required by *NEC Sections 110.10 and 240.1.* The two basic types of overcurrent protective devices that are in common use are fuses and circuit breakers.

1.1.0 Overcurrents

An overcurrent condition may be caused by an **overload** current, a short circuit current, or a

Note: The designations "National Electrical Code," "NE Code," and "NEC," where used in this document, refer to the National Electrical Code®, which is a registered trademark of the National Fire Protection Association, Quincy, MA. *All National Electrical Code (NEC) references in this module refer to the 2002 edition of the NEC.*

Materials

Ensure that you have everything required to teach the course. Check the Materials and Equipment List at the front of this Instructor's Guide.

Classroom

Review the goals of the course.

Explain that terms shown in bold (blue) are defined in the Glossary at the back of this module.

Audiovisual

Show Transparency 1, Course Objectives.

Classroom

List the factors that contribute to overcurrent conditions in distribution systems. Point out that *NEC Article 240* covers overcurrent protection.

Classroom

Define *overload.* List the common causes of overloads.

Define *short circuit* and *AIC rating.* Point out the distinction between clearing a fault level current and clearing an operating current.

ground fault. The overload current is an excessive current relative to normal operating current but one that is confined to the normal conductive paths provided by the conductor and other components and loads of the distribution system. As the name implies, a short circuit current is one that flows outside the normal conducting paths and may be phase-to-phase or phase-to-ground.

1.1.1 Overloads

Overloads are most often between one and six times the normal current level. Usually, they are caused by temporary surge currents that occur when motors are started up or transformers are energized. Such overload currents or transients are normal occurrences. Since they are of brief duration, any temperature rise is momentary and has no harmful effect on the circuit components, so long as the system is designed to accommodate them.

Continuous overloads can result from defective motors (such as worn motor bearings), overloaded equipment, or too many loads on one circuit. Such sustained overloads are destructive and must be cut off by protective devices before they damage the distribution system or system loads. However, since they are of relatively low magnitude compared to short circuit currents, removal of the overload current within a few seconds will generally prevent equipment damage. A sustained overload current results in overheating of conductors and other components and will cause deterioration of insulation, which may eventually result in severe damage and short circuits if not interrupted.

1.1.2 Short Circuits

A short circuit is a conducting connection, whether intentional or accidental, between any of the conductors of an electrical system, either line-to-line or line-to-ground.

The **amperes interrupting capacity (AIC)** rating of a circuit breaker or fuse is the maximum short circuit current that the breaker will safely interrupt. This AIC rating is at rated voltage and frequency.

Whereas overload currents occur at rather modest levels, a short circuit or fault current can be many hundreds of times larger than the normal operating current. A high level fault may be 50,000A (or larger). If not cut off within a matter of a few thousandths of a second, damage and destruction can become rampant—there can be severe insulation damage, melting of conductors, vaporization of metal, ionization of gases, arcing, and fires. Simultaneously, high level short circuit currents can develop huge magnetic field stresses. The magnetic forces between busbars and other conductors can be many hundreds of pounds per lineal foot; even heavy bracing may not be adequate to keep them from being warped or distorted beyond repair.

NEC Section 110.9 clearly states that equipment intended to interrupt current at fault levels (fuses and circuit breakers) must have an interrupting rating sufficient for the nominal circuit voltage and the current that is available at the line terminals of the equipment.

Equipment intended to interrupt current at other than fault levels must have an interrupting rating at nominal circuit voltage sufficient for the current that must be interrupted.

These NEC statements mean that fuses and circuit breakers (and their related components) designed to interrupt fault or operating currents (open the circuit) must have a rating sufficient to withstand such currents. This section emphasizes the difference between clearing fault level currents and clearing operating currents. Protective devices such as fuses and circuit breakers are designed to clear fault currents and therefore must have short circuit interrupting ratings sufficient for fault levels. Equipment such as contactors and safety switches have interrupting ratings for currents at other than fault levels. Thus, the interrupting rating of electrical equipment is now divided into two parts:

- Current at fault (short circuit) levels
- Current at operating levels

Most people are familiar with the normal current-carrying **ampere rating** of fuses and circuit breakers. For example, if an overcurrent protective device is designed to open a circuit when the circuit load exceeds 20A for a given time period, as the current approaches 20A, the overcurrent protective device begins to overheat. If the current barely exceeds 20A, the circuit breaker will open normally or a fuse link will melt after a given period of time with little, if any, arcing. For example, if 40A of current were instantaneously applied to the circuit, the overcurrent protective device would open, but again with very little arcing. However, if a ground fault occurs on the circuit that ran the amperage up to 5,000A, for example, an explosion effect would occur within the protective device. One simple indication of this is the blackened windows of plug fuses.

If this fault current exceeds the interrupting rating of a fuse or circuit breaker, the protective device can be damaged or destroyed; such current can also cause severe damage to equipment and injure personnel. Therefore, selecting overcurrent

Instructor's Notes:

protective devices with the proper interrupting capacity is extremely important in all electrical systems.

There are several factors that must be considered when calculating the required interrupting capacity of an overcurrent protective device. *NEC Section 110.10* states that the overcurrent protective devices, total impedance, component short circuit current ratings, and other characteristics of the circuit to be protected shall be selected and coordinated to permit the circuit protective devices used to clear a fault to do so without extensive damage to the electrical components of the circuit. This fault shall be assumed to be either between two or more of the circuit conductors, or between any circuit conductor and the grounding conductor or enclosing metal raceway.

The component short circuit rating is a current rating given to conductors, switches, circuit breakers, and other electrical components, which, if exceeded by fault currents, will result in extensive damage to the component. The rating is expressed in terms of time intervals and/or current values. Short circuit damage can be the result of heat generated or the electromechanical force of a high-intensity magnetic field.

The NEC's intent is that the design of a system must be such that short circuit currents cannot exceed the short circuit current ratings of the components selected as part of the system. Given specific system components and the level of available short circuit currents that could occur, overcurrent protective devices (mainly fuses and/or circuit breakers) must be used that will limit the energy let-through of fault currents to levels within the withstand ratings of the system components.

1.1.3 Ground Faults

A ground fault is a conducting connection, whether intentional or accidental, between any of the conductors of an electrical system and the conducting material that encloses the conductors or any conducting material that is grounded or may become grounded.

2.0.0 ◆ FUSES

The fuse is a reliable overcurrent protective device. It consists of a fusible link or links encapsulated in a tube and connected to contact terminals. The electrical resistance of the link is so low that it simply acts as a conductor. However, when destructive currents occur, the link quickly melts and opens the circuit to protect conductors and other circuit components and loads. Fuse characteristics are stable and do not require periodic maintenance or testing.

2.1.0 Types of Fuses

NEC Article 240, Part V contains the requirements associated with plug fuses, plug fuseholders, and adapters; *NEC Article 240, Part VI* lists those requirements applying to cartridge fuses and cartridge fuseholders. Plug fuses are normally used in general-purpose branch circuits. Cartridge fuses may be found in branch circuits, feeder circuits, motor circuits, and many other special applications.

Per *NEC Section 240.50(A)*, plug fuses are only permitted in circuits not exceeding 125V between conductors or in circuits supplied by a system

INSIDE TRACK

Fuseholder Safety

Always verify that the fuseholders are not energized before replacing any fuses. Keep in mind that if the fuseholder is wired backfed, the fuseholder will remain energized even though the disconnect is in the OFF position. Also, if a switch is bleeding through, one knife blade may remain touching, keeping it energized.

INSIDE TRACK

Fuse Position

Always make a point of installing fuses right side up, with the label pointing out. When a fuse requires replacement and the label is upside down or backwards, the typical response would be to reach in and twist it around to read the label. This can result in serious shocks or even death.

Classroom

Define *ground fault.*

Teaching Tip

See the Teaching Tip at the end of this module.

Classroom

Discuss various types of fuses and their applications. Provide examples for the trainees to examine.

Refer the trainees to *NEC Article 240, Parts V and VI.*

Classroom

Explain that the voltage rating of a fuse must be equal to or greater than the circuit voltage.

Point out that the ampere rating should not exceed the current-carrying capacity of the circuit. Note the exceptions to this rule.

Emphasize that the interrupting rating must be sufficient to protect against the largest potential short circuit currents.

having a grounded neutral where the line-to-neutral voltage does not exceed 150V.

Plug fuses rated at 15A or less have hexagonal windows or caps so that they may easily be distinguished from fuses of higher ampere ratings. This is a requirement of *NEC Section 240.50(C)*. Plug fuses with an ampere rating above 15A have a round window.

There are two screw-in configurations associated with plug fuses, Edison-base and Type S. An adapter is available to convert an Edison-base socket to a Type S socket. The Edison-base socket is a standard screw-in base, which allows any ampere-rated Edison fuse to be installed. The Type S socket is matched to a specific-rated fuse and will not accept any other fuse except that rating.

NEC Section 240.52 prohibits the use of Edison-base fuseholders in new installations except where they are made to accept Type S fuses using the adapter method.

Per *NEC Section 240.60(A)*, cartridge fuses and fuseholders rated at 300V are only permitted to be used in circuits not exceeding 300V between conductors, or in single-phase, line-to-neutral circuits supplied from a three-phase, four-wire, solidly grounded, neutral source where the line-to-neutral voltage does not exceed 300V.

2.2.0 Voltage Rating

Most low-voltage power distribution fuses have 250V or 600V ratings. The **voltage rating** of a fuse must be at least equal to or greater than the circuit voltage. It can be higher, but never lower. For example, a 600V fuse can be used in a 240V circuit.

The voltage rating of a fuse is a function of or depends upon its capability to open a circuit under an overcurrent condition. Specifically, the voltage rating determines the ability of the fuse to suppress the internal arcing that occurs after a fuse link melts and an arc is produced (**arcing time**). If a fuse is used with a voltage rating lower than the circuit voltage, arc suppression will be impaired and, under some fault current conditions, the fuse may not safely clear the overcurrent. Special consideration is necessary for **semiconductor fuse** applications in which a fuse of a certain voltage rating is used on a lower-voltage circuit.

2.3.0 Ampere Rating

Every fuse has a specific ampere rating. When selecting the ampere rating, consideration must be given to the size and type of load and NEC requirements. The ampere rating of a fuse should normally not exceed the current-carrying capacity of the circuit. For instance, if a conductor is rated to carry 20A, a 20A fuse is the largest that should be used. However, there are specific circumstances in which the ampere rating is permitted to be greater than the current-carrying capacity of the circuit. A typical example is the motor circuit; dual-element fuses are generally permitted to be sized up to 175% and nontime-delay fuses up to 300% of the motor full-load amperes. Generally, the ampere rating of a fuse and switch combination should be selected at 125% of the continuous load current (this usually corresponds to the circuit capacity, which is also selected at 125% of the load current). There are exceptions, such as when the fuse-switch combination is approved for continuous operation at 100% of its rating.

2.4.0 Interrupting Rating

A protective device must be able to withstand the destructive energy of short circuit currents. If a fault current exceeds a level beyond the capability of the protective device, the device may actually rupture, causing additional damage. Therefore, it is important to choose a protective device that can sustain the largest potential short circuit currents. The rating that defines the capacity of a protective

INSIDE TRACK

Fault Current Protection

In a fully rated system, each overcurrent protective device has an interrupting rating that is equal to or greater than the available fault current at the service entrance. Fully rated systems are approved for use in all locations.

In a series rated system, the main upstream breaker has an interrupting capacity equal to or greater than the available fault current at the service entrance, but the downstream protective devices can be rated at lower values and therefore, will have a lower installed cost. Series rated systems have their place in the industry in areas where the tripping of a main breaker would cause a nuisance but would not result in an unsafe condition or the interruption of a critical process. For this reason, series rated systems are not approved for use in all locations. (See also ***NEC Section 110.22.***)

Instructor's Notes:

device to maintain its integrity when reacting to fault currents is called its interrupting rating.

NEC Section 110.9 requires equipment intended to interrupt current at fault levels to have an interrupting rating sufficient for the current that must be interrupted. Interrupting rating and interrupting capacity were covered in your Level Two training; more advanced material is presented in this module.

2.5.0 Selective Coordination

The coordination of protective devices prevents system power outages or blackouts caused by overcurrent conditions. When only the protective device nearest a faulted circuit opens and larger upstream fuses remain closed, the protective devices are selectively coordinated (they discriminate). The word *selective* is used to denote total coordination (isolation of a faulted circuit by the opening of only the localized protective device).

Figure 1 shows the minimum ratios of ampere rating of low peak fuses that are required to provide selective coordination of upstream and downstream fuses.

2.6.0 Current Limitation

If a protective device cuts off a short circuit current in less than one half cycle, before it reaches its total available (and highly destructive) value, the device is a **current-limiting device.** Most modern fuses are current-limiting devices. They restrict fault currents to such low values that a high degree of protection is given to circuit components against even very high short circuit currents. They permit breakers with lower interrupting ratings to be used and can reduce bracing of bus structures. They also minimize the need for other components to have high short circuit current ratings. If not limited, short circuit currents can reach levels of 30,000A or 40,000A or higher in the first half cycle (.008 second at 60Hz) after the start of a short circuit. The heat that can be produced in circuit components by the immense energy of short circuit currents can cause severe insulation damage or even an explosion. At the same time, the huge magnetic forces developed between conductors can crack insulators and distort and destroy bracing structures. Thus, it is extremely important that a protective device limit fault currents before they can reach their full potential level.

A noncurrent-limiting protective device, by permitting a short circuit current to build up to its full value, can let an immense amount of destructive short circuit heat energy through before opening the circuit, as shown in *Figure 2*. On the other hand, a current-limiting device (such as a current-limiting fuse) has such a high speed of response that it cuts off a short circuit long before it can build up to its full peak value, as shown in *Figure 3*.

3.0.0 ◆ OPERATING PRINCIPLES OF FUSES

This section describes the operating characteristics of various types of fuses. These fuses include the nontime-delay and the dual-element, time-delay fuses.

3.1.0 Nontime-Delay Fuses

The basic component of a fuse is the link. Depending upon the ampere rating of the fuse, the single-element, nontime-delay fuse may have one or more links. They are electrically connected to the end blades (or ferrules) and enclosed in a tube or cartridge surrounded by an arc-quenching filler material.

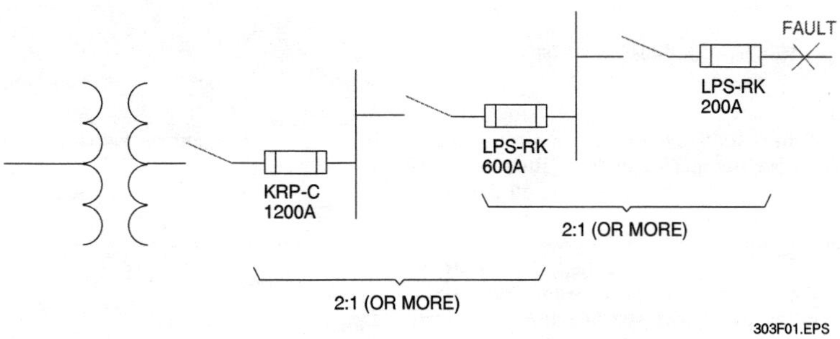

Figure 1 ◆ Selective coordination.

Classroom

Define *selective coordination.*

Audiovisual

Show Transparency 2 (Figure 1).

Classroom

Review the distinction between current-limiting devices and noncurrent-limiting devices.

Audiovisual

Show Transparencies 3 and 4 (Figures 2 and 3).

Classroom

Discuss the operation and applications of nontime-delay fuses.

![Audiovisual]

Show Transparency 5 (Figure 4).

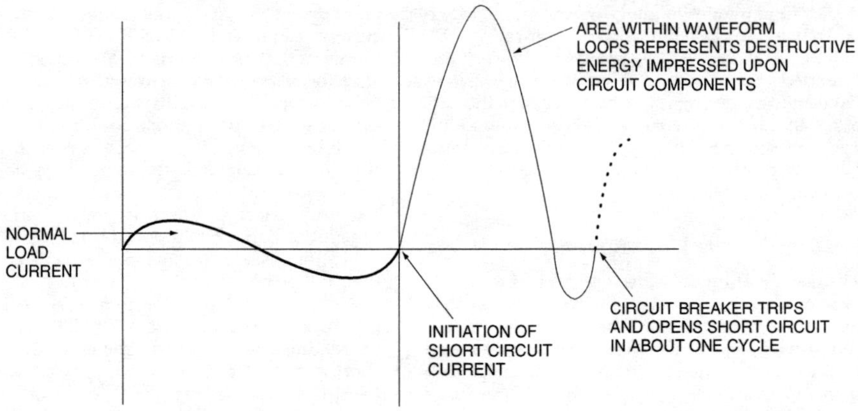

Figure 2 ◆ Characteristics of a noncurrent-limiting protective device.

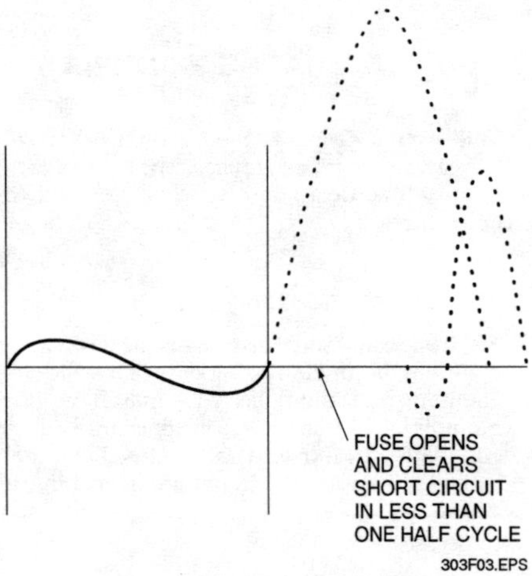

Figure 3 ◆ Characteristics of a current-limiting fuse.

Under normal operation, when the fuse is operating at or near its ampere rating, it simply functions as a conductor. However, as illustrated in *Figure 4*, if an overload current occurs and persists for more than a short interval of time, the temperature of the link eventually reaches a level that causes a restricted segment of the link to melt; as a result, a gap is formed and an electric arc established. As the arc causes the link metal to burn back, the gap becomes progressively larger. The electrical resistance of the arc eventually reaches

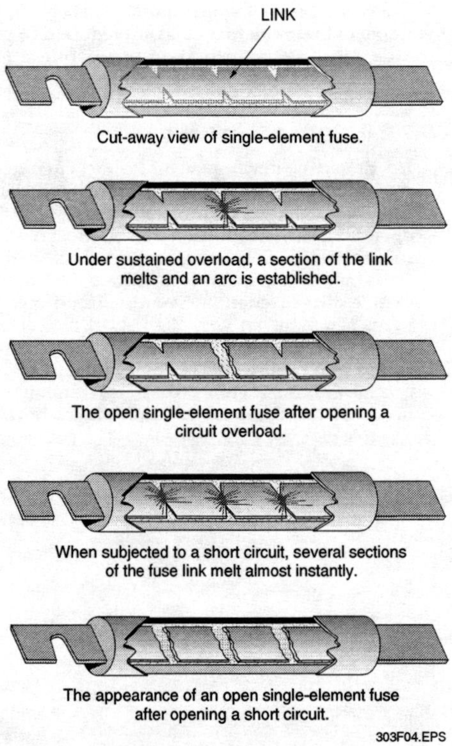

Figure 4 ◆ Characteristics of a single-element fuse.

Instructor's Notes:

such a high level that the arc cannot be sustained and is extinguished; the fuse will have then completely cut off all current flow in the circuit. Suppression or quenching of the arc is accelerated by the filler material.

Overload current normally falls within the region of between one and six times normal current, resulting in currents that are quite high. Consequently, a fuse may be subjected to short circuit currents of 30,000A to 40,000A or higher. The response of current-limiting fuses to such currents is extremely fast. The restricted sections of the fuse link will simultaneously melt within a matter of two-thousandths or three-thousandths of a second in the event of a high-level fault current.

The high resistance of the multiple arcs, together with the quenching effects of the filler particles, results in rapid arc suppression and clearing of the circuit. Again, refer to *Figure 3.* The short circuit current is cut off in less than one half cycle, long before the short circuit current can reach its full value.

3.2.0 Dual-Element, Time-Delay Fuses

Unlike single-element fuses, the dual-element, time-delay fuse can be applied in circuits subject to temporary motor overloads and surge currents to provide both high-performance short circuit and overload protection. Oversizing to prevent nuisance openings is not necessary with this type of fuse. The dual-element, time-delay fuse contains two distinctly separate types of elements. Electrically, the two elements are connected in series. The fuse links (similar to those used in the nontime-delay fuse) perform the short circuit protection function; the overload element provides protection against low-level overcurrents or overloads and will hold an overload that is five times greater than the ampere rating of the fuse for a minimum time of 10 seconds.

As shown in *Figure 5,* the overload section consists of a copper heat absorber and a spring-operated trigger assembly. The heat absorber bar is permanently connected to the heat absorber extension and the short circuit link on the opposite end of the fuse by the S-shaped connector of the trigger assembly. The connector electrically joins the short circuit link to the heat absorber in the overload section of the fuse. These elements are joined by a calibrated fusing alloy. An overload current causes heating of the short circuit link connected to the trigger assembly. The transfer of heat from the short circuit link to the heat absorber begins to raise the temperature of the heat absorber. If the overload is sustained, the temperature of the heat absorber eventually reaches a level that permits the trigger spring to fracture the calibrated fusing alloy and pull the connector free of the short circuit link and the heat absorber. As a result, the short circuit link is electrically disconnected from the heat absorber, the conducting path through the fuse is opened, and the overload current is interrupted. A critical aspect of the fusing alloy is that it retains its original characteristic after repeated temporary overloads without degradation.

The advantages of dual-element fuses are:

- Provide motor overload, ground fault, and short circuit protection
- Permit the use of smaller and less costly switches
- Give a higher degree of short circuit protection (greater current limitation) in circuits in which surge currents or temporary overloads occur
- Simplify and improve blackout prevention (selective coordination)

4.0.0 ◆ UL FUSE CLASSES

Safety is the primary consideration of Underwriters Laboratories (UL). The proper selection, overall functional performance, and reliability of a product are factors that are not within the basic scope of UL activities. However, to develop its safety test procedures, UL does develop the basic performance and physical specifications of standards of a product. In the case of fuses, these standards have culminated in the establishment of distinct **UL classes** of low-voltage (600V or less) fuses. Various UL fuse classes are described in the following paragraphs.

Class R fuses—UL Class R (rejection) fuses are high-performance 1/10A to 600A units, 250V and 600V, having a high degree of current limitation and a short circuit interrupting rating of up to 200,000A (**root-mean-square [rms]** symmetrical). This type of fuse is designed to be mounted in rejection-type fuse clips to prevent older Class H fuses from being installed. Since Class H fuses are not current-limiting devices and are recognized by UL as having only a 10,000A interrupting rating, serious damage could result if a Class H fuse were inserted in a system designed for Class R fuses. Consequently, ***NEC Section 240.60(B)*** requires fuseholders for current-limiting fuses to reject noncurrent-limiting fuses.

Figure 6 shows standard Class H and Class R cartridge fuses. A grooved ring in one ferrule of the Class R fuse provides the rejection feature of the Class R fuse in contrast to the lower interrupting capacity, nonrejection type. *Figure 7* shows fuse rejection clips for Class R fuses.

Classroom

Discuss the operation and applications of dual-element, time-delay fuses. List the advantages of these devices.

Audiovisual

Show Transparency 6 (Figure 5).

Classroom

Discuss UL fuse classes. List the voltage and current ratings for each class.

Audiovisual

Show Transparency 7 (Figure 6).

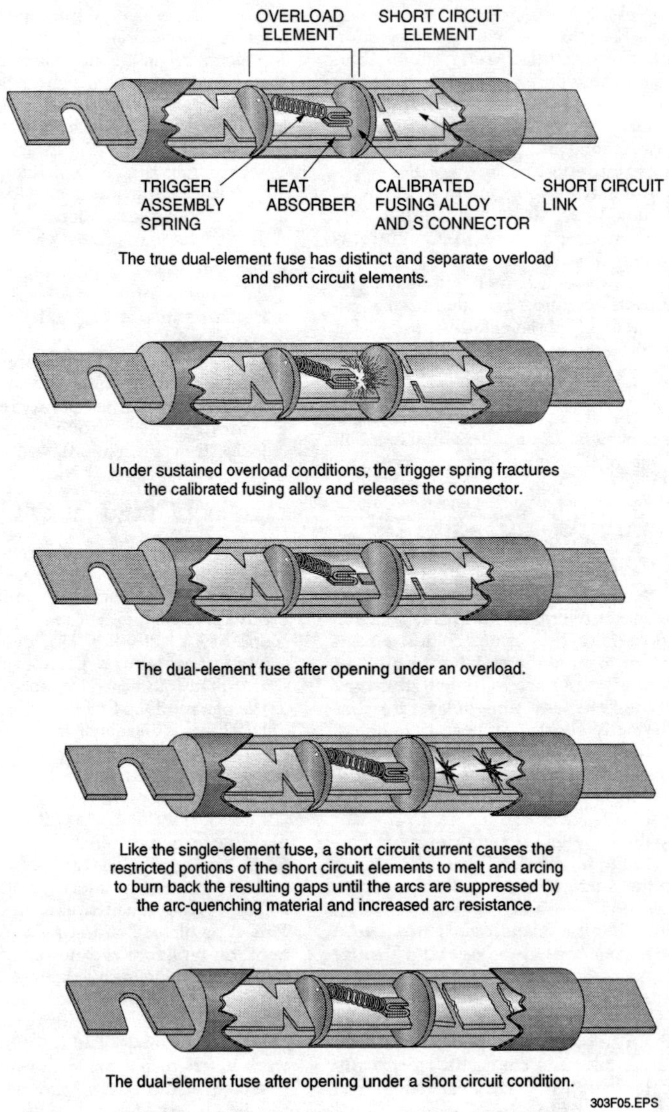

Figure 5 ◆ Characteristics of a dual-element, time-delay fuse.

Instructor's Notes:

Class CC fuses—Class CC fuses are 600V, 200,000A interrupting rating, branch circuit fuses with overall dimensions of 13/32" × 1½". Their design incorporates rejection features that allow them to be inserted into rejection fuseholders and fuse blocks that reject all lower voltage, lower interrupting rating 13/32" × 1½" fuses. They are available from 1/10A through 30A.

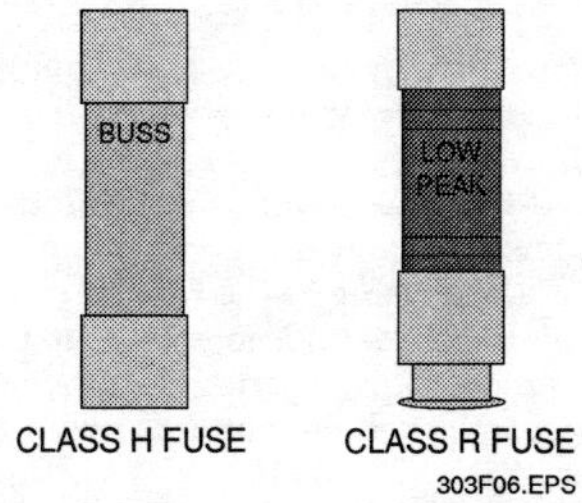

Figure 6 ◆ Comparison of Class H and Class R fuses.

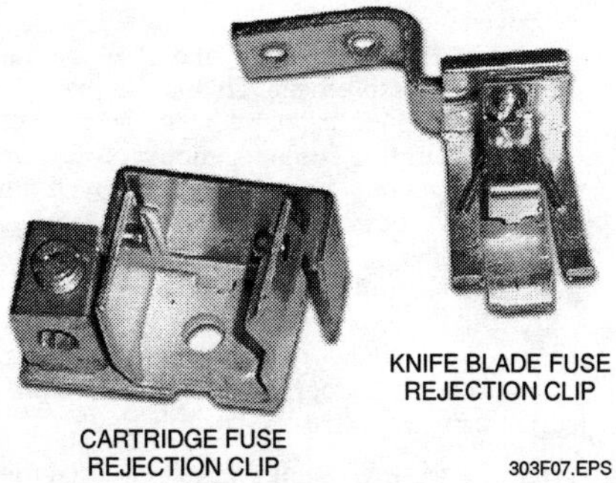

Figure 7 ◆ Fuse rejection clips.

Class G fuses—Class G fuses are 300V, 100,000A interrupting rating branch circuit fuses that are size rejecting to eliminate overfusing. The fuse diameter is 13/32", while the length varies from 1 5/16" to 2¼". They are available in ratings from 1A through 60A.

Class H fuses—Class H fuses are 250V and 600V, 10,000A interrupting rating branch circuit fuses that may be renewable or nonrenewable. They are available in ampere ratings of 1A through 600A.

Class J fuses—Class J fuses are rated to interrupt 200,000A. They are UL labeled as *current limiting*, are rated for 600VAC, and are not interchangeable with other classes.

Class K fuses—Class K fuses are listed by UL as K-1, K-5, or K-9. Each subclass has designated **ampere squared seconds (I^2t)** and **peak let-through (Ip)** maximums. These are dimensionally the same as Class H fuses (**NEC dimensions**), and they can have interrupting ratings of 50,000A, 100,000A, or 200,000A. These fuses are current-limiting devices; however, they are not marked *current limiting* since they do not have a rejection feature.

Class L fuses—Class L fuses are available in ampere ratings of 601A through 6,000A, and are rated to interrupt 200,000A. They are labeled *current limiting* and are rated for 600VAC. They are intended to be bolted into their mountings and are not normally used in clips. Some Class L fuses have time-delay features for all-purpose use.

Class T fuses—Class T fuses are 300V and 600V, with ampere ratings from 1A through 1,200A. They are physically very small and can be applied where space is at a premium. They are fast-acting fuses with an interrupting rating of 200,000A.

INSIDE TRACK

Knife Blade Fuses

In a Class R knife blade fuse, a notch in one blade provides the rejection feature. This picture shows both Class H and Class R knife blade fuses.

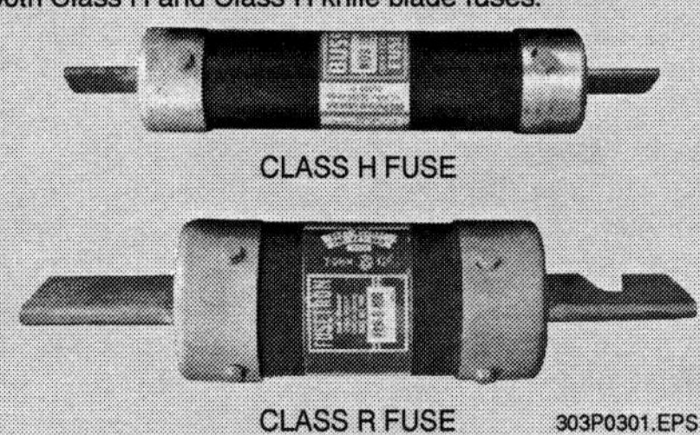

Classroom

Refer the trainees to Figure 7 in the Trainee Module. Provide examples of various fuses and clips for the trainees to examine.

Classroom

List the characteristics of branch circuit listed fuses.

Discuss the classifications of medium-voltage fuses:
- **General-purpose current-limiting fuses**
- **Backup current-limiting fuses**
- **Expulsion fuses**

Review the application considerations for current-limiting fuses. Discuss the relationship between the clearing curve of the fuse being selected and that of any downstream/upstream device(s).

4.1.0 Branch Circuit Listed Fuses

Branch circuit listed fuses are designed to prevent the installation of fuses that cannot provide a comparable level of protection to equipment. The characteristics of branch circuit fuses are as follows:

- They must have a minimum interrupting rating of 10,000A.
- They must have a minimum voltage rating of 125V.
- They must be size rejecting such that a fuse of a lower voltage rating cannot be installed in the circuit.
- They must be size rejecting such that a fuse with a current rating higher than the fuseholder rating cannot be installed in the circuit.

4.2.0 Medium-Voltage Fuses

As defined in *ANSI/IEEE 40-1981,* fuses above 600V are classified as general-purpose, current-limiting; backup current-limiting; or expulsion types.

- *General-purpose current-limiting fuses*—Fuses that are capable of interrupting all currents from the rated interrupting current down to the current that causes melting of the fusible element in one hour.
- *Backup current-limiting fuses*—Fuses that are capable of interrupting all currents from the maximum rated interrupting current down to the rated minimum interrupting current.
- *Expulsion fuses*—Vented fuses in which the expulsion effect of gases produced by the arc and lining of the fuseholder, either alone or aided by a spring, extinguishes the arc.

In the definitions just given, the fuses are defined as either expulsion or current-limiting types. A current-limiting fuse is a sealed, nonventing fuse that, when melted by a current within its interrupting rating, produces arc voltages exceeding the system voltage, which in turn forces the current to zero. The arc voltages are produced by introducing a series of high resistance arcs within the fuse. The result is a fuse that typically interrupts high fault currents within the first half cycle of the fault.

In contrast, an expulsion fuse depends on one arc to initiate the interruption process. The arc acts as a catalyst, causing the generation of de-ionizing gas from its housing. The arc is then elongated either by the force of the gases created or a spring. At some point, the arc elongates far enough to prevent a restrike after passing through a current zero. Therefore, it is not atypical for an expulsion fuse to take many cycles to clear.

4.2.1 Application of Medium-Voltage Fuses

Many of the rules for applying expulsion fuses and current-limiting fuses are the same, but because the current-limiting fuse operates much faster on high fault currents, some additional rules must be applied.

Three basic factors must be considered when applying any fuse: voltage, continuous current-carrying capacity, and interrupting rating.

- *Voltage*—The fuse must have a voltage rating that is equal to or greater than the normal frequency recovery voltage which will be seen across the fuse under all conditions. On three-phase systems, it is a good rule of thumb that the voltage rating of the fuse be greater than or equal to the line-to-line voltage of the system.
- *Continuous current-carrying capacity*—Continuous current values that are shown on the fuse represent the level of current the fuse can carry continuously without exceeding the temperature rises as specified in *ANSI C37.46.* An application that exposes the fuse to a current slightly above its continuous rating but below its minimum interrupting rating may damage the fuse due to excessive heat. This is the main reason overload relays are used in series with backup current-limiting fuses for motor protection.
- *Interrupting rating*—All fuses are given a maximum interrupting rating. This rating is the maximum level of fault current that the fuse can safely interrupt. Backup current-limiting fuses are also given a minimum interrupting rating. When using backup current-limiting fuses, it is important that other protective devices are used to interrupt currents below this level.

When choosing a fuse, it is important that the fuse be properly coordinated with other protective devices located upstream and downstream. To accomplish this, one must consider the **melting time** and **clearing time** characteristics of the devices. Two curves, the minimum melting curve and the total clearing curve, provide this information. To ensure proper coordination, the following rules should be observed:

- The total clearing curve of any downstream protective device must be below a curve representing 75% of the minimum melting curve of the fuse being applied.
- The total clearing curve of the fuse being applied must lie below a curve representing 75% of the minimum melting curve for any upstream protective device.

Instructor's Notes:

4.3.0 Current-Limiting Fuses

To ensure proper application of a current-limiting fuse, it is important that the following additional rules be applied:

- Current-limiting fuses produce arc voltages that exceed the system voltage. Care must be taken to ensure that the peak voltages do not exceed the insulation level of the system. If the fuse voltage rating is not permitted to exceed 140% of the system voltage, there should not be a problem. This does not mean that a higher rated fuse cannot be used, but one must be assured that the system insulation level (BIL) will handle the peak arc voltage produced. BIL stands for basic impulse level, which is the reference impulse insulation strength of an electrical system.
- As with the expulsion fuse, current-limiting fuses must be properly coordinated with other protective devices on the system. For this to happen, the rules for applying an expulsion fuse must be used at all currents that cause the fuse to interrupt in 0.01 second or greater.

When other current-limiting protective devices are on the system, it becomes necessary to use I^2t values for coordination at currents causing the fuse to interrupt in less than 0.01 second. These values may be supplied as minimum and maximum values or minimum melting and total clearing I^2t curves. In either case, the following rules should be followed:

- The minimum melting I^2t of the fuse should be greater than the total clearing I^2t of the downstream current-limiting device.
- The total clearing I^2t of the fuse should be less than the minimum melting I^2t of the upstream current-limiting device.

The fuse selection chart in *Figure 8* should serve as a guide for selecting fuses on circuits of 600V or less. Other valuable information may be found in catalogs furnished by manufacturers of overcurrent protective devices. These are usually obtainable from electrical supply houses or from manufacturers' representatives. You may also write the various manufacturers for a complete list (and price, if any) for all reference materials offered by them.

4.4.0 Fuses for Selective Coordination

The larger the upstream fuse is relative to a downstream fuse (feeder to branch, etc.), the less possibility there is of an overcurrent in the downstream circuit causing both fuses to open. Fast action, nontime-delay fuses require at least a 3:1 ratio between the ampere rating of a large upstream, line-side, time-delay fuse to that of the downstream, load-side fuse in order to be selectively coordinated. In contrast, the minimum selective coordination ratio necessary for dual-element fuses is only 2:1 when used with low peak load-side fuses (*Figure 9*).

The use of dual-element, time-delay fuses affords easy selective coordination, which hardly requires anything more than a routine check of a tabulation of required selectivity ratios. As shown in *Figure 10,* close sizing of dual-element fuses in the branch circuit for motor overload protection provides a large difference (ratio) in the ampere ratings between the feeder fuse and the branch fuse compared to the single-element, nontime-delay fuse.

4.5.0 Fuse Time-Current Curves

When a low-level overcurrent occurs, a long interval of time will be required for a fuse to open (melt) and clear the fault. On the other hand, if the overcurrent is large, the fuse will open very quickly. The opening time is a function of the magnitude of the level of overcurrent. Overcurrent levels and the corresponding intervals of opening times are logarithmically plotted in graph form, as shown in *Figure 11.* Levels of overcurrent are scaled on the horizontal axis, with time intervals on the vertical axis. The curve is therefore called a time-current curve.

The plot in *Figure 11* reflects the characteristics of a 200A, 600V, dual-element fuse. Note that at the 1,000A overload level, the time interval that is required for the fuse to open is 10 seconds. Yet, at approximately the 2,200A overcurrent level, the opening (melt) time of the fuse is only 0.01 second. It is apparent that the time intervals become shorter and shorter as the overcurrent levels become larger. This relationship is called an inverse time-to-current characteristic. Time-current curves are published or are available on most commonly used fuses showing minimum

THINK ABOUT IT

Arcing Fault

In the case of an arcing fault, why might the overcurrent or short circuit protection not clear the fault?

Classroom

List the application considerations for current-limiting fuses.

Show the trainees how to use a typical fuse selection chart. Refer to Figure 8 in the Trainee Module.

Compare the operation of nontime-delay and dual-element, time-delay fuses.

Audiovisual

Show Transparencies 8 and 9 (Figures 9 and 10).

Classroom

Use a time-current curve to trace the operation of a typical fuse.

Audiovisual

Show Transparency 10 (Figure 11).

Teaching Tip

Discuss the "Think About It."

Circuit	Load	Ampere Rating	Fuse Type	Symbol	Voltage Rating (ac)	UL Class	Interrupting Rating (K)	Remarks
Main, Feeder, and Branch (Conventional dimensions)	All type loads (optimum overcurrent protection)	0 to 600A	LOW PEAK® (dual-element time-delay)	LPN-RK LPS-RK	250V 600V	RK1	200	All-purpose fuses. Unequaled for combined short circuit and overload protection.
		601 to 6,000A	LOW PEAK® time-delay	KRP-C	600V	L	200	
	Motors, welders, transformers, capacitor banks (circuits with heavy inrush currents)	0 to 600A	FUSETRON® (dual-element time-delay)	FRN-R FRS-R	250V 600V	RK5	200	Moderate degree of current limitation. Time-delay passes surge currents.
		601 to 4,000A	LIMITRON® (time-delay)	KLU	600V	L	200	All-purpose fuse. Time-delay passes surge currents.
	Non-motor loads (circuits with no heavy inrush currents)	0 to 600A	LIMITRON® (fast-acting)	KTN-R KTS-R	250V 600V	RK1	200	Same short circuit protection as LOW PEAK® fuses but must be sized larger for circuits with surge currents: i.e., up to 300%.
	LIMITRON® fuses particular suited for circuit breaker protection	601 to 6,000A	LIMITRON® (fast-acting)	KTU	600V	L	200	A fast-acting, high-performance fuse.
	All type loads (optimum overcurrent protection)	0 to 600A	LOW PEAK® (dual-element, time-delay)	LPJ	600V	J	200	All-purpose fuses. Unequaled for combined short circuit and overload protection.
	Non-motor loads (circuits with no heavy inrush currents)	0 to 600A	LIMITRON® (quick-acting)	JKS	600V	J	200	Very similar to KTS-R LIMITRON®, but smaller.
		0 to 1,200A	T-TRON™	JJN JJS	300V 600V	T	200	The space saver (1/3 the size of KTN-R/KTS-R).

303F08.EPS

Figure 8 ◆ Fuse selection chart (600V or less).

Instructor's Notes:

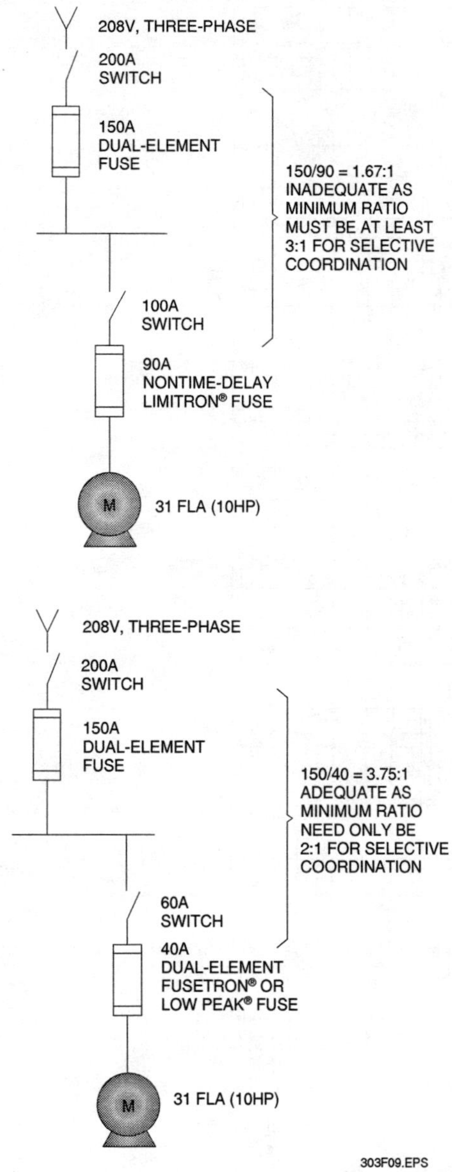

Figure 9 ◆ Fuses used for selective coordination.

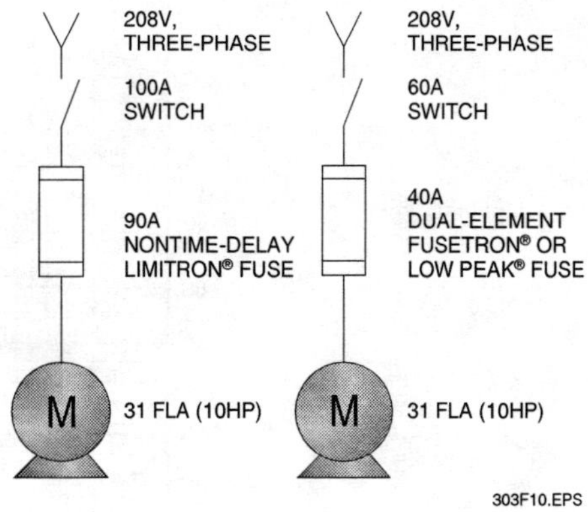

Figure 10 ◆ Comparison of dual-element fuse and single-element, nontime-delay fuse.

melt, average melt, and/or total clear characteristics. Although upstream and downstream fuses are easily coordinated by adhering to simple ampere ratios, these time-current curves permit close or critical analysis of coordination.

4.5.1 Peak Let-Through Charts

Peak let-through charts enable you to determine both the peak let-through current and the apparent prospective rms symmetrical let-through current. Such charts are commonly referred to as current limitation curves. *Figure 12* shows a simplified chart with explanations of the various functions.

> **Note**
> Let-through charts are available in the device manufacturer's literature.

See *Figure 12*. Point 1 shows the system available rms short circuit current of 40,000A. Upward from point 1 is the intersection with the fuse size of 100A (point 2). Follow this left to point 3, the peak let-through current (10,400A). This line represents the point at which the fuse becomes current limiting. From point 3, go down to the X axis and read the apparent symmetrical rms let-through current of 4,000A (point 4).

Classroom

Trace the operation of a typical fuse on a current-limitation curve.

Audiovisual

Show Transparency 11 (Figure 12).

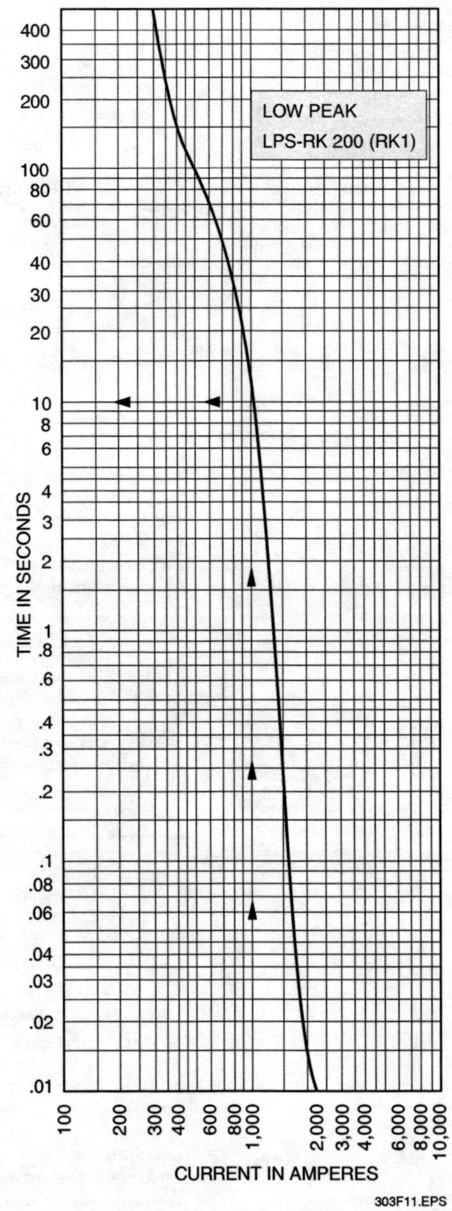

Figure 11 ◆ Typical time-current curve of a fuse.

Instructor's Notes:

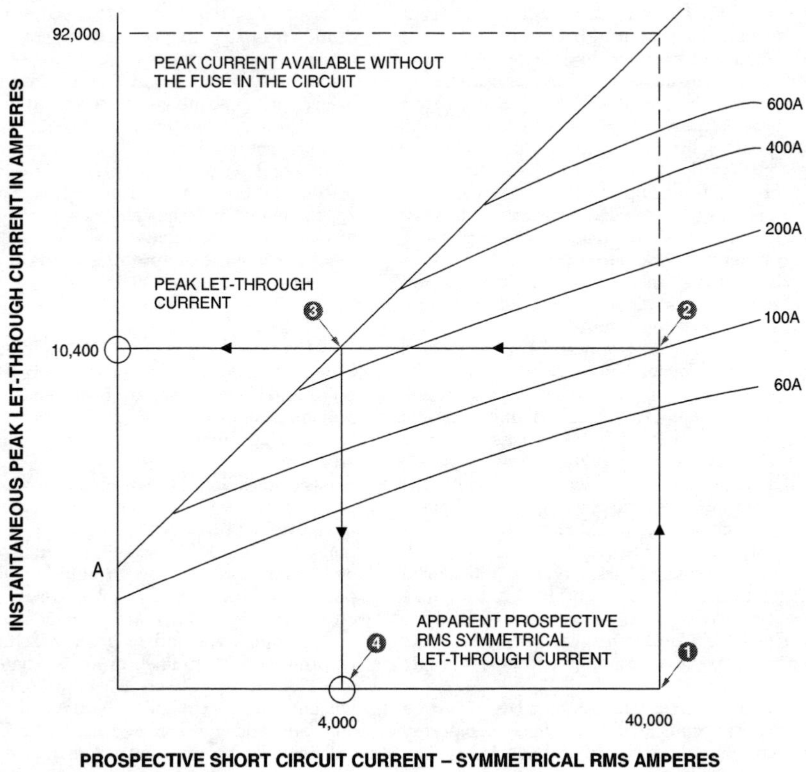

Figure 12 ◆ Principles of forming current limitation curves.

Three factors that affect the let-through performance of a fuse are the short circuit power factor, the point in the sine wave where fault interruption occurs, and the available voltage. The published charts typically represent worst-case scenarios.

5.0.0 ◆ MOTOR OVERLOAD AND SHORT CIRCUIT PROTECTION

When used in circuits with surge currents such as those caused by motors, transformers, and other inductive components, dual-element, time-delay fuses can be sized close to full-load amperes to give maximum overcurrent protection. For example, assume that a 10hp, 208V three-phase motor with integral thermal protection has a full-load current of 31A. *Table 1* shows the fuse type, size, and switch size required by ***NEC Sections 430.32(A)(2) and 430.52*** for a 10hp three-phase motor.

Table 1 shows that a 45A, dual-element fuse will protect the 31A motor compared to the much larger 100A, single-element fuse necessary. It is

Table 1 Fuse and Switch Size for a 10-Horsepower Motor (208V, 3∅, 31 FLA)

Fuse Type	Maximum Fuse Size (Amps)	Required Switch Size (Amps)
Dual-element, time-delay	45A	60A
Single-element, nontime-delay	100A	100A

Classroom

Discuss the specific requirements for motor overload and short circuit protection. Refer the trainees to *NEC Article 430.*

Classroom

Discuss the advantages of using dual-element fuses to provide motor protection.

Audiovisual

Show Transparency 12 (Figure 13).

Classroom

Discuss the operation and applications of various types of circuit breakers. Provide examples for the trainees to examine.

Audiovisual

Show Transparencies 13 and 14 (Figures 14 and 15).

apparent that if a sustained, harmful overload of 300% occurred in the motor circuit, the 100A, single-element fuse would never open and the motor could be damaged. The nontime-delay fuse provides only ground fault and short circuit protection—requiring separate overload protection as per the NEC.

In contrast, the 45A, dual-element fuse provides ground fault and short circuit protection plus a high degree of backup protection against motor burnout from overload or **single phasing** should other overload protective devices fail. If thermal overloads, relays, or contacts should fail to operate, the dual-element fuses will act independently to protect the motor.

Aside from providing only short circuit protection, the single-element fuse also makes it necessary to use larger size switches since a switch rating must be equal to or larger than the ampere rating of the fuse; as a result, the larger switch may cost two or three times more than would be necessary if a dual-element fuse were used (*Figure 13*).

When secondary single phasing occurs, the current in the remaining phases increases to a value of 170% to 200% of the rated full-load current. When primary single phasing occurs, unbalanced voltages that occur in the motor circuit cause excessive current. Dual-element fuses sized for motor overload protection can protect motors against the overload damage caused by single phasing.

The nontime-delay, **fast-acting fuse** must be oversized in circuits in which surge or temporary overload currents occur (**inductive load**). The response of the oversized fuse to short circuit currents is slower. Current builds up to a high level

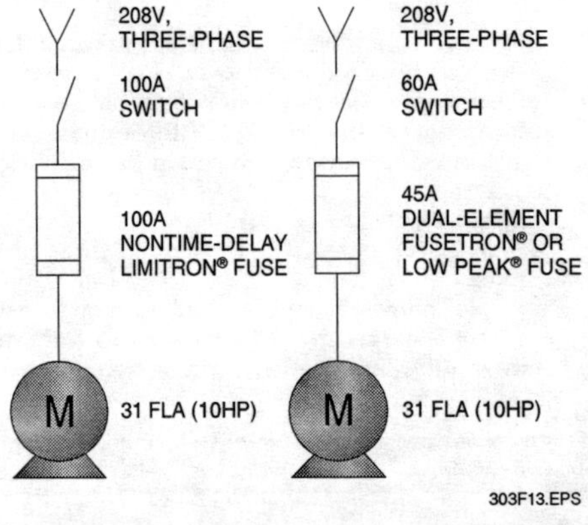

Figure 13 ◆ Dual-element fuses permit the use of smaller and less costly switches.

before the fuse opens, causing the current-limiting action of the oversized fuse to be less than a fuse whose ampere rating is closer to the normal full-load current of the circuit. Consequently, oversizing sacrifices some component protection and although it is permitted by the NEC, the practice is not recommended.

In actual practice, dual-element fuses used to protect motors keep short circuit currents to approximately half the value of the nontime-delay fuses, since the nontime-delay fuses must be oversized to carry the temporary starting current of a motor per *NEC Table 430.52.*

6.0.0 ◆ CIRCUIT BREAKERS

Circuit breakers were covered in your Level Two training. However, some of the more important points are worth repeating here before we cover practical applications of both fuses and circuit breakers.

Basically, a circuit breaker is a device for closing and interrupting a circuit between separable contacts under both normal and abnormal conditions. This is done manually (normal condition) by using its handle to switch it to the ON or OFF positions. However, the circuit breaker is also designed to open a circuit automatically on a predetermined overload or ground fault current without damage to itself or its associated equipment. As long as a circuit breaker is applied within its rating, it will automatically interrupt any fault and is therefore classified as an inherently safe overcurrent protective device.

The internal arrangement of a circuit breaker is shown in *Figure 14,* while its external operating characteristics are shown in *Figure 15.* Note that the handle on a circuit breaker resembles an ordinary toggle switch. On an overload, the circuit breaker opens itself or trips. In a tripped position, the handle jumps to the middle position (*Figure 15*). To reset it, turn the handle to the OFF position and then turn it as far as it will go beyond this position (RESET position); finally, turn it to the ON position.

A standard molded case circuit breaker usually contains:

- A set of contacts
- A magnetic trip element
- A thermal trip element
- Line and load terminals
- Busing used to connect these individual parts
- An enclosing housing of insulating material

The circuit breaker handle manually opens and closes the contacts and resets the automatic trip units after an interruption. Some circuit breakers

Instructor's Notes:

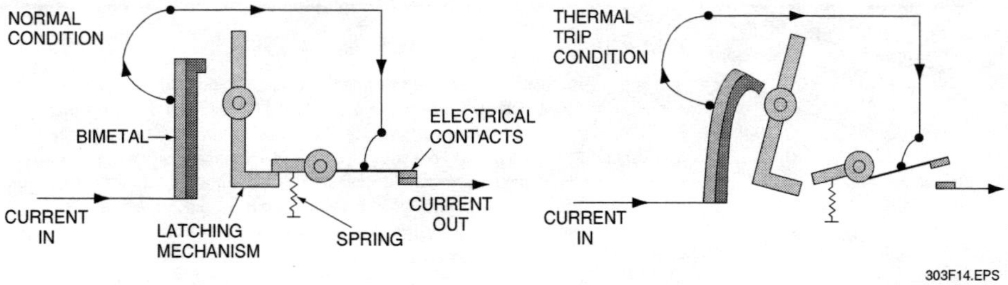

Figure 14 ◆ Internal arrangement of a circuit breaker.

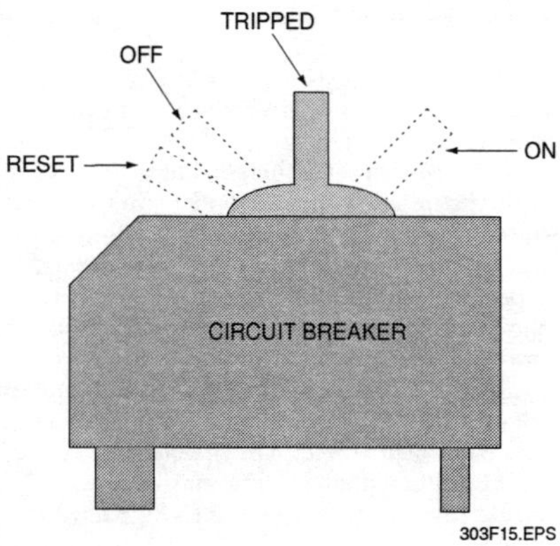

Figure 15 ◆ External characteristics of a circuit breaker.

INSIDE TRACK

Molded Case Circuit Breakers

UL requires that the handles of molded case circuit breakers be trip free. This means that the circuit breaker must open under trip conditions even if the handle is held in the ON position.

Arc Fault Circuit Interrupters (AFCIs)

Arc faults occur as a result of damaged or deteriorated wiring and are a common cause of house fires. AFCIs are special breakers that can detect arcs and eliminate current flow almost immediately. The NEC now requires AFCIs to be installed in all 15A and 20A bedroom receptacle circuits for new construction.

also contain a manually operated push-to-trip testing mechanism.

CAUTION

When a breaker trips, always determine why it tripped before resetting it.

Circuit breakers are grouped for identification according to given current ranges. Each group is classified by the largest ampere rating of its range. These groups are:

- 15A–100A
- 125A–225A
- 250A–400A
- 500A–1,000A
- 1,200A–2,000A

Therefore, they are classified as 100A, 225A, 400A, 1,000A, and 2,000A frames. These numbers are commonly referred to as frame classifications or frame sizes and are terms applied to groups of molded case circuit breakers that are physically interchangeable with each other.

6.1.0 Interrupting Capacity Rating

In most large commercial and industrial installations, it is necessary to calculate available short circuit currents at various points in a system to determine if the equipment meets the requirements of ***NEC Sections 110.9 and 110.10.*** There are a number of methods used to determine the short circuit requirements in an electrical system. Some give approximate values; others require extensive computations and are quite exacting.

The breaker interrupting capacity is based on tests to which the breaker is subjected. There are two such tests; one is set up by UL and the other

Classroom

Point out the new NEC requirement for arc fault circuit interrupters (AFCIs).

Safety

Emphasize the importance of determining why a breaker has tripped before resetting it.

Classroom

Discuss the amperage classifications and typical interrupting capacity ratings of circuit breakers.

Teaching Tip

Discuss the "Case History." Emphasize the importance of wearing appropriate protective equipment and exercising extreme caution when replacing fuses and circuit breakers.

Classroom

Explain that the NEC allows the next larger size protective device to be used when the conductor rating does not correspond to a standard size.

List the amperage classifications of branch circuits. Refer the trainees to *NEC Article 210.*

CASE HISTORY

Arc Flash

A serious injury resulted when a journeyman electrician on an industrial service call attempted to install a three-phase, 400A bolt-in circuit breaker into an energized panel. He carelessly miscalculated the positioning of the bolt-in lugs of the breaker with the breaker handle in the ON position, causing one phase terminal to contact an energized busbar, while contacting the grounded cabinet with another phase terminal. The circuit breaker exploded in his face, in addition to releasing a toxic vapor due to a phenomenon known as *arc flash.* He survived but sustained severe injuries to his eyes and lungs.

The Bottom Line: Working with energized equipment requires the use of properly rated personal protective equipment, including a full-body flash suit, face shield, and glasses.

by NEMA. The NEMA tests are self-certification. UL tests are certified by unbiased witnesses. UL tests have been limited to a maximum of 10,000A in the past, so the emphasis was placed on NEMA tests with higher ratings. UL tests now include the NEMA tests plus other ratings. Consequently, the emphasis is now being placed on UL tests.

The interrupting capacity of a circuit breaker is based on its rated voltage. Where the circuit breaker can be used on more than one voltage, the interrupting capacity will be shown for each voltage level. For example, the LA-type circuit breaker has 42,000A symmetrical interrupting capacity at 240V, 30,000A symmetrical at 480V, and 22,000A symmetrical at 600V.

7.0.0 ◆ CIRCUIT PROTECTION

All conductors in a circuit are to be protected against overcurrents in accordance with their ampacities as set forth in ***NEC Section 240.4.*** They must also be protected against short circuit current damage as required.

INSIDE TRACK

Surge Protection

In addition to the plug-in surge strips that are typically used to protect electronic equipment, some manufacturers are now offering residential surge protection devices that mount directly on the main breaker panel. These devices provide comprehensive protection for AC power in use throughout the residence, and are used in conjunction with surge strips. Surge strips are not rated to handle the higher surges caused by induced lightning, and the comprehensive device reduces the voltage surge to a level that can be handled by the surge protector.

Ampere ratings of overcurrent protective devices must not be greater than the ampacity of the conductor. There is, however, an exception. ***NEC Section 240.4(B)*** states that if such conductor rating does not correspond to a standard size overcurrent protective device, the next larger size overcurrent protective device may be used, provided its rating does not exceed 800A. Likewise, the conductor cannot be part of a multi-outlet branch circuit supplying receptacles for cord- and plug-connected portable loads. When the ampacity of a busway or cablebus does not correspond to a standard overcurrent protective device, the next larger standard rating may be used only if the rating does not exceed 800A (***NEC Sections 368.10 and 370.5***).

7.1.0 Lighting/Appliance Branch Circuits

The branch circuit rating must be classified in accordance with the rating of the overcurrent protective device. Classifications for those branch circuits other than individual loads must be 15A, 20A, 30A, 40A, and 50A, as specified in ***NEC Section 210.3.***

Branch circuit conductors and equipment must be protected by an overcurrent protective device with ampere ratings that conform to ***NEC Section 210.20.*** Basically, the branch circuit conductor and overcurrent protective device must be sized for the actual noncontinuous load plus 125% of the continuous load. The overcurrent protection size must not be greater than the conductor ampacity. Branch circuits rated 15A through 50A with two or more outlets (other than receptacle circuits) must be protected at their rating and the branch circuit conductor sized according to ***NEC Table 210.24.*** This protection is normally installed at the point where the conductors receive their supply.

Instructor's Notes:

7.1.1 Feeder Circuits

The feeder overcurrent protective device ampere rating and feeder conductor ampacity must be as follows:

- *Feeder circuit with no motor load*—The overcurrent protective device size must be at least 125% of the continuous load plus 100% of the noncontinuous load.
- *Feeder circuit with all motor loads*—Size the overcurrent protective device at 100% of the largest branch circuit protective device plus the sum of the full-load current of all other motors.
- *Feeder circuit with mixed loads*—Size the overcurrent protective device at 100% of the largest branch circuit protective device plus the sum of the full-load current of all other motors, plus 125% of the continuous, nonmotor load, plus 100% of the noncontinuous, nonmotor load.

7.1.2 Service Equipment

Each ungrounded service entrance conductor must have an overcurrent device in series with a rating not higher than the ampacity of the conductor. The service overcurrent devices shall be part of the service disconnecting means or be located immediately adjacent to it (***NEC Section 230.91***).

Service disconnecting means can consist of one to six switches or circuit breakers for each service or for each set of service-entrance conductors permitted in ***NEC Section 230.2.*** When more than one switch is used, the switches must be grouped together (***NEC Section 230.72***).

7.1.3 Transformer Secondary Circuits

Field installations indicate nearly 50% of transformers installed do not have secondary protection. The NEC requires overcurrent protection for lighting and appliance panelboards and recommends that secondary conductors be protected from damage by applying the proper overcurrent protective device. For example, the primary overcurrent device protecting a three-wire transformer cannot offer protection to the secondary conductors. ***NEC Section 240.4(F)*** and ***NEC Section 240.21(C)*** discuss protection of transformer secondary conductors.

7.1.4 Motor Circuit Protection

Motors and motor circuits have unique operating characteristics and circuit components. Therefore, these circuits must be dealt with differently from other types of loads. Generally, two levels of overcurrent protection are required for motor branch circuits:

- *Overload protection*—Motor running overload protection is intended to protect the system components and motor from damaging overload currents.
- *Short circuit protection (includes ground fault protection)*—Short circuit protection is intended to protect the motor circuit components such as the conductors, switches, controllers, overload relays, motor, etc., against short circuit currents or grounds. This level of protection is commonly referred to as motor branch circuit protection. Dual-element fuses are designed to provide this protection, as long as they are sized correctly.

There are a variety of ways to protect a motor circuit, depending upon the user's objective. The ampere rating of an overcurrent protective device selected for motor protection depends on whether the overcurrent protective device is of the dual-element, time-delay type or the nontime-delay type.

In general, nontime-delay fuses can be sized at 300% of the motor full-load current for ordinary motors so that the normal motor starting current does not affect the fuse (***NEC Table 430.52***). Dual-element, time-delay fuses are able to withstand normal motor starting current and can be sized closer to the actual motor rating than nontime-delay fuses.

Note

Design E motors require special circuit protection consideration.

A summary of NEC regulations governing overcurrent protection is covered in *Table 2,* while the table in *Figure 16* gives generalized fuse application guidelines for motor branch circuits. *Figure 17* may be used to select dual-element fuses for motor protection.

Note

In many cases, the overcurrent device size is less than that allowed by the NEC. This is an additional safeguard adopted by many manufacturers and installers.

Classroom

Discuss the overcurrent protection sizing requirements for:
- **Feeder circuits**
- **Service equipment**
- **Transformer secondary circuits**
- **Motor circuits**

Refer the trainees to Table 2 in the Trainee Module.

Audiovisual

Show Transparency 15 (Figure 16).

Table 2 NEC Regulations for Overcurrent Protection

Application	Rule	NEC Reference
Scope	Overcurrent protection for conductors and equipment is provided to open the circuit if the current reaches a value that will cause an excessive or dangerous temperature in conductors or conductor insulation. See also ***NEC Sections 110.9 and 110.10*** for requirements for interrupting capacity and protection against fault currents.	***NEC Section 240.1 FPN***
Protection required	Each ungrounded service-entrance conductor must have overcurrent protection in series with each ungrounded conductor.	***NEC Section 230.90(A)***
Number of devices	Up to six circuit breakers or sets of fuses may be considered as the overcurrent device.	***NEC Section 230.90(A), Exception No. 3***
Location in building	The overcurrent device must be part of the service disconnecting means or be located immediately adjacent to it.	***NEC Section 230.91***
Accessibility	In a property comprising more than one building under single management, the ungrounded conductors supplying each building served shall be protected by overcurrent devices, which may be located in the building served or in another building on the same property, provided they are accessible to the occupants of the building served. In a multiple-occupancy building, each occupant shall have access to the overcurrent protective devices.	***NEC Sections 230.92 and 230.72(C), Exception***
Location in circuit	The overcurrent device must protect all circuits and devices, except equipment which may be connected on the supply side, including: (1) Service switch; (2) Special equipment, such as surge arrestors; (3) Circuits for emergency supply and load management (where separately protected); (4) Circuits for fire alarms or fire pump equipment (where separately protected); (5) Meters with all metal housing grounded (600V or less); (6) Control circuits for automatic service equipment, if suitable overcurrent protection and disconnecting means are provided.	***NEC Section 230.94 plus Exceptions***
Installation and use	Listed or labelled equipment shall be used and installed in accordance with any instructions included in the listing or labeling.	***NEC Section 110.3(B)***
Interrupting rating	Equipment intended to interrupt current at fault levels shall have an interrupting rating sufficient for the system voltage and the current which is available at the line terminals of the equipment.	***NEC Section 110.9***
Circuit impedance and other characteristics	The overcurrent protective devices, total impedance, component short circuit current ratings, and other characteristics of the circuit to be protected shall be so selected and coordinated as to permit the circuit protective devices used to clear a fault without the occurrence of extensive damage to the electrical components of the circuit.	***NEC Section 110.10***
General	Bonding shall be provided where necessary to ensure electrical continuity and the capacity to safely conduct any fault current likely to be imposed.	***NEC Section 250.4(A)(4)***
Bonding other enclosures	Metal raceways, cable trays, cable armor, cable sheath, enclosures, frames, fittings, and other metal noncurrent-carrying parts that are to serve as grounding conductors, with or without the use of supplementary equipment grounding conductors, shall be effectively bonded where necessary to ensure electrical continuity and the capacity to safely conduct any fault current likely to be imposed on them. Any nonconductive paint, enamel, or similar coating shall be removed at threads, contact points, and contact surfaces or be connected by means of fittings so designed as to make such removal unnecessary.	***NEC Section 250.96(A)***

Instructor's Notes:

Type of Motor	Dual-Element, Time-Delay Fuses			Nontime-Delay Fuses
	Desired Level of Protection			
	Motor Overload and Short Circuit	**Backup Overload and Short Circuit**	**Short Circuit Only (Based on *NEC Tables 430.147 through 430.150* current ratings)**	**Short Circuit Only (Based on *NEC Tables 430.147 through 430.150* current ratings)**
Service Factor 1.15 or Greater or 40°C Temp. Rise or Less	125% or less of motor nameplate current	125% or next standard size (not to exceed 140%)	150% to 175%	150% to 300%
Service Factor Less Than 1.15 or Greater Than 40°C Temp. Rise	115% or less of motor nameplate current	115% or next standard size (not to exceed 130%)	150% to 175%	150% to 300%

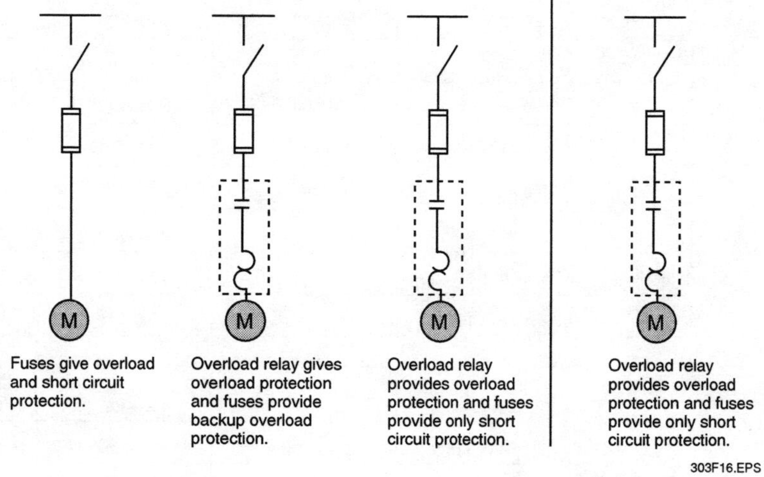

Figure 16 ◆ Fuse application guidelines for motor branch circuits.

Teaching Tip

Provide the trainees with various motor applications and have them determine the size of the overcurrent protection required. Refer to Figure 17 in the Trainee Module.

Dual-Element Fuse Size	Motor Protection (Used without properly sized relays). Motor Full-Load Amps		Backup Motor Protection (Used with properly sized overload relays). Motor Full-Load Amps	
	Motor Service Factor of 1.15 or Greater or With Temp. Rise Not Over 40°C.	Motor Service Factor Less Than 1.15 or With Temp. Rise Not Over 40°C.	Motor Service Factor of 1.15 or Greater or With Temp. Rise Not Over 40°C.	Motor Service Factor of Less Than 1.15 or With Temp. Rise Not Over 40°C.
1/10	0.08–0.09	0.09–0.10	0–0.08	0–0.09
1/8	0.10–0.11	0.11–0.15	0.09–0.10	0.10–0.11
5/100	0.12–0.15	0.14–0.15	0.11–0.12	0.12–0.13
2/10	0.16–0.19	0.18–0.20	0.13–0.16	0.14–0.17
1/4	0.20–0.23	0.22–0.25	0.17–0.20	0.18–0.22
3/10	0.24–0.30	0.27–0.30	0.21–0.24	0.23–0.26
4/10	0.32–0.39	0.35–0.40	0.25–0.32	0.27–0.35
1/2	0.40–0.47	0.44–0.50	0.33–0.40	0.36–0.43
6/10	0.48–0.60	0.53–0.60	0.41–0.48	0.44–0.52
8/10	0.64–0.79	0.70–0.80	0.49–0.64	0.53–0.70
1	0.80–0.89	0.87–0.97	0.65–0.80	0.71–0.87
1 1/8	0.90–0.99	0.98–1.08	0.81–0.90	0.88–0.98
1 1/4	1.00–1.11	1.09–1.21	0.91–1.00	0.99–1.09
1 4/10	1.12–1.19	1.22–1.30	1.01–1.12	1.10–1.22
1 1/2	1.20–1.27	1.31–1.39	1.13–1.20	1.23–1.30
1 6/10	1.28–1.43	1.40–1.56	1.21–1.28	1.31–1.39
1 8/10	1.44–1.59	1.57–1.73	1.29–1.44	1.40–1.57
2	1.60–1.79	1.74–1.95	1.45–1.60	1.58–1.74
2 1/4	1.80–1.99	1.96–2.17	1.61–1.80	1.75–1.96
2 1/2	2.00–2.23	2.18–2.43	1.81–2.00	1.97–2.17

303F17A.EPS

Figure 17 ◈ Selection of fuses for motor protection (1 of 3).

Instructor's Notes:

Dual-Element Fuse Size	Motor Protection (Used without properly sized relays). Motor Full-Load Amps		Backup Motor Protection (Used with properly sized overload relays). Motor Full-Load Amps	
	Motor Service Factor of 1.15 or Greater or With Temp. Rise Not Over 40°C.	Motor Service Factor Less Than 1.15 or With Temp. Rise Not Over 40°C.	Motor Service Factor of 1.15 or Greater or With Temp. Rise Not Over 40°C.	Motor Service Factor of Less Than 1.15 or With Temp. Rise Not Over 40°C.
2$\frac{6}{10}$	2.24–2.39	2.44–2.60	2.01–2.24	2.18–2.43
3	2.40–2.55	2.61–2.78	2.25–2.40	2.44–2.60
3$\frac{2}{10}$	2.56–2.79	2.79–3.04	2.41–2.56	2.61–2.78
3$\frac{1}{2}$	2.80–3.19	3.05–3.47	2.57–2.80	2.79–3.04
4	3.20–3.59	3.48–3.91	2.81–3.20	3.05–3.48
4$\frac{1}{2}$	3.60–3.99	3.92–4.34	3.21–3.60	3.49–3.91
5	4.00–4.47	4.35–4.86	3.61–4.00	3.92–4.35
5$\frac{6}{10}$	4.48–4.79	4.87–5.21	4.01–4.48	4.36–4.87
6	4.80–4.99	5.22–5.43	4.49–4.80	4.88–5.22
6$\frac{1}{4}$	5.00–5.59	5.44–6.08	4.81–5.00	5.23–5.43
7	5.60–5.99	6.09–6.52	5.01–5.60	5.44–6.09
7	6.00–6.39	6.53–6.95	5.61–6.00	6.10–6.52
8	6.40–7.19	6.96–7.82	6.01–6.40	6.53–6.96
9	7.20–7.99	7.83–8.69	6.41–7.20	6.97–7.83
10	8.00–9.59	8.70–10.00	7.21–8.00	7.84 –8.70
12	9.60–11.99	10.44–12.00	8.01–9.60	8.71–10.43
15	12.00–13.99	13.05–15.00	9.61–12.00	10.44–13.04
17$\frac{1}{2}$	14.00–15.99	15.22–17.39	12.01–14.00	13.05–15.21
20	16.00–19.99	17.40–20.00	14.01–16.00	15.22–17.39
25	20.00–23.99	21.74–25.00	16.01–20.00	17.40–21.74
30	24.00–27.99	26.09–30.00	20.01–24.00	21.75–26.09
35	28.00–31.99	30.44–34.78	24.01–28.00	26.10–30.43

303F17B.EPS

Figure 17 ◆ Selection of fuses for motor protection (2 of 3).

Dual-Element Fuse Size	Motor Protection (Used without properly sized relays). Motor Full-Load Amps		Backup Motor Protection (Used with properly sized overload relays). Motor Full-Load Amps	
	Motor Service Factor of 1.15 or Greater or With Temp. Rise Not Over 40°C.	Motor Service Factor Less Than 1.15 or With Temp. Rise Not Over 40°C.	Motor Service Factor of 1.15 or Greater or With Temp. Rise Not Over 40°C.	Motor Service Factor of Less Than 1.15 or With Temp. Rise Not Over 40°C.
40	32.00–35.99	34.79–39.12	28.01–32.00	30.44–37.78
45	36.00–39.00	39.13–43.47	32.01–36.00	37.79–39.13
50	40.00–47.99	43.48–50.00	36.01–40.00	39.14–43.48
60	48.00–55.99	52.17–60.00	40.01–48.00	43.49–52.17
70	56.00–59.99	60.87–65.21	48.01–56.00	52.18–60.87
75	60.00–63.99	65.22–69.56	56.01–60.00	60.88–65.22
80	64.00–71.99	69.57–78.25	60.01–64.00	65.23–69.57
90	72.00–79.99	78.26–86.95	64.01–72.00	69.58–78.26
100	80.00–87.99	86.96–95.64	72.01–80.00	78.27–86.96
110	88.00–99.00	95.65 –108.69	80.01–88.00	86.97–95.65
125	100.00–119.00	108.70–125.00	88.01–100.00	95.66–108.70
150	120.00–139.99	131.30–150.00	100.01–120.00	108.71–30.43
175	140.00–159.99	152.17–173.90	120.01–140.00	130.44–152.17
200	160.00–179.99	173.91–195.64	140.01–160.00	152.18–173.91
225	180.00–199.99	195.65–217.38	160.01–180.00	173.92–195.62
250	200.00–239.99	217.39–250.00	180.01–200.00	195.63–217.39
300	240.00–279.99	260.87–300.00	200.01–240.00	217.40–260.87
350	280.00–319.99	304.35–347.82	240.01–280.00	260.88–304.35
400	320.00–359.99	347.83–391.29	280.01–320.00	304.36–347.83
450	360.00–399.99	391.30–434.77	320.01–360.00	347.84–391.30
500	400.00–479.99	434.78–500.00	360.01–400.00	391.31–434.78
600	480.00–600.00	521.74–600.00	400.01–480.00	434.79–521.74

303F17C.EPS

Figure 17 ◆ Selection of fuses for motor protection (3 of 3).

Instructor's Notes:

Summary

Reliable overcurrent protective devices prevent or minimize costly damage to transformers, conductors, motors, and the many other components and electrical loads that make up the complete electrical distribution system. Consequently, reliable circuit protection is essential to avoid the severe monetary losses that can result from power blackouts and prolonged downtime of various types of facilities. The NEC has set forth various minimum requirements dealing with overcurrent devices and how they should be installed in various types of electrical circuits.

Review Questions

1. Which of the following best describes the maximum short circuit current that a fuse or circuit breaker will safely interrupt?
   a. CIA
   b. AIC
   c. SOL
   d. ICA

2. The minimum interrupting rating of branch circuit listed fuses is ____.
   a. 5,000A
   b. 10,000A
   c. 15,000A
   d. 20,000A

3. The minimum voltage rating of branch circuit fuses is ____.
   a. 24V
   b. 120V
   c. 125V
   d. 240V

4. All of the following are classifications of medium-voltage fuses *except* ____.
   a. general-purpose current-limiting fuses
   b. backup current-limiting fuses
   c. expulsion fuses
   d. proportion fuses

5. An expulsion fuse is best described as a ____.
   a. fuse capable of interrupting all currents from the rated interrupting current down to the current that causes melting of the fusible element in an hour
   b. fuse capable of interrupting all currents from the maximum rated interrupting current down to the rated minimum interrupting current
   c. strap-mounted device
   d. vented fuse in which the expulsion effect of gases produced by the arc and lining of the fuseholder, either alone or aided by a spring, extinguishes the arc

6. Each of the following is a basic factor to consider when applying any fuse *except* ____.
   a. voltage
   b. continuous current-carrying capacity
   c. interrupting rating
   d. manufacturer's brand name

7. The total clearing time of any downstream protective device must be below a curve representing ____ of the minimum melting curve of the fuse being applied.
   a. 10%
   b. 20%
   c. 50%
   d. 75%

8. When a common circuit breaker trips, the handle is in the ____ position.
   a. OFF
   b. ON
   c. middle
   d. RESET

9. Circuit breakers are grouped for identification according to given current ranges. Each group is classified by the ____.
   a. largest ampere rating of its range
   b. smallest ampere rating of its range
   c. absolute lowest ampere rating of its range
   d. overall average ampere rating of its range

10. When circuit breakers are classified as 100A through 2,000A frames, these numbers are normally referred to as the ____.
   a. frame size
   b. overload protective current
   c. maximum voltage allowed on the circuit
   d. physical size of the circuit breaker

Classroom

Have the trainees complete the Review Questions and go over the answers prior to administering the Module Examination.

Examination

Administer the Module Examination. Record the results on Craft Training Report Form 200 and submit the results to the Training Program Sponsor.

GLOSSARY

Trade Terms Introduced in This Module

Ampere rating: The current-carrying capacity of an overcurrent protective device. The fuse or circuit breaker is subjected to a current above its ampere rating; it will open the circuit after a predetermined period of time.

Ampere squared seconds (I^2t): The measure of heat energy developed within a circuit during the fuse's clearing. It can be expressed as *melting* I^2t, *arcing* I^2t, or the sum of them as *clearing* I^2t. I stands for effective let-through current (rms), which is squared, and t stands for time of opening in seconds.

Amperes interrupting capacity (AIC): The maximum short circuit current that a circuit breaker or fuse can safely interrupt.

Arcing time: The amount of time from the instant the fuse link has melted until the overcurrent is interrupted or cleared.

Clearing time: The total time between the beginning of the overcurrent and the final opening of the circuit at rated voltage by an overcurrent protective device. Clearing time is the total of the *melting time* and the *arcing time.*

Current-limiting device: A device that will clear a short circuit in less than one half cycle. Also, it will limit the instantaneous *peak let-through* current to a value substantially less than that obtainable in the same circuit if that device were replaced with a solid conductor of equal impedance.

Fast-acting fuse: A fuse that opens on overloads and short circuits very quickly. This type of fuse is not designed to withstand temporary *overload* currents associated with some electrical loads (*inductive loads*).

Inductive load: An electrical load that pulls a large amount of current—an inrush current—when first energized. After a few cycles or seconds, the current declines to the load current.

Melting time: The amount of time required to melt a fuse link during a specified overcurrent.

NEC dimensions: These are dimensions once referenced in the *National Electrical Code.* They are common to Class H and K fuses and provide interchangeability between manufacturers for fuses and fusible equipment of given *ampere* and *voltage ratings.*

Overload: Can be classified as an overcurrent that exceeds the normal full-load current of a circuit.

Peak let-through (Ip): The instantaneous value of peak current let-through by a current-limiting fuse when it operates in its current-limiting range.

Root-mean-square (rms): The effective value of an AC sine wave which is calculated as the square root of the average of the squares of all the instantaneous values of the current throughout one cycle. Alternating current rms is that value of an alternating current that produces the same heating effect as a given DC value.

Semiconductor fuse: Fuse used to protect solid-state devices.

Short circuit current: Can be classified as an overcurrent that exceeds the normal full-load current of a circuit by a factor many times greater than normal. Also characteristic of this type of overcurrent is that it leaves the normal current-carrying path of the circuit—it takes a shortcut around the load and back to the source.

Single phasing: That condition which occurs when one phase of a three-phase system opens, either in a low-voltage or high-voltage distribution system. Primary or secondary single phasing can be caused by any number of events. This condition results in unbalanced loads in polyphase motors and unless protec-

Instructor's Notes:

tive measures are taken, it will cause overheating and failure.

UL classes: Underwriters Laboratories has developed basic physical specifications and electrical performance requirements for fuses with voltage ratings of 600V or less. These are known as UL standards. If a type of fuse meets with the requirements of a standard, it can fall into that UL class. Typical UL classes are R, K, G, L, H, T, CC, and J.

Voltage rating: The maximum value of system voltage in which a fuse can be used, yet safely interrupt an overcurrent. Exceeding the voltage rating of a fuse impairs its ability to safely clear an *overload* or short circuit.

Additional Resources

This module is intended to present thorough resources for task training. The following reference works are suggested for further study. These are optional materials for continued education rather than for task training.

American Electrician's Handbook. Terrell Croft and Wilfred I. Summers. New York, NY: McGraw-Hill, 1996.

National Electrical Code Handbook. Quincy, MA: National Fire Protection Association, 2002.

Instructor's Notes:

ANSWER KEY

Answers to Review Questions

Answer	Section
1. b	1.1.2
2. b	4.1.0
3. c	4.1.0
4. d	4.2.0
5. d	4.2.0
6. d	4.2.1
7. d	4.2.1
8. c	6.0.0
9. a	6.0.0
10. a	6.0.0

Section 1.1.3 *Overcurrents*

This exercise will help the trainees to understand the terms associated with overcurrent conditions. Allow 15 minutes for this exercise.

1. Draw three columns on the chalkboard/whiteboard. Label each one with a type of fault (overload, short circuit, and ground fault.)
2. Call on trainees to come forward and write out a scenario under any one of the categories. For example, a power drain following a temporary power loss under overload that can occur when several pieces of equipment restart at the same time, a short circuit caused by inadvertently shorting a hot conductor using a screwdriver or other tool, and an electrocution under the grounding column.

Section 4.5.0 *Think About It—Arcing Fault*

An arc fault may not clear a protective device if the duration of the arc is very short or if the arc does not draw enough current to open the device.

CONTREN™ LEARNING SERIES — USER UPDATES

The NCCER makes every effort to keep these textbooks up-to-date and free of technical errors. We appreciate your help in this process. If you have an idea for improving this textbook, or if you find an error, a typographical mistake, or an inaccuracy in NCCER's Contren™ textbooks, please write us, using this form or a photocopy. Be sure to include the exact module number, page number, a detailed description, and the correction, if applicable. Your input will be brought to the attention of the Technical Review Committee. Thank you for your assistance.

Instructors – If you found that additional materials were necessary in order to teach this module effectively, please let us know so that we may include them in the Equipment/Materials list in the Instructor's Guide.

Write: Curriculum Revision and Development Department
National Center for Construction Education and Research
P.O. Box 141104, Gainesville, FL 32614-1104

Fax: 352-334-0932

E-mail: curriculum@nccer.org

Craft ______ Module Name ______

Copyright Date ______ Module Number ______ Page Number(s) ______

Description ______

(Optional) Correction ______

(Optional) Your Name and Address ______

Raceway, Box, and Fitting Fill Requirements

26304-02

Raceway, Box, and Fitting Fill Requirements
Annotated Instructor's Guide

Module 26304-02

MODULE OVERVIEW

This course introduces the Electrical trainee to raceway, box, and fitting fill requirements.

PREREQUISITES

Please refer to the Course Map in the Trainee Module. Prior to training with this module, it is recommended that the trainee shall have successfully completed the following:

Core Curriculum; Electrical Level One; Electrical Level Two; Electrical Level Three, Modules 26301 through 26303.

OBJECTIVES

Upon completion of this module, the trainee will be able to:

1. Size raceways according to conductor fill and NEC installation requirements.
2. Size outlet boxes according to NEC installation requirements.
3. Size and select pull and junction boxes according to NEC installation requirements.
4. Calculate conduit fill using a percentage of the trade size conduit inside diameter (ID).
5. Calculate the required bending radius in boxes and cabinets.

PERFORMANCE TASKS

This is a knowledge-based module; there is no performance examination.

NCCER STANDARDIZED CRAFT TRAINING PROGRAMS

The National Center for Construction Education and Research (NCCER) provides a standardized national program of accredited craft training. Key features of the program include instructor certification, competency-based training, and performance testing. The program provides trainees, instructors, and companies with a standard form of recognition through a National Craft Training Registry. The program is described in full in the *Guidelines for Accreditation*, published by the NCCER. For more information on standardized craft training, contact the NCCER by writing us at P.O. Box 141104, Gainesville, FL 32614-1104, calling 352-334-0911, or e-mailing info@nccer.org. More information may be found at our Web site, www.nccer.org.

HOW TO USE THIS ANNOTATED INSTRUCTOR'S GUIDE

Each page presents two sections of information. The larger section displays each page exactly as it appears in the Trainee Module. The narrow column ties suggested trainee and instructor actions to each page and provides icons to call your attention to material, safety, audiovisual, or testing requirements. The bottom of each page includes space for your notes.

If you see the Teaching Tip icon, that means there is a teaching tip associated with this section. Also refer to the suggested teaching tips at the end of the module.

SAFETY CONSIDERATIONS

Ensure that the trainees are equipped with appropriate personal protective equipment.

PREPARATION

Before teaching this module, you should review the Module Outline, Objectives, and the Materials and Equipment List. Be sure to allow ample time to prepare your own training or lesson plan and gather all required equipment and materials.

MATERIALS AND EQUIPMENT LIST

Materials:
Transparencies
Markers/chalk
Copy of the latest edition of the *National Electrical Code®*
Various types of raceways
Various examples of outlet, pull, and junction boxes
Short lengths of conductors of various wire sizes
Raised device covers
Extension rings
Examples of switches, receptacles, clamps, and other devices
Module Examinations*

Equipment:
Overhead projector and screen
Whiteboard/chalkboard
Appropriate personal protective equipment

*Located in the Test Booklet packaged with this Annotated Instructor's Guide.

ADDITIONAL RESOURCES

This module is intended to present thorough resources for task training. The following reference works are suggested for both instructors and motivated trainees interested in further study. These are optional materials for continued education rather than for task training.

American Electrician's Handbook. Terrell Croft and Wilfred I. Summers. New York, NY: McGraw-Hill, 1996.

National Electrical Code Handbook. Quincy, MA: National Fire Protection Association, 2002.

TEACHING TIME FOR THIS MODULE

An outline for use in developing your lesson plan is presented below. Note that each Roman numeral in the outline equates to one session of instruction. Each session has a suggested time period of 2½ hours. This includes 10 minutes at the beginning of each session for administrative tasks and one 10-minute break during the session. Approximately 12½ hours are suggested to cover *Raceway, Box, and Fitting Fill Requirements.* You will need to adjust the time required for hands-on activity and testing based on your class size and resources.

Topic	Planned Time
Session I. Introduction; Conduit Fill Requirements	
A. Introduction	________
B. Conduit Fill Requirements	________
1. Conduit Fill Calculations	________
Session II. Conduit Fill Requirements	
A. Conduit Fill Requirements	________
1. Practice Calculations	________
Session III. Conduit Bodies, Pull Boxes, and Junction Boxes	
A. Conduit Bodies, Pull Boxes, and Junction Boxes	________
1. Sizing Pull and Junction Boxes	________
2. Cabinets and Cutout Boxes	________
Session IV. Outlet Boxes	
A. Outlet Boxes	________
1. Sizing Outlet Boxes	________
Session V. Review; Module Examination	
A. Review	________
B. Module Examination	________
1. Trainees must score 70% or higher to receive recognition from the NCCER.	
2. Record the testing results on Craft Training Report Form 200 and submit the results to the Training Program Sponsor.	

Module 26304-02

Raceway, Box, and Fitting Fill Requirements

Course Map

This course map shows all of the modules in the third level of the Electrical curriculum. The suggested training order begins at the bottom and proceeds up. Skill levels increase as you advance on the course map. The local Training Program Sponsor may adjust the training order.

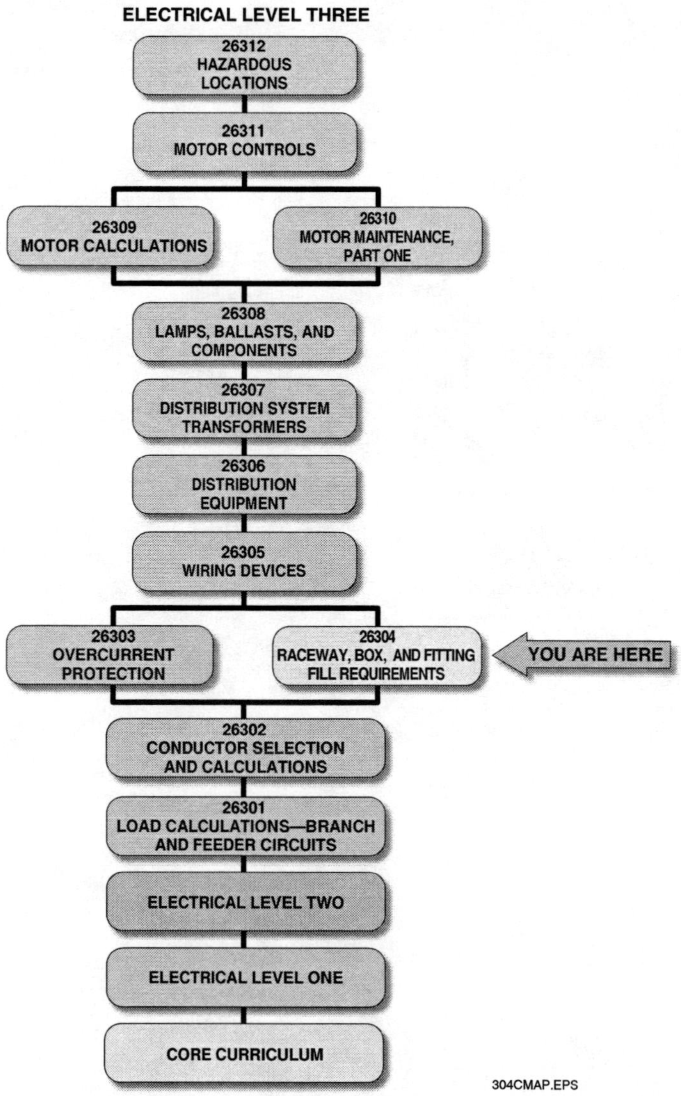

304CMAP.EPS

Homework

Assign reading of Module 26304.

MODULE 26304 CONTENTS

Figures

Table

MODULE 26304

Raceway, Box, and Fitting Fill Requirements

Objectives

When you have completed this module, you will be able to do the following:

1. Size raceways according to conductor fill and NEC installation requirements.
2. Size outlet boxes according to NEC installation requirements.
3. Size and select pull and junction boxes according to NEC installation requirements.
4. Calculate conduit fill using a percentage of the trade size conduit inside diameter (ID).
5. Calculate the required bending radius in boxes and cabinets.

Prerequisites

Before you begin this module, it is recommended that you successfully complete the following: Core Curriculum; Electrical Level One; Electrical Level Two; Electrical Level Three, Modules 26301 through 26303.

Required Trainee Materials

1. Pencil and paper
2. Appropriate personal protective equipment
3. Copy of the latest edition of the *National Electrical Code*

1.0.0 ◆ INTRODUCTION

A raceway is any channel which is designed and used for the sole purpose of holding wires, **cables,** or busbars. Raceways are constructed of either metal or insulating material.

Types of raceways include:

- Rigid metal conduit (RMC)
- Intermediate metal conduit (IMC)
- Rigid nonmetallic conduit (RNC)
- Flexible metal conduit (FMC)
- Liquid-tight flexible metal conduit (LFMC)
- Liquid-tight flexible nonmetallic conduit (LFNC)
- Electrical metallic tubing (EMT)
- Electrical nonmetallic tubing (ENT)
- Underfloor raceways
- Cellular metal floor raceways
- Cellular concrete floor raceways
- Surface metal raceways, wireways, and auxiliary gutters

Raceways provide mechanical protection for the conductors that run in them and prevent accidental damage to insulation and the conducting metal. Depending on the application, some raceways are also designed to protect conductors from the harmful chemical attack of corrosive atmospheres, and reduce hazards to life and property by confining arcs and flame caused by faults in the wiring system.

One of the most important functions of metal raceways is to provide a path for the flow of fault current to ground, thereby preventing voltage buildup on conductor and equipment enclosures. This feature helps to minimize shock hazards to personnel and damage to electrical equipment. To maintain this feature, it is extremely important that all metal raceway systems be securely **bonded** together into a continuous conductive path and properly connected to an electrical grounding system.

Note: The designations "National Electrical Code," "NE Code," and "NEC," where used in this document, refer to the National Electrical Code®, which is a registered trademark of the National Fire Protection Association, Quincy, MA. *All National Electrical Code (NEC) references in this module refer to the 2002 edition of the NEC.*

Materials

Ensure that you have everything required to teach the course. Check the Materials and Equipment List at the front of this Instructor's Guide.

Classroom

Review the goals of the course.

Explain that terms shown in bold (blue) are defined in the Glossary at the back of this module.

Audiovisual

Show Transparency 1, Course Objectives.

Classroom

Define *raceway*. List the common types of raceways and provide examples for the trainees to examine.

Explain that the primary function of a raceway is to provide protection for the conductors. Point out that metal raceways also provide a grounding path.

List the required locations of boxes/fittings in raceway systems.

Discuss conduit fill requirements. Refer the trainees to *NEC Chapter 9, Table 1.*

Show Transparency 2 (Figure 1).

Explain that raceway sizes must be adjusted to account for the number of conductors and the ambient temperature.

In a conduit raceway system, a box or fitting must be installed at:

- Each conductor splice point
- Each outlet, switch point, or junction point
- Each pull point for the connection of conduit and other raceways

Furthermore, boxes or other fittings are required when a change is made from conduit to open wiring. Electrical workers also install pull boxes in raceway systems to facilitate the pulling of conductors.

The NEC specifies specific maximum fill requirements for raceways, outlet boxes, pull boxes, and **junction boxes.** Fill requirements specify the area of conductors in relation to the box, fitting, or raceway system. This module is designed to cover these NEC requirements and apply these rules to practical applications.

2.0.0 ◆ CONDUIT FILL REQUIREMENTS

NEC Chapter 9, Table 1 summarizes rules on the maximum number of conductors permitted in raceways. In conduits, for either new work or rewiring of existing raceways, the maximum **conduit fill** must not exceed the percentages listed in *NEC Chapter 9, Table 1.* In all such cases, fill is based on using the actual cross-sectional areas of the particular types of conduit *(NEC Chapter 9, Table 4)* and conductors *(NEC Chapter 9, Table 8)* used. This is better illustrated in *Figure 1*, using ½" EMT conduit as an example.

Jamming is the wedging of three cables lying side by side in a conduit, resulting in a failed conductor installation. *NEC Chapter 9, Table 1, FPN No. 2* states that when pulling three conductors or cables into a raceway, if the ratio of the inside diameter of the raceway to the outside diameter of the conductor or cable is between 2.8 and 3.2, jamming can occur.

Other derating rules are specified in *NEC Article 310.* For example, if more than three conductors are used in a single conduit, a reduction in **ampacity** is required. Ambient temperature is another consideration that may call for derating of wires below the values given in NEC tables. This is covered in the Electrical Level Three module *Conductor Selection and Calculations.*

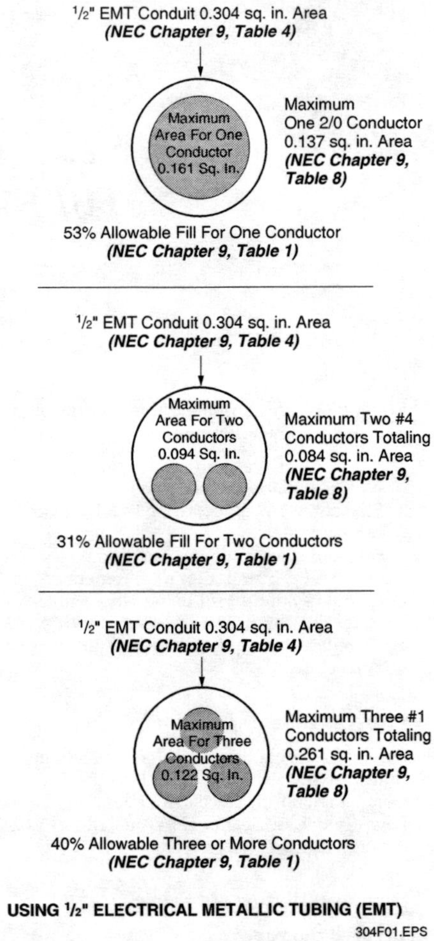

Figure 1 ◆ Conduit fill using ½" EMT.

INSIDE TRACK

Jamming

Be aware that bends in conduit, particularly EMT, may distort the inside diameter of the conduit enough to cause jamming, regardless of whether or not the correct fill has been applied. When pulling three conductors into a raceway, apply the formula found in ***NEC Chapter 9, Table 1, FPN No. 2*** to see if the next larger size raceway should be used.

Instructor's Notes:

2.1.0 Conduit Fill Calculations

The allowable number of conductors in a raceway system is calculated as percentage of fill, as per *NEC Chapter 9, Table 1.* When using this table, remember that equipment grounding or bonding conductors, where installed, must be included when calculating conduit fill or tubing fill. The actual dimensions of the equipment grounding or bonding conductor (insulated or bare) must be used in the calculation.

Most installations require three or more conductors in a conduit, therefore the typical percentage of fill to remember is 40%. This means that you must know the total area of the conduit to be used and the combined areas of all conductors that are to be installed in the conduit. *Figure 2* illustrates this comparison.

> **Note**
>
> If all of the conductors are of the same size, refer to ***NEC Annex C, Tables C1 through C12,*** being careful to identify the type of raceway and conductor being used. If you are using compact aluminum conductors, refer to ***NEC Tables C1(A) through C12(A).***

To calculate how many No. 6 THW wires you can install in a 2" EMT conduit, proceed as follows:

Step 1 Refer to *NEC Annex C, Table C1.*

Step 2 Locate the section of the table listing THW wire.

Step 3 Move right across the table until you reach the 2" *Trade Size* column.

Step 4 Move down this column until you are lined up with conductor size No. 6.

Step 5 The total quantity of No. 6 THW wire in a 2" EMT conduit is determined to be 18.

You can also use these tables if you know the size and type of wire and need to know the size of a raceway required.

To find the size of rigid metallic conduit required for four 500 kcmil THHN conductors, proceed as follows:

Step 1 Refer to *NEC Annex C, Table C8.*

Step 2 Locate the section of the table listing THHN insulation.

Step 3 Move down the table until you reach 500 kcmil.

Step 4 Move right across the table from 500 kcmil until you reach the column with the number 4, indicating 4 conductors. Look up to the top of this column to find that you would need a trade diameter of 3". (If the quantity of conductors is not listed in the row, you must use the next higher listed quantity.)

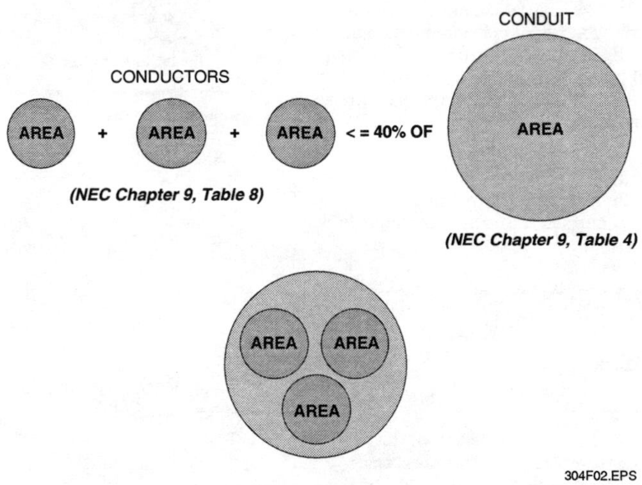

Figure 2 ◆ Conductor area compared to conduit area.

Classroom

Point out that when three or more conductors are installed in a raceway, the maximum fill is 40%.

Audiovisual

Show Transparency 3 (Figure 2).

Classroom

Refer the trainees to *NEC Annex C.* Demonstrate how to determine the maximum number of conductors that may be installed in a raceway of a given size. Provide practice exercises for the trainees.

Show the trainees how to determine the size of the raceway to be used for a given size and type of wire. Provide practice exercises for the trainees.

Classroom

Refer the trainees to *NEC Chapter 9, Tables 4 and 5* and explain how to determine the conduit size and fill requirements for installations with more than one size of conductor. Provide example problems for the trainees to solve.

Audiovisual

Show Transparency 4 (Figure 3).

If you have more than one size of conductor installed in the same raceway, use ***NEC Chapter 9, Tables 4 and 5*** to determine the proper conduit size and fill requirements. Use the following procedure.

Step 1 Using ***NEC Chapter 9, Table 5***, select the type and size of conductors being used.

Step 2 Move right to the *Approx. Area Sq. In.* column to find the area required for each conductor size.

Step 3 Multiply the number of each size of conductors by the quantity to determine the total sq. in. required for each conductor size. Add the total sq. in. required for each conductor size together to determine the total area required.

Step 4 Determine the type and size of raceway being used (e.g., EMT, IMC, rigid, etc.).

Step 5 Refer to ***NEC Chapter 9, Table 4*** to locate the type of raceway.

Step 6 Once the appropriate raceway is identified, move right across the row to the column that corresponds to the number of conductors in the raceway. Typically, the *Over 2 Wires* column will be used.

Step 7 Go down this column until you reach a value that is equal to or greater than the amount calculated in Step 3.

Step 8 Move left across the table to find the trade size of conduit required.

For example, take a situation in which a 1,000A service-entrance conductor is to be installed using three parallel rigid metal conduits, each containing three 500 kcmil THHN ungrounded conductors and one 350 kcmil THHN grounded conductor (neutral), as shown in *Figure 3*. What size of rigid conduit is required?

Step 1 Refer to ***NEC Chapter 9, Table 5*** to determine the area (in sq. in.) of a 500 kcmil THHN conductor. The area is found to be 0.7073 sq. in.

INSIDE TRACK

Pocket-Size References

There are many pocket-sized references that list these conduit fill tables and steps. They are often available at electrical supply houses free of charge.

Step 2 Repeat Step 1 to find the area for 350 kcmil THHN. The area is found to be 0.5242 sq. in.

Step 3 Multiply the number of conductors of each size by their corresponding areas.

$0.7073 \times 3 = 2.1219$ (total area required for 500 kcmil conductors)

$0.5242 \times 1 = .5242$ (total area required for 350 kcmil conductor)

Step 4 Add the totals together to obtain the total area required for all conductors.

$2.1219 + 0.5242 = 2.6461$ sq. in.

Step 5 Refer to ***NEC Chapter 9, Table 4*** and locate *Rigid Metal Conduit*.

Step 6 Looking in the *Over 2 Wires* column, scan down until you find a number equal to or greater than 2.6461. The value found is 3.000. Now scan back to the left to determine that the conduit size required is 3".

Therefore, each of the three conduits containing three 500 kcmil THHN conductors and one 350 kcmil THHN conductor must be at least 3" trade size to comply.

Note

You may have noticed that the internal diameter of a raceway does not match the trade size in all cases, nor does it match the internal diameter of different raceways in the same trade size. The trade size is a size given to a raceway with an internal diameter close to that size. In recognizing the difference in internal diameters, the NEC has expanded ***NEC Chapter 9, Table 4*** and added the tables in ***Annex C.***

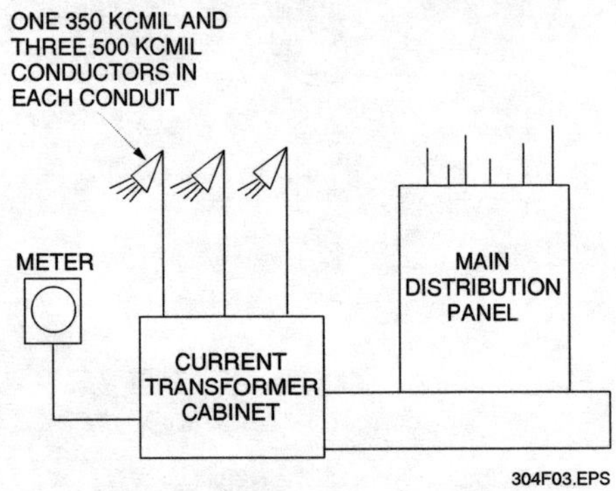

Figure 3 ◆ Power riser diagram of a 1,000A service-entrance conductor.

Instructor's Notes:

3.0.0 ◆ CONDUIT BODIES, PULL BOXES, AND JUNCTION BOXES

NEC Section 300.15 states that at each splice point or pull point for the connection of conduit, cables, or other raceways, a box or fitting must be installed. The NEC has specific requirements for conduit bodies, pull boxes, and junction boxes. For installation rules, see *Table 1*.

Conduit bodies provide access to the wiring through removable covers. Typical examples are Types T, C, X, L, and LB. Conduit bodies enclosing No. 6 or smaller conductors must have an area twice that of the largest conduit to which they are attached, but the number of conductors within the body must not exceed that allowed in the conduit.

When conduit bodies or boxes are used as junction boxes or pull boxes, a minimum size box is required to allow conductors to be installed without undue bending. The calculated dimensions of the box depend on the type of conduit arrangement and the size of the conduit involved.

Table 1 Installation Requirements for Conduit Bodies, Pull Boxes, and Junction Boxes

Application	Installation Requirements	NEC Reference
Conduit bodies	Conduit bodies enclosing No. 6 conductors or smaller must have a cross-sectional area twice that of the largest conduit or tubing to which it is attached. The maximum number of conductors allowed in the conduit body must not exceed the allowable fill for the attached conduit. Conduit bodies must not contain splices, taps, or devices unless they are durably and legibly marked by the manufacturer with their cubic inch capacity. Conduit bodies must be supported in a rigid and secure manner.	***NEC Section 314.16(C)***
Minimum sizes	Boxes and conduit bodies used as pull or junction boxes must comply with (a) through (d) below: (a) For raceways containing conductors of No. 4 or larger, the minimum dimensions of pull or junction boxes installed in a raceway or cable run must comply with the following: • In straight pulls, the length of the box must not be less than eight times the trade diameter of the largest conduit. • Where angle or U-pulls are made, the distance between each raceway entry inside the box and the opposite wall of the box must not be less than six times the trade diameter of the largest raceway in a row. This distance must be increased for additional entries by the amount of the sum of the diameters of all other raceway entries in the same row on the same wall of the box. Each row must be calculated individually and the single row that provides the maximum distance must be used. • The distance between raceway entries enclosing the same conductor must be less than six times the trade diameter of the largest raceway.	***NEC Section 314.28(A)***
	(b) If pull boxes or junction boxes have any dimension over 6', all conductors must be cabled or racked up in an approved manner.	***NEC Section 314.28(B)***
	(c) All boxes and fittings must be provided with covers compatible with the box, and if metal, must comply with ***NEC Section 250.110.***	***NEC Section 314.28(C)***
	(d) Where permanent barriers are installed in a box, each section shall be considered a separate box.	***NEC Section 314.28(D)***
Accessibility	Junction, pull, and outlet boxes must be accessible without removing any part of a building.	***NEC Section 314.29***
Over 600V	Special requirements apply to boxes used on systems of over 600V.	***NEC Article 314, Part V***

INSIDE TRACK

Sizing Conduit

Although the NEC specifies minimum size requirements for raceway system components (conduit, conduit bodies, boxes, etc.), it is often wise to oversize these components to allow for future system expansion, to provide additional protection against conductor damage, or to help facilitate the pulling of conductors on long or complicated conduit runs.

Classroom

Review the NEC requirements for conduit bodies, pull boxes, and junction boxes. Refer the trainees to Table 1 in the Trainee Module.

Teaching Tip

Discuss the "Case History." Emphasize the importance of properly sizing and selecting pull boxes, particularly in lengthy or complex installations.

Discuss the "What's Wrong with This Picture?" See the TeachingTip at the end of this module.

Wireways and Conduit Bodies

INSIDE TRACK

Both wireways and conduit bodies must provide access to the wiring through removable covers, as shown here.

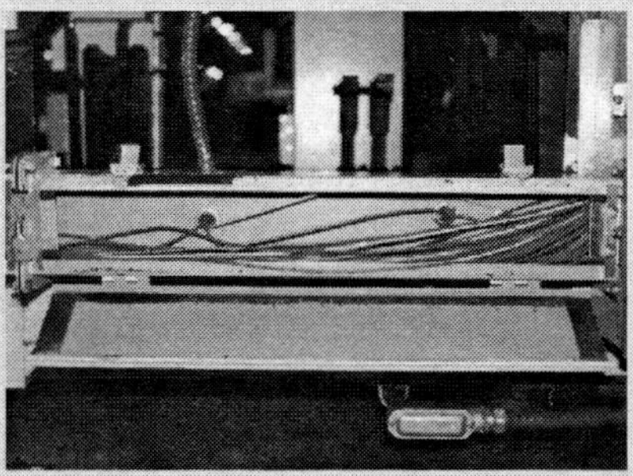

WIREWAY

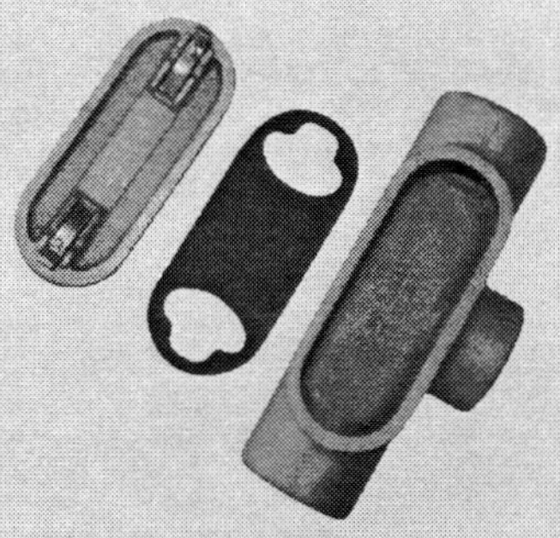

CONDUIT BODY

304P0401.EPS

Pulling Boxes

CASE HISTORY

In the early 1970s while employed as plant electrician for a crude oil refinery, an electrical contractor installed three parallel runs of four 500 kcmil THW conductors in 4" rigid metal conduit at the refinery. This was permitted according to the NEC conduit fill calculations; however, these runs were lengthy and changed directions several times. Standard LB conduit fittings were used instead of pull boxes. When the new system was energized, a very loud boom was heard and the pipe rack that held the conduits shook. It was later learned that the insulation on one of the ungrounded 500 kcmil conductors had been damaged in pulling around the LB and it shorted to the conduit. All conductors in that conduit had to be removed and discarded as scrap copper, the damaged conduit replaced, and four new 500 kcmil THW conductors pulled into the conduit. An expensive lesson was learned about pulling large conductors through lengthy runs.

The Bottom Line: Situations such as this one have led to changes in the code requiring an increase in the minimum wire bending space for conduit bodies.

What's Wrong with This Picture?

304P0402.EPS

Instructor's Notes:

3.1.0 Sizing Pull and Junction Boxes

Figure 4 shows a junction box with several conduits entering it. Since 4" conduit is the largest size in the group, the minimum length required for the box can be determined by the following calculation:

4" (trade size of conduit) × 8 [per ***NEC Section 314.28(A)(1)***] = 32" (minimum length of box)

Therefore, this particular pull box must be at least 32" in length. The width of the box, however, need be only of sufficient size to enable locknuts and bushings to be installed on all the conduits or connectors entering the enclosure.

Junction or pull boxes in which the conductors are pulled at an angle, as shown in *Figure 5,* must have a distance of not less than six times the trade diameter of the largest conduit. The distance must be increased for additional conduit entries by adding it to the sum of the diameter of all other conduits entering the box on the same side (the wall of the box). The distance between raceway entries enclosing the same conductors must not be less than six times the trade diameter of the largest conduit.

Since the 4" conduit is the largest in this case:

$$L_1 = 6 \times 4 + (3 + 2) = 29"$$

Since the same number and sizes of conduit are located on the adjacent wall of the box, L_2 is calculated in the same way; therefore, $L_2 = 29$.

The distance (D) = 6 × 4 or 24". This is the minimum distance permitted between conduit entries enclosing the same conductor.

The depth of the box need only be of sufficient size to permit locknuts and bushings to be properly installed. In this case, a 6"-deep box would suffice.

If the conductors are smaller than No. 4, the length restriction does not apply.

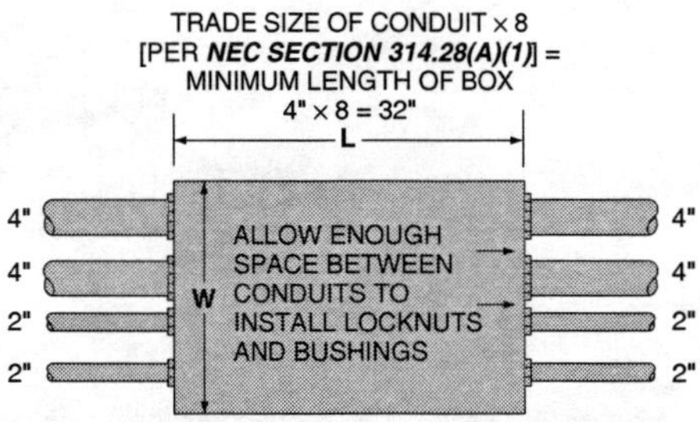

Figure 4 ◆ Pull box used on straight pulls.

Figure 6 shows another straight pull box. What is the minimum length if the box has one 3" conduit and two 2" conduits entering and leaving the box? Again, refer to *NEC Section 314.28(A)(1)* and find that the minimum length is eight times the largest conduit size. In this case, it is:

$$8 \times 3" = 24"$$

Let's review the installation requirements for pull or junction boxes with angular or U-pulls [*NEC Section 314.28(A)(2)*]. Two conditions must be met in order to determine the length and width of the required box:

- The minimum distance to the opposite side of the box from any conduit entry must be at least six times the trade diameter of the largest raceway.
- The sum of the diameters of the raceways on the same wall must be added to this figure.

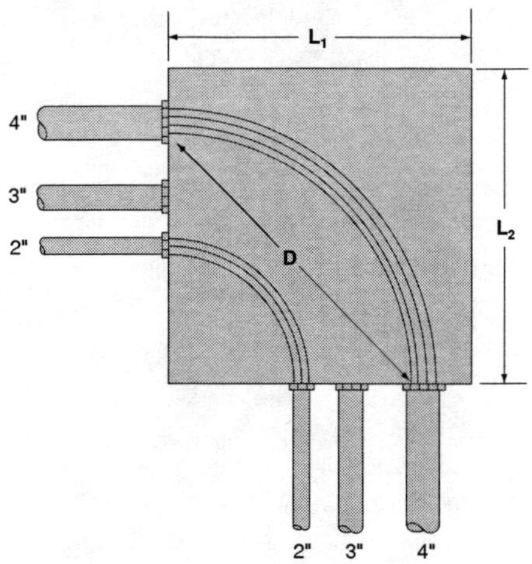

Figure 5 ◆ Junction box with conduit runs entering at right angles.

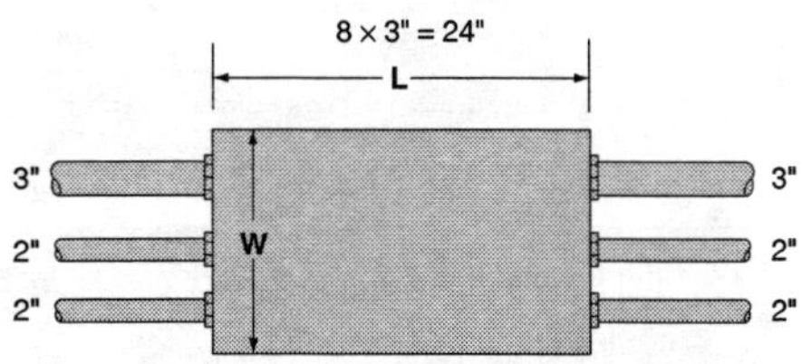

Figure 6 ◆ Typical straight pull box.

Classroom

Explain how to size a junction box. Point out the difference between sizing a box for a straight pull and sizing a box for an angled pull.

Provide example problems for the trainees to solve.

Audiovisual

Show Transparencies 5 through 7 (Figures 4 through 6).

Classroom

Explain how to calculate the required distances between conduit entries for a box to be used in an angled pull.

Audiovisual

Show Transparencies 8 and 9 (Figures 7 and 8).

Figure 7 shows the minimum length of a box with two 3" conduits, two 2" conduits, and two 1½" conduits in a right-angle pull. The minimum length based on this configuration is:

$$\begin{aligned} 6 \times 3" &= 18" \\ 1 \times 3" &= 3" \\ 2 \times 2" &= 4" \\ 2 \times 1\tfrac{1}{2}" &= \underline{3"} \\ & \quad 28" \end{aligned}$$

Since the number and size of conduits on the two sides of the box are equal, the box is square and has a minimum size dimension of 28". However, the distance between conduit entries must now be checked to ensure that all NEC requirements are met; that is, the spacing (D) between conduits enclosing the same conductor must not be less than six times the conduit diameter. Again, refer to *Figure 7* and note that the 1½" conduits are the closest to the left-hand corner of the box. Therefore, the distance (D) between conduit entries must be:

$$6 \times 1\tfrac{1}{2}" = 9"$$

The next group, the two 2" conduits, is calculated in a similar fashion:

$$6 \times 2" = 12"$$

The remaining raceways in this example are the two 3" conduits, and the minimum distance between the 3" conduit entries must be:

$$6 \times 3" = 18"$$

A summary of the conduit entry distances is shown in *Figure 8*. However, some additional math is required to obtain the spacing (w) between the conduit entries. For example, the distance from the corner of the pull box to the center of the conduits (w) may be found by the following equation:

$$\text{Spacing} = \frac{\text{diagonal distance (D)}}{\sqrt{2}}$$

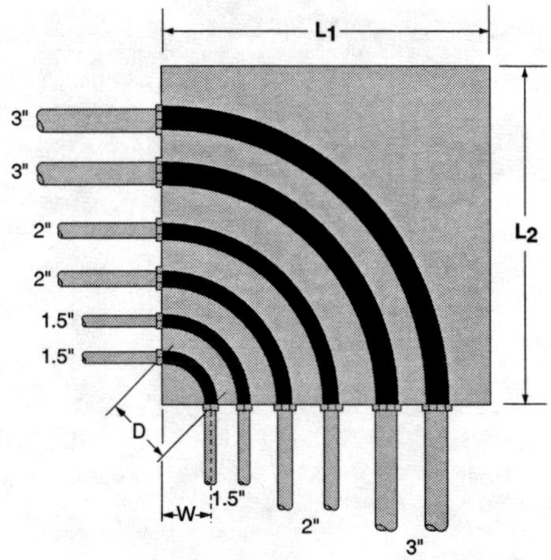

Figure 7 ◆ Minimum size pull box for angle conduit entries.

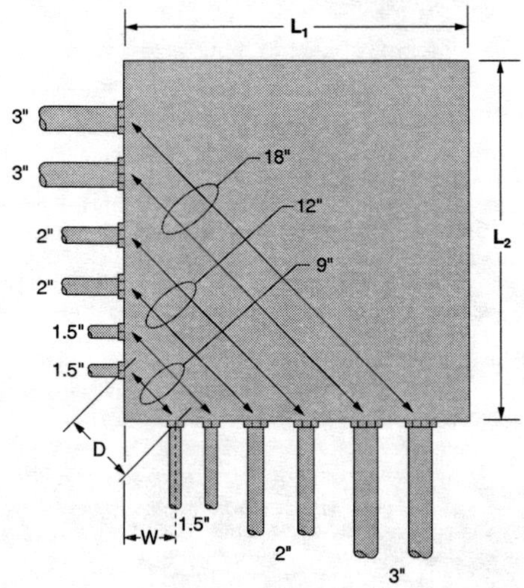

Figure 8 ◆ Required distances between conduit entries.

INSIDE TRACK

Locknut Spacing

A rule of thumb is to allow a minimum of ½" clearance between locknuts when coming into an enclosure.

Instructor's Notes:

Consequently, the spacing (w) for the 1½" conduit may be determined using the following equation:

$$\frac{9}{\sqrt{2}} = \frac{9}{1.414} = 6.4"$$

Therefore, the spacing (w) is 6.4". This distance is measured from the left lower corner of the box in each direction, both vertically and horizontally, to obtain the center of the first set of 1½" conduits. This distance must be added to the spacing of the other conduits, including locknuts or bushings.

Using all information calculated thus far, the required measurements of the pull box may be further calculated as follows:

Step 1 Calculate space (w):

$$D = 6 \times 1½" = 9"$$

Step 2 Divide this number (9") by the square root of 2 (1.414) and make the following calculation:

$$w = \frac{9"}{1.414} = 6.4"$$

Step 3 Measure from the left, lower corner of the pull box over 6.4" to obtain the center of the **knockout** for the first 1½" conduit. Measure up (from the lower left corner) to obtain the center of the knockout for this same cable run on the left side of the pull box.

Step 4 Since there are two 3" (trade size) conduits, each with a measured outside diameter (including locknuts) of approximately 4.25", the space for these two conduits can be found using the following equation:

$$2 \times 4.25" = 8.5"$$

Step 5 The space required for the two 2" (trade size) conduits, each with a measured outside diameter (including locknuts) of approximately 3.25", may be determined in a similar manner:

$$2 \times 3.25" = 6.5"$$

Step 6 The space required for the two 1.5" (trade size) conduits, each with a measured outside diameter (including locknuts) of approximately 2.75", may be determined using the same equation:

$$2 \times 2.75" = 5.5"$$

Step 7 To find the required space for locknuts and bushings, multiply 0.5" by the total number of conduit entries on one side of the box. Since there are a total of six conduit entries, use the following equation:

$$6 \times 0.5" = 3.0"$$

Step 8 Add all figures obtained in Steps 2 through 7 together to obtain the total required length of the pull box.

Clear space (w)	= 6.4"
1.5" conduits	= 5.5"
2" conduits	= 6.5"
3" conduits	= 8.5"
Space between locknuts	= 3.0"
Total length of box	= 29.9"

Since the same number and size of conduits enter on the bottom side of the pull box and leave, at a right angle, on the left side of the pull box, the box will be square. Although a box exactly 29.9" will suffice for this application, the next larger standard size is 30"; this should be the size pull box selected. Even if a custom pull box is made in a sheet metal shop, the workers will still probably make it an even 30" unless specifically ordered otherwise.

THINK ABOUT IT

Sizing Pull Boxes

This box has conduit exiting both in a straight pull and at an angle. How would you size it?

304P0403.EPS

Classroom

Calculate the total required length of a pull box.

Provide example problems for the trainees to solve.

Teaching Tip

Discuss the "Think About It."

Classroom

List the code requirements for cabinets and cutout boxes. Refer the trainees to *NEC Article 312.*

Audiovisual

Show Transparency 10 (Figure 9).

Classroom

Discuss the fill requirements for auxiliary gutters. Refer the trainees to *NEC Article 366.*

3.2.0 Cabinets and Cutout Boxes

NEC Article 312 deals with the installation requirements for cabinets, cutout boxes, and meter sockets. In general, where cables are used, each cable must be secured to the cabinet or cutout box by an approved method. Furthermore, the cabinets or cutout boxes must have sufficient space to accommodate all conductors installed in them without crowding.

NEC Table 312.6(A) gives the minimum wire bending space at terminals, along with the width of sizing gutter in inches.

Figure 9 gives a summary of NEC requirements for the installation of cabinets and cutout boxes.

Other basic NEC requirements for cabinets and cutout boxes are as follows:

- *NEC Table 312.6(A)* applies when the conductor does not enter or leave the enclosure through the wall opposite its terminal.
- *NEC Section 312.6(B), Exception No. 1* states that a conductor must be permitted to enter or leave an enclosure through the wall opposite its terminal, provided the conductor enters or leaves the enclosure where the gutter joins an adjacent gutter that has a width that conforms to *NEC Table 312.6(B)* for that conductor.
- *NEC Section 312.6(B)(2), Exception No. 2* states that a conductor not larger than 350 kcmil must be permitted to enter or leave an enclosure containing only a meter socket(s) through the wall opposite its terminal, provided the terminal is a lay-in type where either: the terminal is directly facing the enclosure wall, and the offset is not greater than 50% of the bending space specified in *NEC Table 312.6(A)*; or the terminal is directed toward the opening in the enclosure and is within a 45° angle of directly facing the enclosure wall.
- *NEC Table 312.6(B)* must apply where the conductor enters or leaves the enclosure through the wall opposite its terminal.
- *NEC Section 376.23(B)* requires that where insulated conductors No. 4 or larger enter a wireway through a raceway or cable, the distance between those raceway and cable entries shall not be less than six times the trade size of the largest raceway in a row for angle or U pulls and not less than eight times the trade size of the largest raceway or cable for straight pulls.

NEC Article 366 covers the installation requirements for auxiliary gutters, which are permitted to supplement wiring spaces at meter centers, distribution centers, and similar points of wiring systems and may enclose conductors or busbars, but must not be used to enclose switches, overcurrent devices, appliances, or other similar equipment.

In general, auxiliary gutters must not contain more than 30 current-carrying conductors at any cross-section. The sum of the cross-sectional areas of all contained conductors at any cross-section of an auxiliary gutter must not exceed 20% of the interior cross-sectional area of the auxiliary gutter. We discussed earlier that conductors installed in conduits and tubing must not exceed 40% fill. Auxiliary gutters are limited to only 20%.

When dealing with auxiliary gutters, always remember the number 30. This is the maximum number of current-carrying conductors allowed in any auxiliary gutter. It is permitted to install more than 30 current-carrying conductors if the ampacities are reduced per *NEC Table 310.15(B)(2)(a)*.

INSIDE TRACK

Minimum Bending Radius

Adhering to the NEC minimum bending radius requirements helps to ensure a good installation by providing:

- Adequate room to make connections
- Adequate wire training space to prevent damage to the conductors
- Room for future expansion

Instructor's Notes:

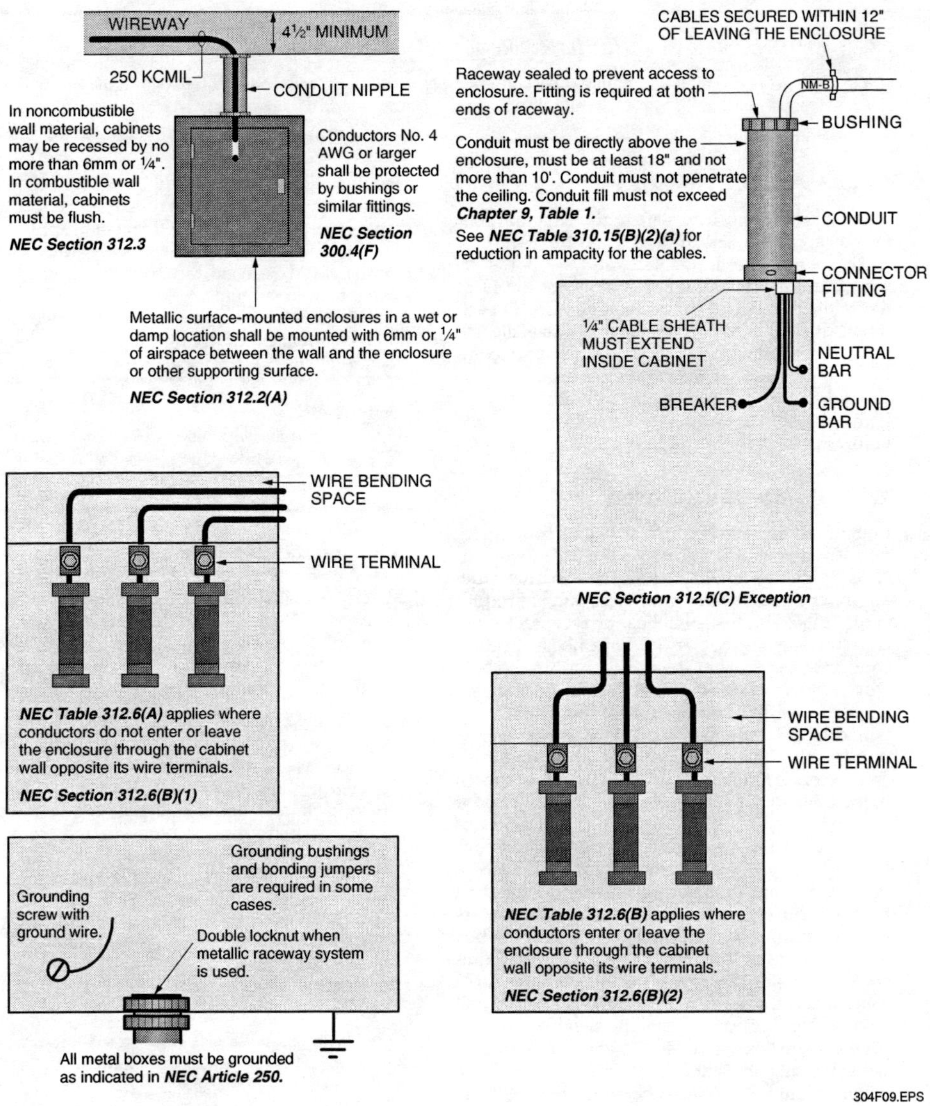

Figure 9 ◆ Installation requirements for cabinets and cutout boxes.

Classroom

Discuss outlet box sizing. Point out that the maximum number of conductors permitted in standard boxes is listed in *NEC Table 314.16(A)*.

Audiovisual

Show Transparency 11 (Figure 10).

> *Cable Racks*
>
> INSIDE TRACK
>
> Per ***NEC Section 314.28(B)***, all pull boxes or junction boxes having any dimension over 6' must have all conductors cabled or racked up in an approved manner.

4.0.0 ◆ OUTLET BOXES

On every job, many boxes are required for outlets, switches, pull boxes, and junction boxes. All of these must be sized, installed, and supported to meet current NEC requirements. Since the NEC limits the number of conductors allowed in each outlet or switch box according to its size, electricians must install boxes large enough to accommodate the number of conductors that must be spliced in the box or fed through. Therefore, a knowledge of the various types of boxes and the volume of each is essential.

4.1.0 Sizing Outlet Boxes

In general, the maximum number of conductors permitted in standard outlet boxes is listed in ***NEC Table 314.16(A)***. The figures in this table apply where no fittings or devices such as fixture studs, cable clamps, switches, or receptacles are contained in the box and where no grounding conductors are part of the wiring within the box. Obviously, in all modern residential wiring systems, there will be one or more of these items contained in the outlet box. Therefore, where one or more of these items are present, the number of conductors is reduced by one less than that shown in the table for each type of fitting and by two for each device strap. For example, a deduction of two conductors must be made for each strap containing a device such as a switch or duplex receptacle; a further deduction of one conductor shall be made for one or more grounding conductors entering the box. For example, a 3" × 2" × 3½" box is listed in the table as allowing a maximum number of eight No. 12 wires. If the box contains cable clamps and a duplex receptacle, three wires will have to be deducted from the total of eight—providing for only five No. 12 wires. If a ground wire is used, only four No. 12 wires may be used, which might be the case when a three-wire cable with ground is used to feed a three-way wall switch.

Figure 10 illustrates a wiring configuration for outlet boxes and the maximum number of conductors permitted in them as governed by ***NEC Section 314.16***. This example shows two single-gang switch boxes joined or ganged together to hold a single-pole toggle switch and a duplex receptacle. This type of arrangement is likely to be found above kitchen countertops, where the duplex receptacle is provided for small appliances and the single-pole switch (on a separate branch circuit) could be used to control a garbage disposal. This arrangement is also useful above a workbench—the receptacle for small power tools and the switch to control lighting over the bench.

> **WARNING!**
>
> Stamped metal boxes often have sharp edges and should be handled with care to avoid cuts.

Suppose the boxes pictured in *Figure 10* are 3" × 2" × 2½". Since ***NEC Table 314.16(A)*** gives the capacity of one 3" × 2" × 2½" device box as 12.5 cu. in., the total capacity of both boxes in *Figure 10* is 25 cu. in. These two boxes have the capacity to allow 10 No. 12 **American Wire Gauge (AWG)** conductors less the following deductions.

- Per ***NEC Section 314.16(B)(4)***, two conductors must be deducted for each strap-mounted device based on the largest size conductor connected to the device. Since there is one duplex receptacle and one single-pole toggle switch, four conductors must be deducted from the total number stated previously.
- Per ***NEC Section 314.16(B)(2)***, the combined boxes contain one or more cable clamps. Another conductor must be deducted based on the largest size conductor in the box.

> **Note**
>
> Only one deduction is made for similar clamps, regardless of the number. However, any unused clamps may be removed to facilitate the electrical worker's job (i.e., to allow for more work space).

- Per ***NEC Section 314.16(B)(5)***, the equipment grounding conductors, regardless of the number, count as one conductor only. Therefore, deduct one conductor based on the largest size grounding conductor in the box.

Instructor's Notes:

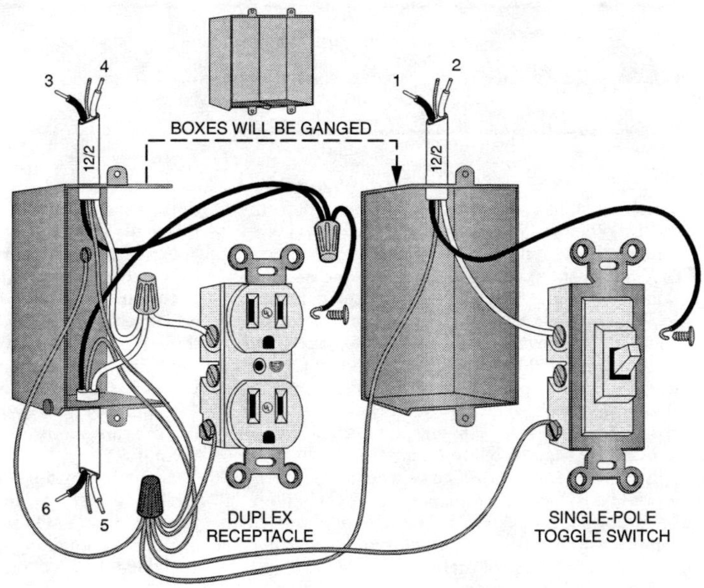

Figure 10 ◆ Box sizing example.

Therefore, to comply with the NEC, and considering the combined deduction of six conductors, only four No. 12 AWG conductors may be installed in the outlet box configuration in *Figure 10.*

Figure 10 shows three nonmetallic-sheathed (NM) cables, designated 12/2 with ground, entering the ganged outlet boxes. This is a total of six current-carrying conductors. This arrangement is in violation of the NEC because the total number of conductors exceed the NEC limits. One alternative is to go to 3" × 2" × 2¾" device boxes, which would then have a total of 28 cu. in. for the two boxes and allow a maximum of 12 No. 12 conductors.

Also note the pigtails in *Figure 10.* Conductors that both originate and terminate in the same outlet box are exempt from being counted against the allowable capacity of an outlet box [***NEC Section 314.16(B)(1)***]. Because these wires originate and terminate in the same set of ganged boxes, they are not counted against the total number of conductors. By the same token, the three grounding conductors extending from the wire nut to the individual grounding screws on the devices originate and terminate in the same set of boxes. These conductors are also exempt from being counted.

A pictorial definition of stipulated conditions as they apply to ***NEC Section 314.16*** is shown in the following illustrations. *Figure 11* shows a raised device cover and a box extension ring. These components, when combined with the appropriate outlet boxes, serve to increase the usable work space. Each type is marked with its cu. in. capacity, which may be added to the figures in ***NEC Table 314.16(A)*** to calculate the increased number of conductors allowed.

Figure 11 ◆ Raised box cover and box extension ring.

Classroom

Point out that raised device covers can be used to increase outlet box capacity.

Classroom

Explain how to size an outlet box for a typical application.

Audiovisual

Show Transparencies 12 and 13 (Figures 12 and 13).

Extension Rings

INSIDE TRACK

If you add extension rings to a box, all conductors must be long enough to extend a minimum of 3" beyond the device without a splice.

Figure 12 shows typical wiring configurations, which must be counted as conductors when calculating the total capacity of outlet boxes. A wire passing through the box without a splice or tap is counted as one conductor. Therefore, a cable containing two wires that passes in and out of an outlet box with a splice or tap is counted as two conductors. However, wires that enter a box and are either spliced or connected to a terminal, and then exit again, are counted as two conductors. In the case of two cables, the total conductors charged will be four. Wires that enter and terminate in the same box are charged as individual conductors and in this case, the total charge would be two conductors. Remember, when one or more grounding wires enter the box and are joined, a deduction of only one conductor is required, regardless of their number.

Further components that require deduction adjustments from those specified in ***NEC Table 314.16(A)*** include fixture studs, hickeys, and fixture stud extensions [***NEC Section 314.16(B)(3)***]. One conductor must be deducted from the total for each type of fitting used. Two conductors must be deducted for each strap-mounted device, such as duplex receptacles and wall switches; a deduction of one conductor is made when one or more internally mounted cable clamps are used.

Figure 13 shows components that may be used in outlet boxes without affecting the total number of conductors. Such items include grounding clips and screws, wire nuts, and cable connectors when the latter are inserted through knockout holes in the outlet box and secured with locknuts. Prewired fixture wires are not counted against the total number of allowable conductors in an outlet box, nor are conductors originating and ending in the box.

To better understand how outlet boxes are sized, we will take two No. 12 AWG conductors installed in ½" EMT and terminating into a metallic outlet box containing one duplex receptacle. What size outlet box will meet NEC requirements?

Step 1 Use ***NEC Section 314.16*** to calculate the total number of conductors and equivalents.

One receptacle = 2
Two No. 12 conductors = 2
Total No. 12 conductors = 4

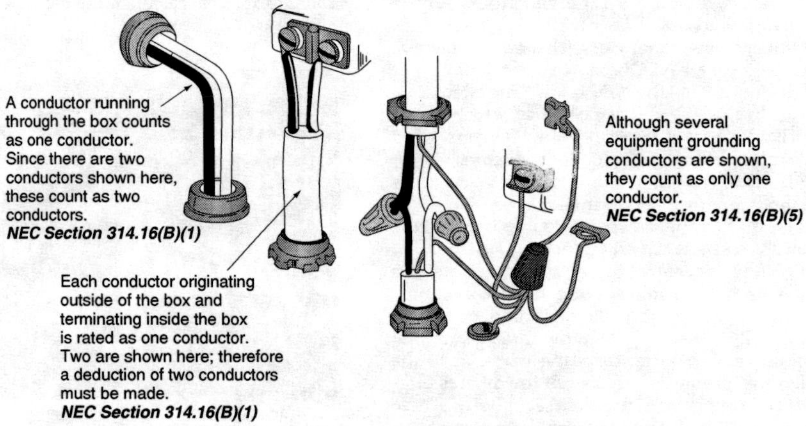

Figure 12 ◆ Wiring configurations that must be counted as conductors when calculating box capacity.

Instructor's Notes:

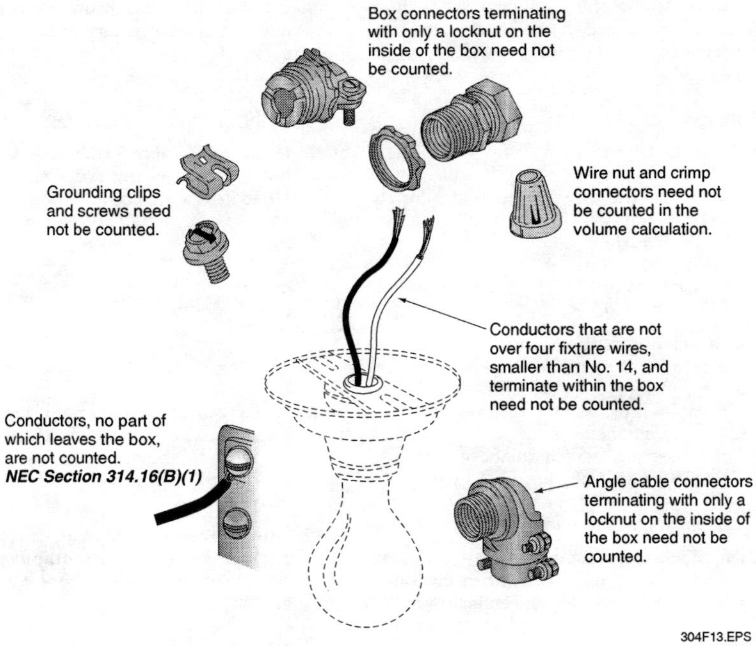

Figure 13 ◆ Components that do not affect the capacity of an outlet box.

Step 2 Determine the amount of space required for each conductor. ***NEC Table 314.16(B)*** gives the box volume required for each conductor:

No. 12 AWG = 2.25 cu. in.

Step 3 Calculate the outlet box space required by multiplying the number of cu. in. required for each conductor by the number of conductors found in Step 1.

$4 \times 2.25 = 9.00$ cu. in.

Once you have determined the required box capacity, again refer to ***NEC Table 314.16(A)*** and note that a 3" × 2" × 2¼" box comes closest to our requirements. This box size is rated for 10.5 cu. in.

Where four No. 12 conductors enter the box, two additional No. 12 conductors must be added to our previous count for a total of six conductors.

$6 \times 2.25 = 13.5$ cu. in.

Again, refer to ***NEC Table 314.16(A)*** and note that a 3" × 2" × 2¾" device box with a rated capacity of 14.0 cu. in. is the closest device box that meets NEC requirements. Of course, any box with a larger capacity is permitted.

Summary

The NEC specifies certain fill requirements for raceways, outlet boxes, pull and junction boxes, cabinets, cutout boxes, auxiliary gutters, and similar conductor-containing housings. In some cases, NEC tables may be used to determine the proper size of housing; in other cases, calculations are required in conjunction with tables and manufacturers' specifications.

This module should provide a firm foundation for determining the proper size of raceway, box, or fitting for any given application. However, it is recommended that you carefully review the NEC requirements pertaining to all such fill requirements.

Review Questions

1. Any channel used for holding wires, cables, or busbars is called a ____.
   a. knockout
   b. raceway
   c. junction box
   d. conduit fill

Teaching Tip

See the Teaching Tip at the end of this module.

Classroom

Have the trainees complete the Review Questions and go over the answers prior to administering the Module Examination.

Examination

Administer the Module Examination. Record the results on Craft Training Report Form 200 and submit the results to the Training Program Sponsor.

2. According to the NEC, when three or more conductors are installed in a conduit, the total area of the conductors must not exceed ____ of the conduit's total area.
   a. 31%
   b. 40%
   c. 53%
   d. 75%

3. Where in the NEC would you find information on sizing pull and junction boxes?
   a. *NEC Article 314*
   b. *NEC Article 430*
   c. *NEC Article 440*
   d. *NEC Article 450*

4. What special procedures must be taken with cables installed in a pull box that has a dimension of 6' in length or over?
   a. The cables must be racked in the box in an approved manner.
   b. The cables must terminate within the box.
   c. The cables must not terminate within the box.
   d. Cable clamps of any kind are not allowed.

5. When calculating pull box size, ____ is the recommended minimum amount of clearance that should be allowed between locknuts.
   a. 1"
   b. ½"
   c. ¼"
   d. ⅛"

6. If a straight pull box contains two 4" conduits entering one end and leaving the other end, what is the minimum length of the box?
   a. 12"
   b. 24"
   c. 32"
   d. 36"

7. If a straight pull box contains one 4" conduit and one 3" conduit (both sizes entering and leaving from opposite ends), what is the minimum length of the box?
   a. 24"
   b. 32"
   c. 36"
   d. 38"

8. The maximum number of current-carrying conductors allowed in any auxiliary gutter without using derating is ____.
   a. 30
   b. 20
   c. 40
   d. 25

9. What is the maximum percentage of fill allowed in an auxiliary gutter?
   a. 10%
   b. 20%
   c. 30%
   d. 40%

10. Using *NEC Table 312.6(B)*, what is the minimum wire bending space that must be provided in a cabinet or cutout box in which there are three No. 1/0 THWN conductors per terminal?
    a. 7"
    b. 12"
    c. 14"
    d. 18"

11. What effect do extension rings have on outlet box fill capacity?
    a. The box capacity remains the same.
    b. The box capacity decreases.
    c. The box capacity increases.
    d. This only applies to NM boxes, and the box capacity decreases.

12. How many conductor deductions must be allowed for three bare equipment grounding conductors in an outlet box?
    a. 1
    b. 2
    c. 3
    d. 4

13. How many conductor deductions must be allowed for an angle cable connector terminating with only a locknut on the inside of the outlet box?
    a. 6
    b. 4
    c. 2
    d. 0

14. How many conductor deductions must be counted for three No. 16 AWG fixture wires that terminate within an outlet box?
    a. 1
    b. 2
    c. 0
    d. 3

15. How many conductor deductions must be counted for two wire nuts used for splicing conductors inside an outlet box?
    a. 0
    b. 1
    c. 2
    d. 3

Instructor's Notes:

GLOSSARY

Trade Terms Introduced in This Module

American wire gauge (AWG): The U.S. standard for wire measurement.

Ampacity: The current-carrying capacity of conductors or equipment, expressed in amperes.

Bonded: The permanent joining of metallic parts to form an electrically conductive path that will ensure electrical continuity and the capacity to conduct safely any current likely to be imposed.

Cable: An assembly of two or more conductors which may be insulated or bare.

Conduit fill: Amount of cross-sectional area used in a raceway.

Junction box: A group of electrical terminals housed in a protective box or container.

Knockout: A portion of an enclosure designed to be easily removed for raceway installation.

Additional Resources

This module is intended to present thorough resources for task training. The following reference works are suggested for further study. These are optional materials for continued education rather than for task training.

American Electrician's Handbook. Terrell Croft and Wilfred I. Summers. New York, NY: McGraw-Hill, 1996.

National Electrical Code Handbook. Quincy, MA: National Fire Protection Association, 2002.

Instructor's Notes:

ANSWER KEY

Answers to Review Questions

Answer	Section
1. b	1.0.0
2. b	2.1.0
3. a	3.0.0; Table 1
4. a	3.0.0; Table 1
5. b	3.1.0
6. c	3.1.0
7. b	3.1.0
8. a	3.2.0
9. b	3.2.0
10. a	3.2.0
11. c	4.1.0
12. a	4.1.0
13. d	4.1.0; Figure 13
14. c	4.1.0; Figure 13
15. a	4.1.0

Section 3.0.0 — *What's Wrong with This Picture?*

Point out that these cables must be suitably installed in a raceway if the ceiling area is used as a return air plenum. Also, they must be installed in a "neat and workmanlike manner" per *NEC Section 110.12,* and securely attached to the building structure to allow access to any equipment above the suspended ceiling.

Section 3.1.0 — *Think About It—Sizing Pull Boxes*

Calculate the straight pull dimension, then calculate the angle pull dimension. Use the dimension that results in the largest box size.

Section 4.1.0 — *Sizing Outlet Boxes*

This exercise allows the trainees to practice sizing outlet boxes for various applications. For this exercise, you will need several types of outlet boxes, along with short lengths of conductors and various devices. Allow 30 minutes for this exercise.

CAUTION: Warn the trainees to handle the sharp edges of any metal boxes carefully to avoid cuts.

1. Divide the trainees into groups. Provide each group with an outlet box, one or more strap-mounted devices, conductors, clamps, raised device covers, extension rings, and so on.
2. Have the trainees calculate whether the box size is adequate for the given number of devices and conductor size.
3. Provide additional configurations as time permits.

CONTREN™ LEARNING SERIES — USER UPDATES

The NCCER makes every effort to keep these textbooks up-to-date and free of technical errors. We appreciate your help in this process. If you have an idea for improving this textbook, or if you find an error, a typographical mistake, or an inaccuracy in NCCER's Contren™ textbooks, please write us, using this form or a photocopy. Be sure to include the exact module number, page number, a detailed description, and the correction, if applicable. Your input will be brought to the attention of the Technical Review Committee. Thank you for your assistance.

Instructors – If you found that additional materials were necessary in order to teach this module effectively, please let us know so that we may include them in the Equipment/Materials list in the Instructor's Guide.

Write: Curriculum Revision and Development Department
National Center for Construction Education and Research
P.O. Box 141104, Gainesville, FL 32614-1104

Fax: 352-334-0932

E-mail: curriculum@nccer.org

Craft ____________ Module Name ____________

Copyright Date ____________ Module Number ____________ Page Number(s) ____________

Description

____

____

____

(Optional) Correction

____

____

(Optional) Your Name and Address

____

____

Wiring Devices

26305-02

MODULE OVERVIEW

This module introduces the Electrical trainee to the methods and procedures used when sizing and selecting wiring devices.

PREREQUISITES

Please refer to the Course Map in the Trainee Module. Prior to training with this module, it is recommended that the trainee shall have successfully completed the following:

Core Curriculum; Electrical Level One; Electrical Level Two; Electrical Level Three, Modules 26301 through 26304.

OBJECTIVES

Upon completion of this module, the trainee will be able to:

1. Select wiring devices according to the National Electrical Manufacturers' Association (NEMA) classifications.
2. Size wiring devices in accordance with NEC requirements.
3. Discuss the NEMA enclosure classifications.
4. Follow NEC regulations governing the installation of wiring devices.
5. Explain the types and purposes of grounding wiring devices.
6. Determine the maximum load allowed on specific wiring devices.

PERFORMANCE TASKS

Under supervision of the instructor, the trainee should be able to:

1. Identify different types of receptacles.
2. Identify different types of switches.
3. Select the appropriate receptacle for a given application.

NCCER STANDARDIZED CRAFT TRAINING PROGRAMS

The National Center for Construction Education and Research (NCCER) provides a standardized national program of accredited craft training. Key features of the program include instructor certification, competency-based training, and performance testing. The program provides trainees, instructors, and companies with a standard form of recognition through a National Craft Training Registry. The program is described in full in the *Guidelines for Accreditation*, published by the NCCER. For more information on standardized craft training, contact the NCCER by writing us at P.O. Box 141104, Gainesville, FL 32614-1104, calling 352-334-0911, or e-mailing info@nccer.org. More information may be found at our Web site, www.nccer.org.

HOW TO USE THIS ANNOTATED INSTRUCTOR'S GUIDE

Each page presents two sections of information. The larger section displays each page exactly as it appears in the Trainee Module. The narrow column ties suggested trainee and instructor actions to each page and provides icons to call your attention to material, safety, audiovisual, or testing requirements. The bottom of each page includes space for your notes.

If you see the Teaching Tip icon, that means there is a teaching tip associated with this section. Also refer to the suggested teaching tips at the end of the module.

SAFETY CONSIDERATIONS

Ensure that the trainees are equipped with appropriate personal protective equipment.

PREPARATION

Before teaching this module, you should review the Module Outline, Objectives, Performance Tasks, and the Materials and Equipment List. Be sure to allow ample time to prepare your own training or lesson plan and gather all required equipment and materials.

MATERIALS AND EQUIPMENT LIST

Materials:
Transparencies
Markers/chalk
Copy of the latest edition of the *National Electrical Code®*
Examples of devices, including relays, contactors, and conductors
Various types of receptacles
Various types of wall switches
Various safety switches, including:
- General-duty
- Heavy-duty
- Double-pole, double-throw

Module Examinations*
Performance Profile Sheets*

Equipment:
Overhead projector and screen
Whiteboard/chalkboard
Appropriate personal protective equipment

*Located in the Test Booklet packaged with this Annotated Instructor's Guide.

ADDITIONAL RESOURCES

This module is intended to present thorough resources for task training. The following reference works are suggested for both instructors and motivated trainees interested in further study. These are optional materials for continued education rather than for task training.

American Electrician's Handbook, Terrell Croft and Wilfred I. Summers. New York, NY: McGraw-Hill, 1996.

National Electrical Code Handbook. Quincy, MA: National Fire Protection Association, 2002.

TEACHING TIME FOR THIS MODULE

An outline for use in developing your lesson plan is presented below. Note that each Roman numeral in the outline equates to one session of instruction. Each session has a suggested time period of 2½ hours. This includes 10 minutes at the beginning of each session for administrative tasks and one 10-minute break during the session. Approximately 10 hours are suggested to cover *Wiring Devices.* You will need to adjust the time required for hands-on activity and testing based on your class size and resources.

Topic	Planned Time
Session I. Introduction; Receptacles; Locating Receptacles; Laboratory	
A. Introduction	________
B. Receptacles	________
1. Receptacle Characteristics	________
2. Mounting Receptacles	________
3. Types of Receptacles	________
C. Locating Receptacles	________
1. Residential Occupancies	________
2. Commercial Applications	________

D. Laboratory

Have the trainees identify various receptacles and their applications.

Note the proficiency of each trainee. __________

Session II. Switches; NEC Requirements for Switches

- A. Switches __________
  1. Switch Configurations __________
  2. Switch Identification __________
- B. NEC Requirements for Switches __________

Session III. Safety Switches; Laboratory

- A. Safety Switches __________
  1. Heavy-Duty Switches __________
     - a. Switch Blade and Jaws __________
     - b. Fuse Clips __________
     - c. Terminal Lugs __________
     - d. Insulating Material __________
     - e. Operating Mechanism and Cover Latching __________
     - f. Enclosures __________
     - g. Interlocked Receptacles __________
     - h. Accessories __________
  2. General-Duty Switches __________
     - a. Switch Blades and Jaws __________
     - b. Fuse Clips __________
     - c. Terminal Lugs __________
     - d. Insulating Material __________
     - e. Operating Mechanism and Cover Latching __________
     - f. Enclosures __________
  3. Double-Throw Safety Switches __________
  4. NEC Safety Switch Requirements __________
- B. Laboratory __________

  Have the trainees identify various switches and their applications.

  Note the proficiency of each trainee.

Session IV. Review; Module Examination; Performance Testing

- A. Review __________
- B. Module Examination __________
  1. Trainees must score 70% or higher to receive recognition from the NCCER.
  2. Record the testing results on Craft Training Report Form 200 and submit the results to the Training Program Sponsor.
- C. Performance Testing __________
  1. Trainees must perform each task to the satisfaction of the instructor to receive recognition from the NCCER.
  2. Record the testing results on Craft Training Report Form 200 and submit the results to the Training Program Sponsor.

Module 26305-02

Wiring Devices

Course Map

This course map shows all of the modules in the third level of the Electrical curriculum. The suggested training order begins at the bottom and proceeds up. Skill levels increase as you advance on the course map. The local Training Program Sponsor may adjust the training order.

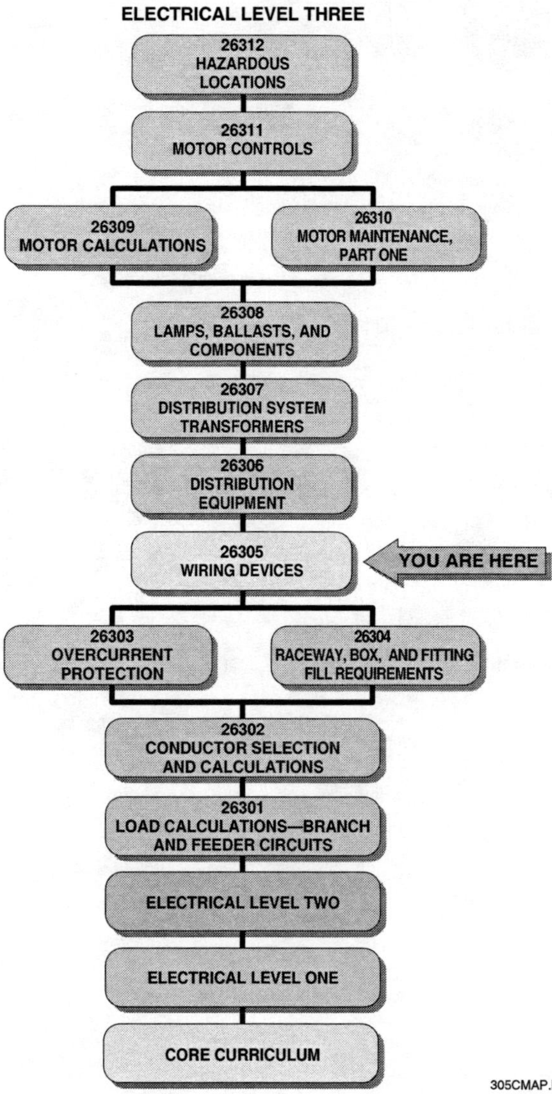

305CMAP.EPS

Homework

Assign reading of Module 26305.

MODULE 26305 CONTENTS

Figures

Table

Instructor's Notes:

MODULE 26305

Wiring Devices

Objectives

When you have completed this module, you will be able to do the following:

1. Select wiring devices according to the National Electrical Manufacturers' Association (NEMA) classifications.
2. Size wiring devices in accordance with NEC requirements.
3. Discuss the NEMA enclosure classifications.
4. Follow NEC regulations governing the installation of wiring devices.
5. Explain the types and purposes of grounding wiring devices.
6. Determine the maximum load allowed on specific wiring devices.

Prerequisites

Before you begin this module, it is recommended that you successfully complete the following: Core Curriculum; Electrical Level One; Electrical Level Two; Electrical Level Three, Modules 26301 through 26304.

Required Trainee Materials

1. Pencil and paper
2. Appropriate personal protective equipment
3. Copy of the latest edition of the *National Electrical Code*

1.0.0 ◆ INTRODUCTION

The NEC defines a **device** as "a unit of an electrical system that is intended to carry but not utilize electric energy." This covers a wide assortment of system components that include (but are not limited to) the following:

- **Switches**
- Relays
- Contactors
- **Receptacles**
- Conductors

This module covers switching devices and receptacles. Both are commonly known as *wiring devices*. Other devices, such as relays and contactors, are covered in other modules.

2.0.0 ◆ RECEPTACLES

A receptacle is a contact device installed at an outlet of a branch circuit for the connection of a single **attachment plug.** Several types and configurations are available for use with many different attachment plugs, each designed for a specific application (*Figure 1*). For example, receptacles

THINK ABOUT IT

Relay and Contactor Coils

Can the coils of relays and contactors be considered devices based on the definition of devices in the NEC?

Note: The designations "National Electrical Code," "NE Code," and "NEC," where used in this document, refer to the National Electrical Code®, which is a registered trademark of the National Fire Protection Association, Quincy, MA. *All National Electrical Code (NEC) references in this module refer to the 2002 edition of the NEC.*

Materials

Ensure that you have everything required to teach the course. Check the Materials and Equipment List at the front of this Instructor's Guide.

Classroom

Review the goals of the course.

Explain that terms shown in bold (blue) are defined in the Glossary at the back of this module.

Audiovisual

Show Transparency 1, Course Objectives.

Show Transparency 2, Performance Tasks.

Classroom

Define *device.* List the common types of devices and provide examples for the trainees to examine.

Discuss receptacle plug configurations. Refer the trainees to Figure 1 in the Trainee Module.

Teaching Tip

Discuss the "Think About It."

Teaching Tip

Provide several receptacles and have the trainees identify the number of wires/poles and determine whether or not the device is a grounding type.

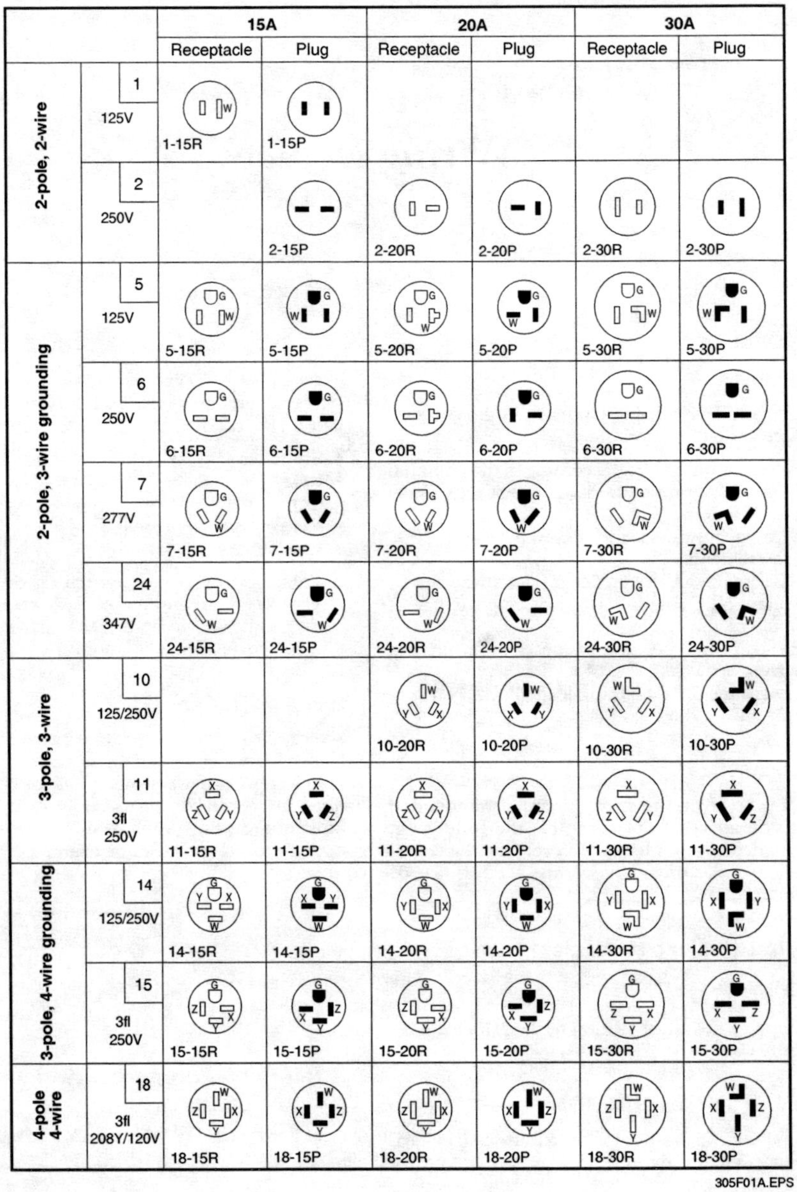

Figure 1 ◆ Common receptacles and plugs (1 of 2).

5.2 ELECTRICAL LEVEL THREE—TRAINEE MODULE 26305

Instructor's Notes:

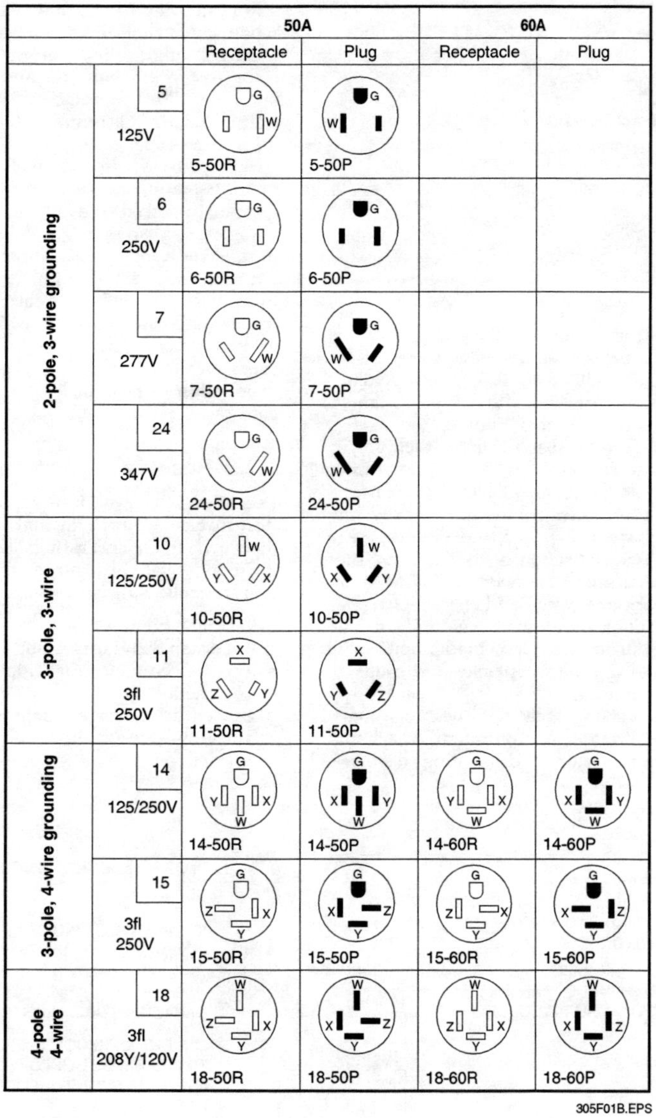

Figure 1 ◆ Common receptacles and plugs (2 of 2).

![Classroom]

Explain that receptacles are rated according to their voltage and amperage capacity. Note that the unique configuration assigned to each rating prevents interchangeability.

Point out the difference between system grounds and equipment grounds.

Explain that *NEC Article 406* covers the installation of receptacles, cord connectors, and cord caps.

List the types of information found on a typical receptacle.

Audiovisual

Show Transparency 3 (Figure 2).

INSIDE TRACK

Factory-Installed Power Cords with Plugs

Never change a factory-installed power cord on an appliance or equipment to make it match a receptacle. Receptacles and branch circuit ratings must be matched to the appliance or equipment.

are available for two-wire, 120V, 15A and 20A circuits; others are designed for use on two-wire and three-wire, 240V/20A, 30A, 40A, and 50A circuits. There is also a variety of other types, many of which are discussed in this module.

Receptacles are rated according to their voltage and amperage capacity. This rating determines the number and configuration of the contacts, both on the receptacle and the receptacle mating plug. The chart in *Figure 1* was developed by the Wiring Device Section of the National Electrical Manufacturers' Association (NEMA) and illustrates various voltage and current characteristics. Note that all configurations in *Figure 1* are for general-purpose, straight blade, nonlocking devices. Locking-type receptacles and plugs are covered later in this module.

Unsafe interchangeability has been eliminated by assigning a unique configuration to each voltage and current rating. All dual ratings have been eliminated, and interchangeability exists only where it does not present an unsafe condition.

Each configuration is designated by a number composed of the chart line number, the amperage, and either R for receptacle or P for plug cap. For example, a 5-15R is found on line 5 and represents a 15A receptacle.

A clear distinction is made between system grounds and equipment grounds. System grounds, referred to as *grounded conductors,* normally carry current at ground potential, and terminals for such conductors are marked *W* for white in the chart. Equipment grounds, referred to as *grounding conductors,* carry current only during ground fault conditions. The terminals for such conductors are marked *G* for grounding or green in the chart. ***NEC Article 406*** covers the installation of receptacles, cord connectors, and cord caps.

2.1.0 Receptacle Characteristics

Receptacles have various symbols and information inscribed on them that help to determine their proper use and ratings. For example, *Figure 2* shows a standard duplex receptacle and contains the following printed inscriptions:

- Testing laboratory label
- Canadian Standards Association (CSA) label
- Type of conductor for which the terminals are designed
- Current and voltage ratings listed by maximum amperage, maximum voltage, and current restrictions

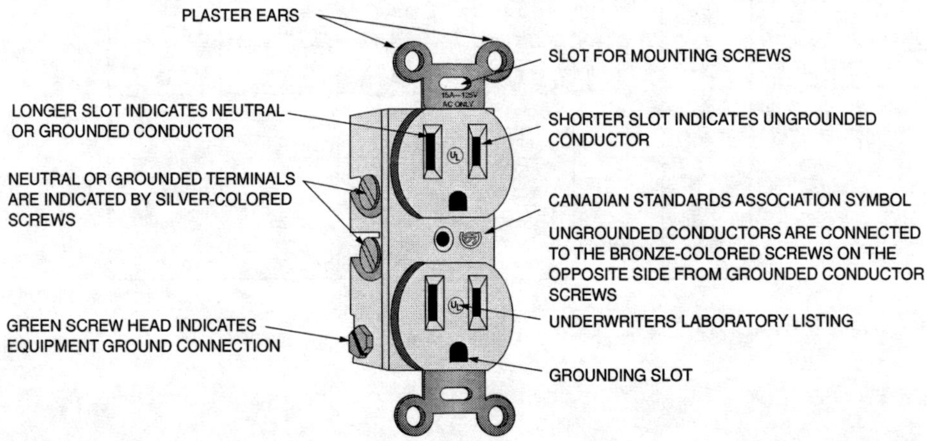

Figure 2 ◆ Typical duplex receptacle.

305F02.EPS

Instructor's Notes:

The testing laboratory label is an indication that the device has undergone extensive testing by a nationally recognized testing lab and has met certain minimum safety requirements. The label does not indicate quality. The receptacle in *Figure 2* is marked with the UL label, which indicates that the device type was tested by Underwriters Laboratories, Inc. of Northbrook, IL. ETL Testing Laboratories, Inc. of Cortland, NY is another nationally recognized testing laboratory. These laboratories provide labeling, listing, and follow-up service for the safety testing of electrical products to nationally recognized safety standards or specifically designated requirements of jurisdictional authorities. The CSA label is an indication that the material or device has undergone a similar testing procedure by the Canadian Standards Association and is acceptable for use in Canada.

Current and voltage ratings are listed by maximum amperage, maximum voltage, and current restriction. On the device shown in *Figure 2,* the maximum current rating is 15A. The maximum voltage allowed on this device is 125V.

Conductor markings are also usually found on duplex receptacles. Receptacles with quick-connect wire clips will be marked *USE No. 14 SOLID WIRE ONLY.* If the inscription CO/ALR is marked on the receptacle, either copper, aluminum, or copper-clad aluminum wire may be used. The letters ALR stand for *aluminum revised.* Receptacles marked with the inscription CU/AL should be used for copper only, although they were originally intended for use with aluminum also. However, such devices frequently failed when connected to 15A or 20A circuits. Consequently, devices marked with CU/AL are no longer acceptable for use with aluminum conductors.

The remaining markings on duplex receptacles may include the manufacturer's name or logo, the words *WIRE RELEASE* inscribed under the wire release slots, and the letters GR beneath or beside the green grounding screw.

The screw terminals on receptacles are color-coded. For example, the terminal with the green screw head is the equipment ground connection and is connected to the U-shaped slots on the receptacle. The silver-colored terminal screws are for connecting the grounded or neutral conductors and are associated with the longer of the two vertical slots on the receptacle. The brass-colored terminal screws are for connecting the ungrounded or hot conductors and are associated with the shorter vertical slots on the receptacle.

> **Note**
> The long vertical slot accepts the grounded or neutral conductor. The shorter vertical slot accepts the ungrounded or hot conductor.

2.2.0 Mounting Receptacles

Although no actual NEC requirements exist for mounting heights and positioning of receptacles, other than the prohibition against mounting receptacles face up on countertops and similar work surfaces, there are certain NEC guidelines regarding receptacle placement. For example, the NEC states that receptacles shall be not more than 20" above the countertop per ***NEC Section 210.52(C)(5).*** Also, where allowed, receptacles may not be located more than 12" below the countertop surface. See ***NEC Section 210.52(C)(5) Exception.*** In addition to these NEC guidelines, certain installation methods have become standard in the electrical industry. *Figure 3* shows common mounting heights of duplex receptacles used on conventional residential and small commercial installations. However, these dimensions are frequently varied to suit the building structure. For example, ceramic tile might be placed above a kitchen or bathroom countertop. If the dimensions in *Figure 3* put the receptacle part of the way out of the tile (i.e., half in and half out), the mounting height should be adjusted to place the receptacle either completely in the tile or completely out of the tile, as shown in *Figure 4.*

Refer again to *Figure 3* and note that the mounting heights are given to the bottom of the outlet box. Many dimensions on electrical drawings are given to the center of the outlet box or receptacle. However, during the actual installation, workers installing the outlet boxes can mount them more accurately (and in less time) by lining up the bottom of the box with a chalk mark rather than trying to eyeball this mark to the center of the box.

A decade or so ago, most electricians mounted receptacle outlets 12" from the finished floor to the center of the outlet box. However, a recent survey taken of over 500 homeowners shows that they prefer a mounting height of 15" from the finished floor to the bottom of the outlet box. It is easier to plug and unplug the **cord** assemblies at this height. However, always check the working drawings, written specifications, and details of

Classroom

Point out the connection screws on a typical receptacle.

Discuss the NEC guidelines for the placement of receptacles.

Audiovisual

Show Transparencies 4 and 5 (Figures 3 and 4).

Classroom

Discuss the NEC requirements for mounting outlet boxes.

Audiovisual

Show Transparency 6 (Figure 5).

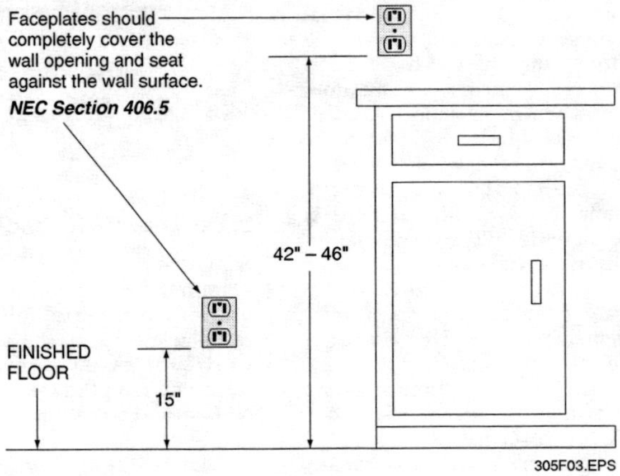

Figure 3 ◆ Mounting heights of duplex receptacles.

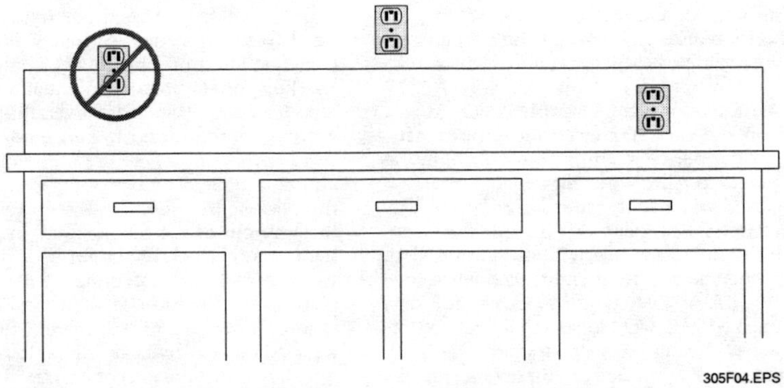

Figure 4 ◆ Adjusting mounting heights.

construction for measurements that may affect the mounting height of a particular receptacle outlet. For example, those confined to wheelchairs may require more specific receptacle height locations to fit their individual needs.

> **Note**
> Be sure to check local codes as well as the job specifications for specific receptacle mounting requirements.

NEC Section 314.20 requires all outlet boxes installed in walls or ceilings of concrete, tile, or other noncombustible material, such as plaster or drywall, to be installed in such a manner that the front edge of the box or fitting is not set back from the finished surface by more than ¼". Where walls and ceilings are constructed of wood or other combustible materials, outlet boxes and fittings must be flush with the finished surface of the wall. See *Figure 5*.

Wall surfaces such as drywall or plaster that contain wide gaps or are broken, jagged, or otherwise damaged, must be repaired so there will be no gaps or open spaces greater than ⅛" between the outlet box and the wall material. These repairs should be made prior to installing the faceplate. Such repairs are best made using a noncombustible caulking or spackling compound. See *Figure 6*.

Instructor's Notes:

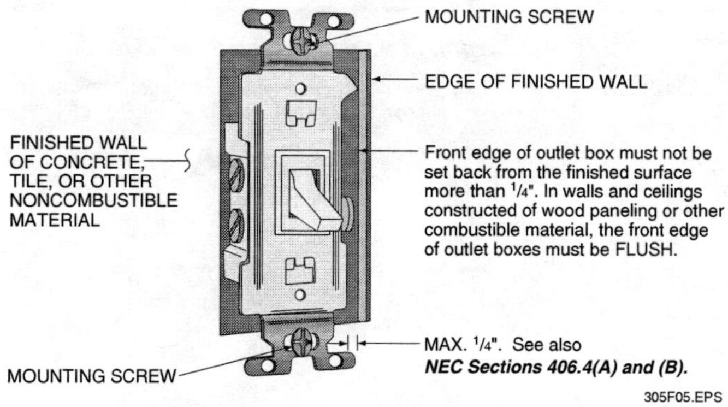

Figure 5 ◆ NEC requirements for mounting outlet boxes in walls or ceilings.

Figure 6 ◆ Gaps or openings around outlet boxes must be repaired.

2.3.0 Types of Receptacles

There are many types of receptacles. For example, the duplex receptacles discussed previously are the straight-blade type, which accepts a straight blade connector or plug. This is the most common type of receptacle in the U.S. and can be found on virtually all electrical projects from residential to large industrial installations.

- *Twist lock receptacles*—Twist lock receptacles are designed to accept a slightly curved blade connector or plug. The plug/connector and the receptacle lock together with a slight twist. This prevents accidental unplugging of the equipment.
- *Pin-and-sleeve receptacles*—Pin-and-sleeve devices have a unique locking feature. These receptacles are made with an extremely heavy-duty plastic housing. They are manufactured with long brass pins for long life and are color-coded according to voltage for easy identification.
- *Low-voltage receptacles*—These receptacles are designed for both AC and DC systems where the maximum potential is 50V. Receptacles used for low-voltage systems must have a minimum current-carrying capacity of 15A.
- *480V receptacles*—Portable electrical equipment operating at 460V to 480V is common on many industrial installations. This includes welders, battery chargers, and other types of portable equipment. Special 480V plugs and receptacles are used to connect and disconnect such equipment from a power source. Equipment grounding is required in all cases.

WARNING!

Make certain that the plug-and-cord assembly is compatible with both the equipment and receptacle before connecting to any receptacle. Polarity and equipment grounding checks on the plug-and-cord assembly should be made on a monthly basis or more often if subjected to hard use.

Classroom

Discuss common types of receptacles and their applications. Provide examples for the trainees to examine.

Safety

Emphasize that any plug-and-cord assembly used must be compatible with both the equipment and the receptacle.

Classroom

Identify the NEC requirements for receptacles in residential occupancies.

Audiovisual

Show Transparencies 7 through 9 (Figures 7 through 9).

INSIDE TRACK

Hazardous Location Receptacles

Arc-eliminating receptacles like the one shown here are typically found in petrochemical facilities and other classified locations.

305P0501.EPS

3.0.0 ◆ LOCATING RECEPTACLES

Several NEC sections specify requirements for locating receptacles in all types of installations. This section presents a summary of these requirements.

3.1.0 Residential Occupancies

NEC Section 210.52 should be referred to when laying out outlets for residential and some commercial installations. This section details the general provisions along with small appliance circuit requirements, laundry requirements, unfinished basements, attached garages, and other areas of the home. *Figure 7* illustrates various NEC requirements for locating receptacles.

In general, every dwelling—regardless of its size—must have receptacles located in each habitable area so that no point along the floor line in any wall space (2' wide or wider) is more than 6' from an outlet in that space. The purpose of this requirement is to prevent the need for extension cords and to minimize the use of cords across doorways, fireplaces, and similar openings.

In addition, a minimum of two 20A small appliance branch circuits are required to serve all receptacle outlets, including refrigeration equipment, in the kitchen, pantry, breakfast room, dining room, or similar area of the dwelling unit. Such circuits, whether two or more are used, must have no other outlets connected to them other than in the rooms listed.

At least one receptacle is required in each laundry area, supplied by a separate 20A circuit, on the outside of the building at the front and back (GFCI protected), in each basement, in each attached and detached garage, in each hallway 10' or more in length, and at an accessible location for servicing any HVAC equipment. *Figures 8 and 9* summarize these and other NEC requirements regarding the installation of receptacles in dwelling units.

Instructor's Notes:

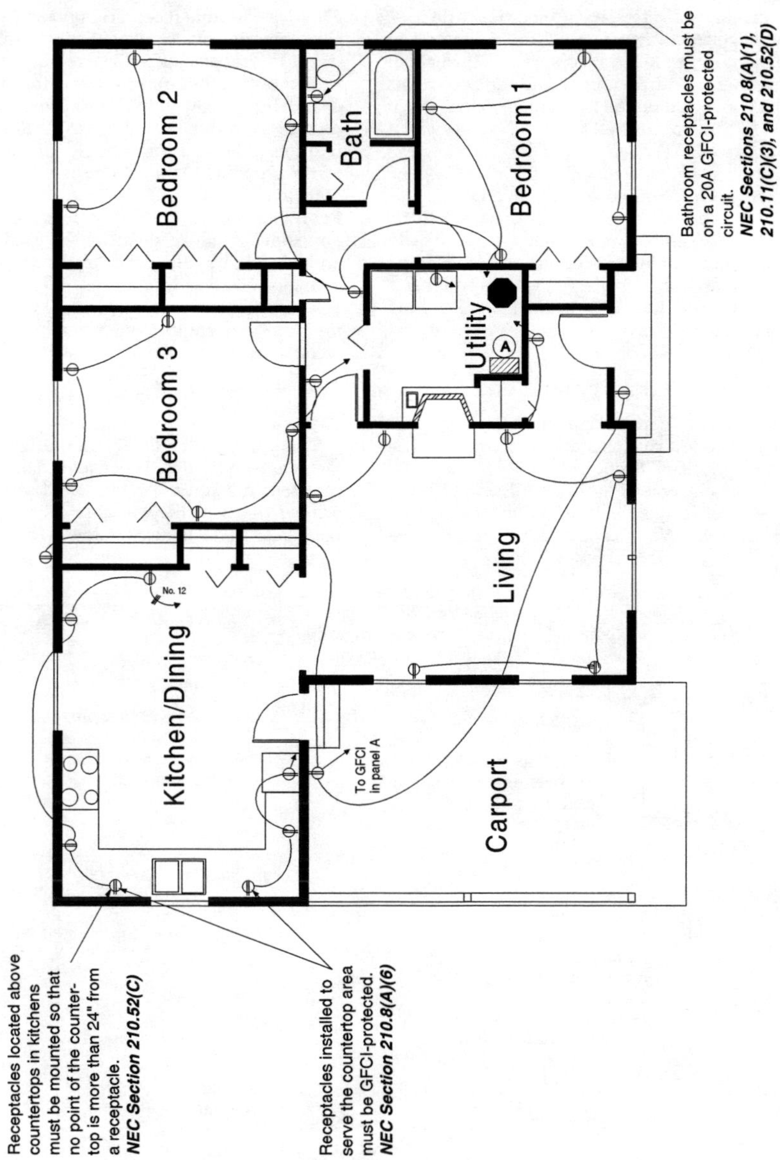

305F07.EPS

Figure 7 ◆ NEC requirements for locating receptacles.

Homework

Have the trainees review the receptacle locations in their homes and report on whether or not they meet the NEC requirements.

Classroom

Note that the NEC allows the use of GFCI receptacles to replace existing ungrounded receptacles, but they must be suitably marked.

Audiovisual

Show Transparency 10 (Figure 10).

Classroom

Identify the NEC requirements for receptacles in commercial occupancies.

Explain how to calculate the receptacle requirements for show windows.

Audiovisual

Show Transparency 11 (Figure 11).

When upgrading existing electrical systems, the NEC permits the use of a GFCI receptacle in place of an ungrounded receptacle. With such an arrangement, additional grounded receptacles may be connected on the load side connections of the GFCI, but must be marked *No Equipment Ground* per ***NEC Section 406.3(D)(3)***, as shown in *Figure 10*.

Other receptacles and related circuits are provided as needed according to the load to be served. For example, receptacles are normally provided in residential occupancies for electric ranges, clothes dryers, and similar appliances. Most operate on 120/240V branch circuits using 30A to 60A receptacles.

3.2.0 Commercial Applications

Receptacles for commercial installations have only a few code-driven installation requirements. For example, ***NEC Section 210.60(A)*** states that guest rooms in hotels, motels, and similar occupancies must have receptacle outlets installed in accordance with ***NEC Section 210.52. NEC Section 210.60(B)*** permits receptacle outlets to be located conveniently for permanent furniture layout. At least two receptacles shall be readily accessible.

The only other requirement for commercial installations deals with the placement of receptacle outlets in show windows. ***NEC Section 210.62*** requires at least one receptacle for each 12 linear feet of show window area measured horizontally at its maximum width. See *Figure 11*. To calculate the number of receptacles required at the top of any show window, measure the total linear feet and then divide this figure by 12. Any remainder or major fraction thereof requires an additional receptacle. For example, the show window in *Figure 11* is 18' in length. Consequently, the number of receptacles required may be calculated as follows:

$$\frac{18'}{12'} = 1.5 \text{ receptacles}$$

To comply with ***NEC Section 210.62***, two receptacles are required in this area. Had the calculation resulted in a figure of 1.01, the local inspection authorities would probably require only one receptacle in the show window.

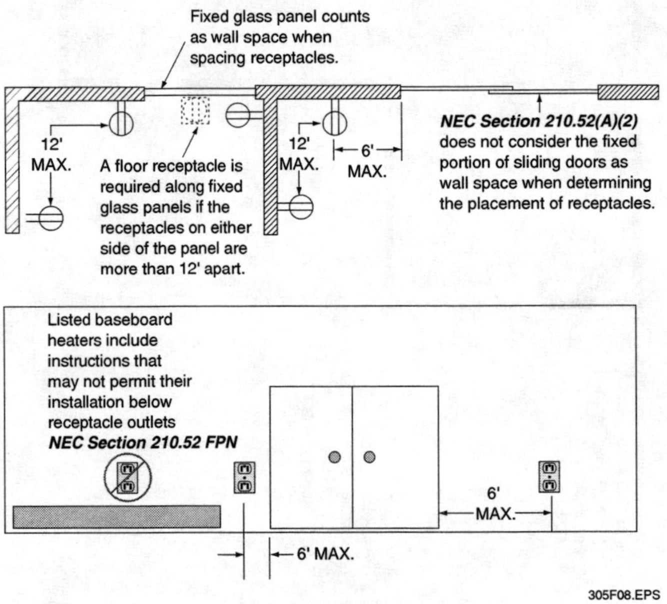

Figure 8 ◆ NEC requirements for dwelling unit receptacles.

Instructor's Notes:

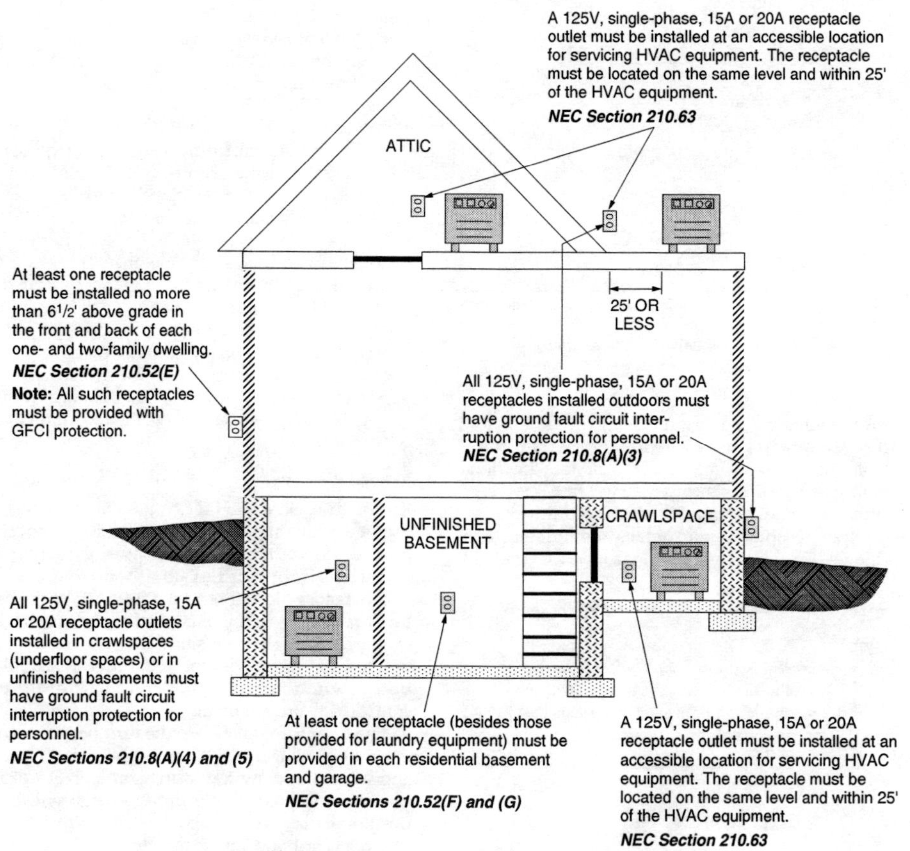

Figure 9 ◆ NEC requirements for placement of dwelling unit receptacles. 305F09.EPS

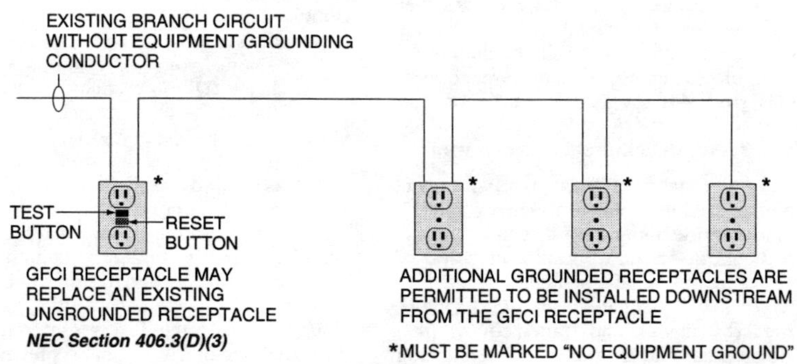

Figure 10 ◆ A GFCI may replace an ungrounded receptacle. 305F10.EPS

Laboratory

Have the trainees identify various receptacles and their applications. Note the proficiency of each trainee.

Classroom

Discuss common types of switches and their applications. Provide examples for the trainees to examine.

Point out that *NEC Article 404* covers the installation and use of switches.

Audiovisual

Show Transparency 12 (Figure 12).

Classroom

Define *pole* and *throw.* List various switch configurations.

Audiovisual

Show Transparency 13 (Figure 13).

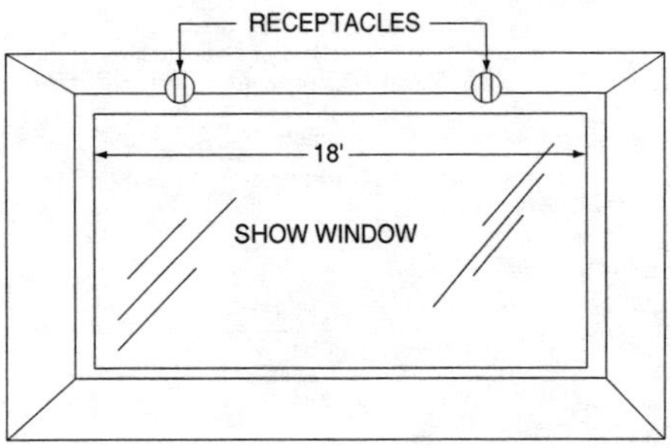

Figure 11 ◆ Commercial show window receptacle placement.

Of course, GFCIs are required on all 15A and 20A receptacles installed in commercial bathrooms, commercial garages, crawlspaces, kitchens, rooftops, boathouses (other than shore-powered outlets), and all receptacles installed outdoors.

Other receptacles and related circuits are provided as needed according to the load to be served.

> **Note**
> Per ***NEC Section 220.3(B)(9),*** the number of general-purpose receptacles permitted on a branch circuit is calculated at not less than 180VA for each receptacle on a yoke.

4.0.0 ◆ SWITCHES

The purpose of a switch is to make and break an electrical circuit in a safe and convenient manner. In doing so, a switch may be used to manually control lighting, motors, fans, and various other items connected to an electrical circuit. Switches may also be activated by light, heat, chemicals, motion, and electric energy for automatic operation. *NEC Article 404* covers the installation and use of switches.

The NEC switch definitions are as follows:

- *Bypass isolation switch*—This is a manually operated device used in conjunction with a transfer switch to provide a means of directly connecting load conductors to a power source and of disconnecting the transfer switch.
- *General-use switch*—A switch intended for use in general distribution and branch circuits. It is rated in amperes and is capable of interrupting its rated current at its rated voltage.
- *General-use snap switch*—A form of general use switch that is designed for installation in device boxes or on outlet box covers, or otherwise used in conjunction with wiring systems recognized by the NEC.
- *Isolation switch*—A switch intended for isolating an electric circuit from the source of power. It has no interrupting rating and is intended to be operated only after the circuit has been opened by some other means (see *Figure 12*).
- *Motor circuit switch*—A switch that is rated in horsepower and is capable of interrupting the maximum operating overload current of a motor of the same horsepower rating as the switch at its rated voltage.
- *Transfer switch*—A device used to transfer one or more load conductor connections from one power source to another. This type of switch may be either automatic or nonautomatic.

4.1.0 Switch Configurations

Although basic switch terms are covered to some extent in earlier modules, a brief review is warranted here. In general, the major terms used to identify the characteristics of switches are *pole* and *throw.*

The term pole refers to the number of conductors that the switch will control in the circuit. For example, a single-pole switch breaks the connection to only one conductor in the circuit. A double-pole switch breaks the connection to two conductors, and so forth.

The term throw refers to the number of internal operations that a switch can perform. *Figure 13* shows common switch configurations. For example, a single-pole, single-throw switch will make one conductor when thrown in one direction (the ON position) and break the circuit when thrown in the opposite direction (the OFF position). The common ON/OFF toggle switch is a single-pole, single-throw (SPST) switch. The single-pole, double-throw (SPDT) switch, also known as the **three-way switch,** is used to control a single load, such as a lamp, from two locations. A double-pole, single-throw (DPST) switch opens or closes two conductors at the same time. Both conductors are either open or closed, that is, in the ON or OFF position. A double-pole, double-throw (DPDT) switch is used to direct a two-wire circuit through one of two different paths. One application of a double-pole, double-throw switch is an electrical transfer switch in which certain circuits may be energized from either the main electric service or from an emergency standby generator. The double-pole, double-throw switch makes the circuit from one or the other and prevents the circuit from being energized from both sources at once.

Instructor's Notes:

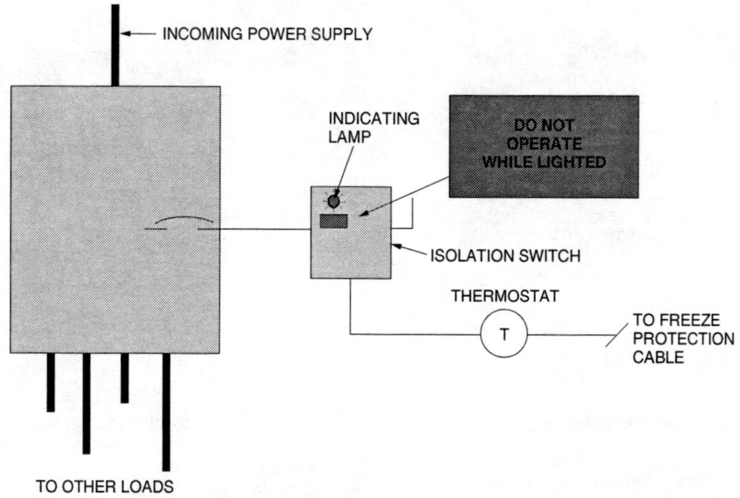

Figure 12 ◆ Isolation switch.

INSIDE TRACK

Three-Way Switches

This photo shows a three-way switch that has been disassembled. Notice the contact arrangement inside the switch housing. The black-colored screw on the bottom right is permanently and electrically connected to the two brass-colored arms that contact either one of the two upper screw terminals as the white switch handle or toggle is moved from one position to the other.

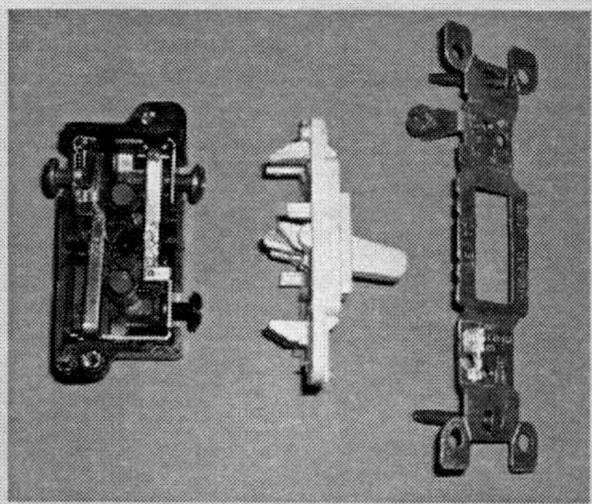

305P0502.EPS

Classroom

Discuss the internal configuration of a three-way switch.

Classroom

Emphasize that the switch selected must be suitable for the application.

Point out the connection screws on a typical switch.

List the types of information found on a typical switch.

Audiovisual

Show Transparency 14 (Figure 14).

Classroom

Point out that the type and grade of switch for a particular installation may be listed in the electrical specifications for the job.

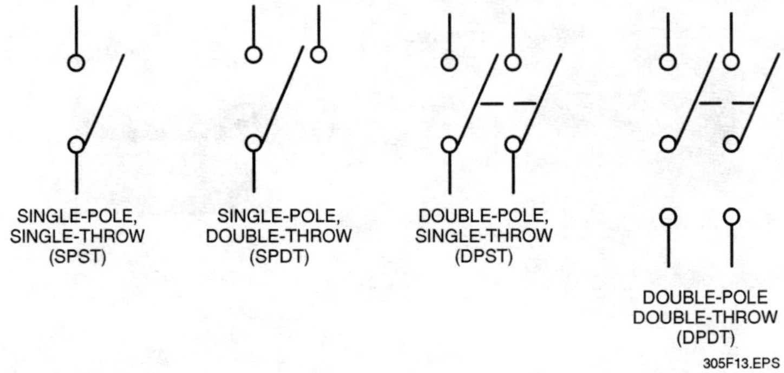

Figure 13 ◆ Common switch configurations.

4.2.0 Switch Identification

Switches vary in grade, capacity, and purpose. It is very important that the proper type of switch be selected for the given application. For example, most single-pole toggle switches used for the control of lighting are restricted to AC use only and are not suitable for use on DC circuits, such as a 32VDC emergency lighting circuit. A switch rated for AC only will not extinguish a DC arc quickly enough. Not only is this a dangerous practice (causing arcing and heating of the device), but the switch contacts would probably burn up after only a few operations of the handle, if not the first time.

Figure 14 shows a typical single-pole toggle switch—the type most often used to control AC lighting in all installations. Note the identifying marks. They are similar to those on the duplex receptacle discussed previously.

Screw terminals are also color-coded on conventional toggle switches. Switches are typically constructed with a ground screw attached to the metallic strap of the switch. The ground screw is usually a green-colored hex head screw. This screw is for connecting the equipment grounding conductor to the switch. On three-way switches, the common or pivot terminal usually has a black or bronze screw head.

The switch shown is the type normally used for residential construction. Heavy-duty switches are normally used on commercial wiring. Some heavy-duty switches are rated for use on 277V circuits with current-carrying ratings up to 30A. It is important to check the rating of each switch before it is installed.

The exact type and grade of switch to be used on a specific installation is often dictated by the project drawings or written specifications. Sometimes wall switches are specified by manufacturer and catalog number; other times they are specified by type, grade, voltage, current rating, etc., leaving the contractor or electrician to select the manufacturer. The naming of a certain brand of switch for a particular project does not necessarily mean that this brand must be used. A typical paragraph from an electrical specification (concerning the substitution of materials) may read as follows:

The naming of a certain make or manufacturer in the Specifications is to establish a quality standard for the article desired. The Contractor is not restricted to the use of the specific brand of the manufacturer named unless so indicated in the Specifications. However, where a substitution is requested, a substitution will be permitted only with the written approval of the Architect-Engineer. No substitute material or equipment shall be ordered, fabricated, shipped, or processed in any manner prior to the approval of the Architect-Engineer. The Contractor shall assume all responsibility for additional expenses as required in any way to meet changes from the original material or equipment specified. If notice of substitution is not furnished to the Architect-Engineer within ten days after the contract is awarded, the equipment and materials named in the Specifications are to be used.

Electrical specifications dealing with wall switches are covered in at least two sections of the specifications:

- 16100, *Basic Materials and Methods*
- 16500, *Lighting*

Brief excerpts from these two sections follow.

SECTION 16B—BASIC MATERIALS AND WORKMANSHIP

SWITCH OUTLET BOXES: Wall switches shall be mounted approximately 54" above the finished floor (AFF) unless otherwise noted. When the switch is

Instructor's Notes:

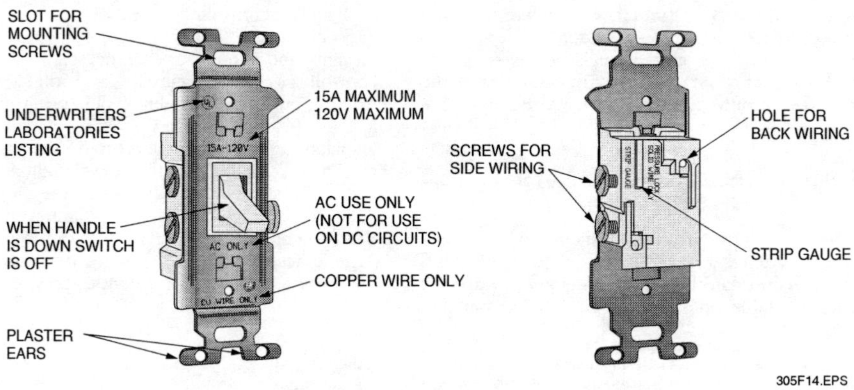

Figure 14 ◆ Identifying marks on a single-pole, single-throw toggle switch.

mounted in a masonry wall, the bottom of the outlet box shall be in line with the bottom of a masonry unit. Where more than two switches are located, the switches shall be mounted in a ***gang switch*** *outlet box with gang cover. Dimmer switches shall be individually mounted unless otherwise noted. Switches with pilot lights, switches with overload motor protection, and other special switches that will not conveniently fit under gang wall plates may be individually mounted.*

EQUIPMENT AND INSTALLATION WORKMANSHIP

a. *All equipment and material shall be new and shall bear the manufacturer's name and trade name. The equipment and material shall be essentially the standard product of a manufacturer regularly engaged in the production of the required type of equipment and shall be the manufacturer's latest approved design.*

b. *The Electrical Contractor shall receive and properly store the equipment and material pertaining to the electrical work. The equipment shall be tightly covered and protected against dirt, water, chemical or mechanical injury, and theft. The manufacturer's directions shall be followed completely in the delivery, storage, protection, and installation of all equipment and materials.*

c. *The Electrical Contractor shall provide and install all items necessary for the complete installation of the equipment as recommended or as required by the manufacturer of the equipment or required by code without additional cost to the Owner, regardless of whether the items are shown on the plans or covered in the Specifications.*

d. *It shall be the responsibility of the Electrical Contractor to clean the electrical equipment, make necessary adjustments, and place the equipment into operation before turning the equipment over to the Owner. Any paint that was scratched during construction shall be touched up with factory color paint to the satisfaction of the Architect. Any items that were damaged during construction shall be replaced.*

WIRING DEVICES

a. *GENERAL: The wiring devices specified below with Arrow Hart numbers may also be the equivalent wiring device as manufactured by Bryant Electric, Harvey Hubbell, or Pass & Seymour. All other items shall be as specified.*
(1) *Single Pole AH#1991*
(2) *Three-Way AH#1993*
(3) *Four-Way AH#1994*
(4) *Switch with pilot light AH#2999-R*
(5) *Motor Switch—Surface AH#6808*
(6) *Motor Switch—Flush AH#6808-F*

b. *WALL SWITCHES: Where more than one flush wall switch is indicated in the same location, the switches shall be mounted in gangs under a common plate.*

c. *WALL PLATE: Stainless steel wall plates with satin finish minimum .030 inches shall be provided for all outlets and switches.*

In general, the preceding electrical specifications give the grade of materials to be used on the project and the manner in which the electrical system must be installed. Most specification writers use an abbreviated language; although it is relatively difficult for beginners to understand, experience makes possible a proper interpretation

Classroom

Review the portions of a specification relating to switches.

Classroom

Identify the NEC requirements for switches in residential occupancies.

Audiovisual

Show Transparency 15 (Figure 15).

with little difficulty. However, electricians involved with any project should make certain that everything is clear. If it is not, contact the architectural or engineering firm and clarify the problem prior to performing the work, not after a system has been completed.

5.0.0 ◆ NEC REQUIREMENTS FOR SWITCHES

There are many NEC requirements for installing light switches (*Figure 15*). For example, wall switch-controlled lighting outlets are required in each habitable room of all residential occupancies. Wall switch-controlled lighting is also required in each bathroom, hallways, stairways, attached garages, and at outdoor entrances. A wall switch-controlled receptacle may be used in place of the lighting outlet in habitable rooms other than the kitchen and bathrooms. Providing a wall switch for room lighting is intended to prevent an occupant's groping in the dark for table lamps or pull chains. In stairways with six or more steps, the stairway lighting must be controlled at two locations—at both the top and bottom of the stairway. This is accomplished by using two three-way switches, as discussed in previous modules.

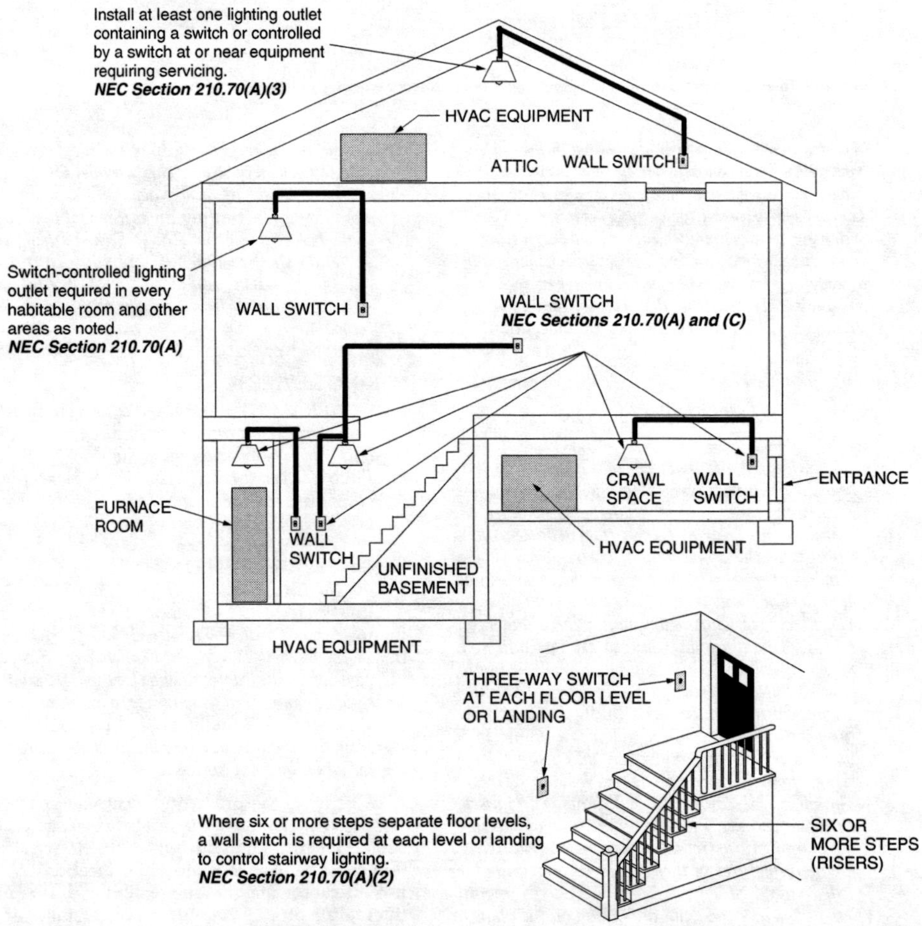

Figure 15 ◆ NEC requirements for residential wall switches.

Instructor's Notes:

Finding Location Requirements for Switched Lighting in the NEC

Even though ***NEC Article 404*** covers some of the requirements of switch installations, you must also refer to ***NEC Section 210.70*** to find the location requirements for switched lighting.

Lighting outlets are also required in attics, crawlspaces, utility rooms, and basements when these spaces are used for storage or contain equipment requiring servicing, such as HVAC equipment. Again, if the basement or attic stairs have six or more steps, a three-way switch is required at each landing.

At least one wall switch-controlled lighting outlet is required in each guest room in hotels, motels, or similar locations, as shown in *Figure 16.* In areas other than kitchens, bathrooms, and hallways, a wall switch-controlled receptacle is permitted in lieu of the lighting outlet.

In many commercial installations, circuit breakers in panelboards that are listed and marked as such are permitted by *NEC Section 404.11* and *NEC Section 240.83(D)* to control main area lighting where the areas are constantly illuminated during operating hours. Consequently, wall switches are not required in these areas. However, wall switches are normally installed at outdoor entrances, entrances to storerooms, small offices, toilets, and similar locations.

Wiring diagrams of switch circuits (single-pole, three-way, and **four-way switches**) were thoroughly discussed in previous modules and these diagrams will not be repeated here. However, it is recommended that the trainee review these diagrams at this time if deemed necessary.

6.0.0 ◆ SAFETY SWITCHES

Enclosed single-throw safety switches are manufactured to meet industrial, commercial, and residential requirements. See *Figure 17.* The two basic types of safety switches are:

- General-duty
- Heavy-duty

Double-throw switches are also manufactured with enclosures and features similar to the general-duty and heavy-duty single-throw designs.

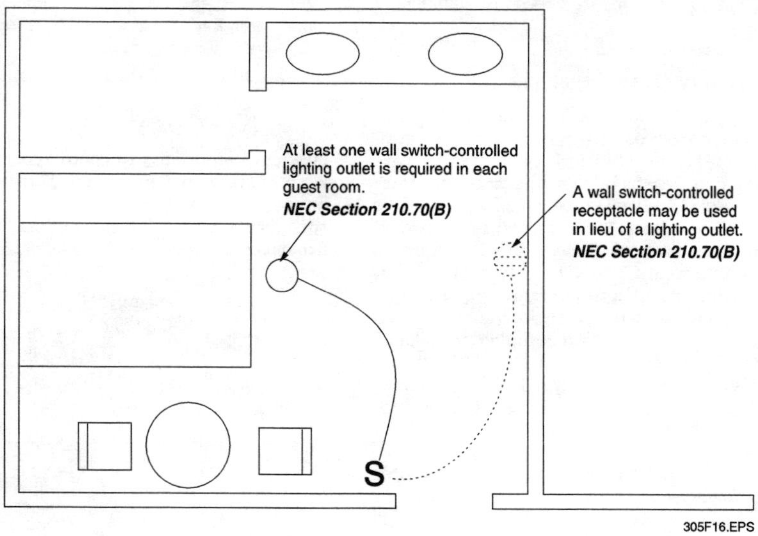

Figure 16 ◆ Wall switch requirements for guest rooms.

Classroom

Identify the NEC requirements for switches in commercial occupancies.

Audiovisual

Show Transparency 16 (Figure 16).

Classroom

Discuss safety switches and their applications. Provide an example for the trainees to examine.

Classroom

Discuss the applications and NEMA classifications of heavy-duty switches.

Describe the components of a heavy-duty switch:
- **Plated blades and jaws**
- **Fuse clips with steel springs**
- **Screw-type terminal lugs**
- **Arc suppressors**
- **Spring-driven operating mechanism**

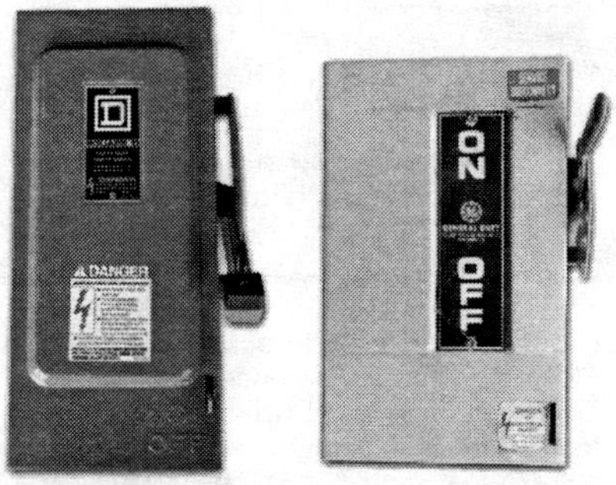

Figure 17 ◆ Typical safety switches.

The majority of safety switches have visible blades and safety handles. The switch blades are in full view when the enclosure door is open and there is visually no doubt when the switch is off. The only exceptions are NEMA Type 7 and 9 enclosures; these do not have visible blades. The switch handles on all types of enclosures are an integral part of the box, not the cover, so that the handle is in control of the switch blades under normal conditions. *NEC Table 430.91* provides a partial listing of enclosure types and what they provide.

6.1.0 Heavy-Duty Switches

Heavy-duty switches are intended for applications where ease of maintenance, rugged construction, and continued performance are primary concerns. They can be used in atmospheres where general-duty switches would be unsuitable, and are therefore widely used in industrial applications. Heavy-duty switches are rated 30A through 1,200A and 240V to 600VAC or VDC. Switches with horsepower ratings are capable of opening a circuit up to six times the rated current of the switch. When equipped with Class J or Class R fuses for 30A through 600A switches, or Class L fuses in 800A and 1,200A switches, many heavy-duty safety switches are UL listed for use on systems with up to 200,000 rms symmetrical amperes available fault current. This, however, is about the highest short circuit rating available for any heavy-duty safety switch. Applications include use where the required enclosure is NEMA Type 1, 3R, 4, 4X, 7, 9, 12, or 12K.

6.1.1 Switch Blades and Jaws

Two types of switch contacts are used in today's safety switches: the butt contact and the knife blade and jaw. On switches with knife blade construction, the jaws distribute a uniform clamping pressure on both sides of the blade contact surface. In the event of a high-current fault, the electromagnetic forces that develop squeeze the jaws tightly against the blade. In the butt-type contact, only one side of the blade's contact surface is held in tension against the conducting path. Electromagnetic forces due to high current faults force the contacts apart, causing them to burn severely. Consequently, the knife blade and jaw construction is the preferred type for use on all heavy-duty switches. The action of the blades moving in and out of the jaws aids in cleaning the contact surfaces. All current-carrying parts of these switches are plated to reduce heating by keeping oxidation to a minimum. Switch blades and jaws are made of copper for high conductivity. Spring-clamped blade hinges are another feature that helps ensure clean contact surfaces and cool operation. Visible blades are used to provide visual evidence that the circuit has been opened.

WARNING!
Before changing fuses or performing maintenance on any safety switch, always visually check the switch blades and jaws to ensure that they are in the OFF position, and verify with a voltage tester.

6.1.2 Fuse Clips

Fuse clips are plated to control corrosion and to keep heating to a minimum. All fuse clips on heavy-duty switches have steel reinforcing springs for increased mechanical strength and firmer contact pressure. See *Figure 18.*

Figure 18 ◆ Fuse clip.

Instructor's Notes:

6.1.3 Terminal Lugs

Most heavy-duty switches have front removable, screw-type terminal lugs. Most switch lugs are suitable for copper or aluminum wire except NEMA Type 4, 4X, 12, and 12K switches, which have all-copper current-carrying parts and lugs designated for use with copper wire only. Listed equipment with terminal lugs is suitable for the wire sizes and number of wires per pole as stated in *NEC Tables 312.6(A) and (B).*

6.1.4 Insulating Material

As the voltage rating of switches is increased, arc suppression becomes more difficult, and the choice of insulation material becomes more critical. Arc suppressors are usually made of insulation material and magnetic suppressor plates when required. All arc suppressor materials must provide proper control and extinguishing of arcs.

6.1.5 Operating Mechanism and Cover Latching

Most heavy-duty safety switches have a spring-driven, quick-make, quick-break mechanism. A quick-break action is necessary to safely switch the mechanism OFF under a heavy load.

The spring action, in addition to making the operation quick-make, quick-break, firmly holds the switch blades in the ON or OFF position. The operating handle is an integral part of the switching mechanism and is in direct control of the switch blades under normal conditions.

The switching mechanism should include a one-piece crossbar connected to all switch blades. This adds to the overall stability and integrity of

INSIDE TRACK

Mounting Single-Throw Knife Switches

Single-throw knife switches should always be mounted so that gravity tends to open the switch and not close it.

305P0503.EPS

CASE HISTORY

Case History: Visually Verifying Disconnection

A near-fatality occurred when an electrician received a 480V electrical shock while replacing damaged fuses in a pump motor switch. Although he placed the switch in the OFF position, the switch failed to disconnect the power to the fuses. It was discovered that the switch failed because of a missing screw in the switching arm linkage. If the switch had operated correctly, the fuse clips would have been de-energized, and the grounding bar would have effectively grounded all three fuses.

The Bottom Line: You must confirm the absence of power before grabbing fuses.

Teaching Tip

Discuss the "Case History." Emphasize the importance of confirming that the power is off prior to replacing fuses.

Classroom

Point out that dual-cover interlocks prevent the switch from being opened while the handle is in the ON position.

Explain that heavy-duty switch enclosures are available to suit various environments. List the NEMA classifications for switch enclosures.

Point out that some switches incorporate interlocked receptacles to allow loads to be connected directly to the switch.

the switching assembly by promoting proper alignment and uniform switch blade operation.

Dual cover interlocks are standard on most heavy-duty switches where the NEMA enclosure permits. However, NEMA Type 7 and 9 enclosures have bolted covers and obviously cannot contain dual cover interlocks. The purpose of a dual cover interlock is to prevent the enclosure door from being opened when the switch handle is in the ON position and to prevent the switch from being turned ON while the door is open. A means of bypassing the interlock is provided to allow an energized switch to be inspected by qualified personnel. However, this practice is extremely dangerous and should be avoided if at all possible. Heavy-duty switches can be padlocked in the OFF position with up to three padlocks.

6.1.6 Enclosures

Heavy-duty switches are available in a variety of enclosures that have been designed to conform to specific industry requirements based upon the intended use. Sheet metal enclosures (NEMA Type 1) are constructed from cold-rolled steel, which is usually phosphatized and finished with an electrode-deposited enamel paint. NEMA Type 3R rainproof and Type 12 and 12K dust-tight enclosures are manufactured from sheet steel and painted to provide better weather protection. NEMA Type 4 or 4X enclosures are made of corrosion-resistant stainless steel or nonmetallic materials. NEMA Type 7 and 9 enclosures are cast from copper-free aluminum and finished with an enamel paint. NEMA Type 1 switches are general-purpose devices designed for use indoors to protect personnel from live parts and the enclosed equipment from dirt. Switches rated through 200A are provided with ample knockouts; 400A through 1,200A switches are provided without knockouts.

The following are the NEMA enclosure types that will be encountered most often. Always make certain that the proper enclosure is chosen for the application.

NEMA Type 3R switches are designated rainproof and are designed for use outdoors. NEMA Type 3R enclosures for switches rated through 200A have provisions for interchangeable bolt-on hubs at the top endwall, as illustrated in *Figure 19*. NEMA Type 3R switches rated higher than 200A have blank top endwalls. Knockouts are provided (below live parts only) on enclosures for 200A and smaller Type 3R switches. Type 3R switches are available in ratings through 1,200A.

NEMA Type 4 and 4X switches are designated dust-tight, watertight, and corrosion-resistant, and are designed for indoor and outdoor use. Common applications include commercial kitchens, dairies, canneries, and other types of food processing facilities, as well as areas where mildly corrosive liquids are present. All NEMA Type 4 and 4X enclosures are provided without knockouts. The use of watertight hubs is required. Available switch ratings are 30A through 600A.

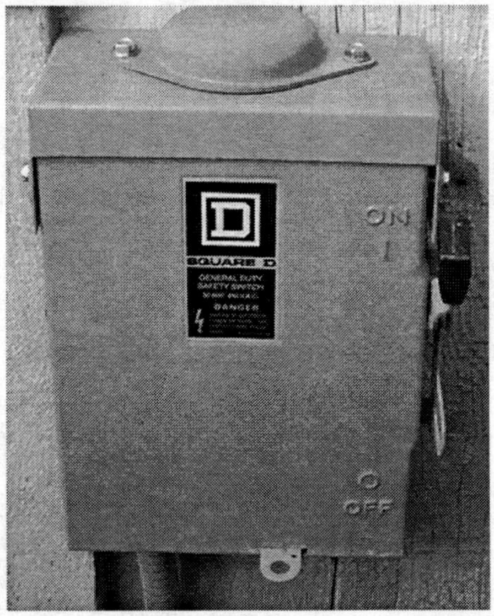

305F19.EPS

Figure 19 ◆ NEMA Type 3R enclosure.

NEMA Type 12 and 12K switches are designated dust-tight (except at knockout locations on Type 12K) and are designed for indoor use. Common applications include heavy industrial use where the switch must be protected from dust, lint, flying material, oil seepage, etc. NEMA 12K switches have knockouts in the bottom and top endwalls only. Available switch ratings are 30A through 600A in Type 12 and 30A through 200A in Type 12K.

6.1.7 Interlocked Receptacles

Heavy-duty, 60A NEMA Type 1 and 12 switches with an interlocked receptacle are also available. This receptacle provides a means of connecting and disconnecting loads directly to the switch. A non-defeating interlock prevents the insertion or removal of the receptacle plug while the switch is in the ON position. It also prevents operation of the switch if an incorrect plug is used.

Instructor's Notes:

6.1.8 Accessories

Accessories for field installation include Class R fuse kits, fuse pullers, insulated neutrals with grounding provisions, equipment grounding kits, watertight hubs for use with NEMA Type 4 or 4X switches, and interchangeable bolt-on hubs for Type 3R switches.

An electrical interlock consists of auxiliary contacts for use where control or monitoring circuits need to be switched in conjunction with the safety switch operation. Kits can be either factory-installed or field-installed, and they contain either one normally open and one normally closed contact or two normally open and two normally closed contacts. The electrical interlock is actuated by a pivot arm that operates directly from the switch mechanism. The electrical interlock is designed so that its contacts disengage before the blades of the safety switch open and engage after the safety switch blades close.

6.2.0 General-Duty Switches

General-duty switches for residential and light commercial applications are used where operation and handling are moderate and where the available fault current is 10,000 rms symmetrical amperes or less. Some general-duty safety switches, however, exceed this specification in that they are UL listed for use on systems having up to 100,000 rms symmetrical amperes of available fault current when Class R fuses and Class R fuse kits are used. Class T fusible switches are also available in 400A, 600A, and 800A ratings. These switches accept 300VAC Class T fuses only. Some examples of general-duty switch applications include residential, farm, and small business service entrances and light-duty branch circuit disconnects.

General-duty switches are rated up to 600A at 240VAC in general-purpose (Type 1) and rainproof (Type 3R) enclosures. Some general-duty switches are horsepower-rated and capable of opening a circuit up to six times the rated current of the switch; others are not. Always check the switch specifications under a horsepower-rated condition before use.

6.2.1 Switch Blades and Jaws

All current-carrying parts of general-duty switches are plated to minimize oxidation and reduce heating. Switch jaws and blades are made of copper for high conductivity. Where required, a steel reinforcing spring increases the mechanical strength of the jaws and contact pressure between the blade and jaw. Good pressure contact maintains the blade-to-jaw resistance at a minimum, which in turn promotes cool operation. All general-duty switch blades feature visible blade construction. With the door open, there is visually no doubt when the switch is OFF. However, you should always verify this with a voltage tester.

6.2.2 Fuse Clips

Fuse clips are normally plated to control corrosion and keep heating to a minimum. Where required, steel reinforcing springs are provided to increase the mechanical strength of the fuse clip. The result is a firmer, cooler connection to the fuses, as well as superior fuse retention.

6.2.3 Terminal Lugs

Most general-duty safety switches are furnished with mechanical set screw lugs that are suitable for aluminum or copper conductors.

6.2.4 Insulating Material

Switch and fuse bases are made of a strong, noncombustible, moisture-resistant material that provides the required phase-to-phase and phase-to-ground insulation.

6.2.5 Operating Mechanism and Cover Latching

Although not required by either the UL or NEMA standards, some general-duty switches have spring-driven, quick-make, quick-break operating mechanisms. Operating handles are an integral part of the operating mechanism and are not mounted on the enclosure cover. The handle provides an indication of the status of the switch. When the handle is up, the switch is ON. When the handle is down, the switch is OFF. A padlocking bracket is provided that allows the switch handle to be locked in the OFF position. Another bracket is provided that allows the enclosure to be padlocked closed.

6.2.6 Enclosures

General-duty safety switches are available in either a NEMA Type 1 enclosure for general-purpose indoor applications or a NEMA Type 3R enclosure for rainproof outdoor applications. Enclosure types were discussed in detail earlier in this module.

Classroom

Discuss the types of accessories used with heavy-duty switches.

Compare the components of a general-duty switch to those of a heavy-duty switch. Point out the NEMA classifications of general-duty switches.

Classroom

Discuss the operation and applications of double-throw safety switches. Provide an example for the trainees to examine.

Audiovisual

Show Transparency 17 (Figure 20).

Classroom

Point out the NEC requirements for safety switches. Refer the trainees to Table 1 in the Trainee Module.

Teaching Tip

Discuss the "Think About It."

6.3.0 Double-Throw Safety Switches

Double-throw safety switches are used as manual transfer switches and are not intended for use as motor circuit switches; thus, horsepower ratings are generally unavailable. Double-throw switches are available as either fused or nonfusible devices. Two general types of switch operation are available:

- Quick-make, quick-break (for use where fast action is the most desirable feature)
- Slow-make, slow-break (for use where the application might expose the switch to voltage fluctuations or current surges on startup)

Figure 20 shows a practical application of a double-throw safety switch used as a transfer switch in conjunction with a standby emergency generator system.

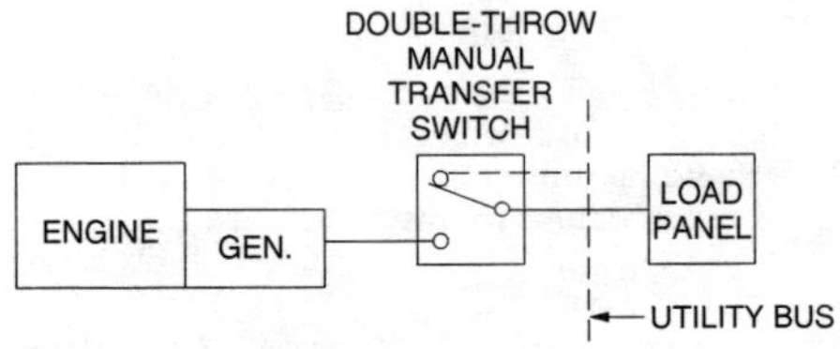

Figure 20 ◆ Practical application of a double-throw safety switch.

6.4.0 NEC Safety Switch Requirements

Safety switches, in both fusible and nonfusible types, are used as a disconnecting means for services, feeders, and branch circuits. Installation requirements involving safety switches are found in several places throughout the NEC, including:

- ***NEC Article 312***
- ***NEC Article 404***
- ***NEC Article 430, Part VIII***
- ***NEC Article 440, Part II***
- ***NEC Section 450.8(C)***

When used as a service disconnecting means, the major installation requirements are listed in *Table 1.*

THINK ABOUT IT

Identifying Devices

Which component in the picture is not considered a device by NEC definition?

Instructor's Notes:

Table 1 NEC Installation Requirements Governing Switches Used for Service Disconnects

Application	Installation Requirements	NEC Reference or Comment
General	A means must be provided to disconnect all conductors serving the premises from service-entrance conductors. Each disconnecting means must be permanently identified. Disconnecting means must be installed at a readily accessible location nearest the service-entrance conductors. Service disconnecting means cannot be installed in bathrooms. Each service disconnecting means must be suitable for the prevailing conditions.	***NEC Section 230.70***
Number of disconnects	Service disconnecting means can consist of no more than six switches for each service.	***NEC Section 230.71***
	The two to six disconnects permitted above shall be grouped. Each disconnect shall be identified as to the load served. *Exception:* Disconnect for the water pump of a fire pump protection system.	***NEC Section 230.72***
Service disconnecting means	The service disconnecting means shall not be rated less than the load to be carried: *One circuit*—Not less than 15A *Two circuits*—Not less than 30A *Single-family dwelling*—Not less than three-wire, 100A *All others*—Not less than 60A	***NEC Sections 230.79(A), (B), (C), and (D)***
Working space	Requirements for any electrical equipment apply; that is, the minimum head room of working spaces around service equipment must be 6.5' or more. The dimensions of working space in the direction of access to live parts operating at 600V nominal or less to ground and likely to require examination, adjustment, servicing, or maintenance while energized must not be less than indicated in ***NEC Table 110.26(A).***	***NEC Section 110.26***
Type of disconnect	Disconnects may include either a manually operated or power-operated safety switch, provided the switch can be opened by hand in the event of a power supply failure.	***NEC Section 230.76***
Disconnection of grounded conductor	Where the service disconnecting means does not disconnect the grounded conductor from the premises wiring, other means shall be provided for this purpose in the service equipment.	***NEC Section 230.75***
Connections to terminals	Service conductors must be connected to the disconnecting means by pressure connectors, clamps, or other approved means. Soldered connections are forbidden.	***NEC Section 230.81***
More than one building on the same premises	In industrial establishments under single management, the disconnecting means for several buildings may be conveniently located if NEC conditions are met.	***NEC Section 225.32, Exception 1***
Service disconnecting means with more than one switch or circuit breaker	Where service disconnecting means consist of more than one switch or circuit breaker, the combined ratings of all the circuit breakers or switches shall not be rated at less than the load to be carried.	***NEC Section 230.80***

Laboratory

Have the trainees identify various switches and their applications. Note the proficiency of each trainee.

Classroom

Have the trainees complete the Review Questions and go over the answers prior to administering the Module Examination.

Examination

Administer the Module Examination. Record the results on Craft Training Report Form 200 and submit the results to the Training Program Sponsor.

Performance Testing

Administer the Performance Test and fill out Performance Profile Sheets for each trainee. If desired, trainee proficiency noted during laboratory sessions may be used to complete the Performance Test. Be sure to record the results on Craft Training Report Form 200 and submit the results to the Training Program Sponsor.

Summary

The NEC defines a device as a unit of an electrical system that is intended to carry or help carry but not utilize electric energy. This covers a wide assortment of system components that include, but are not limited to, the following:

- Switches
- Relays
- Contactors
- Receptacles
- Conductors

The purpose of a switch is to make and break an electrical circuit in a safe and convenient manner. *NEC Article 404* covers most of the installation requirements for switches.

A receptacle is a contact device installed at the outlet of a circuit for the connection of a single attachment plug. A single receptacle is a single contact device with no other contact device on the same yoke. A multiple receptacle is a single device containing two or more receptacles, with the most common being the duplex receptacle. *NEC Sections 210.50, 210.52, 210.60, 210.62, 220.3(B)(9), and 220.3(B)(10)* cover many of the requirements for receptacles.

Review Questions

1. Which of the following statements is true concerning duplex receptacles?
   a. CU/AL markings indicate the use of both copper and aluminum wire.
   b. CU/AL markings indicate the use of aluminum or copper-clad wire.
   c. CO/ALR markings indicate the use of both copper and aluminum wire.
   d. CO/ALR markings indicate the use of copper wire only.
2. The ____ in a duplex receptacle is for connection to a grounded conductor.
   a. long slot
   b. short slot
   c. grounding slot
   d. slot connected to the green screw
3. Which of the following wire sizes may usually be used in quick-connect wire clips in 15A receptacles and switches?
   a. No. 14 AWG
   b. No. 12 AWG
   c. No. 4 AWG
   d. No. 2 AWG
4. What is the maximum distance that outlet boxes may be set back from the finished surface in concrete walls?
   a. 1/16"
   b. 1/8"
   c. 1/4"
   d. 1/2"
5. What is the maximum distance that outlet boxes may be set back from the finished surface in wood-paneled walls?
   a. 1/16"
   b. 1/8"
   c. 1/4"
   d. 0"
6. The intent of *NEC Section 210.52(A)(1)* is to ____.
   a. prevent the need for extension cords
   b. ensure ample amperage for cord-and-plug appliances
   c. prevent the use of more appliances than the circuit can handle
   d. enable extension cords longer than 6' to be used
7. Receptacles installed to serve ____ must be protected with a GFCI.
   a. crawlspaces above ground level
   b. bedrooms
   c. living rooms
   d. kitchen countertop areas
8. The NEC requires at least one receptacle for every ____ of show window in a store building.
   a. 3'
   b. 6'
   c. 12'
   d. 18'
9. In a set of electrical specifications, you would most likely find information concerning wall switches for lighting control in ____.
   a. Section 15408
   b. Section 16500
   c. Section 16700
   d. Section 16800
10. The ____ is the preferred contact for use in heavy-duty switches.
    a. butt-type
    b. knife blade and jaw
    c. clip-type
    d. thermocouple

Instructor's Notes:

GLOSSARY

Trade Terms Introduced in This Module

Attachment plug: The male connector for electrical cords.

Cord: A small, flexible conductor assembly, usually jacketed.

Device: An item intended to carry or help carry but not utilize electric energy.

Four-way switch: A device that, when used in conjunction with two three-way switches, offers control of an electrical outlet (usually lighting) at three or more locations.

Gang switch: A unit of two or more switches that allows control of two or more circuits from one location. The entire mechanism is mounted in one box under one cover.

Receptacle: A contact device installed at an outlet for the connection of an attachment plug and flexible cord to supply portable equipment.

Switch: A device used to open, close, or change the connection of a circuit.

Three-way switch: A switch used to control a light or set of lights from two different locations.

Additional Resources

This module is intended to present thorough resources for task training. The following reference works are suggested for further study. These are optional materials for continued education rather than for task training.

American Electrician's Handbook. Terrell Croft and Wilfred I. Summers. New York, NY: McGraw-Hill, 1996.

National Electrical Code Handbook. Quincy, MA: National Fire Protection Association, 2002.

Instructor's Notes:

ANSWER KEY

Answers to Review Questions

Answer	Section
1. c	2.1.0
2. a	2.1.0; Figure 2
3. a	2.1.0
4. c	2.2.0
5. d	2.2.0
6. a	3.1.0
7. d	3.1.0; Figure 7
8. c	3.2.0
9. b	4.2.0
10. b	6.1.1

TEACHING TIPS—MODULE 26305-02

Section 1.0.0 *Think About It—Relay and Contactor Coils*

The coils of contactors and relays use electrical energy to operate; therefore, they would not be devices according to the NEC definition. Contactors and relays without the coils installed are considered devices.

Section 6.4.0 *Think About It—Identifying Devices*

The light bulb is not considered a device because it uses electrical energy.

CONTREN™ LEARNING SERIES — USER UPDATES

The NCCER makes every effort to keep these textbooks up-to-date and free of technical errors. We appreciate your help in this process. If you have an idea for improving this textbook, or if you find an error, a typographical mistake, or an inaccuracy in NCCER's Contren™ textbooks, please write us, using this form or a photocopy. Be sure to include the exact module number, page number, a detailed description, and the correction, if applicable. Your input will be brought to the attention of the Technical Review Committee. Thank you for your assistance.

Instructors – If you found that additional materials were necessary in order to teach this module effectively, please let us know so that we may include them in the Equipment/Materials list in the Instructor's Guide.

Write: Curriculum Revision and Development Department
National Center for Construction Education and Research
P.O. Box 141104, Gainesville, FL 32614-1104

Fax: 352-334-0932

E-mail: curriculum@nccer.org

Craft ________________ Module Name ________________

Copyright Date ________ Module Number ________ Page Number(s) ________

Description

____

____

____

____

(Optional) Correction

____

____

____

(Optional) Your Name and Address

____

____

____

Distribution Equipment

26306-02

Distribution Equipment
Annotated Instructor's Guide

Module 26306-02

MODULE OVERVIEW

This module introduces the Electrical trainee to the methods and procedures related to distribution equipment, including grounding, switchboard testing and maintenance, ground fault sensing, and interpreting electrical drawings.

PREREQUISITES

Please refer to the Course Map in the Trainee Module. Prior to training with this module, it is recommended that the trainee shall have successfully completed the following:

Core Curriculum; Electrical Level One; Electrical Level Two; Electrical Level Three, Modules 26301 through 26305.

OBJECTIVES

Upon completion of this module, the trainee will be able to:

1. Describe the purpose of switchgear.
2. Describe the four general classifications of circuit breakers and list the major circuit breaker ratings.
3. Describe switchgear construction, metering layouts, wiring requirements, and maintenance.
4. List NEC requirements pertaining to switchgear.
5. Describe the visual and mechanical inspections and electrical tests associated with low-voltage and medium-voltage cables, metal-enclosed busways, and metering and instrumentation.
6. Describe a ground fault relay system and explain how to test it.

PERFORMANCE TASKS

This is a performance-based module; there is no performance examination.

NCCER STANDARDIZED CRAFT TRAINING PROGRAMS

The National Center for Construction Education and Research (NCCER) provides a standardized national program of accredited craft training. Key features of the program include instructor certification, competency-based training, and performance testing. The program provides trainees, instructors, and companies with a standard form of recognition through a National Craft Training Registry. The program is described in full in the *Guidelines for Accreditation*, published by the NCCER. For more information on standardized craft training, contact the NCCER by writing us at P.O. Box 141104, Gainesville, FL 32614-1104, calling 352-334-0911, or e-mailing info@nccer.org. More information may be found at our Web site, www.nccer.org.

HOW TO USE THIS ANNOTATED INSTRUCTOR'S GUIDE

Each page presents two sections of information. The larger section displays each page exactly as it appears in the Trainee Module. The narrow column ties suggested trainee and instructor actions to each page and provides icons to call your attention to material, safety, audiovisual, or testing requirements. The bottom of each page includes space for your notes.

If you see the Teaching Tip icon, that means there is a teaching tip associated with this section. Also refer to the suggested teaching tips at the end of the module.

SAFETY CONSIDERATIONS

Ensure that the trainees are equipped with appropriate personal protective equipment.

PREPARATION

Before teaching this module, you should review the Module Outline, Objectives, and the Materials and Equipment List. Be sure to allow ample time to prepare your own training or lesson plan and gather all required equipment and materials.

MATERIALS AND EQUIPMENT LIST

Materials:
Transparencies
Markers/chalk
Copy of the latest edition of the *National Electrical Code*®
Manufacturer's literature for various types of distribution equipment
Module Examinations*

Equipment:
Overhead projector and screen
Whiteboard/chalkboard
Appropriate personal protective equipment

*Located in the Test Booklet packaged with this Annotated Instructor's Guide.

ADDITIONAL RESOURCES

This module is intended to present thorough resources for task training. The following reference works are suggested for both instructors and motivated trainees interested in further study. These are optional materials for continued education rather than for task training.

American Electrician's Handbook, Terrell Croft and Wilfred I. Summers. New York, NY: McGraw-Hill, 1996.

National Electrical Code Handbook. Quincy, MA: National Fire Protection Association, 2002.

TEACHING TIME FOR THIS MODULE

An outline for use in developing your lesson plan is presented below. Note that each Roman numeral in the outline equates to one session of instruction. Each session has a suggested time period of 2½ hours. This includes 10 minutes at the beginning of each session for administrative tasks and one 10-minute break during the session. Approximately 12½ hours are suggested to cover *Distribution Equipment.* You will need to adjust the time required for hands-on activity and testing based on your class size and resources.

Topic	Planned Time
Session I. Introduction; Voltage Classifications; Switchboards; Switchgear	
A. Introduction	________
B. Voltage Classifications	________
C. Switchboards	________
1. Applications	________
2. General Description	________
3. Switchboard Frame Heating	________
4. Low-Voltage Spacing Requirements	________
5. Cable Bracing	________
D. Switchgear	________
1. Switchgear Construction	________
2. Control and Metering Wire Standards	________
3. Wiring System	________
a. Door-Mounted Wiring Restrictions	________
b. Terminal Connections	________
4. Metering Current and Potential Transformers	________
5. Switchgear Handling, Storage, and Installation	________

Session II. Switchboard Testing and Maintenance; NEC Requirements

A. Switchboard Testing and Maintenance
   1. General Maintenance Guidelines ____________
   2. Typical Guidelines ____________
      a. Thermographic Survey ____________
      b. Metal-Enclosed Switchgear and Switchboards ____________
      c. Low-Voltage Cables (600V Maximum) ____________
      d. Medium-Voltage Cables (15kV Maximum) ____________
      e. Metal-Enclosed Busways ____________
      f. Metering and Instrumentation ____________

B. NEC Requirements ____________
   1. Requirements for Electrical Installations ____________
   2. Requirements for Conductors ____________
   3. Grounding ____________
   4. Switchboards and Panelboards ____________

Session III. Ground Faults; HVL Switches; Bolted Pressure Switches

A. Ground Faults ____________
   1. Ground Fault Systems ____________
   2. Sensing Operation ____________
   3. Zero Sequencing Sensor Mounting ____________
   4. Relay Mounting ____________
   5. Connections ____________
   6. Relay Settings ____________
      a. Coordination with Downstream Circuit Breakers ____________
      b. Instantaneous Trip Feature ____________
   7. Generic National Electrical Testing Association Ground Fault System Test ____________
      a. Procedures ____________

B. HVL Switches ____________
   1. Ratings ____________
   2. Variations ____________
   3. Opening Operation ____________
   4. Closing ____________
   5. Maintenance ____________
   6. Sluggish Operation ____________

C. Bolted Pressure Switches ____________
   1. Ground Fault ____________
   2. Phase Failure ____________
   3. Blown Main Fuse Detector ____________
   4. Maintenance ____________

Session IV. Transformers; Instrument Transformers; Circuit Breakers

A. Transformers
   1. Transformer Theory
      a. No-Load Operation
      b. Load Operation
   2. Transformer Types
   3. Dry Transformers (Air-Cooled)
   4. Sealed Dry Transformers
   5. Transformer Nameplate Data
   6. Transformer Case Inspections
   7. Transformer Tests

B. Instrument Transformers
   1. Potential Transformers
   2. Current Transformers
   3. Instrument Transformer Maintenance

C. Circuit Breakers
   1. Circuit Breaker Ratings

Session V. Electrical Drawing Identification; Electrical Prints; Interpreting One-Line Diagrams; Panelboards; Review; Module Examination

A. Electrical Drawing Identification
   1. Electrical Diagram Symbology

B. Electrical Prints
   1. Single-Line Diagrams
   2. Elementary Diagrams
   3. Interconnection Diagrams
   4. Connection Diagrams
      a. Point-to-Point Method
      b. Cable Method

C. Interpreting One-Line Diagrams
   1. Interpreting Secondary One-Line Diagrams
   2. Shop Drawings

D. Panelboards
   1. Lighting and Power Panelboards
   2. Panelboard Construction
      a. Identification of Conductors
      b. Number of Circuits
   3. Panelboard Protective Devices
   4. Branch Circuit Protective Devices

E. Review

F. Module Examination
   1. Trainees must score 70% or higher to receive recognition from the NCCER.
   2. Record the testing results on Craft Training Report Form 200 and submit the results to the Training Program Sponsor.

Module 26306-02

Distribution Equipment

Course Map

This course map shows all of the modules in the third level of the Electrical curriculum. The suggested training order begins at the bottom and proceeds up. Skill levels increase as you advance on the course map. The local Training Program Sponsor may adjust the training order.

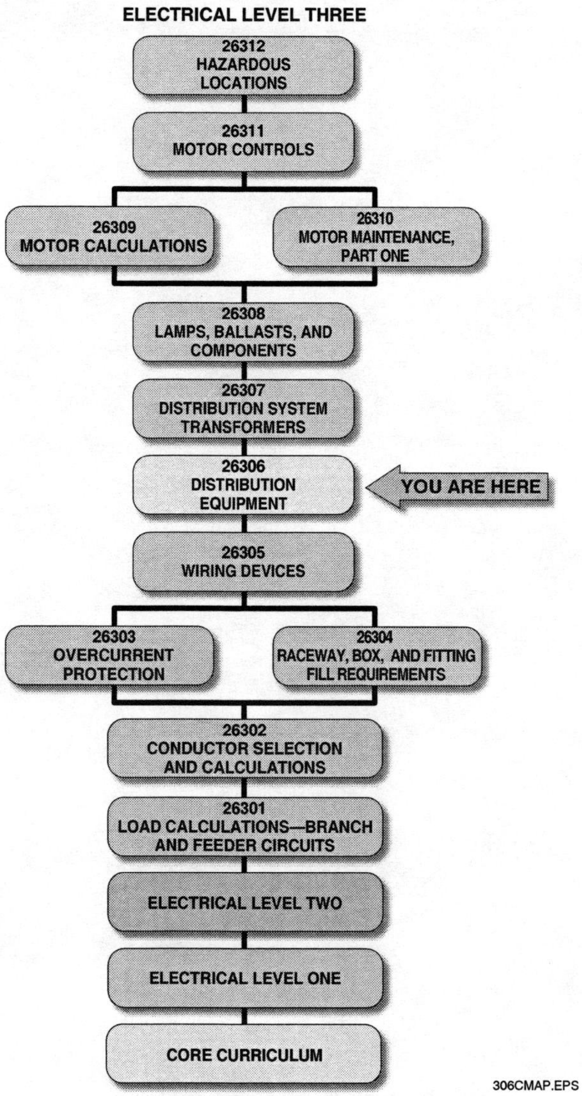

306CMAP.EPS

Assign reading of Module 26306.

MODULE 26306 CONTENTS

Instructor's Notes:

Figures

Tables

Instructor's Notes:

MODULE 26306

Distribution Equipment

Objectives

When you have completed this module, you will be able to do the following:

1. Describe the purpose of switchgear.
2. Describe the four general classifications of circuit breakers and list the major circuit breaker ratings.
3. Describe switchgear construction, metering layouts, wiring requirements, and maintenance.
4. List NEC requirements pertaining to switchgear.
5. Describe the visual and mechanical inspections and electrical tests associated with low-voltage and medium-voltage cables, metal-enclosed busways, and metering and instrumentation.
6. Describe a ground fault relay system and explain how to test it.

Prerequisites

Before you begin this module, it is recommended that you successfully complete the following: Core Curriculum; Electrical Level One; Electrical Level Two; Electrical Level Three, Modules 26301 through 26305.

Required Trainee Material

1. Pencil and paper
2. Appropriate personal protective equipment
3. Copy of the latest edition of the *National Electrical Code*

1.0.0 ◆ INTRODUCTION

An electrical power system consists of several subsystems on both the utility (supply) side and the customer (user) side. Electricity generated in power plants is stepped up to transmission voltage and fed into a nationwide grid of transmission lines. This power is then bought, sold, and dispatched as needed. Local utility companies take power from the grid and reduce the voltage to levels suitable for subtransmission and distribution through various substations to the customer. This may range from the common 200A, 120/240V residential service to hundreds of thousands of amps at voltages from 480V to 69kV in an industrial facility.

From the point of service, customers must control, distribute, and manage the power to supply their electrical needs. This module will discuss how this is done using a typical industrial facility as an example. We will discuss the various components of the distribution system and their interdependence. An understanding of single-line diagrams will allow analysis of a facility's distribution system.

> **Note**
> The voltage conventions used in this module are industry standards for distribution systems.

2.0.0 ◆ VOLTAGE CLASSIFICATIONS

Electrical equipment is usually classified by voltage. The various voltage levels are classified as low, medium, high, extra high, and ultra high:

- *Low voltage (LV)*—Low voltage is considered to be 600V and below. It is typically used to supply nominal voltage directly to electrical loads.

Note: The designations "National Electrical Code," "NE Code," and "NEC," where used in this document, refer to the National Electrical Code®, which is a registered trademark of the National Fire Protection Association, Quincy, MA. *All National Electrical Code (NEC) references in this module refer to the 2002 edition of the NEC.*

Materials

Ensure that you have everything required to teach the course. Check the Materials and Equipment List at the front of this Instructor's Guide.

Classroom

Review the goals of the course.

Explain that terms shown in bold (blue) are defined in the Glossary at the back of this module.

Audiovisual

Show Transparency 1, Course Objectives.

Classroom

Describe the components in a typical distribution system.

List the five voltage classifications.

Classroom

Define *switchboard.* Point out the difference between service-entrance equipment and distribution system equipment.

Describe the components in a typical switchboard.

Point out that switchboards have specific bus spacing requirements to prevent overheating of the frame. Refer the trainees to Table 1 in the Trainee Module.

Common voltages in this category are 120V, 208V, 240V, and 480V.

- *Medium voltage (MV)*—Medium voltage ranges from 601V to 15kV. It is mainly used for distribution purposes and for supplying large electrical loads.
- *High voltage (HV)*—High voltage ranges from 15kV to 230kV. It is mainly used for transmission purposes.
- *Extra high voltage (EHV) or very high voltage (VHV)*—Extra high voltage, or very high voltage, ranges from 230kV to 800kV. It is used only for transmission purposes.
- *Ultra high voltage (UHV)*—Ultra high voltage is any voltage greater than 800kV. Presently, the common voltages in this category range between 1,100kV and 1,500kV.

3.0.0 ◆ SWITCHBOARDS

According to the National Electrical Code, the term **switchboard** may be defined as a large single panel, frame, or assembly of panels on which switches, overcurrent and other protective devices, **buses,** and instruments may be mounted, either on the face or back or both. Switchboards are generally accessible from both the rear and from the front and are not intended to be installed in cabinets.

3.1.0 Applications

Switchboards are used in modern distribution systems to subdivide large blocks of electrical power. One location for switchboards is typically where the main power enters the building. In this location, the switchboard is referred to as **service-entrance equipment.** The other location common for switchboards is downstream from the service-entrance equipment. In the downstream location, the switchboard is commonly referred to as **distribution system equipment.**

3.2.0 General Description

A switchboard consists of a stationary structure that includes one or more freestanding units of uniform height that are mechanically and electrically joined to make a single coordinated installation. These cubicles contain circuit-interrupting devices. They take up less space in a plant, have more eye appeal, and eliminate the need for a separate room to protect personnel from contact with lethal voltages.

The main portion of the switchboard is formed from heavy-gauge steel welded with members across the top and bottom to provide a rigid enclosure. Most switchboard enclosures are divided into three sections: the front section, the bus section, and the cable section. These three sections are physically separated from one another by metal partitions. This confines any damage that may occur to any one section and keeps it from affecting the other sections.

Typical switchboard components include:

- Circuit breakers
- Fuses
- Motor starters
- Ground fault systems
- Instrument transformers
- Switchboard metering
- Control power transformers
- Busbars

Electrical ratings include three-phase, three-wire and three-phase, four-wire systems with voltage ratings up to 600V and current ratings up to 4,000A.

A switchboard enclosure is described as a dead front panel, which means that no live parts are exposed on the opening side of the equipment; however, it contains energized breakers. Busbars can be a standard size or customized. Standard sizes are usually made of silver-plated or tin-plated copper or tin-plated aluminum. Conventional bus sizing is .25" × 2" through .375" × 7". Copper provides an ampacity of 1,000A/sq. in. of cross-sectional area. When using aluminum, the ampacity is 750A/sq. in.

When two busbars are bolted together using Grade S hardware with the proper torque, the ampacity of the connection is 200A/sq. in. of the lapped portion for aluminum or copper bussing. Bussing joints must be bolted together to the specified torque and include Belleville washers or Keps nuts. Aluminum busbars must be tin-plated, and copper busbars over 600A must be plated with tin or silver.

3.3.0 Switchboard Frame Heating

Table 1 shows guidelines that should be observed in order to keep heat losses in the iron switchboard frame members to a safe minimum. The dimensions are recommended values and should be adhered to whenever possible.

> **Note**
> Some switchboard frames are engineered differently and will have values other than those shown in *Table 1.*

Instructor's Notes:

Table 1 Switchboard Frame Heating Guidelines

Amperes	Minimum Distance from Phase Bus to Closest Steel Member	Minimum Distance from Neutral Bus to Closest Steel Member
3,000	4"	2"
4,000	6"	3"
5,000 and over	12"	see below
5,000 to 6,000	An aluminum or nonmagnetic material should be used in place of steel frame sections. Wherever possible, you must maintain 12" to steel members and 6" to aluminum or nonmagnetic members. Neutral spacing can be 6" and 3", respectively. If the main bus is tapered, it is permissible (at 4,000A and below) to use steel frames for those sections containing the tapered bus.	
6,000 and over	You must use an aluminum or nonmagnetic material for frame sections and maintain 12" to steel members and 6" to aluminum or nonmagnetic members. Neutral spacing can be 6" and 3", respectively. The use of any steel frame members is discouraged. If the main bus is tapered, it is permissible (at 4,000A and below) to use steel frames for those sections containing the tapered bus.	

Note: For amperages above 8,000A, the neutral spacing must be 12" wherever possible.

Classroom

Describe the bus spacing requirements for a typical three-phase system.

Audiovisual

Show Transparency 2 (Figure 1).

3.4.0 Low-Voltage Spacing Requirements

To minimize tracking or arcing from energized parts to ground, switchboard construction includes spacing requirements. These spacing requirements are measured between live parts of opposite polarity and between live parts and grounded metal parts. *Figure 1* illustrates typical switchboard spacing requirements.

An isolated dead metal part, such as a screwhead or washer, interposed between uninsulated

VOLTAGE INVOLVED		MINIMUM SPACING BETWEEN LIVE PARTS OF OPPOSITE POLARITY		MINIMUM SPACING THROUGH AIR AND OVER SURFACE BETWEEN LIVE PARTS AND GROUNDED METAL PARTS
GREATER THAN	MAX.	THROUGH AIR	OVER SURFACE	BOTH THROUGH AIR AND OVER SURFACE
0	125	1/2"	3/4"	1/2"
125	250	3/4"	1 1/4"	1/2"
250	600	1"	2"	1"*

* A through air spacing of not less than 1/2" is acceptable (1) at a molded-case circuit breaker or a switch other than a snap switch, (2) between uninsulated live parts of a meter mounting or grounded dead metal, and (3) between grounded dead metal and the neutral of a 480Y/277V, three-phase, four-wire switchgear section.

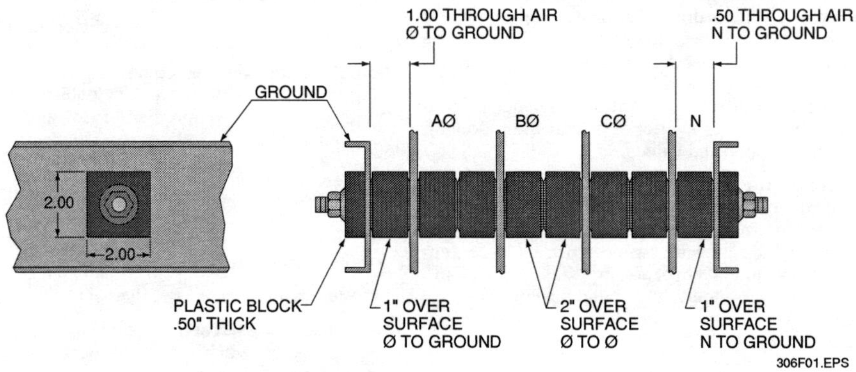

Figure 1 ◆ Typical busbar spacing requirements.

Classroom

Discuss the cable bracing requirements of various types of systems.

Define *switchgear.* List the four classifications of switchgear. Refer the trainees to Figures 2 through 4 in the Trainee Module.

INSIDE TRACK

Busbars

Busbars have very specific spacing requirements. Note the red spacer blocks on the switchgear shown here.

306P0601.EPS

live parts of opposite polarity or between an uninsulated live part and grounded dead metal is considered to reduce the spacing by an amount equal to the dimension of the interposed part along the path of measurement.

When measuring over-surface spacing, any slot, groove, and the like that is 0.013" (0.33mm) wide or less and in the contour of the insulating material is to be disregarded.

When measuring spacing, an air space of 0.013" or less between a live part and an insulating surface is to be disregarded, and the live part is to be considered in contact with the insulating material. A pressure wire connector shall be prevented from any turning motion that would result in less than the minimum acceptable spacings. The means used to ensure turn prevention must be reliable, such as a shoulder or boss. A lock washer alone is not acceptable.

A means of turn prevention need not be provided if spacings are not less than the following minimum accepted values:

- When the connector and any connector of opposite polarity have each been turned 30 degrees toward the other
- When the connector has been turned 30 degrees toward other live parts of opposite polarity and toward grounded dead metal parts

3.5.0 Cable Bracing

All construction using conductors and having a short circuit current rating greater than 50,000 rms symmetrical amperes requires a cable brace positioned as close to the supply lugs as possible. The cable brace is intended to be mounted in the same area that is allotted for wire bending. It is not necessary to provide additional mounting height to accommodate the cable brace.

The cable brace requirement does not apply to load-side cables, main breakers, or switches. It only applies when cables are connected directly to an unprotected line-side bus. The bus restrictions for a line-side bus include:

- There can be no splice in edgewise bus mounting of 2,100A or less rated at 50,000 rms symmetrical amperes.
- There can be no splice in flatwise bus mounting of 600A or less rated over 50,000 rms symmetrical amperes.

Note

This does not apply to connections made from the through bus to a switch or circuit breaker.

The cable restrictions for a line-side bus include:

- Busing of 600A or less that is rated over 50,000 rms symmetrical amperes cannot use cables; it must be bus connected.
- If cabling is required, 800A minimum busing must be used.

Cable bracing requirements may be excluded if the busing is able to fully withstand the total available short circuit current.

4.0.0 ◆ SWITCHGEAR

Switchgear is a general term used to describe switching and interrupting devices and assemblies of those devices containing control, metering, protective, and regulatory equipment, along with the associated interconnections and supporting structures. Switchgear performs two basic functions:

- Provides a means of switching or disconnecting power system apparatus
- Provides power system protection by automatically isolating faulty components

Switchgear can be classified as:

- **Metal-enclosed switchgear** (low voltage)
- Metal-clad switchgear (low and medium voltage)
- Metal-enclosed interrupters
- Unit substations

Instructor's Notes:

The low-voltage and medium-voltage switchgear assemblies are completely enclosed on all sides and topped with sheet metal, except for ventilating openings and inspection windows. They contain primary power circuit switching or interrupting devices, buses, connections, and control and auxiliary devices. *Figure 2* shows typical low-voltage, metal-clad switchgear.

The station-type cubicle switchgear consists of indoor and outdoor types with power circuit breakers rated from 14.4kV to 34.5kV, 1,200A to 5,000A, and 1,500kVA to 2,500kVA interrupting **capacity.** Equipment can be special ordered and built at higher kVA ratings.

4.1.0 Switchgear Construction

Switchgear consists of a stationary structure that includes one or more freestanding units of uniform height that are mechanically and electrically joined to make a single coordinated installation. These units, commonly referred to as *cubicles,* contain circuit-interrupting devices such as circuit breakers.

Switchgear enclosures are formed from heavy-gauge sheet steel that has been welded or bolted together. Structural members across the top, sides, and bottom provide a rigid enclosure. Metal-clad switchgear enclosures are divided into three sections: the front section, the bus section, and the cable or termination section.

These three sections are physically separated from one another by metal partitions. This confines any damage that may occur to any one section and keeps it from affecting the other sections. It also separates power between the sections for ease and safety of maintenance.

306F02.EPS

Figure 2 ◆ Typical low-voltage, metal-clad switchgear.

The rigid enclosure provides the primary structural strength of the switchgear assembly and the means by which the switchgear is fastened to its foundation. The strength of the enclosure and its mounting system will vary depending on its intended use. For example, switchgear used in a nuclear application must meet certain seismic qualifications.

The enclosure also provides the required supports and mounts for items to be located in the switchgear and provides for the necessary interconnections between the switchgear and other plant systems. The number of sections and physical makeup of switchgear varies depending on the voltage and current ratings, plant specifications, and specific manufacturer.

Figures 3 and *4* show external and internal views, respectively, of medium-voltage, metal-clad switchgear. This equipment is available in voltages from 4.76kV to 27kV and current ranges from 1,200A through 3,000A.

4.2.0 Control and Metering Wire Standards

Switchboard control and meter wiring standards must meet the requirements of *Machine Tool Wires*

306F03.EPS

Figure 3 ◆ Medium-voltage, metal-clad switchgear (exterior view).

Classroom

Describe the construction of switchgear.

Teaching Tip

Take the trainees on a tour of a local commercial facility and allow them to examine the switchgear.

Classroom

Explain that switchgear must conform to the requirements of various standards.

List some of the basic requirements for switchgear.

Emphasize that the wiring in switchgear must be neatly organized and securely harnessed.

Figure 4 ◆ Medium-voltage, metal-clad switchgear (interior view).

and Cables Safety Standard UL 1063, Switchboard Safety Standard UL 891, and *Service Equipment Safety Standard UL 869,* including the following highlights:

- Stranded copper conductor with thermoplastic insulation, UL and CSA approved
- Insulation thickness of .030" (1/32") for No. 16–No. 10 AWG size conductor
- 600V, 90°C minimum rating for dry locations; 60°C minimum rating when exposed to oil and moisture
- Conductors no smaller than No. 14 AWG for control wiring

4.3.0 Wiring System

The NEC requires wiring to be supported mechanically to keep the wiring in place. Wire harnessing is generally used within the switchboard with the following restrictions:

- Each bundle or cable of wires must be run in a vertical or horizontal direction, securing the harness by means of plastic cable ties or cable clips.
- Plastic wire cable clamps shall be placed at strategic locations along the harnessing to hold the harness firmly in place to prevent interference with the control components' required electrical, mechanical, and arcing clearances.
- Apply wire ties to the harnessed wiring every 3" to 4" with self-adhesive cable ties spaced at every 12".
- Some precautions to be observed when wiring the switchboard electrical components are:
  - –Keep control wires at least ½" from moving parts.
  - –Avoid running wires across sharp metal edges. To protect the wiring from mechanical damage, use approved cable protectors, such as a nylon clip cable guard, a wire guard for edge protection, or special edge protection molding.
  - –Wires must not touch exposed bare electrical parts of opposite polarity.
  - –Wires must not interfere with the adjustment or replacement of components.
  - –Wires should be as straight and as short as possible.
  - –Wires shall not be spliced.

Instructor's Notes:

–To eliminate possible strain on the control wire, a certain amount of slack should be given to the individual or harnessed conductor terminated at a component connection.
–The equipment ground busbar shall not be used as a portion of the control or metering circuits.
–Do not use pliers for bending control wiring. Use your hands or an approved wire bending device.

4.3.1 Door-Mounted Wiring Restrictions

No incoming wiring connections may be made directly to the door-mounted devices. Wires from the door-mounted equipment to the panel terminal block should be a minimum of 19-strand wire.

Wires from the door must be neatly cabled so that the door can be opened easily without placing excessive strain on the wire terminal connections. In some cases, the cable must be separated into two bundles to accomplish this. Insulated sleeving, tubing, or vinyl tape must be used to bundle and protect the flexible wires.

4.3.2 Terminal Connections

All control or metering wiring entering or leaving the switchboard should terminate at terminal blocks, leaving one side of the terminal block free for the user's connections. No factory connections are allowed on the user's terminal connection point. For factory wiring, allow a maximum of two control wires on the same side of a terminal block. No more than three connections are allowed on terminals of control transformers, meters, meter selector switches, and metering equipment.

Since bolted pressure switches or any 100% current-rated, molded-case circuit breaker's line and load power terminals are allowed a higher maximum operating temperature than the recommended insulated conductor's operating temperature, the control wires cannot be placed directly on the 100%-rated disconnect device's line and load connections.

In all cases, control wires cannot touch any exposed part of opposite electrical polarity.

4.4.0 Metering Current and Potential Transformers

Ground connections on a potential transformer (PT) or current transformer (CT) secondary terminal must be connected to the ground bus. CT secondary terminals must be shorted if no metering equipment is connected to the current transformer.

PTs are required to have primary and secondary fusing. If protective circuits, such as ground fault or phase failure protective systems, are placed in the secondary circuit of the potential transformer, no secondary fusing is required.

Metering circuit connections made directly to the incoming bus must be provided with current-limiting fuses that are equal in rating to the available interrupting capacity.

> **Note**
> CTs and PTs will be discussed later.

4.5.0 Switchgear Handling, Storage, and Installation

The following is a basic guideline for the handling of switchgear. It is important to emphasize that these recommendations only supplement the manufacturer's instructions. Manufacturers include instruction books and drawings with their equipment. It is absolutely imperative that you read and understand these documents before handling any equipment.

- *Switchgear handling*—Immediately upon receipt of switchgear, an inspection for damage during transit should be performed. If any damage is noted, the transportation company should be notified immediately.
- *Switchgear rigging*—Instructions for switchgear should be found in the manufacturer's instruction books and drawings. Verify that the rigging is suitable for the size and weight of the equipment.
- *Switchgear storage*—Indoor switchgear that is not being installed should be stored in a clean, dry location. The equipment should be level and protected from the environment if construction is proceeding. The longer equipment is in storage, the more care is required for protection of the equipment. If a temporary cover is used to protect the equipment, this cover should not prevent air circulation. If the building is not heated or temperature controlled, heaters should be used to prevent moisture/condensation buildup. Outdoor switchgear that cannot be installed immediately must be provided with temporary power. This power will allow operation of the space heaters provided with the equipment.

Classroom

Discuss the restrictions on wiring door-mounted equipment in switchgear.

List the requirements for making switchgear terminal connections.

Provide an overview of the wiring requirements for potential and current transformers.

Discuss general switchgear handling, storage, and installation practices. Emphasize the importance of strictly adhering to the manufacturer's instructions.

Teaching Tip

Invite a switchgear manufacturer's representative to present an overview of a typical installation.

Safety

Emphasize the importance of following established safety procedures when working on any electrical equipment.

Classroom

Explain how a keyed interlock operates.

- *Bus connections*—The main bus that is usually removed during shipping should be reconnected. Ensure that the contact surfaces are clean and pressure is applied in the correct manner. The conductivity of the joints is dependent on the applied pressure at the contact points. The manufacturer's torque instructions should be referenced.
- *Cable connections*—When making cable connections, verify the phasing of each cable. This procedure is done in accordance with the connection diagrams and the cable tags. When forming and mounting cables, ensure that the cables are tightened per the manufacturer's instructions.
- *Grounding*—Any sections of ground bus that were previously disconnected for shipping should be reconnected when the units are installed. All secondary wiring should be connected to the switchgear ground bus. The ground bus should be connected to the system ground with as direct a connection as possible. If it is to be run in metal conduit, adequate bonding to the circuit is required. The ground connection is necessary for all switchgear and should be of sufficient ampacity to handle any abnormal condition.

5.0.0 ◆ SWITCHBOARD TESTING AND MAINTENANCE

This section covers general switchboard testing and maintenance.

> **WARNING!**
> When working on switchboards or any piece of electrical equipment, you must always be aware of and follow all applicable safety procedures. If you are not sure if the procedures you are following are safe, contact your supervisor. It is better to check the procedure than to go ahead and follow unsafe work practices.
>
> Always follow established procedures for grounding and locking out a circuit, and ensure that you have control over the switchboard voltage source(s). It is a good practice to bond neutrals to one another and to the grounding conductor when testing in order to ensure a stable reference point to ground.

INSIDE TRACK

Keyed Interlocks

Keyed interlocks, such as the one shown here, ensure that qualified personnel perform operations in the required sequence by preventing or allowing the operation of one part only when another part is locked in a predetermined position. These devices can be used for a variety of safety applications, such as preventing personnel from accessing a high-voltage compartment before opening the disconnect switch.

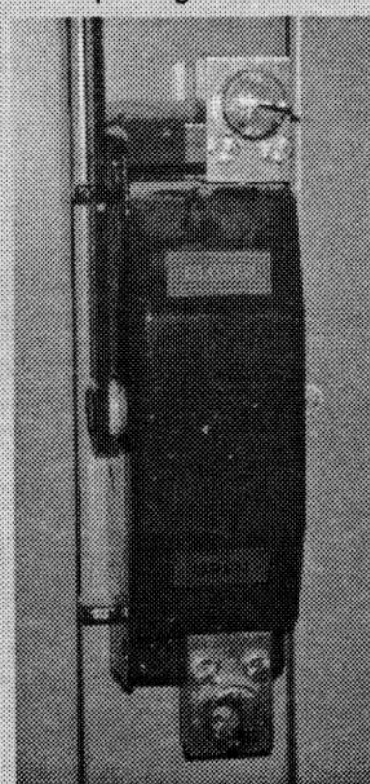

306P0602.EPS

Instructor's Notes:

> **Note**
> Test values will differ depending on whether you are performing an acceptance test or a maintenance test.

5.1.0 General Maintenance Guidelines

To perform a visual inspection:

Step 1 Check the exterior for the proper fit of doors and covers, paint, etc.

Step 2 Check the interior, particularly the current-carrying parts, including:
- Inspect the busbars for dirt, corrosion, and/or overheating.
- If necessary, perform an infrared or thermographic test. Note any discoloration that would represent a poor bus joint.
- Check the busbar supports for cracks.
- Check for correct electrical spacing.
- Verify the integrity of all bolted connections.

To clean the switchboard:

Step 1 Vacuum the interior (do not use compressed air).

Step 2 Wipe down the interior using a clean, lint-free cloth. Use nonconductive, nonresidue solution, such as contact cleaner or denatured alcohol.

To check equipment operation:

Step 1 Manually open and close circuit breakers and switches.

Step 2 Electrically operate all components, such as ground fault detectors, sure trip metering, current transformers, test blocks, ground lights, blown main fuse detectors, and phase failure detectors.

To perform a megger test:

Step 1 Isolate the bus by opening all circuit breakers and switches.

Step 2 Disconnect any devices, such as relays and transformers, that may be connected to the busbars.

Step 3 Make sure all personnel are clear of the switchboard.

Step 4 Use a 1,000V megger to check the phase-to-phase and phase-to-ground resistance. Megger readings should reflect the values listed in the equipment manufacturer's instructions. Typical values are shown in *Table 2*.

INSIDE TRACK

Meggers

To test for potential insulation breakdown, phase-to-phase shorts, or phase-to-ground shorts in switchgear, you need to apply a much higher potential than that supplied by the battery of an ohmmeter. A megohmmeter, or *megger*, is commonly used for these tests. The megger is a portable instrument consisting of a hand-driven DC generator, which supplies the level of voltage for making the measurement, and the instrument portion, which indicates the value of the resistance being measured.

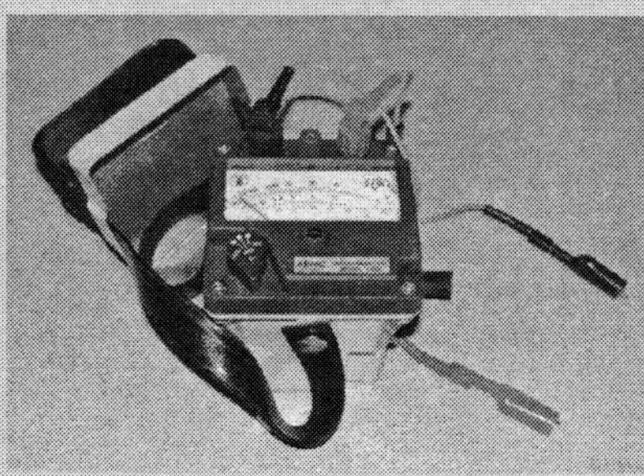

306P0603.EPS

Classroom

Provide an overview of the general maintenance procedures performed on switchgear:
- **Visual inspection**
- **Cleaning**
- **Operating test**
- **Resistance test**

Point out the typical resistance values shown in Table 2 in the Trainee Module.

Classroom

Explain the purpose of a thermographic survey. Refer the trainees to Figure 5 in the Trainee Module.

Safety

Point out that thermographic surveys are conducted while the equipment is energized and may only be performed by qualified personnel.

Classroom

Explain how to perform a visual and mechanical inspection on switchgear.

Safety

Emphasize that only authorized personnel may perform tests on switchgear, and the voltage must be verified as OFF prior to any testing procedures.

Table 2 Typical Insulation Resistance Tests on Electrical Apparatus and Systems at 68°F

Minimum Voltage Rating of Equipment	Minimum Test Voltage (VDC)	Recommended Minimum Insulation Resistance (in Megohms)
2–250V	500	50
251–600V	1,000	100
601–5,000V	2,500	1,000
5,001–15,000V	2,500	5,000
15,001–39,000V	5,000	20,000

5.2.0 Typical Guidelines

This section provides typical guidelines for performing various tests on distribution equipment.

5.2.1 Thermographic Survey

A thermographic survey (*Figure 5*) involves checking switches, busways, open buses, switchgear, cable and bus connections, circuit breakers, rotating equipment, and load tap changers.

> **WARNING!**
> This test is performed while the equipment is energized and the covers are removed. This test may only be performed by qualified personnel under the appropriate safe work plan or permit.

Infrared surveys should be performed during periods of maximum possible loading and not at less than 40% of the rated load of the electrical equipment being inspected. Negative test results include:

- Temperature gradients of 1°C to 3°C indicate a possible deficiency and require investigation.
- Temperature gradients of 4°C to 15°C indicate a deficiency. Repair as time permits.
- Temperature gradients of 16°C and above indicate a major deficiency. Secure power and repair as soon as possible.

5.2.2 Metal-Enclosed Switchgear and Switchboards

> **WARNING!**
> You must be certified and authorized to perform these tests; care should be taken to ensure that there is no voltage present.

To perform a visual and mechanical inspection:

Step 1 Inspect the physical, electrical, and mechanical condition of the equipment.

Step 2 Compare the equipment nameplate information with the latest single-line diagram, and report any discrepancies.

Step 3 Check for proper anchorage, required area clearances, physical damage, and proper alignment.

Step 4 Inspect all doors, panels, and sections for missing paint, dents, scratches, fit, and missing hardware.

Step 5 Inspect all bus connections for high resistance. Use a low-resistance ohmmeter or check tightness of bolted bus joints using a calibrated torque wrench.

Step 6 Test all electrical and mechanical interlock systems for proper operation and sequencing:
- A closure attempt must be made on all locked-open devices. An opening attempt must be made on all locked-closed devices.
- A key exchange must be made with all devices operated in normally off positions.

Step 7 Clean the entire switchgear using the manufacturer's approved methods and materials.

Step 8 Inspect insulators for evidence of physical damage or contaminated surfaces.

Step 9 Inspect the lubrication:
- Verify appropriate contact lubricant on moving current-carrying parts.
- Verify appropriate lubrication of moving and sliding surfaces.
- Exercise all active components.
- Inspect all indicating devices for proper operation.

Instructor's Notes:

306F05.EPS

Figure 5 ◆ Infrared imager used in thermographic surveys.

WARNING!
Electrical testing may produce hazardous voltages and may only be performed by qualified personnel under the appropriate safe work plan or permit. Prepare the area to avoid any accidental contact with the system under test, and wear appropriate personal protective equipment.

To perform electrical testing:

Step 1 Perform ratio and polarity tests on all current and voltage transformers.

Step 2 Perform ground resistance tests.

Step 3 Perform insulation resistance tests on each bus section (phase-to-phase and phase-to-ground) for one minute. Refer to the specific manufacturer's guidelines, an example of which is shown in *Table 2*.

Step 4 Perform an overpotential test on each bus section (phase-to-ground) for one minute. Refer to specific manufacturer's guidelines, an example of which is shown in *Table 3*.

Table 3 Overpotential DC Test Voltages for Electrical Apparatus Other Than Inductive Equipment

	DC Test Voltage Max.	
Nominal Voltage Class	**New**	**Used**
250V	2,500VDC	1,500VDC
600V	3,500VDC	2,000VDC
5,000V	18,000VDC	11,000VDC
15,000V	50,000VDC	30,000VDC

Note
The values shown in *Tables 2* and *3* are typical acceptance values. Maintenance values will vary by manufacturer.

Step 5 Perform an insulation resistance test on the control wiring. Do not perform this test on wiring connected to solid-state components.

Step 6 Perform a phasing check on double-ended switchgear to ensure proper bus phasing from each source.

Any values of insulation resistance less than those listed in the manufacturer's literature should be investigated. Overpotential tests should not proceed until insulation resistance levels are raised above minimum values.

Overpotential test voltages must be applied in accordance with the manufacturer's literature. Test results are evaluated on a go/no-go basis by slowly raising the test voltage to the required value. The final test voltage is applied for one minute.

5.2.3 Low-Voltage Cables (600V Maximum)

To perform a visual and mechanical inspection:

Step 1 Inspect cables for physical damage and proper connection in accordance with the single-line diagram.

INSIDE TRACK

Manufacturer's Data

Never assume anything when it comes to equipment operation, testing, or maintenance. Always refer to the manufacturer's installation, operating, and maintenance instructions for the equipment in use. These materials provide important data that explain the warranty requirements, appropriate test procedures, and specific maintenance and test points.

Classroom

Explain how to perform electrical testing on switchgear.

Safety

Emphasize that only authorized personnel may perform tests on switchgear.

Classroom

Point out the overpotential test voltages shown in Table 3 in the Trainee Module.

Explain how to inspect and test low-voltage cables (up to 600V).

Classroom

Explain how to inspect and test medium-voltage cables (up to 15kV).

Describe how to inspect and test metal-enclosed busways.

Explain how to inspect and test meters and instruments.

Step 2 Verify the integrity of all bolted connections.

Step 3 Check color-coded cable against the applicable engineer's specifications and NEC standards.

To perform electrical testing:

Step 1 Perform an insulation resistance test on each conductor with respect to ground and adjacent conductors. The applied potential should be 1,000VDC for one minute.

Step 2 Perform a continuity test to ensure proper cable connection. The minimum insulation resistance values must not be less than two megohms.

5.2.4 Medium-Voltage Cables (15kV Maximum)

To perform a visual and mechanical inspection:

Step 1 Inspect exposed sections for physical damage.

Step 2 Inspect for shield grounding, cable support, and termination.

Step 3 Inspect for proper fireproofing in common cable areas.

Step 4 If cables are terminated through window-type CTs, make an inspection to verify that neutrals and grounds are properly terminated for normal operation of the protective devices.

Step 5 Visually inspect the jacket and insulation condition.

Step 6 Inspect for proper phase identification and arrangement.

5.2.5 Metal-Enclosed Busways

To perform a visual and mechanical inspection:

Step 1 Inspect the bus for physical damage.

Step 2 Inspect for proper bracing, suspension, alignment, and enclosure.

Step 3 Check the tightness of bolted joints using a calibrated torque wrench.

Step 4 Check for proper physical orientation per the manufacturer's labels to ensure proper cooling. Perform continuity tests on each conductor to verify that proper phase relationships exist.

Step 5 Check outdoor busways for removal of weep-hole plugs if applicable and also for the proper installation of a joint shield.

To perform electrical testing:

Step 1 Perform an insulation resistance test. Measure the insulation resistance on each bus run (phase-to-phase and phase-to-ground) for one minute.

Step 2 Perform AC or DC overpotential tests on each bus run, both phase-to-phase and phase-to-ground.

Step 3 Perform a contact resistance test on each connection point of the uninsulated bus. On an insulated bus, measure the resistance of the bus section and compare values with adjacent phases.

Step 4 Insulation resistance test voltages and resistance values must be in accordance with the manufacturer's specifications.

Step 5 Apply overpotential test voltages in accordance with the manufacturer's specifications.

5.2.6 Metering and Instrumentation

To perform a visual and mechanical inspection:

Step 1 Examine all devices for broken parts, indication of shipping damage, and wire connection tightness.

Step 2 Verify that meter connections are in accordance with appropriate diagrams.

To perform electrical testing:

Step 1 Check the calibration of meters at all cardinal points.

Step 2 Calibrate watt-hour meters to one-half of one percent (0.5%).

Step 3 Verify all instrument multipliers.

Instructor's Notes:

6.0.0 ◆ NEC REQUIREMENTS

This section is designed to provide a brief description of the NEC articles that are applicable to switchboard construction, installation, and accessories.

6.1.0 Requirements for Electrical Installations

NEC requirements for electrical installations include the following:

- *Interrupting rating*—The interrupting rating is the maximum current a device is intended to interrupt under standard test conditions. *NEC Section 110.9* defines the equipment interrupting rating as sufficient to interrupt the current that is available at the line-side terminals of the equipment.
- *Deteriorating agents*—*NEC Section 110.11* provides for the protection of equipment and conductors from environments that could cause deterioration, such as gases, vapors, liquids, or moisture, unless specifically designed for such environments.
- *Mechanical execution of work*—*NEC Section 110.12(A)* states that electrical equipment is to be installed in a neat and professional manner. Any openings provided by the equipment manufacturer or at the time of installation that are not being used must be sealed equivalent to the structure wall. This section also forbids the use of electrical equipment with damaged parts that may affect the safe operation or mechanical strength of the equipment.
- *Mounting and cooling*—*NEC Section 110.13* states that electrical equipment shall be securely fastened to its mounting surface by mechanical fasteners, excluding wooden plugs driven into concrete, masonry, plaster, or similar materials. Equipment shall be located so as not to restrict air flow required for convection or forced-air cooling.
- *Electrical connections*—Due to the resistive oxidation created when dissimilar metals are connected, splicing devices and pressure connectors must be identified for the conductor material with which they are to be used (*NEC Section 110.14*). Dissimilar metal conductors may not be mixed in terminations or splices. Fluxes, solders, and antioxidation compounds must be suitable for use and must not adversely affect conductors, installation, or equipment. Terminals for use with more than one conductor or aluminum must be identified as such.
- *Markings*—The manufacturer's trademark or logo and system ratings, including voltage, current, and wattage, must be permanently attached to the equipment (*NEC Sections 110.16 and 110.21*).
- *Disconnect identification*—Each disconnecting means, such as circuit breakers, fused switches, **feeders,** or unfused disconnects, must be clearly marked as to its purpose at its point of origin unless located in such a manner that its purpose is evident (*NEC Section 110.22*).

CASE HISTORY

How* Not *to Build a Better Mousetrap

The expensive switchgear shown here had an unused opening that was left uncovered after the installation. A rodent entered the compartment, shorting out one of the busbars and causing extensive damage.

The Bottom Line: ***NEC Section 110.12(A)*** requires that any unused openings must be sealed equivalent to the structure wall.

306P0604.EPS

Classroom

Review the NEC requirements relating to electrical installations. Refer the trainees to *NEC Article 110.*

Teaching Tip

Discuss the "Case History." Emphasize the consequences of failing to adhere to code requirements.

Classroom

Review the NEC requirements relating to conductors. Refer the trainees to the applicable sections in the NEC.

- *Working space*—Suitable access and working space shall be maintained around electrical equipment to permit safe operation and maintenance (*NEC Section 110.26*). Minimum clearances in front of all electrical enclosures must conform to those specified in *NEC Table 110.26(A);* in all cases, space must be adequate to allow doors or hinged parts to open to a 90-degree angle. In special case installations, the clearances in *NEC Table 110.26(A)* must be adhered to. Storage of any kind is not permitted within the clearance area. In accordance with *NEC Section 110.26(C)(1),* at least one entrance of ample size must be provided to access the work area. In cases of services over 1,200A and over 6' wide, two entrances are required. The work space must be adequately illuminated.
- *Flash protection*—*NEC Section 110.16* states that switchboards, panelboards, industrial control panels, and motor control centers in other-than-dwelling occupancies that are likely to require examination, adjustment, servicing, or maintenance while energized shall be field marked to warn qualified persons of potential electric arc flash hazards. The marking shall be located so as to be clearly visible to qualified persons before examination, adjustment, servicing, or maintenance of the equipment.

6.2.0 Requirements for Conductors

NEC Section 200.6 covers requirements associated with identifying grounded conductors. It includes the following:

- *Neutrals*—Grounded conductors (neutrals) size No. 6 AWG and smaller are color coded with a solid white or gray marking for the entire length of the conductor. Conductors size No. 6 and larger may be color coded with a solid white marking tape at termination points at the time of installation. Where different electrical systems are run together, each system's grounded conductor must be distinctively identified [*NEC Section 200.6(D)*].
- *Protection*—**Branch circuit** conductors must be protected by overcurrent devices, as specified in *NEC Section 240.4.*
- *Loading*—*NEC Section 210.19(A)* states that protective device calculations for continuous duty circuits are calculated at 125% of the continuous load. This equates to an 80% loading factor on the branch circuit.
- *Tap rules*—Tap conductors are conductors that are tapped onto the line-side bus of the switchboard to feed control circuits, control power transformers, and metering devices. Overcurrent devices (typically fuses) are connected where the conductor to be protected receives its supply. Per *NEC Section 240.21,* tap conductors do not require protection if the following conditions are met:
  - –The length of the conductor is not over 10'.
  - –The ampacity of the conductor is not less than the combined loads supplied by the conductor.
  - –The conductors do not extend beyond the switchboard.
  - –The conductors are enclosed in a raceway except at the point of connection to the bus.
  - –For field installations where the tap conductors leave the enclosure or vault in which the tap is made, the rating of the overcurrent device on the line side of the tap conductors does not exceed 10 times the tap conductor's ampacity.
- *Markings*—All conductors and cables shall be permanently marked to indicate the manufacturer, voltage, AWG size, and insulation type (*NEC Sections 110.15, 110.21, 310.11, and 310.12*).
  - –Grounded conductors (neutrals) size No. 6 and smaller shall have a continuous marking of white or gray for the entire length of the conductor. Larger conductors may be marked at each termination with white marking tape.
  - –Grounding conductors (equipment grounding wires) shall be permitted to be bare wire. In cases of insulated grounding conductors, the conductor will have a continuous marking of green for the entire length of the conductor. Larger conductors may be marked at each end and every point where the conductor is accessible.
  - –Ungrounded conductors (phase wires) must be distinguishable from grounded or grounding conductors with colors other than white, gray, or green. Typical ungrounded conductor identification colors are black, red, blue, brown, orange, and yellow. Conductors size No. 6 or smaller must have a continuous marking. Larger cables may be marked at each termination.
  - –In switchboards fed by a four-wire, delta system in which one phase is grounded at its midpoint, the phase having the higher voltage must be marked with an orange color according to *NEC Section 110.15.*
- *Ampacities*—The ampacities of cable are determined by the cable size and insulation type (*NEC Section 310.15*).

Instructor's Notes:

6.3.0 Grounding

NEC grounding requirements include the following:

- *Grounding*—*NEC Section 250.20(B)* states that AC systems between 50V and 1,000V must be grounded when any of the following conditions are met:
  - –Where the system can be grounded in such a way that the maximum phase-to-ground voltage does not exceed 150V
  - –When the system is three-phase, four-wire, wye-connected and the neutral is used as a circuit conductor
  - –When the system is three-phase, four-wire, delta-connected and the midpoint of a phase is used as a conductor (developed neutral)
- *Grounding electrode conductor*—*NEC Sections 250.24 and 250.66* cover the requirements of grounding electrode conductors, including proper sizing of the equipment grounding conductors to the service equipment enclosures. *NEC Section 250.28* states that for grounded systems (delta or wye), an unspliced main bonding jumper in the service equipment must be used to connect the grounding conductor and the service disconnect enclosure to the grounded conductor of the system within the enclosure.

Note
Some systems are ungrounded and will not blow fuses.

- *Electrodes*—*NEC Sections 250.52, 250.53, and 250.56* require that when rod or pipe electrodes are used, they must extend a minimum of 8' into the soil. The electrode must be no less than ¾" in diameter for pipe and ⅝" in diameter for rods. It must be galvanized metal or copper-coated to resist corrosion. Underground structures, such as water piping systems, may also be used as an electrode. Underground gas piping systems must not be used. Aluminum electrodes are not permitted. Rod, pipe, or plate electrodes must maintain a resistance of no more than 25Ω to ground. If the resistance is above 25Ω, an additional electrode is required to maintain the minimum resistance.
- *Grounding of ground wire conduits*—*NEC Section 250.64(E)* states that a grounding conductor or its enclosure must be securely mounted to the surface along which it runs. In cases where the conductor is enclosed, the enclosure must be electrically continuous and firmly grounded.
- *Ground connection surfaces*—Nonconducting coatings, such as paint, enamel, or insulating materials, must be thoroughly removed at any point where a grounding connection is made (*NEC Section 250.12*).

6.4.0 Switchboards and Panelboards

NEC requirements for switchboards and panelboards include the following:

- *Dedicated space*—*NEC Sections 110.26(E) and (F) and 408.8* state that panelboards and switchboards may only be installed in spaces specifically designed for such purposes. No other piping, ducts, or devices may be installed or pass through such areas, except equipment that is necessary to the operation of the electrical equipment.
- *Inductive heating*—*NEC Section 408.3(B)* states that busbars and conductors must be arranged so as to avoid overheating due to inductive forces.
- *Phasing*—*NEC Section 408.3(E)* states that phasing in switchboards must be arranged A, B, C from front to back, top to bottom, and left to right, respectively, when facing the front of the switchboard. In systems containing a high leg, the B phase must be the phase conductor having a higher voltage to ground.
- *Wire bending space*—*NEC Section 408.3(F)* states that the wire bending space must be in accordance with *NEC Tables 312.6(A) and (B).*
- *Minimum spacing*—*NEC Section 408.36* states that the spacing between bare metal parts and conductors must be as specified in *NEC Table 408.36.* Conductors entering the bottom of switchboards must have the clearances specified in *NEC Section 408.10.*
- *Conductor insulation*—Insulated conductors within switchboards must be listed as flame-retardant and rated at not less than the voltage applied to them or any adjacent conductors they may come in contact with (*NEC Section 408.9*).

7.0.0 ◆ GROUND FAULTS

Ground faults exist when an unintended current path is established between an ungrounded conductor and ground. These faults can occur due to deteriorated insulation, moisture, dirt, rodents, foreign objects, such as tools, and careless installation.

Ground faults are usually high arcing and low level in nature, which conventional breakers will not detect. Ground fault protection is used to protect equipment and cables against these low-level faults.

Classroom

Review the NEC requirements for grounding. Refer the trainees to *NEC Article 250.*

Point out the NEC requirements relating to switchboards and panelboards. Refer the trainees to the applicable sections in the NEC.

Discuss the purpose of ground fault protection.

Teaching Tip

Discuss the "What's Wrong with This Picture?"

Classroom

List the three basic methods of sensing ground faults.

Audiovisual

Show Transparency 3 (Figure 6).

What's Wrong with This Picture?

306P0605.EPS

Ground fault protection is required per the NEC on solidly-grounded wye services of more than 150V to ground but not exceeding a phase-to-phase voltage of 600V with each service disconnecting means of 1,000A or more.

7.1.0 Ground Fault Systems

The three basic methods of sensing ground faults include:

- Ground-return method
- Residual method
- Zero sequence method

The ground-return method incorporates a sensing coil around the grounding electrode conductor. The residual method uses three individual sensing coils to monitor the current on each phase conductor. The zero sequence method requires a single, specially designed sensor to monitor all the phases and the neutral conductor of a system at the same time, as shown in *Figure 6*.

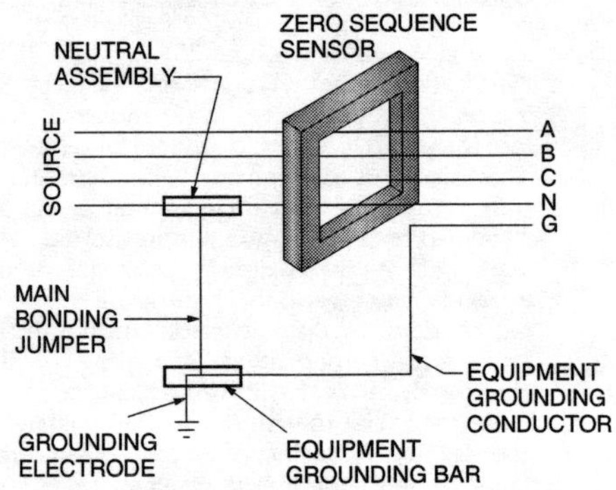

306F06.EPS

Figure 6 ◆ Zero sequencing diagram.

6.16 ELECTRICAL LEVEL THREE—TRAINEE MODULE 26306

Instructor's Notes:

7.2.0 Sensing Operation

When circuit conditions are normal, the currents from all the phase and neutral (if used) conductors add up to zero, and the sensor current transformer produces no signal. When any ground fault occurs, the currents add up to equal the ground fault current, and the sensor produces a signal proportional to the ground fault. This signal provides power to the ground fault relay, which trips the circuit breaker.

A ground fault lasting for less than the time-delay period will not pick up the ground trip coil, thus eliminating nuisance tripping of self-clearing faults.

The ground fault relay is a high-reliability device due to its solid-state construction. The use of redundant, self-protecting, and high-reliability components further improves the performance. Self-protection against failure is provided through an internal fuse that will blow and result in a tripping function if the solid-state circuitry fails during a ground fault situation.

7.3.0 Zero Sequencing Sensor Mounting

The sensor current transformer (sensor) should be mounted so that all phase and neutral (if used) conductors pass through the core window once. The ground conductor (if used) must not pass through the core window. The neutral conductors must be free of all grounds after passing through the core window (see *Figure 6*).

When so specified by the system design engineer, the sensor may be mounted so that only the conductor connecting the neutral to ground at the service equipment passes through the core window. In such cases, the sensor must provide power to the particular ground fault relay that is associated with the main circuit breaker.

Maintain at least two inches of clearance from the iron core of the sensor to the nearest busbar or cable to avoid false tripping. Cable conductors should be bundled securely and braced to hold them at the center of the core window. The sensor should be mounted within an enclosure and protected from mechanical damage.

7.4.0 Relay Mounting

The ground fault relay should be mounted in a vertical position within an enclosure with the terminal block at the lower end. The location of the relay should be such that the trip setting knob is accessible without exposing the operator to contact with live parts or arcing from disconnect operation.

7.5.0 Connections

Connections for standard applications should be made in accordance with the wiring diagrams in the manufacturer's literature. An example of one circuit is shown in *Figure 7*. Wires from the sensor to the ground fault relay should be no longer than 25' and no smaller than No. 14 AWG wire. Wires from the ground fault relay to the trip coil should be no longer than 50' and no smaller than No. 14 AWG wire. All wires should be protected from arcing fault and physical damage by barriers, conduit, armor, or location in an equipment enclosure. Do not disconnect or short circuit wires to the circuit breaker trip coil at any time when the power is turned on.

7.6.0 Relay Settings

The ground fault relay has an adjustable trip setting. The amount of time delay is factory set and is available in nominal time delays of 0.1, 0.2, 0.3, and 0.5 second. When ground fault protection is used in downstream steps, the feeder should have the next lower time-delay curve than the main, the branch the next lower curve than the feeder, and so on.

High trip settings on main and feeder circuits are desirable to avoid nuisance tripping. High settings usually do not reduce the effectiveness of the protection if the ground path impedance is reasonably low. Ground faults usually quickly reach a value of 40% or more of the available short circuit current in the ground path circuit.

7.6.1 Coordination with Downstream Circuit Breakers

It is recommended that the magnetic trips of any downstream circuit breakers that are not equipped with ground fault protection be set as low as possible. Likewise, the ground fault relay trip settings for main or feeder circuits should be

> INSIDE TRACK
>
> ***Ground Fault Trip Settings***
>
> Even minor ground faults will usually arc, causing immediate damage. Major damage can occur in a matter of a second. Ideally, the ground fault relay should respond in less than 30 cycles (½ second in a 60Hz system).

Classroom

Describe the operation of a ground sensing system. Cover:
- **Sensor mounting**
- **Relay mounting**
- **Connections**
- **Relay settings**

Audiovisual

Show Transparency 4 (Figure 7).

Classroom

Describe how to inspect and test a ground fault system.

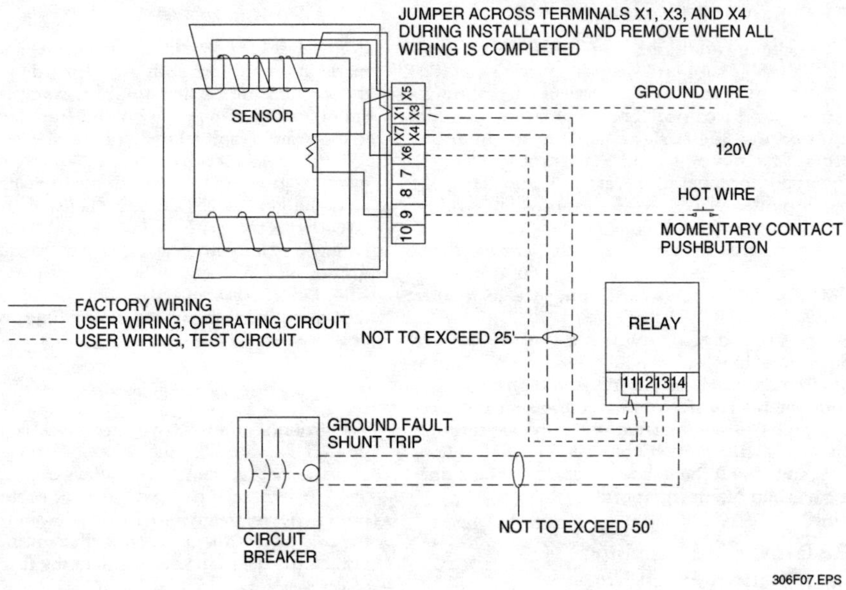

Figure 7 ◆ Typical wiring diagram.

higher than the magnetic trip settings for unprotected downstream breakers where possible. This will minimize nuisance tripping of the main or feeder breaker for ground faults occurring on downstream circuits.

7.6.2 Instantaneous Trip Feature

Standard ground-powered ground fault relays have a built-in instantaneous trip feature. This instantaneous trip has a fixed time delay of approximately 1½ cycles, and the fixed trip setting is higher than found on most feeder or branch breakers to avoid nuisance tripping. Its purpose is to interrupt very high-current ground faults on main disconnects as quickly as possible and to protect the ground fault relay components.

7.7.0 Generic National Electrical Testing Association Ground Fault System Test

This section provides an overview of a generic visual inspection and electrical test for ground faults. Always follow the procedures specified by the equipment manufacturer for the system being tested.

7.7.1 Procedures

Perform a visual inspection:

Step 1 Inspect the components for physical damage.

Step 2 Determine if a ground sensor was located properly around the appropriate conductor(s):
- Zero sequence sensing requires all phases and the neutral to be encircled by the sensor(s).
- Ground return sensing requires the sensor to encircle the main bonding jumper.

Step 3 Inspect the main bonding jumper to ensure:
- Proper size
- Termination on the line side of the neutral disconnect link
- Termination on the line side of the sensor on zero sequence systems

Instructor's Notes:

Step 4 Inspect the grounding electrode conductor to ensure:
- Proper size
- Correct switchboard termination

Step 5 Inspect the ground fault control power transformer for proper installation and size. When the control transformer is supplied from the line side of the ground fault protection circuit interrupting device, overcurrent protection and a circuit disconnecting means must be provided.

Step 6 Visually inspect the switchboard neutral bus downstream of the neutral disconnect line to verify the absence of ground connections.

Perform electrical tests as required by *NEC Section 230.95(C):*

Step 1 Check for proper ground fault system performance, including correct response of the circuit interrupting device confirmed by primary/secondary ground sensor current injection:
- Measure the relay pickup current.
- Ensure that the relay time delay is measured at two values above the pickup current.

Step 2 Test system operation at 57% of the rated voltage.

Step 3 Functionally check the operation of the ground fault monitor panel for:
- Trip test
- No-trip test
- Nonautomatic reset

Step 4 Verify proper sensor polarity on the phase and neutral sensors for residual systems.

Step 5 Measure the system neutral insulation resistance downstream of the neutral disconnect link to verify the absence of grounds.

Step 6 Test systems (zone interlock/time coordinates) by simultaneous ground sensor current injection, and monitor for the proper response.

Test result evaluation:

- The system neutral insulation resistance should be above 100Ω and preferably one megohm or greater.
- The maximum pickup setting of the ground fault protection must be 1,200A and the maximum time delay must be one second for ground fault currents equal to or greater than 3,000A, according to ***NEC Section 230.95(A).***
- The relay pickup current should be within 10% of the manufacturer's calibration marks or fixed setting.
- The relay timing should be in accordance with the manufacturer's published time-current characteristics.

8.0.0 ◆ HVL SWITCH

Figure 8 shows the general appearance of an HVL (high-voltage limiting) switch. The HVL switch is a switching device for primary circuits up to the full interrupting current of the switch. The switches are single-throw devices designed for use on 2.4kV to 34.5kV systems.

HVL switches may provide both switching and overcurrent protection. HVL switches are commonly used as a service disconnect in unit substations and for sectionalizing medium voltage feeder systems. The HVL switch is designed to conform to ANSI standards for metal enclosed switchgear.

8.1.0 Ratings

Switch ratings are as follows:

- *Switch kV*—The design voltage for the switch. Of course, nominal system voltage is the normal application method; thus, a 5kV switch may be used for nominal system voltages of 2.4kV or 4.16kV, etc.

306F08.EPS

Figure 8 ◆ High-voltage limiting switch.

Classroom

Discuss HVL switches and their applications. Refer the trainees to Figure 8 in the Trainee Module.

List the ratings associated with HVL switches.

Classroom

List the six types of HVL switches.

Explain how an HVL switch operates.

- ***Basic impulse insulation level,*** *or BIL (kV)*—The maximum voltage pulse that the equipment will withstand.
- *Frequency (Hertz)*—All HVL switches may be used in either 50Hz or 60Hz power systems.
- *Withstand (kV)*—The maximum 60Hz voltage that can be applied to the switch for one minute without causing insulation failure.
- *Capacitor switching (kVAR)*—The maximum capacitance expressed in kVAR that can be switched with the HVL.
- *Fault close*—The maximum, fully offset fault current that the switch can be closed into without sustaining damage. The term fully offset means that the fault current will have a delaying DC component in addition to the AC component.
- *Short time current*—The amount of current that the switch will carry for 10 seconds without sustaining any damage.
- *Continuous current (amps)*—The amount of current that the switch will carry continuously.
- *Interrupting current (amps)*—The maximum amount of current that the switch will safely interrupt.

8.2.0 Variations

There are six main types of switches:

- *Upright*—The upright switch design is the most common type. The upright construction of the service entry, jaws, and arc chutes are located near the top of the cubicle. The hinge point is below the jaws and arc chutes.
- *Inverted*—The inverted switch design has the terminals, jaws, and arc chutes located near the bottom of the cubicle. The hinge point is above the jaws and arc chutes. This type of switch is used primarily as a main switch to a lineup of other switches. Its handle operation is identical to that of an upright switch; to close the switch, the handle is moved up, and to open it, the handle is moved down.
- *Fused/unfused*—HVL switches are available in both fused and unfused models. If equipped with fuses, the entire HVL switch has the fault interrupting capacity of the fuse and therefore provides fault protection. Either current-limiting or boric acid fuses may be used in the HVL switch.
- *Duplex*—A duplex switch is actually two switches, each in its own bay. The bays are mechanically connected and the switches are electrically connected on the load side. This switch may be used to supply power to a single load from two different sources.
- *Selector*—A selector allows an HVL switch to have double-throw characteristics. The selector switch is a single switch with a load connected to the moving or switch mechanism. Throwing the switch to one side connects the load to one source, while throwing it the other way connects it to a second source. The selector switch will be interlocked with another switch to prevent the selector switch from interrupting current flow. The selector serves a purpose similar to the duplex switch. However, the selector switch is not an interrupter; it is a disconnect.
- *Motor-operated*—This type of switch is most commonly used as the major component in an automatic transfer scheme. It can also be used when open and close functions are to be initiated from remote locations.

8.3.0 Opening Operation

In the closed position, the main switch blade is engaged on the stationary interrupting contacts. The circuit current flows through the main blades.

As the switch operating handle is moved toward the open position, the stored energy springs are charged. After the springs become fully charged, they toggle over the dead center position, discharging force to the switch operating mechanism.

The action of the switch operating mechanism forces the movable main blade off the stationary main contacts while the interrupting contacts are held closed, momentarily carrying all the current without arcing. Once the main contacts have separated well beyond the striking distance, the interrupting blade contact that was held captive has charged the interrupter blade hub spring, and the interrupter blade is suddenly forced free and flips open.

The resulting arc drawn between the stationary and movable interrupting contacts is elongated and cooled as the plastic arc chute absorbs heat and generates an arc-extinguishing gas to break up and blow out the arc. The combination of arc stretching, arc cooling, and extinguishing gas causes a quick interruption with only minor erosion of the contacts and arc chutes. The movable main and interrupting contacts continue to the fully open position and are maintained there by spring pressure.

8.4.0 Closing

When the switch operating handle is moved towards the closed position, the stored energy

Instructor's Notes:

springs are being charged and the main blades begin to move. As the main and interrupter blades approach the arc chute, the stored energy springs become fully charged and toggle over the dead center position.

When the main and movable blades approach the main stationary contacts, a high voltage arc leaps across the diminishing air gap in an attempt to complete the circuit. The arc occurs between the tip of the stationary main contacts and a remote corner of the movable main blades. This arc is short and brief because the fast-closing blades minimize the arcing time.

The spring pressure and momentum of the fast-moving main blades completely close the contacts. The force is great enough to cause the contacts to close even against repelling short circuit magnetic forces if a fault exists. At the same time, the interrupter blade tip is driven through the twin stationary interrupting contacts, definitely latching and preparing them for an interrupting operation when the switch is opened.

WARNING!
Maintenance and testing may only be performed by qualified personnel under the appropriate safe work plan or permit.

8.5.0 Maintenance

Maintenance tasks for an HVL switch include the following:

Step 1 The HVL switch should be operated several times. Observe the mechanism and check for binding.

Step 2 Inspect the interrupting and main blades every 100 operations for excessive wear or damage; replace as necessary. Also, inspect the arc chutes for damage.

Step 3 Clean the switch and its compartment thoroughly. Use a clean cloth and avoid solvents.

Step 4 Lubricate the switch. The pivot points on the switch should be greased. The switch contacts should also be lubricated with a light film of grease after being cleaned.

Step 5 Final maintenance checks include phase-to-ground and phase-to-phase megger testing. If the results are satisfactory, then a DC high-potential test is performed.

8.6.0 Sluggish Operation

A switch that is operating sluggishly hesitates on the opening cycle. This contrasts with the normal snapping action. Observing the interrupter blade during the opening operation is the proper way to determine sluggish operation. Sluggishness must be repaired to prevent the switch from locking up completely. Perform the following procedure:

Step 1 Tease the switch closed and then open again while watching the interrupter blades closely. Sluggishness on close will be noted by the main blade's being engaged behind the contacts of the arc chute. On opening, the interrupter blades may hesitate momentarily.

Step 2 Disconnect the links from the operating shaft. Never operate the switch with the links off as this may break the handle crank casting. This is because the main spring energy is absorbed by the handle crank rather than the main blades.

Step 3 Rotate the handle approximately 45 degrees, and hold it in this position while trying to operate the switch by hand. Excessive binding will prevent rotation of the shaft.

Step 4 Check the contact adjustment at the jaw and hinge.

Step 5 Check for binding between the interrupter blade and the arc chute.

Step 6 Remove the front panel over the operating mechanism and disconnect the spring yoke from the cam. Check for binding between the spring pivot and the sides of the operator. Check the spring for breaks.

9.0.0 ◆ BOLTED PRESSURE SWITCHES

Bolted pressure switches (*Figure 9*) are used frequently on service-entrance feeders in switchgear such as that shown in *Figure 10*. They are often used in lieu of circuit breakers because they are inexpensive. Bolted pressure switches can be manually operated or motor operated. However, unlike a circuit breaker, they can only be automatically tripped by three events: a ground fault, a phase failure, or a blown main fuse detector.

9.1.0 Ground Fault

Under normal conditions, the currents in all conductors surrounded by the ground fault CT equal

Classroom

Discuss the maintenance tasks associated with HVL switches.

Safety

Caution the trainees that maintenance and testing may only be performed by qualified personnel.

Classroom

Discuss bolted pressure switches and their applications. Refer the trainees to Figures 9 and 10 in the Trainee Module.

Classroom

Explain the purpose of a phase failure relay.

Discuss the function of a blown main fuse detector.

Point out that switch failures are often caused by lack of maintenance.

Discuss the maintenance procedures performed on switches.

Safety

Emphasize the importance of following established safety procedures when working on any electrical equipment.

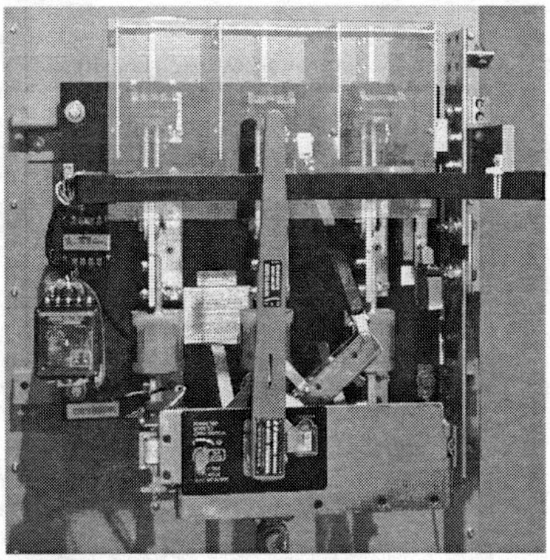

306F09.EPS

Figure 9 ◆ Manually operated bolted pressure switch.

306F10.EPS

Figure 10 ◆ Switchgear.

zero. When a ground fault occurs, this sensed current increases, eventually reaching the ground fault relay pickup point and causing the bolted pressure switch to trip.

The ground fault system may also be tested. By pressing the test button, a green test light will illuminate, indicating correct circuit operation. To actually test the switch, press the TEST and RESET buttons simultaneously. This sends an actual trip signal through the current sensor, thus tripping the switch. Whenever a bolted pressure switch is tripped, a red light or a red flag will trip. Additionally, the ground fault relay must be reset before the switch can be reclosed.

9.2.0 Phase Failure

If a phase failure relay is installed, it will cause a trip of the bolted pressure switch if a phase is lost. This could occur when a tree limb knocks a line down. Under this condition, the phase failure relay will sense the lost phase and trip the bolted pressure switch, preventing a single-phasing condition.

9.3.0 Blown Main Fuse Detector

If one of the in-line main fuses were to blow, the blown main fuse detector would detect it and cause a trip of the bolted pressure switch. The trip signal generated comes from a capacitor trip unit. This ensures that power is always available to trip the switch.

9.4.0 Maintenance

These switches have a high failure rate due to lack of maintenance. All manufacturers of bolted pressure switches recommend annual maintenance. Lack of annual maintenance will eventually result in a switch that is stuck shut. Since these switches are often used as service-entrance equipment, a stuck switch can pose immediate personnel safety hazards, as well as equipment failures.

> **WARNING!**
> When performing any maintenance, always follow the safety procedures of your company.

Due to the high interrupting capacity of the switch when operated under load, the grease that is used on the movable blades deteriorates over time and eventually turns into an adhesive. Even when the switch is not operated on a recurring basis, the grease still deteriorates due to the high temperatures associated with the current drawn by the phase. The deterioration of this grease has been shown to cause the switch to stick shut. The grease must be cleaned off yearly with denatured alcohol and replaced.

> **Note**
> Regular electrical grease cannot be used; use only the grease specifically recommended by the switch manufacturer.

Instructor's Notes:

Additionally, infrared scanning of in-service bolted pressure switches has revealed a marked heating concern in switches. A digital low-resistance ohmmeter (DLRO) is used to ensure that all three phases carry similar current loads. DLRO readings should never be greater than 75 microhms, and there should not be more than a 5% difference between the phases.

Typical annual maintenance includes:

Step 1 De-energize the switch, lock and tag it, and perform a preliminary operational check.

Step 2 Record pre-maintenance DLRO readings.

Step 3 With the switch open, disassemble the crossbar to free all three phases.

Step 4 Clean off all old grease with denatured alcohol or a similar solvent.

Step 5 Inspect the arc tips and arc chutes for damage.

Step 6 Adjust all pivotal connections on each blade to within the manufacturer's recommended tolerances.

Step 7 Apply an appropriate grease to the movable blades and the area where the blades come in contact with the stationary assembly.

Step 8 Check the pullout torque on each individual blade prior to crossbar reassembly. It should be in accordance with the manufacturer's prescribed limits. Too much torque will result in a switch that will be unable to open under load.

Step 9 Record the DLRO readings.

Step 10 Reassemble the crossbar assembly.

Step 11 Close and open the switch manually several times. Ensure that no phases hang up on the arc chute assembly.

Step 12 Megger the switch.

Step 13 Energize and test all accessories, such as the ground fault detector, phase failure detector, and blown main fuse detector.

Remember, if the switch is physically stuck shut, de-energize the switch from the incoming power supply, and take extra precautions when trying to unstick the switch. It may be necessary to pry the blades open, but beware of the excessive outward force that will result from a charged opening spring. To alleviate this, discharge the opening spring before commencing any work on the switch.

10.0.0 ◆ TRANSFORMERS

Transformers are used to step voltage up and down in the power transmission and distribution system.

The reason for such high transmission voltages is twofold. First, as a transformer increases transmission voltage, the required current decreases in the same proportion; therefore, larger amounts of power can be transmitted and line losses reduced. Second, to send large amounts of power over long distances at a high current and a low voltage requires a very large diameter wire. The reduction in current reduces the conductor size, which results in a cost reduction.

A transformer is an electrical device that uses the process of electromagnetic induction to change the levels of voltage and current in an AC circuit without changing the frequency and with very little loss of power.

10.1.0 Transformer Theory

As current flows through a conductor, a magnetic field is produced around the conductor. This magnetic field begins to form at the instant current begins to flow and expands outward from the conductor as the current increases in magnitude.

When the current reaches its peak value, the magnetic field is also at its peak value. When the current decreases, the magnetic field also decreases.

Alternating current (AC) changes direction twice per cycle. These changes in direction or alternation create an expanding and collapsing magnetic field around the conductor.

If the conductor is wound into a coil, the magnetic field expanding from each turn of the coil cuts across other turns of the coil. When the source current starts to reverse direction, the magnetic field collapses, and again the field cuts across the other turns of the coil.

The result in both cases is the same as if a conductor is passed through a magnetic field. An electromotive force (EMF) is induced in the conductor. This EMF is called a self-induced EMF because it is induced in the conductor carrying the current.

The direction of this induced EMF is always opposite the direction of the EMF that caused the

Classroom

Provide an overview of the annual maintenance procedure for a bolted pressure switch.

Discuss the purpose of a transformer and explain how it operates.

Classroom

Explain the difference between no-load operation and load operation.

Discuss power transformers and distribution transformers.

Describe the operation and applications of dry transformers. Refer the trainees to Figure 11 in the Trainee Module.

current to flow initially. This principle is known as Lenz's law:

- An induced EMF always has such a direction as to oppose the action that produced it.
- For this reason, the EMF induced is also known as a counter-electromotive force (CEMF).

The counter-electromotive force reaches a value nearly equal to the applied voltage; thus, the primary current is limited when the secondary is open circuited.

10.1.1 No-Load Operation

The operation of a transformer is based on the principle that electrical energy can be transferred efficiently by mutual induction from one winding to another. When the primary winding is energized from an AC source, an alternating magnetic flux is established in the transformer core. This flux links the turns of the primary with the secondary, thereby inducing a voltage in them. Since the same flux cuts both windings, the same voltage is induced in each turn of both windings. Whenever the secondary of a transformer is left disconnected (or open), there is no current drawn by the secondary winding. The primary winding draws the amount of current required to supply the magnetomotive force, which produces the transformer core flux. This current is called the exciting or magnetizing current.

The exciting current is limited by the CEMF in the primary and a small amount of resistance, which cannot be avoided in any current-carrying conductor.

10.1.2 Load Operation

When a load is connected to the secondary winding of a transformer, the secondary current flowing through the secondary turns produces a counter-magnetomotive force. According to Lenz's law, this magnetomotive force is in a direction that opposes the flux that produced it. This opposition tends to reduce the transformer flux and is accompanied by a reduction in the CEMF in the primary. Since the primary current is limited by the internal impedance of the primary winding and the CEMF in the winding, whenever the CEMF is reduced, the primary current continues to increase until the original transformer flux reaches a state of equilibrium.

10.2.0 Transformer Types

Transformers can be divided into two main categories: power transformers and **distribution transformers.** Power transformers handle large amounts of power and are generally used at transmission level voltages. Distribution transformers are designed to handle larger currents at lower voltage levels. Distribution transformers have smaller kVA ratings and are physically much smaller than power transformers. Power transformers often have an auxiliary means of cooling, such as fans and radiators. Distribution transformers are usually self-cooled, using no fans or other cooling methods. Where distribution transformers may be pole-mounted or pad-mounted, power transformers are always pad-mounted.

Although there is some overlap between power and distribution transformers, a transformer that is rated at more than 500kVA and/or 34.5kV is generally a power transformer. A transformer rated below these values can be considered a distribution transformer. Remember, there is an overlap in kVA capacity and voltage depending on the system and power requirements.

10.3.0 Dry Transformers (Air-Cooled)

Many transformers do not use an insulating liquid to immerse the core and windings. Dry or air-cooled transformers are used for many jobs where small, low-kVA transformers are required. Large distribution transformers are usually oil filled for better cooling and insulating. However, for installations in buildings and other locations where the oil in oil-filled transformers would be a serious fire hazard, dry transformers are used. These transformers are generally of the core form. The core and coils are similar to those of other transformers. A three-phase, dry-type transformer is shown in *Figure 11*. The enclosing side plates on the high-voltage side have been removed to show the baffles that control the direction of air circulation.

The case is made of sheet metal and provided with ventilating louvers for the circulation of cooling air. To increase the output, fans can be installed to draw cooling air through the coils at a faster rate than is possible with natural circulation.

Either Class B or Class H insulation is used for the windings. Class B insulation may be operated safely at a hot-spot temperature of 130°C. Class H insulation may be operated safely at a hot-spot temperature of 180°C. The use of these materials makes it possible to manufacture smaller transformers. Both Class B and Class H insulation consist of mica, asbestos, fiberglass, and similar inorganic material. Temperature-resistant organic varnishes are used as the binder for Class B insulation. Silicone or fluorine compounds or similar materials are used as the binder for Class H insulation. Such transformers use high-temperature

Instructor's Notes:

306F11.EPS

Figure 11 ◆ Dry-type transformer.

insulation only in locations where the high temperature requires such insulation.

10.4.0 Sealed Dry Transformers

Hermetically sealed dry transformers are constructed in large sizes for voltages above 15kV. They are used for installations in buildings and other locations where oil-filled transformers would be a serious fire hazard, but they may also be used for lower voltages and kVA ratings and for water-submersible transformers in locations subject to floods. Nitrogen is typically used for the insulation and cooling of sealed dry transformers.

10.5.0 Transformer Nameplate Data

Transformer nameplate data includes the following:

- *Electrical ratings*—The information relating to the transformer electrical parameters can be found on the nameplate.
- *Voltage ratings*—The voltage rating identifies the nominal root mean square (rms) voltage value at which the transformer is designed to operate. A transformer can operate within a ±5% range of its rated primary voltage. If the primary voltage is increased to more than +5%, the windings of the transformer can overheat. Operation of the transformer at more than −5% decreases its power output proportionally to the percent voltage reduction. Transformer windings are rated as follows:
  - –Phase-to-phase and phase-to-neutral for wye windings, such as 480Y/277VAC
  - –Phase-to-phase for delta windings, such as 480VAC
  - –Dual-voltage windings, such as 480VAC × 240VAC

  When transformers are equipped with a tap changer, the voltage ratings in the nameplate indicate the nominal voltages.
- *BIL*—This identifies the maximum impulse voltage the winding insulation can withstand without failure.
- *Phase*—The phase information indicates the number of phase windings contained in a transformer tank.
- *Frequency*—The frequency rating of a transformer is the normal operating system frequency. When a transformer is operated at a lower frequency, the reactance of the primary winding decreases. This causes a higher exciting current and an increase in flux density. In addition, there is an increase in core loss, which results in overall heating.
- *Class*—Transformers are classified by the type of cooling they employ.
- *Temperature rise*—The temperature rise rating is the maximum elevation above ambient temperature that can be tolerated without causing insulation damage.
- *Capacity*—The capacity of a transformer to transfer energy is related to its ability to dissipate the heat produced in the windings. The capacity rating is the product of the rated voltage and the current that can be carried at that voltage without exceeding the temperature rise limitation.
- *Impedance*—Impedance identifies the opposition of a transformer to the passage of short circuit current.
- *Phasor diagrams*—Phasor diagrams show phase and polarity relationships of the high and low windings. They can be used with the schematic connection diagram to provide test connection points and to provide proper external system connections.

10.6.0 Transformer Case Inspections

When inspecting the inside of a dry-type, air-cooled transformer case, look for the following:

- Bent, broken, or loose parts
- Debris on the floor or in the coils
- Corrosion of any part
- Worn or frayed insulation

Classroom

List the types of information found on transformer nameplates.

Explain how to conduct an inspection of a transformer case.

Classroom

Describe the various tests performed on transformers.

List the functions performed by instrument transformers.

Discuss the operation and applications of potential transformers.

- Shifted core members
- Damaged tap changer mounts or mechanisms
- Misaligned core spacers and loose coil elements
- Broken or loose blocking

Upon the completion of the inspection, replace the covers and bolt securely. All information should be recorded on appropriate inspection sheets.

10.7.0 Transformer Tests

The following tests are the recommended minimum tests that should be included as part of a maintenance program. These tests are conducted to determine and evaluate the present condition of the transformer. From the results of these tests, a determination is made as to whether the transformer is suitable for service. All tests should be performed using the standards and procedures provided by the transformer manufacturer.

- *Continuity and winding resistance test*—There should be a continuity check of all windings. If possible, measure the winding resistance and compare it to the factory test values. An increase of more than 10% could indicate loose internal connections.
- *Insulation resistance test*—To ensure that no grounding of the windings exists, a 1,000V insulation resistance test should be made.
- *Ratio test*—A turns ratio test should be made to ensure proper transformer ratios and to ensure that all connections were made. If equipped with a tap changer, all positions should be checked.
- *Core ground*—This test is performed in the same way as the insulation resistance test, except the measurement is made from the core to the frame and ground bus. Remove the core ground strap before the test.
- *Heat scanning*—After the transformer is energized, a heat scan test should be done to detect loose connections. This test is performed using an infrared scanning device that shows or indicates hot spots.

11.0.0 ◆ INSTRUMENT TRANSFORMERS

For all practical purposes, the voltages and currents used in the primary circuits of substations are much too large to be used to provide operating quantities to relaying or metering circuits. In order to reduce voltage and currents to usable levels, instrument transformers are employed. Instrument transformers are used to:

- Protect personnel and equipment from the high voltages and/or currents used in electric power transmission and distribution
- Provide reasonable use of insulation levels and current-carrying capacity in relay and metering systems and other control devices
- Provide a means to combine voltage and/or current phasors to simplify relaying or metering

Instrument transformers are manufactured with a multitude of different ratios to provide a standard output for the many different system primary voltage levels and load currents. There are two major classifications of instrument transformers: potential (voltage) transformers and current transformers.

11.1.0 Potential Transformers

Potential transformers are designed to reduce primary system voltages down to usable levels for metering and are often referred to as voltage transformers or VTs. Potential transformers are often used where the system's primary voltage exceeds 600V and sometimes on 240V and 480V systems.

The standard secondary circuit voltage level for a potential transformer circuit is 120V for circuits below 25kV and 115V for circuits above 25kV at the potential transformer's rated primary voltage. These voltages correspond to typical transformation ratios of standard transmission voltages. The current flowing in the secondary of the potential transformer circuit is very low under normal operating conditions, typically less than one ampere.

Potential transformers are constructed as lightly loaded distribution transformers with the design emphasis on winding ratio accuracy rather than thermal ratings. Potential transformer construction can be air-insulated dry, case epoxy-insulated, oil-filled, or SF_6-insulated, depending upon the primary circuit voltage level.

The standard output voltage of potential transformers is either 120V or 69.3V, depending on

INSIDE TRACK

SF_6 Insulation

Sulfur hexafluoride (SF_6) is a colorless, odorless, nontoxic, nonflammable gas that is used as an insulating gas in electrical equipment. SF_6 is used as a gaseous dielectric for transformers, condensers, and circuit breakers.

Instructor's Notes:

whether its primary uses phase-to-phase or phase-to-neutral connections. Understanding the operation of a potential or voltage transformer is simplified by the inspection of its equivalent circuit.

Potential transformers must have their secondary circuits grounded for safety reasons in the event that a short circuit develops between the primary and secondary windings and to negate the effects of parasitic capacitance between the primary and the secondary. *Figure 12* shows the connection of an ideal potential transformer circuit.

11.2.0 Current Transformers

A current transformer is designed to reduce high primary system currents down to usable levels. Current transformers are used whenever system primary voltage isolation is required. The standard secondary circuit current for a current transformer circuit is 5A with full-rated current flowing in the primary circuit.

> **WARNING!**
> The voltage level across a current transformer's secondary terminals can rise to a very dangerous level if the secondary circuit opens while the primary circuit is energized.

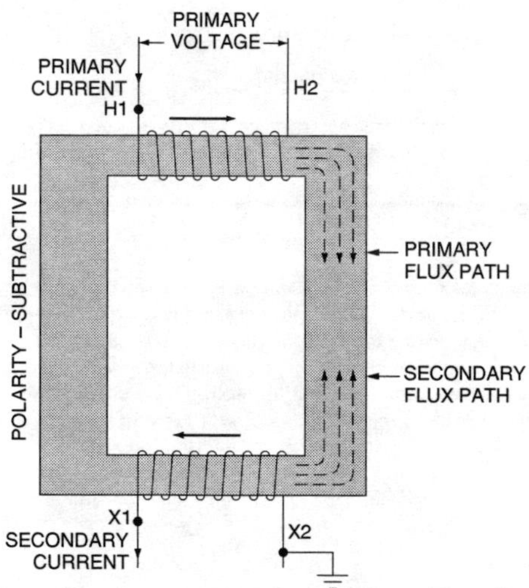

Figure 12 ◆ Potential transformer construction.

The primary considerations in current transformer design are the current-carrying capability and saturation characteristics. Insulation systems are of the same generic types as potential transformers; however, SF_6 insulation is infrequently used in current transformer construction.

Current transformers are manufactured in four basic types: oil-filled (for example, donut type), bar, window, and **bushing.** The bushing-type transformer is normally applied on circuit breakers or power transformers. The other types are used for the remaining indoor and outdoor installations. *Figure 13* illustrates some common types of current transformer construction.

The major criteria for the selection of the current transformer for relaying are its primary current rating, maximum burden, and saturation characteristics. Saturation is particularly important in relaying due to the fact that many relays are called upon to operate only under fault conditions.

Current transformer circuits operate at a very low voltage. Connected loads (burdens) range from 0.2Ω to 2Ω. These small impedances,

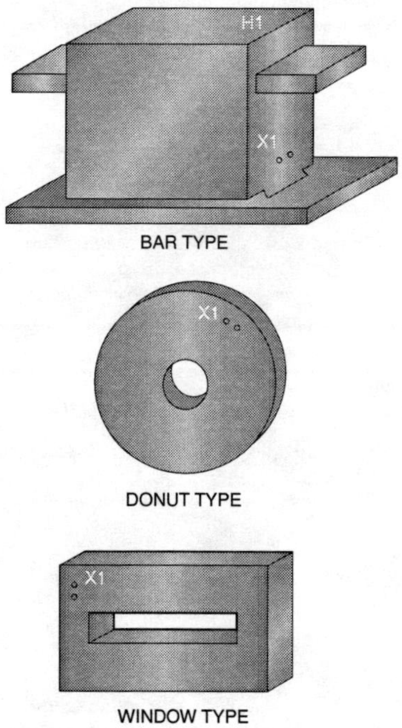

Figure 13 ◆ Types of current transformer construction.

Audiovisual

Show Transparency 5 (Figure 12).

Classroom

Discuss the operation and applications of current transformers.

Audiovisual

Show Transparency 6 (Figure 13).

Classroom

Describe the inspection procedure for an instrument transformer.

Teaching Tip

Discuss the "Think About It."

Classroom

List the four classifications of circuit breakers.

Teaching Tip

Discuss the "Case History." Emphasize the consequences of failing to verify that the power has been de-energized prior to performing any type of service or maintenance.

together with a maximum continuous current of up to 5A, keep these circuits at low potentials. The voltage can become high momentarily during faults when large secondary currents flow. This voltage is a function of the current, burden, and transformer VA capability.

As with potential transformers, current transformers must also have their secondary windings grounded in the event of an insulation breakdown between the primary and secondary and to negate the effects of parasitic capacitance.

11.3.0 Instrument Transformer Maintenance

Instrument transformers require regular inspection and maintenance. A typical inspection includes:

Step 1 Inspect for physical damage, and check the nameplate information for compliance with instructions and specification requirements.

Step 2 Verify the proper connection of transformers against the system requirements.

Step 3 Verify the tightness of all bolted connections, and ensure that adequate clearances exist between the primary circuits and the secondary circuit wiring.

Step 4 Verify that all required grounding and shorting connections provide good contact.

Step 5 Test for proper operation of the transformer withdrawal mechanism (trip out) and grounding operation when applicable.

> THINK ABOUT IT
>
> *Transformers*
>
> What is the difference between an instrument (potential) transformer and a control transformer? Why can't one device serve both functions?

12.0.0 ◆ CIRCUIT BREAKERS

Circuit breakers are the only circuit interrupting devices that combine a full fault current interruption rating and the ability to be manually or automatically opened or closed. A circuit breaker is defined as a mechanical switching device that is capable of making, carrying, and breaking currents under normal circuit conditions and also making, carrying (for a specified time), and breaking currents under specified abnormal circuit conditions, such as a short circuit (according to IEEE). The four general classifications of circuit breakers are:

- **Air circuit breakers** (ACBs)
- Oil circuit breakers (OCBs)
- Vacuum circuit breakers (VCBs)
- Gas circuit breakers (GCBs)

Circuit breakers may conveniently be divided into low-voltage, medium-voltage, and high-voltage classes. Although there is considerable overlap among these classes, each one has certain characteristic features.

12.1.0 Circuit Breaker Ratings

Circuit breaker ratings are given on the breaker nameplate. The information from the nameplate

> CASE HISTORY
>
> *OCB Circuit Breaker Testing*
>
> An assistant engineer and an electrician had the responsibility of testing the insulation values of the bushings on an OCB. The assistant engineer pulled down the handle and opened the breaker, visually determining that the contacts were open and safe. He instructed the electrician to move the test leads from one bushing to the next as the assistant engineer read the values. On the third and last bushing, the electrician was electrocuted as he placed the test leads on the bushing. The last contact had not opened due to a broken porcelain insulator skirt, which allowed the cap to remain stationary while the rest of the mechanism rotated as the handle was pulled down. This was not noticeable from the assistant engineer's eye level when he determined all contacts were opened.
>
> **The Bottom Line:** ***Never*** come in contact with switchgear such as this without first testing each and every contact point with the proper meter, and ***never*** rely on someone else's word that the power is off.

Instructor's Notes:

should be reviewed when considering any breaker selection problem. The same rating information should be included in any documentation for breaker applications. The rating information includes:

- *Rated voltage*—The rated voltage is the maximum voltage for which the circuit breaker is designed.
- *Rated current*—This is the continuous current that the circuit breaker can carry without exceeding a standard temperature rise (usually 55°C).
- *Interrupting rating*—This is the maximum value of current at rated voltage that the circuit breaker is required to successfully interrupt for a limited number of operations under specified conditions. The term is usually applied to abnormal or emergency conditions.

13.0.0 ◆ ELECTRICAL DRAWING IDENTIFICATION

Before looking at actual plant diagrams, it is necessary to understand the symbology used to condense electrical drawings. The designer uses symbols and abbreviations as a type of shorthand. This section will present the standard symbols, abbreviations, and device numbers that make up the designer's shorthand. For IEEE device numbers, see the *Appendix*.

13.1.0 Electrical Diagram Symbology

It is imperative that every line, symbol, figure, and letter in a diagram have a specific purpose and that the information be presented in its most concise form. For example, when the rating of a current transformer is given, a transformer symbol is shown, and an abbreviation such as CT is not needed; the information is implied by the symbol itself. Writing the unit of measure (amp) in this case is also unnecessary, since a current transformer is always rated in amperes. Thus, the numerical rating and the transformer symbol are sufficient. The key to reading and interpreting electrical diagrams is to understand and use the electrical legend. The legend shows the symbols used in the diagram and also contains general notes and other important information. Most electrical legends are very similar; however, there are some variations between the legends developed by different companies. Only the legend specifically designed for a given set of drawings should be used for those drawings.

The legend prevents the necessity of memorizing all the symbols presented on a diagram and can be used as a reference for unfamiliar symbols. Typically, the legend will be found in the bottom right corner of a print or on a separate drawing. In addition to symbols, abbreviations are an important part of the designer's shorthand. For example, a circle can be used to symbolize a meter, relay, motor, or indicating light. A circle's application can generally be distinguished by its location in the circuit; however, the designer uses a set of standard abbreviations to make the distinction clear. The following abbreviations are used to represent meters:

A	Ammeter
AH	Ampere-hour meter
CRO	Oscilloscope
DM	Demand meter
F	Frequency meter
GD	Ground detector
OHM	Ohmmeter
OSC	Oscillograph
PF	Power factor meter
PH	Phase meter
SYN	Synchroscope
TD	Transducer
V	Voltmeter
VA	Volt-ammeter
VAR	VAR meter
VARH	VAR hour meter
W	Wattmeter
WH	Watt-hour meter

As mentioned earlier, indicating lamps may also be represented by a circle. The following abbreviations are used to represent indicating lamps:

A	Amber
B	Blue
C	Clear
G	Green
R	Red
W	White

Relays are another component commonly represented by a circle. The following abbreviations are used for relays:

CC	Closing coil
CR	Closing/control relay
TC	Trip coil
TR	Trip relay
TD	Time-delay relay
TDE	Time-delay energize
TDD	Time-delay de-energize
X	Auxiliary relay

Still another component that is commonly represented by a circle is the motor. Motors usually have the horsepower rating in or near the circle

Classroom

Explain that circuit breakers are rated by voltage, current, and interrupting capacity.

Discuss the symbols and abbreviations used on electrical diagrams. Point out that plant diagrams may also refer to IEEE device numbers. Refer the trainees to the *Appendix* in the Trainee Module.

Classroom

Explain how contacts are depicted on electrical drawings.

Audiovisual

Show Transparencies 7 through 9 (Figures 14 through 16).

Classroom

List the types of information found on single-line (one-line) diagrams.

Audiovisual

Show Transparency 10 (Figure 17).

representing them. The abbreviation for horsepower is hp. Any other piece of equipment represented by a circle will be identified in the legend, notes, or spelled out on the diagram itself.

Contacts and switches are also identified using standard abbreviations. The following is a list of these abbreviations:

A	Breaker A contact
B	Breaker B contact
BAS	Bell alarm switch
BLPB	Backlighted pushbutton
CS	Control switch
FS	Flow switch
LS	Limit switch
PB	Pushbutton
PS	Pressure switch
PSD	Differential pressure switch
TDO	Time-delay open
TDC	Time-delay closed
TS	Temperature switch
XSH	Auxiliary switch

The following figures illustrate examples of these abbreviations and symbols. *Figure 14* shows A and B contacts in their normally de-energized state. If relay CR is de-energized, contact A is open and contact B is shut. When relay CR is energized, contact A is shut and contact B is open.

Figure 15 illustrates a control switch and its associated contacts. Contacts 1 through 4 open and close as a result of the operation of control switch 1 (CS1).

In the stop position, contact 2 is shut, and the red indicating lamp is lit. In the start position, contacts 3 and 4 are shut, energizing the M coil and the amber indicating lamp, respectively. When the switch handle is released, the spring returns to the run position, and contact 4 opens, de-energizing the amber lamp and closing contact 1 to energize the green lamp.

There are many abbreviations used on electrical drawings. The designer makes an effort to use standard abbreviations; however, you will encounter nonstandard abbreviations. Nonstandard abbreviations will typically be defined in the diagram notes or legend. *Figure 16* defines abbreviations commonly used in wiring prints and specifications. The symbols further illustrate descriptions of the abbreviations.

14.0.0 ◆ ELECTRICAL PRINTS

This section will cover the specific types of electrical prints with which you need to be familiar in order to install and maintain electrical systems.

14.1.0 Single-Line Diagrams

Analyzing and reading complex electrical circuits can be very difficult. Diagrams are simplified to single-line (one-line) diagrams to aid in reading the prints.

A single-line diagram is defined as a diagram that indicates by means of single lines and standard symbology the paths, interconnections, and component parts of an electric circuit or system of circuits. This type of drawing uses a single line to represent all conductors (phases) of the system. All components of power circuits are represented by symbols and notations. One-line diagrams are valuable tools for system visualization during planning, installation, operation, and maintenance, and they provide a basic understanding of

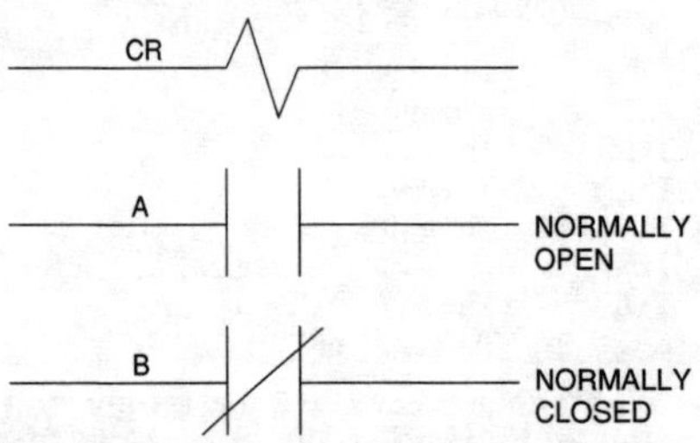

Figure 14 ◆ Contact symbols.

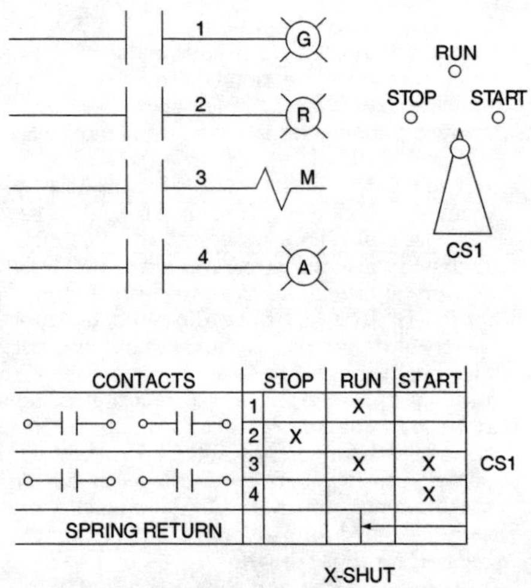

Figure 15 ◆ Switch development.

Instructor's Notes:

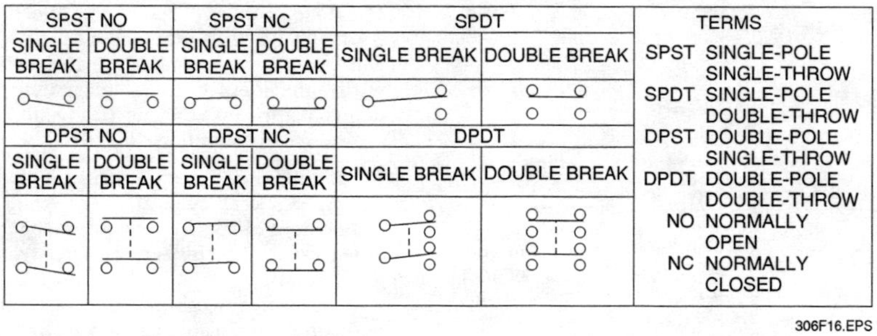

Figure 16 ◆ Supplementary contact symbols.

how a portion of the electrical system functions in terms of the physical components of the circuit.

Two types of single-line diagrams include the overall plant single-line diagram and the project single-line diagram. Overall plant single-line diagrams show the electrical power distribution system, in simplified form, from the utility or generated supply to the load side of the substation protection devices. The overall plant single-line diagrams do not include distribution panels, motor control centers, motors, or similar electrical equipment located on the load side of the substation. Project single-line diagrams show the electrical power distribution and utilization for a particular project or local plant area. These diagrams are the continuation of overall plant single-line diagrams and indicate the power distribution from the load side of the substation protective device to the final point of utilization on a branch circuit. An example of a single-line diagram is shown in *Figure 17*.

14.2.0 Elementary Diagrams

An elementary diagram is a drawing that falls between one-line diagrams and schematics in terms of complexity. An elementary diagram is a wiring diagram showing how each individual conductor is connected. *Figure 18* is an example of an elementary diagram.

Elementary diagrams, interconnection diagrams, and connection diagrams all illustrate individual conductors. Elementary diagrams are used to show the wiring of instrument and electrical control devices in an elementary ladder or schematic form. The elementary diagram reflects the control wiring required to achieve the operation and sequence of operations described in the logic diagram. When presented in ladder form, the vertical lines in each ladder diagram represent the hot and neutral wires of a 120VAC circuit. If a number of schemes are connected to the same 120VAC circuit, the vertical lines are continuous

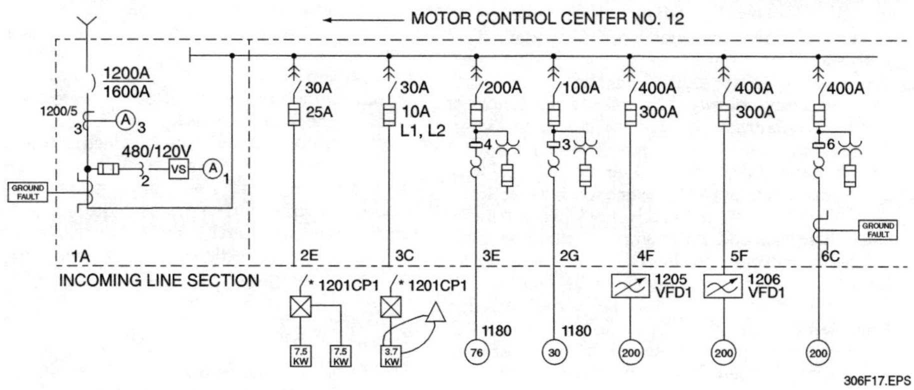

Figure 17 ◆ One-line diagram.

Classroom

Discuss the types of information found on elementary diagrams.

Audiovisual

Show Transparency 11 (Figure 18).

Classroom

List the types of information found on interconnection diagrams.

Point out that interconnection diagrams show the actual wiring connections between unit assemblies or equipment, while connection diagrams show the internal connections of the unit.

Discuss the two types of connection diagrams:
- **Point-to-point method**
- **Cable method**

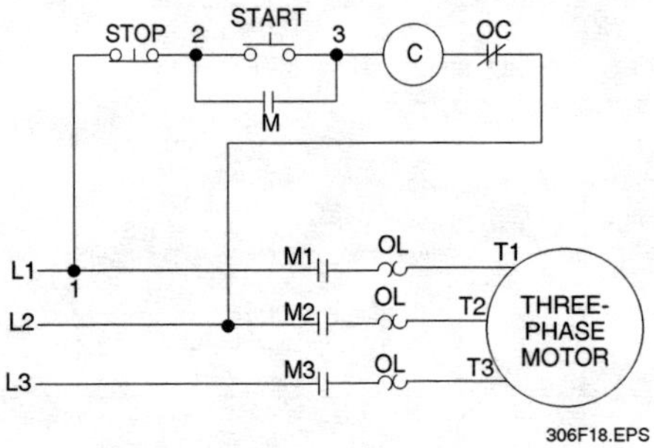

Figure 18 ◆ Elementary diagram.

from the top to the bottom of the ladder. The hot wire is always shown on the left side of the ladder and the neutral on the right. A ground symbol is normally not shown on the neutral wire.

The hot and neutral wire numbers are shown at the top of the vertical lines on each ladder diagram. The circuit identification number and source of the 120VAC circuit are also shown at the top center of each ladder. If two or more 120VAC circuits are represented in a single ladder, the vertical lines are broken, and the wire numbers and circuit identification are entered at the top of each ladder segment. Each horizontal line in a ladder diagram represents a circuit path. All devices shown on a single horizontal line represent a series circuit path; parallel circuit paths are shown on two or more horizontal lines.

14.3.0 Interconnection Diagrams

When troubleshooting electrical circuits, you may use an elementary circuit diagram to determine the cause of a failure; however, since elementary diagrams are drawn without regard to physical locations, connection diagrams should be used to aid in locating faulty components. Interconnection and connection diagrams are structured in such a way that they present all the wires that were shown in the elementary drawing in their actual locations. These drawings show all electrical connections within an enclosure, with each wire labeled to indicate where each end of the wire is terminated.

The interconnection diagram is made to show the actual wiring connections between unit assemblies or equipment. Internal wiring connections within unit assemblies or equipment are usually omitted. The interconnection diagrams will appear adjacent to the schematic diagram or on a separate drawing, depending upon the format chosen when making the schematic diagram. The development of the interconnection diagram is integrated with that of the schematic diagram and only the equipment, terminal blocks, and wiring pertinent to the accompanying schematic diagram appear in the interconnection diagram. A typical interconnection diagram will contain:

- An outline of the equipment involved in its relative physical location
- Terminal blocks in the equipment that are concerned with the wiring illustrated on the schematic
- Wire numbers, cable sizes, cable numbers, cable routing, and cable tray identification (should not be repeated on the interconnection diagram except where necessary)
- Wiring between equipment (normally shown as individual cables but may be combined on complex drawings)
- Equipment identification information

14.4.0 Connection Diagrams

The connection diagram shows the internal wiring connections between the parts that make up an apparatus. It will contain as much detail as necessary to make or trace any electrical connections involved. A connection diagram generally shows the physical arrangement of component electrical connections. It differs from the interconnection diagram by excluding external connections between two or more unit assemblies or pieces of equipment.

The schematic diagram shows the arrangement of a circuit with the components represented by conventional symbols. Its intent is to show the function of a circuit. The schematic, like the elementary drawing, is not laid out with respect to physical locations.

A wiring diagram also shows the physical locations of all electrical equipment and/or components with all interconnecting wiring. It shows the actual connection point of every wire and the color of the wires connected to each terminal of every component. It allows the electrician to easily locate terminals and wires. A wiring diagram in conjunction with a schematic greatly aids in troubleshooting a given piece of equipment. Connection diagrams can be shown in various forms.

The following sections illustrate two types of connection diagrams.

14.4.1 Point-to-Point Method

The point-to-point method is used for the simpler diagrams where sufficient space is available to show each individual wire without sacrificing the

Instructor's Notes:

clarity of the diagram. *Figure 19* is a point-to-point connection diagram.

14.4.2 Cable Method

In complex diagrams, individual wires are cabled so as to conserve drawing space. *Figure 20* is a cable connection diagram.

15.0.0 ◆ INTERPRETING ONE-LINE DIAGRAMS

The one-line diagram in *Figures 21* and *22* is typical of those used to show how an electrical system is to be installed. In general, a one-line diagram is never drawn to scale. Such drawings show the major components in an electrical system and then use

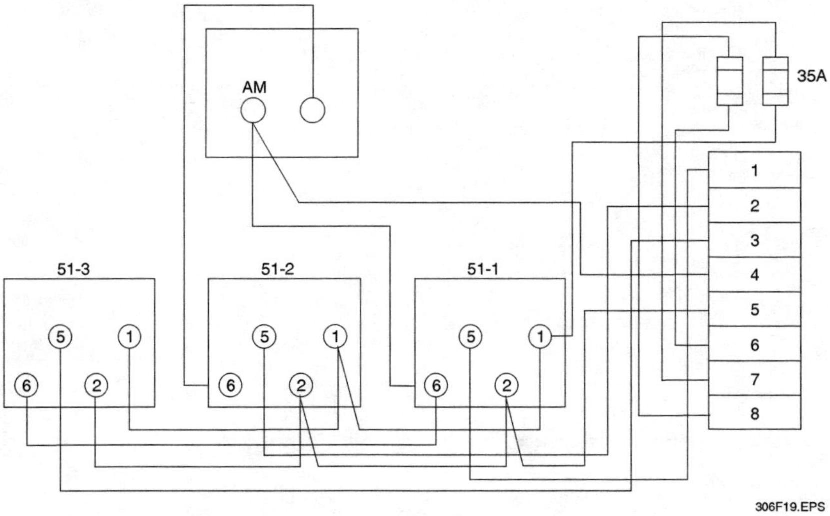

Figure 19 ◆ Point-to-point connection diagram.

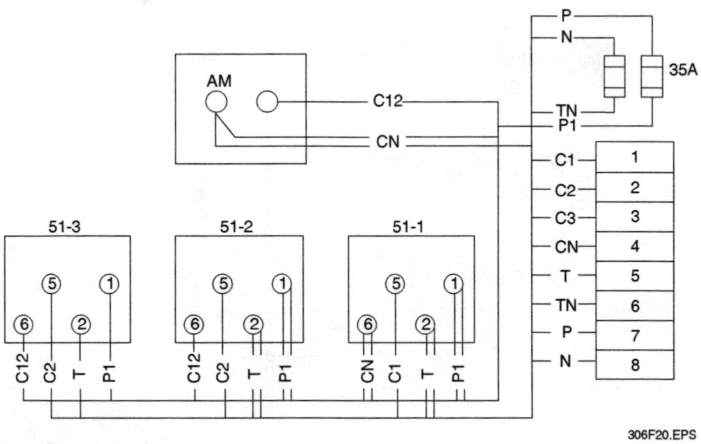

Figure 20 ◆ Cable connection diagram.

Audiovisual

Show Transparencies 12 and 13 (Figures 19 and 20).

Classroom

Demonstrate how to interpret a one-line diagram.

Audiovisual

Show Transparencies 14 and 15 (Figures 21 and 22).

Teaching Tip

Have the trainees identify various components on the diagram.

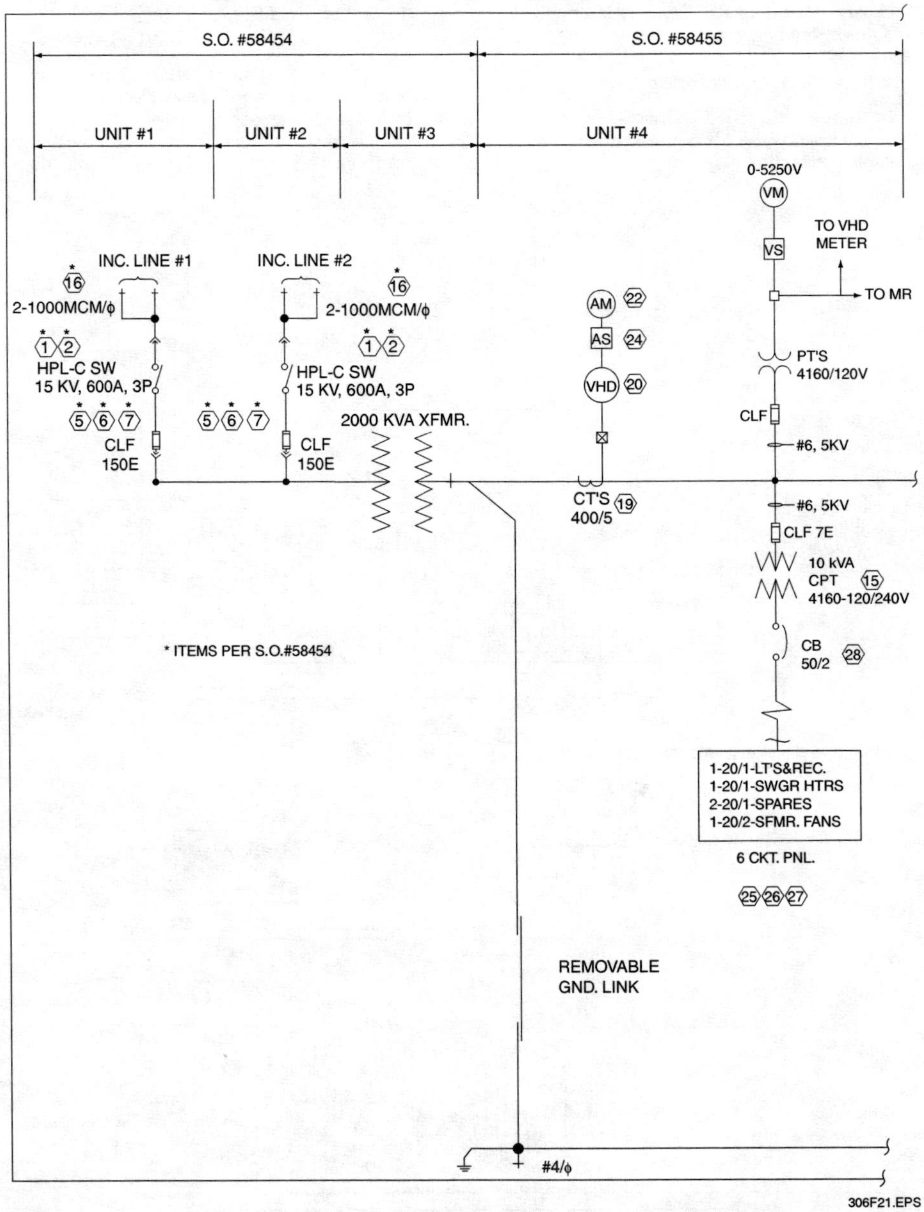

Figure 21 ◈ One-line diagram of a 2,000kVA substation (left side).

6.34 ELECTRICAL LEVEL THREE—TRAINEE MODULE 26306

Instructor's Notes:

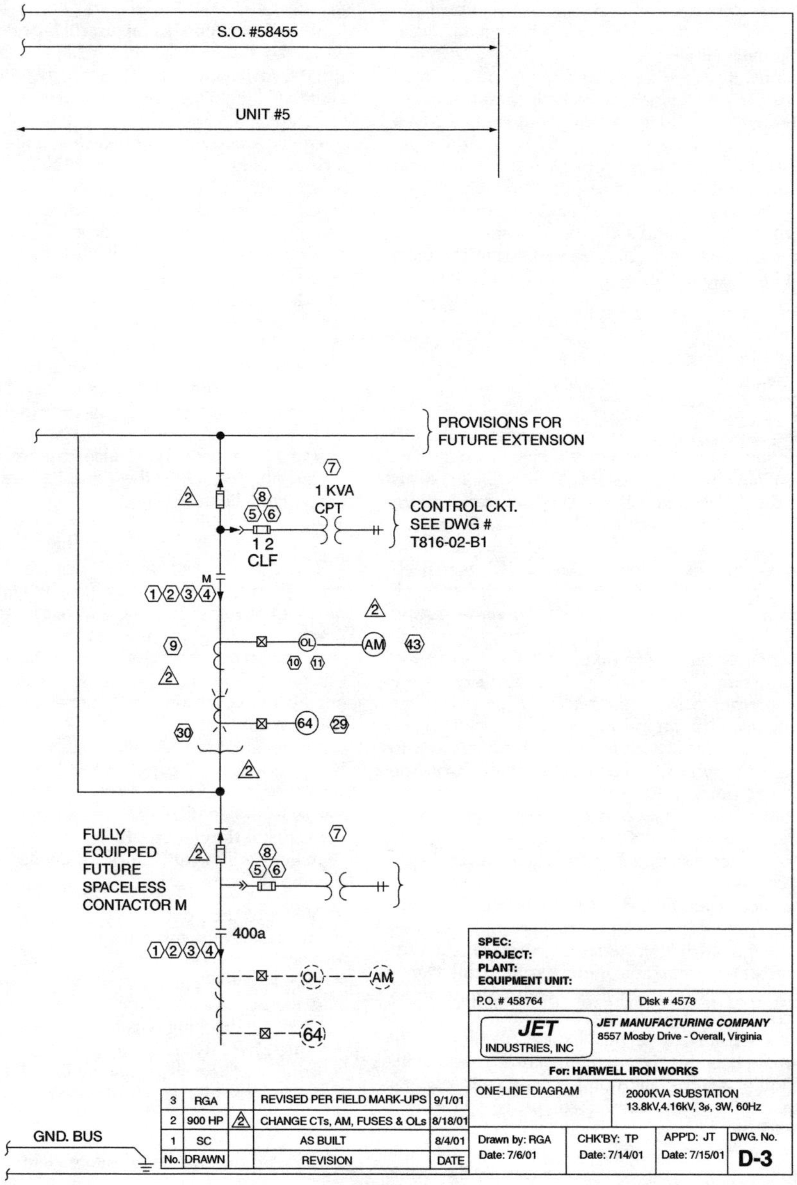

306F22.EPS

Figure 22 ◆ One-line diagram of a 2,000kVA substation (right side).

Teaching Tip

Have the trainees identify various components on the diagram.

Classroom

Explain how to interpret the incoming power connections as depicted on a one-line diagram.

Audiovisual

Show Transparency 16 (Figure 23).

Classroom

Discuss the types of information found in the drawing's title block.

Audiovisual

Show Transparency 17 (Figure 24).

only one drawing line to indicate the connections between these components. Even though only one line is used between components, this single line may indicate a raceway of two, three, four, or more conductors. Notes, symbols, tables, and detailed drawings are used to supplement and clarify a one-line diagram. Refer again to *Figures 21* and *22.* These drawings were prepared by an electrical manufacturing company to give workers at the job site an overview of a 2,000kVA substation utilizing a 13.8kV primary and a 4.16kV, three-phase, three-wire, 60Hz secondary. Note that this drawing sheet is divided into the following sections:

- Service order (S.O.) numbers
- Unit numbers
- One-line diagram
- Title block
- Revision notes

Service order numbers are arranged at the top of the drawing sheet in a time sequence, bar-chart type arrangement. For example, S.O. #58454 deals with the primary side of the 2,000kVa transformer, including the transformer itself. This section includes a high-voltage switchgear with an indoor/outdoor enclosure. The switchgear itself consists of two HLP-C interrupter switches, each rated at 15kV, 600A, with 150E current-limiting fuses (CLF).

Service order #58455 deals with the wiring and related components on the secondary side of the transformer and begins with a low-voltage switchgear with an indoor/outdoor enclosure.

Service order #58454 is further subdivided into three units which are indicated as such on the drawing immediately under the S.O. number. Unit #1 deals with incoming line #1, unit #2 deals with incoming line #2, and unit #3 covers the 2,000kVA transformer and its related connections and components.

Service order #58455 is further subdivided into two units: unit #4 and unit #5. Basically, unit #4 covers grounding, the installation of current transformers, various meters, potential transformers, a 10kVA, 4,160/240V transformer, and a six-circuit panel, all derived from a 600A, 4,160V, three-wire, 60Hz main bus. Unit #5 continues with the main bus and covers the installation and connection of a complete motor control center, along with another fully-equipped future space, less **contactors.**

The one-line diagram takes up most of the drawing sheet and gives an overview of the entire installation. We will begin at the left side of the drawing where incoming line #1 is indicated. This section of the drawing is shown in *Figure 23.*

Incoming line #1 (partially abbreviated on the drawings as INC. LINE #1) consists of two, 1,000MCM (kcmil) conductors per phase, as indicated by note 2-1000MCM/Ø (i.e., parallel 1,000kcmil conductors). Since this is a three-phase system, a total of six 1,000kcmil conductors are used.

The single line continues to engage separable connectors at the single-throw, 15kV, 600A, three-pole switch. Overcurrent protection is provided by current-limiting fuses as indicated by the fuse symbol combined with a note. *Figure 21* shows that the single line continues to the high-voltage bus, which connects to the primary side of the 2,000kVA transformer. Incoming line #2 (partially abbreviated on the drawings as INC. LINE #2) is identical to line #1. This line also connects to the high-voltage bus, which connects to the primary side of the 2,000kVA transformer. Notice the numerals, each enclosed by a hexagon, placed near various components in these two high-voltage primaries. Note also that an asterisk is placed above each of these marks. A note on the drawing indicates:

**ITEMS PER S.O. #58454*

These marks appear in a supplemental schedule known as the *Bill of Materials,* which describes the marked items, lists the number required, manufacturer, catalog number, and a brief description of each. Such schedules are extremely useful to estimators, job superintendents, and workers to ensure that each required item is accounted for and installed.

Every electrical drawing should have a title block, normally located in the lower right-hand corner of the drawing sheet; the size of the block varies with the size of the drawing and also with the information required.

In general, the title block for an electrical drawing should contain:

- Name of the project
- Address of the project
- Name of the owner or client
- Name of the person or firm who prepared the drawing
- Date the drawing was made
- Scale(s), if any
- Initials of the drafter, checker, designer, and engineer with dates under each
- Job number
- Drawing sheet number
- General description of the drawing

The title block for the project in question is shown in *Figure 24.*

Sometimes, electrical drawings will have to be partially redrawn or modified during the plan-

Instructor's Notes:

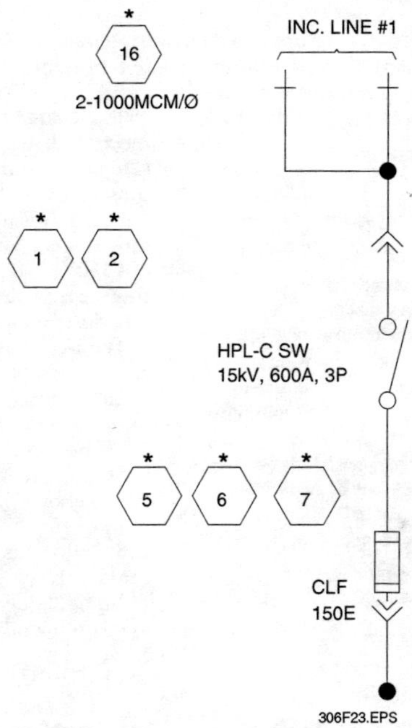

Figure 23 ◆ Incoming high-voltage line.

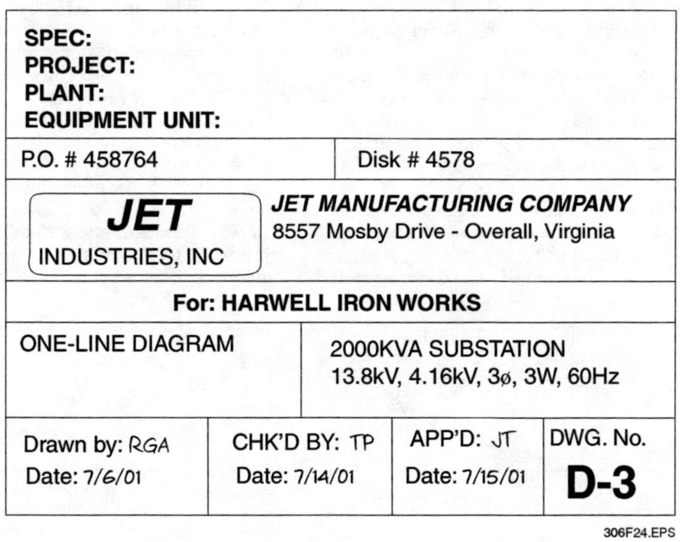

Figure 24 ◆ Typical drawing title block.

Classroom

Emphasize the importance of retaining the latest version of the drawings for any job.

Audiovisual

Show Transparency 18 (Figure 25).

Classroom

Explain how to relate the material found on the one-line diagram to the Bill of Materials for the job. Refer the trainees to Figure 26 in the Trainee Module.

ning or construction of a project. It is extremely important that such modifications are noted and dated on the drawings to ensure that all workers have an up-to-date set of drawings. In some situations, sufficient space is left near the title block for the dates and descriptions of revisions, as shown in *Figure 25.*

> **CAUTION**
>
> When a set of electrical drawings has been revised, always make certain that the most up-to-date set is used for all future layout work. Either destroy the obsolete set of drawings, or else clearly mark *Obsolete Drawing—Do Not Use* on the affected drawing sheets. Also, when working with a set of working drawings and written specifications for the first time, thoroughly check each page to see if any revisions or modifications have been made to the originals. Doing so can save much time and expense for all concerned with the project.

15.1.0 Interpreting Secondary One-Line Diagrams

Referring again to *Figure 21,* note that a 600A, 4,160V, three-phase, three-wire, 60Hz main bus is used to feed the remaining secondary elements. Also note the removable ground link between the transformer and the grounding bus connection. The drawing shows this conductor as a No. 4/0 AWG. Looking back at the 2,000kVA transformer, note that the main bus continues in a horizontal line to the right of the transformer symbol. The first equipment encountered is the metering section. Note the current transformers (CTs), which are designated by both symbol and note. The 400/5 note indicates that the CTs have a ratio of 400 to 5; that is, if 400A are flowing in the main bus, only 5A will flow to the meters. Again, numerals enclosed in hexagons are placed at each component in this section. Referring to the Bill of Materials schedule in *Figure 26,* we see a description of Item (Mark) #19 as CTs 400/5 Type JAF-0. Two are required; the catalog number is 750X10G304, and the manufacturer is GE (General Electric). Continuing from the CTs down to Item #20, the schedule describes this as a three-phase, three-wire, watt-hour meter with a 15-minute demand. It is designed to register with CTs with a ratio of 400/5 and PTs with a primary/secondary at 4,160V/240V. Locate the remaining numerals in this group, and find their descriptions in the schedule in *Figure 26.*

The two taps from the main bus in the drawing shown in *Figure 22* are for feeding two motor control centers (MCC); one is to be put into use immediately, while the other is a fully-equipped MCC (less contactors) for future use. First, look at the complete MCC. An enlarged view of this section is shown in *Figure 27.* This feeder is provided with overcurrent protection by means of current-limiting fuses (CLF), which are fuse type EJ-2 rated at 170A (see Item #8, *Figure 26*).

Immediately beneath this device, note that a tap is taken from the main line, fused with 5kV MTG fuses (Item #5) and also 5kV, CLF, 2E fuses (Item #6) before terminating at a 1kVA, 4,160V/120V CPT transformer (Item #7). This transformer is provided to accommodate a 120V control circuit. Since motor controls and motor control circuits are covered later in your training, the 120V control circuit will not be explained in this module.

Now backtrack to the main feeder, and continue downward to a contactor before another group of current transformers are installed in the circuit. These CTs are accompanied by notes and Item #9. Referring to the schedule in *Figure 26* for a description of Item #9, we see that these three CTs have a ratio of 150/5; that is, when the circuit is drawing 150A, the metering devices will receive only 5A, but the meter itself will indicate 150A. This circuit continues to a 200-line block overload relay with three elements and then on to an ammeter with a range of 0A to 150A.

3	RGA		Revised per field mark-ups	9/1/01
2	900HP	△2	Change CTs, AM, fuses & OLs	8/18/01
1	SC		As built	8/4/01
No.	**Drawn**		**Revision**	**Date**

306F25.EPS

Figure 25 ◆ Typical drawing revision block.

Instructor's Notes:

Mark	Req'd	Cat. No.	Mfg.	Description.
1	2	IC2957B103C	GE	Disc. Handle & Elec. Interlock ASM.
				(400A) (CAT#116C9928G1)
2	2	IC2957B108E	GE	Vert. Bus (CAT#195B4010G1)(400A)
				Shutter ASM. (CAT#116C9927G1)(400A)
3	2	IC2957B10BF	GE	Coil Finger ASM. (CAT#194A6949G1)(400A)
				Safety Catch (CAT#194A6994G1)(400A)
				Stab Fingers (CAT#232A6635G)(400A)
4	1		Toshiba	5kV, 300A, 3P, Vacuum Contactor
				120VAC Rectified Control
				Type CV461J-GAT2
5	4	2033A73G03	W	5kV Fuse MTG (2/CPT)
6	4	677C592G09	W	5kV, CLF, 2E Fuses Type CLE-PT
7	2	HN1K0EG15	Micron	1kVA, 4160-120 CPT
8	3	9F60LJD809	GE	CLF Size 9R (170A) Type EJ-2 (600HP)
9	3	615X3	GE	CT'S 150/5A Type JCH-0
10	1	CR224C610A	GE	200 Line Block O.L. Rly. 3 Elements
				Ambient Compensated W/INC. Contact
11	0	CR123C3.56A	GE	O.L. HTR (600HP)
11A	3	CR123C3.26A	GE	O.L. HTR . (2.79A)(700HP)
12	1	7022AB	AG	Off Delay R.Y .5-5 SEC.
13	0	CR2810A14A	GE	Machine Tool RLY. INO&INC 120VAC (MR)
14	1	CR294OUM301	GE	Emergency Stop PB (Push to Stop Pull to Reset) W/NP
15	1	9T28Y5611	GE	10kVA CPT. 4160-120/240V
16	2	643X92	GE	PT'S 4160/120V Type JVM-3/2FU
17	2	9F60CED007	GE	CLF 7E, 4.8kV Type EJ-1
18	2	9F61BNW451	GE	Fuse Clips Size C

306F26A.EPS

Figure 26 ◆ Bill of Materials schedule (1 of 2).

The next item on this main vertical feeder is a ground sensor that is connected to a solid-state ground guard relay. The feeder is then terminated at a 4,160V/480V power transformer (not shown in *Figure 27*) with the secondary conductors of this transformer entering and connecting to the busbars in a motor control center (MCC) enclosure, as shown in *Figure 28*. The remaining feeder in this one-line wiring diagram is for future use and is similar to the circuit just described.

Classroom

Discuss the wiring connections for a motor control center as depicted on a one-line diagram.

Audiovisual

Show Transparencies 19 and 20 (Figures 27 and 28).

Mark	Req'd	Cat. No.	Mfg.	Description.
19	2	750X10G304	GE	CT'S 400/5 Type JAF-0
20	1	700X64G885	GE	DWH-Meter 3ϕ, 3W, 60HZ, Type DSM-63 W/15MIN. Demand Register CT'S Ratio 400/5 & PT'S 4160-120V
21	1	50-103021P	GE	VM Scale 0-5250V Type AB-40
22	1	50-103131L	GE	AM Scale 0-400A Type AB-40
23	1	10AA004	GE	VS Type SBM
24	1	10AA012	GE	AS Type SBM
25	1	TL612FL	GE	6 CKT. PNL.
26	4	TQL1120	GE	20/1 C/B Type TQL.
27	1	TQL2120	GE	20/2 C/B Type TQL.
28	1	TEB12050WL	GE	50/2 C/B Type TEB.
29	1	3512C12H02	W	Type GR Groundgard RLY. Solid State
30	1	3512C13H03	W	GRD. Sensor
31	2	H	Smout Hollman	1/2 LT. REC.
32	2	7604-1	GE	LT. SW. & Receptacle
33	2	4D846G20	GE	120VAC, 250W HTR
34	1		Econo	Econo Lift for Contactor
35	11	Lot	Cook	NP/Schedule DWG. 58455-A1
36	3	Hold	T & B	Lug
37	0	50250440LSPK	GE	AM Scale 0-100A PNL. Type 2%
				ACC. Type 250 4-1/2 Case
38	1	NON10	Bus	10A, 250V Fuse
39	1	CP232	AH	2P, 250V Pull-Apart Fuse Block
40	1		Cook	SWGR NP S.O.#58455

306F26B.EPS

Figure 26 ◆ Bill of Materials schedule (2 of 2).

Instructor's Notes:

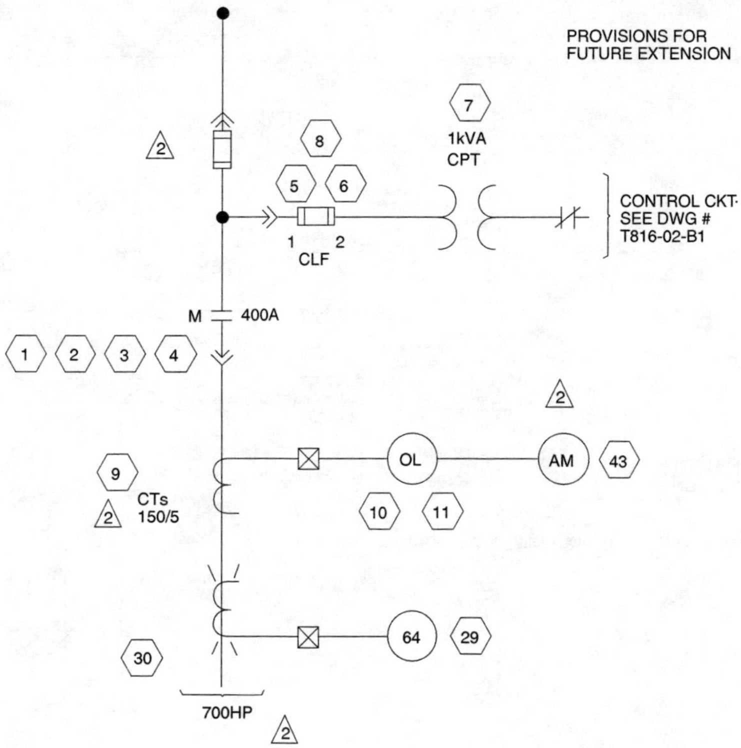

306F27.EPS

Figure 27 ◆ Enlarged view of the MCC feeder.

Teaching Tip

Have the trainees identify various components on the diagram.

Teaching Tip

Have the trainees identify various components on the diagram.

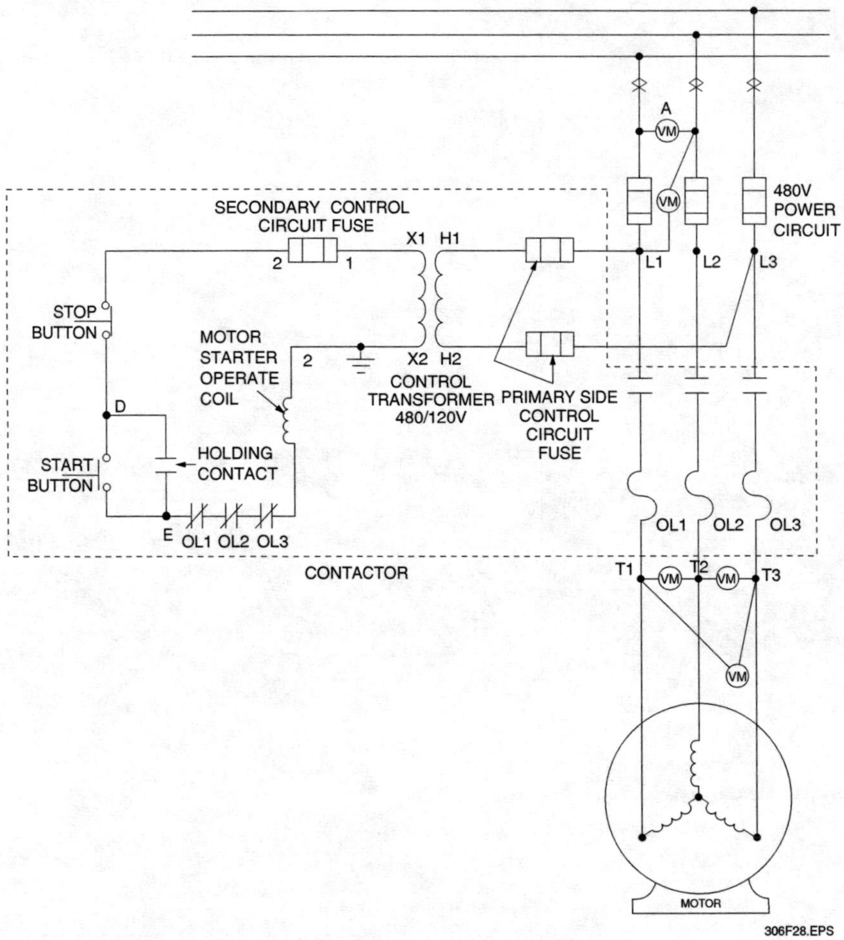

Figure 28 ◆ Motor control circuit diagram.

Instructor's Notes:

15.2.0 Shop Drawings

When large pieces of electrical equipment are needed, such as high-voltage switchgear and motor control centers, most are custom built for each individual project. In doing so, shop drawings are normally furnished by the equipment manufacturer prior to shipment to ensure that the equipment will fit the location at the shop site and to instruct the workers about preparing for such equipment as rough-in conduit and cable trays.

The drawing in *Figure 29* is one page of a shop drawing showing a pictorial view of the enclosure.

Shop drawings will also usually include connection diagrams for all components that must be field wired or connected.

As-built drawings, including detailed factory-wired connection diagrams, are also included to assist workers and maintenance personnel in making the final connections and then in troubleshooting problems once the system is in operation.

Typical drawings are shown in *Figures 30* and *31*.

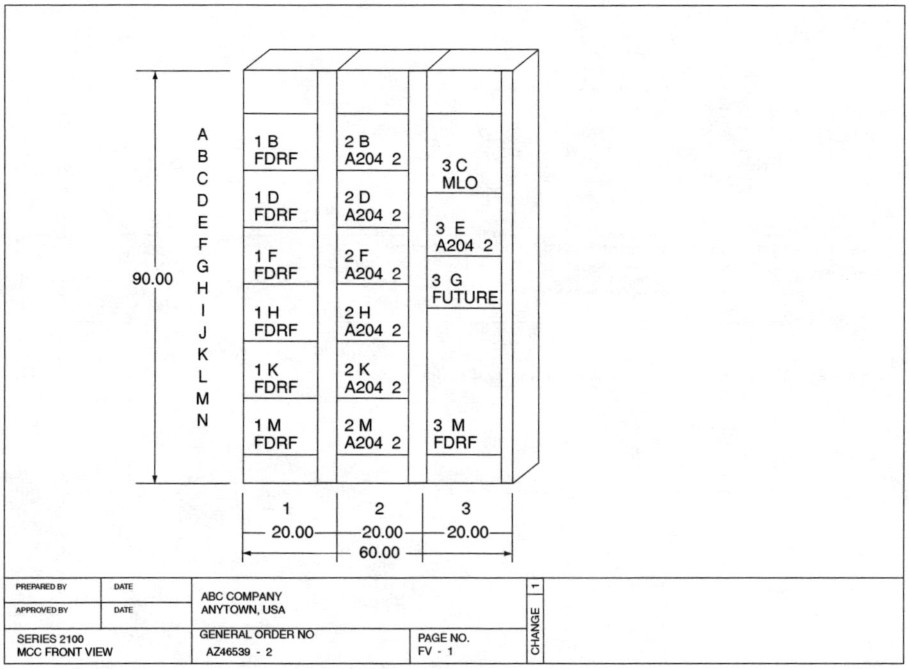

306F29.EPS

Figure 29 ◆ View of a motor control center.

Classroom

Discuss the purpose of shop drawings. Point out that they are normally supplied with large pieces of equipment.

Audiovisual

Show Transparency 21 (Figure 29).

Classroom

Discuss the purpose of as-built drawings. Point out that they are valuable tools for future system service and maintenance.

Audiovisual

Show Transparencies 22 and 23 (Figures 30 and 31).

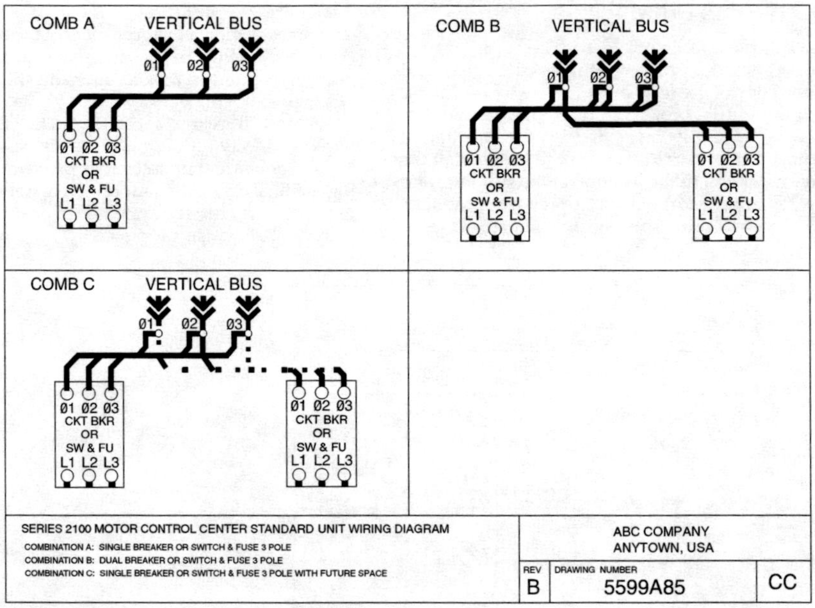

306F30.EPS

Figure 30 ◆ Motor control center standard unit wiring diagram.

Instructor's Notes:

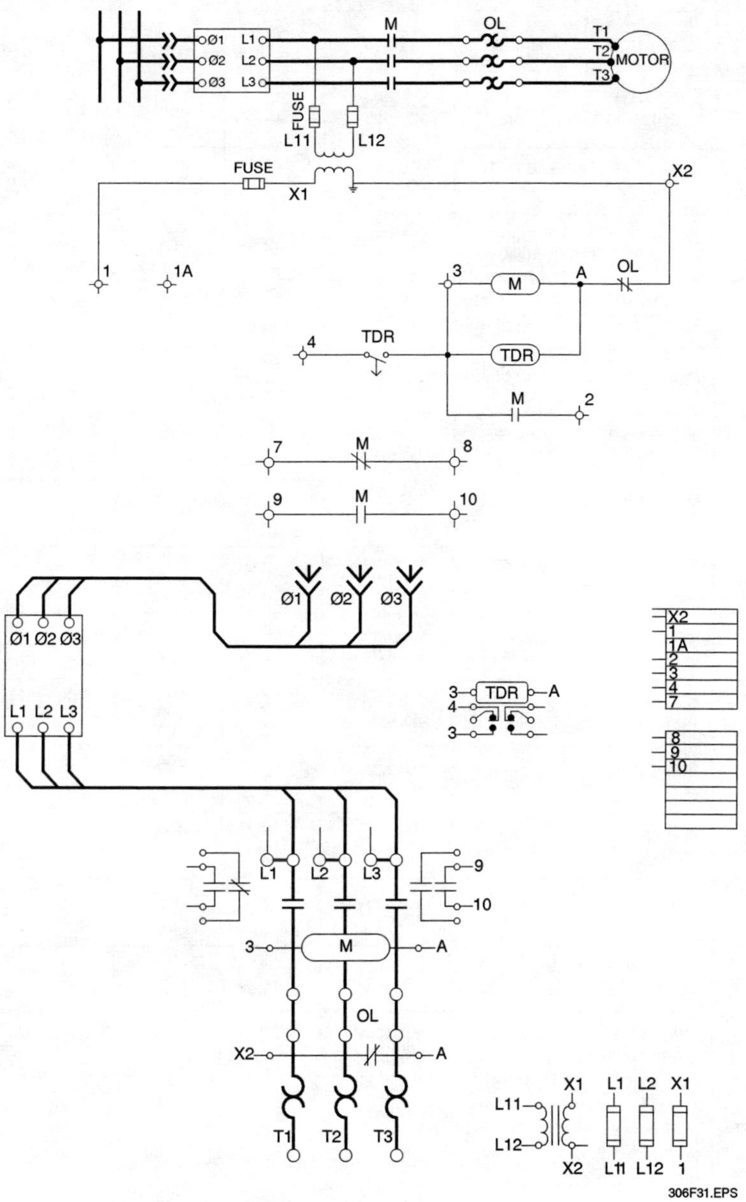

Figure 31 ◆ Unit diagrams for motor control center.

Classroom

Discuss the types of information found on a panelboard schedule. Refer the trainees to Figure 32 in the Trainee Module.

PANEL NO.	LOCATION	MAINS	VOLTAGE RATING	NO. OF CIRCUITS	BREAKER RATINGS	POLES	PURPOSE
P-1	BASEMENT N. CORRIDOR	BREAKER 100A	208/120V 3Ø, 4W	19 2 5	20A 20A 20A	1 2 1	LIGHTING AND RECEPTACLES SPARES
P-2	BASEMENT N. CORRIDOR	BREAKER 100A	208/120V 3Ø, 4W	24 2 0	20A 20A	1 2	LIGHTING AND RECEPTACLES SPARES
P-3	2ND FLOOR N. CORRIDOR	BREAKER 100A	208/120V 3Ø, 4W	24 2 0	20A 20A	1 2	LIGHTING AND RECEPTACLES SPARES
P-4	BASEMENT S. CORRIDOR	BREAKER 100A	208/120V 3Ø, 4W	24 2 0	20A 20A	1 2 1	LIGHTING AND RECEPTACLES SPARES
P-5	1ST FLOOR S. CORRIDOR	BREAKER 100A	208/120V 3Ø, 4W	23 2 1	20A 20A 20A	1 2 1	LIGHTING AND RECEPTACLES SPARES
P-6	2ND FLOOR S. CORRIDOR	BREAKER 100A	208/120V 3Ø, 4W	22 2 2	20A 20A 20A	1 2 1	LIGHTING AND RECEPTACLES SPARES
P-7	MFG. AREA S. WALL E.	BREAKER 100A	208/120V 3Ø, 4W	5 7 2	20A 20A 20A	1 1 1	LIGHTING AND RECEPTACLES SPARES
P-8	MFG. AREA S. WALL W.	BREAKER 100A	208/120V 3Ø, 4W	5 7 2	20A 20A 20A	1 1 1	LIGHTING AND RECEPTACLES SPARES
P-9	MFG. AREA S. WALL E.	BREAKER 100A	208/120V 3Ø, 4W	5 7 2	50A 20A 20A	1 1 1	LIGHTING AND RECEPTACLES SPARES
P-10	MFG. AREA S. WALL W.	BREAKER 100A	208/120V 3Ø, 4W	5 7 2	50A 20A 20A	1 1 1	LIGHTING AND RECEPTACLES SPARES
P-11	MFG. AREA EAST WALL	LUGS ONLY 225A	208/120V 3Ø, 4W	6	20A	3	BLOWERS AND VENTILATORS
P-12	BOILER ROOM	BREAKER 100A	208/120V 3Ø, 4W	10 4	20A 20A	1 1	LIGHTING AND RECEPTACLES SPARES
P-13	BOILER ROOM	LUGS ONLY 225A	208/120V 3Ø, 4W	6	20A	3	OIL BURNERS AND PUMPS
P-14	MFG. AREA EAST WALL	LUGS ONLY 400A	208/120V 3Ø, 4W	3 2 1	175A 70A 40A	3 3 3	CHILLERS FAN COIL UNITS FAN COIL UNITS
P-15	MFG. AREA WEST WALL	LUGS ONLY 600A	208/120V 3Ø, 4W	5	100A	3	TROLLEY BUSWAY AND ELEVATOR

306F32.EPS

Figure 32 ◆ Schedule of electric panelboards for an industrial building.

Instructor's Notes:

16.0.0 ◆ PANELBOARDS

This section covers panelboard construction and protective devices.

16.1.0 Lighting and Power Panelboards

Circuit control and overcurrent protection must be provided for all circuits and the power-consuming devices connected to these circuits. Lighting and power panels located throughout large buildings being supplied with electrical energy provide this control and protection. *Figure 32* (see p. 6.46) shows a schedule of fifteen panelboards provided in a typical industrial building to feed electrical energy to the various circuits.

16.2.0 Panelboard Construction

In general, panelboards are constructed so that the main feed busbars run the height of the panelboard. The buses to the branch circuit protective devices are connected to the alternate main buses. In an arrangement of this type, the connections directly across from each other are on the same phase, and the adjacent connections on each side are on different phases. As a result, multiple protective devices can be installed to serve the 208V equipment. An example of a panelboard is shown in *Figure 33.*

16.2.1 Identification of Conductors

The ungrounded conductors may be any color except green (or green with a yellow stripe), which is reserved for grounding purposes only, or white or gray, which are reserved for the grounded circuit conductor. See ***NEC Section 200.6.***

NEC Section 210.4(D) requires that where different voltages exist in a building, the ungrounded conductors for each system must be identified at each accessible location. Identification may be by color-coding, marking, tape, tagging, or other approved means. The means of identification must be permanently posted at each branch circuit panelboard.

For example, this situation may occur when the building is served with 277/480V and step-down transformers are used to provide 120/208V for lighting and receptacle outlets. Examples of panelboard wiring connections are shown in *Figures 34, 35, 36,* and *37.*

306F33.EPS

Figure 33 ◆ Typical panelboard.

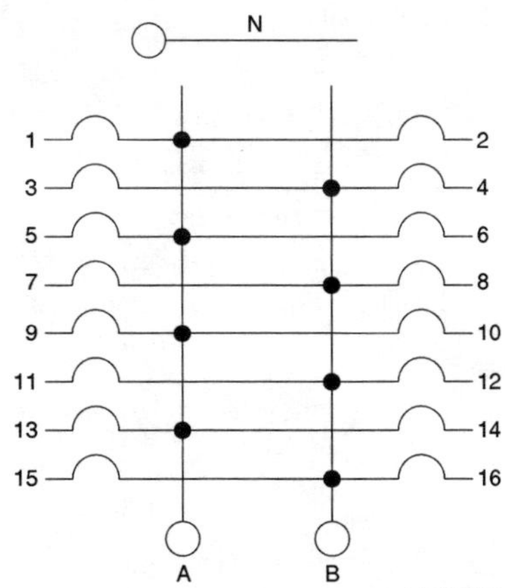

306F34.EPS

Figure 34 ◆ Lighting and appliance branch circuit panelboard—single-phase, three-wire connections.

Classroom

Describe the construction of a typical panelboard. Refer the trainees to Figure 33 in the Trainee Module.

Explain how to identify the conductors in a panelboard.

Audiovisual

Show Transparencies 24 through 27 (Figures 34 through 37).

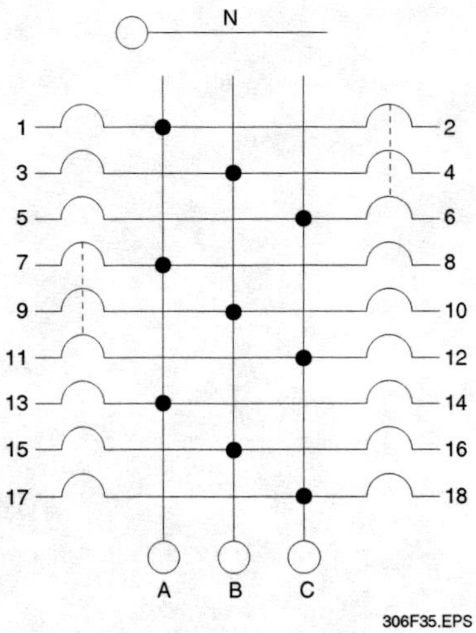

Figure 35 ◆ Lighting and appliance branch circuit panelboard—three-phase, four-wire connections.

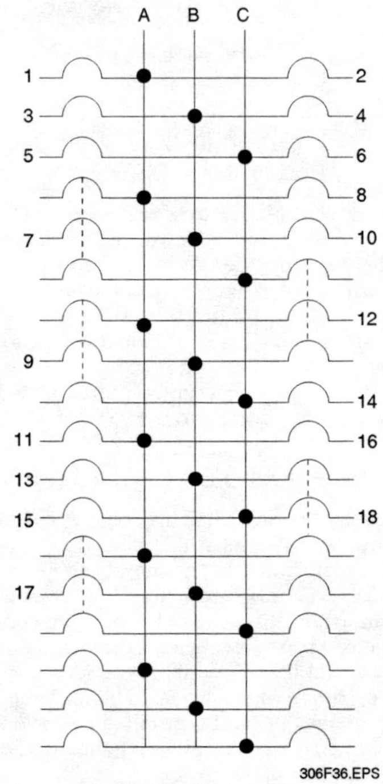

Figure 36 ◆ Bakery panelboard circuit showing alternate numbering scheme.

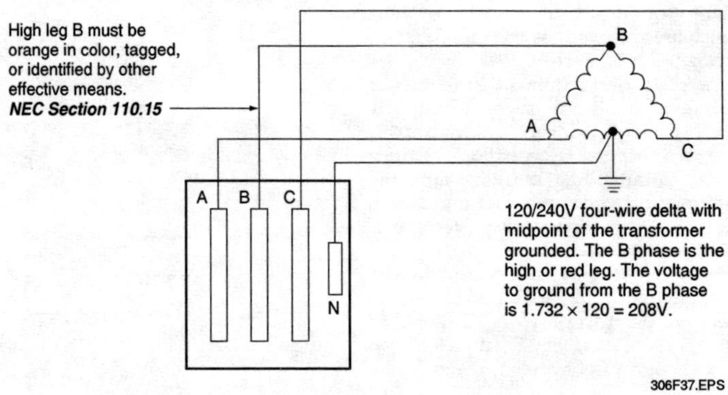

Figure 37 ◆ Panelboards and switchboards supplied by four-wire, delta-connected system.

Instructor's Notes:

16.2.2 *Number of Circuits*

The number of overcurrent devices in a panelboard is determined by the needs of the area being served. Using the bakery panelboard in *Figure 36* as an example, there are 13 single-pole circuits and five three-pole circuits. This is a total of 28 poles. When using a three-phase supply, the incremental number is six (a pole for each of the three phases on both sides of the panelboard). The minimum number of poles that could be specified for the bakery is 30. This would limit the power available for growth and would not permit the addition of a three-pole lead. The reasonable choice is to go to 36 poles, which provides flexibility for growth loads.

16.3.0 Panelboard Protective Devices

The main protective device for a panelboard may be either a fuse or a circuit breaker. This section concentrates on the use of circuit breakers. The selection of the circuit breaker should be based on the necessity to:

- Provide the proper overload protection
- Ensure a suitable voltage rating
- Provide a sufficient interrupting current rating
- Provide short circuit protection
- Coordinate the breaker(s) with other protective devices

The choice of the overload protection is based on the rating of the panelboard. The trip rating of the circuit breaker cannot exceed the amperage capacity of the busbars in the panelboard. The number of branch circuit breakers is generally not a factor in the selection of the main protective device except in a practical sense. It is a common practice to have the total amperage of the branch breakers greatly exceed the rating of the main breaker; however, it makes little sense for a single branch circuit breaker to be the same size as, or larger than, the main breaker.

The voltage rating of the breaker must be higher than that of the system. Breakers are usually rated at 250V to 600V.

The importance of the proper interrupting rating cannot be overstressed. You should recall that if there is ever any question as to the exact value of the short circuit current available at a point, the circuit breaker with the higher interrupting rating is to be installed.

Many circuit breakers used as the main protective device are provided with an adjustable magnetic trip (*Figure 38*). Adjustments of this trip determine the degree of protection provided by the circuit breaker if a short circuit occurs. The manufacturer of this device provides exact information about the adjustments to be made. In general, a low setting may be 10 or 12 times the overload trip rating.

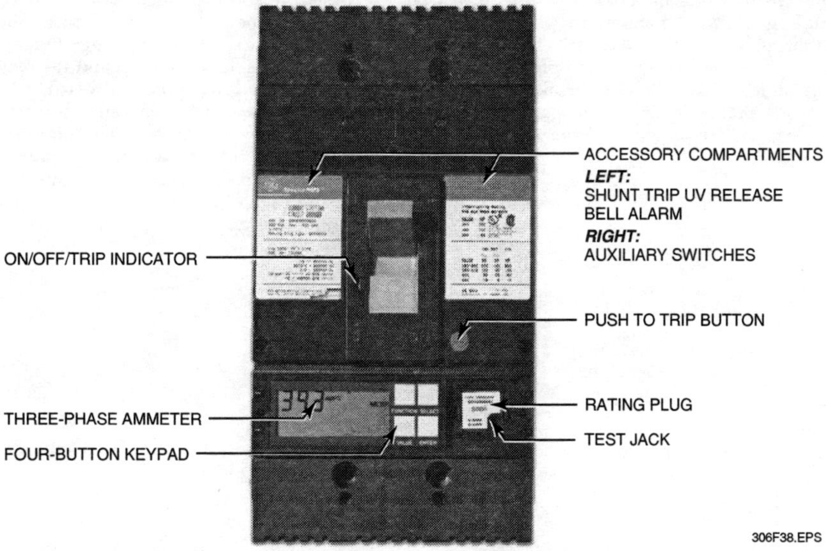

Figure 38 ◆ Circuit breaker with adjustable magnetic trip.

Classroom

Explain how to calculate the number of overcurrent devices required in a panelboard.

Discuss the factors involved in circuit breaker selection.

Describe the operation of magnetic trip circuit breakers. Refer the trainees to Figure 38 in the Trainee Module.

Classroom

Explain that a double-pole breaker takes up the same space as two single-pole breakers. Refer the trainees to Figure 39 in the Trainee Module.

Two rules should be followed whenever the magnetic trip is set:

- The magnetic trip must be set to the minimum practical setting.
- The setting must be lower than the value of the short circuit current available at that point.

If subfeed lugs are used, ensure that the lugs are suitable for making multiple breaker connections, as required by ***NEC Section 110.14(A).*** In general, this means that a separate lug is to be provided for each conductor being connected.

If taps are made to the subfeeder, they can be reduced in size according to ***NEC Section 240.21.*** This specification is very useful in cases such as that of panel P-12 in *Figure 32.* For this panel, a 100A main breaker is fed by a 350MCM conductor. Within the distances given in ***NEC Section 240.21(B)(1),*** a conductor with a 100A rating may be tapped to the subfeeder and connected to the 100A main breaker in the panel.

Per ***NEC Section 110.14(C),*** the temperature rating of conductors must be selected and coordinated so as not to exceed the lowest temperature rating of any connected termination, conductor, or device.

16.4.0 Branch Circuit Protective Devices

The schedule of panelboards for the industrial building (*Figure 32*) shows that lighting panels P-1 through P-6 have 20A circuit breakers, including double-pole breakers to supply special receptacle outlets. A double-pole breaker requires the same installation space as two single-pole breakers. Breakers are shown in *Figure 39.*

306F39.EPS

Figure 39 ◆ Branch circuit protective devices.

Summary

This module explained the purpose of switchgear. Switchgear construction, metering layouts, wiring requirements, and maintenance were discussed. It also explained NEC requirements for these systems and provided a basic understanding of how to apply them. Circuit breakers, their four general classifications, and the major circuit breaker ratings were also addressed. Additionally, ground fault relay systems and the testing of such systems were explained. This module also covered visual and mechanical inspections and electrical tests associated with low-voltage and medium-voltage cables, metal-enclosed busways, and metering and instrumentation.

Instructor's Notes:

Review Questions

1. Voltages classified as medium voltage are in the range of ____.
   a. 120V to 240V
   b. 120V to 600V
   c. 601V to 15kV
   d. 15kV to 230kV

2. The term *interrupting rating* refers to the ____.
   a. trip setting of a circuit breaker
   b. voltage rating of a fuse
   c. highest voltage level a device can withstand
   d. maximum current a device will safely interrupt at rated voltage

3. Which of the following is a color that can be used to designate an ungrounded conductor?
   a. Green
   b. White
   c. Gray
   d. Red

4. When an unintended path is established between an ungrounded conductor and ground, it is called a(n) ____.
   a. phase fault
   b. open circuit
   c. ground fault
   d. overload

5. A device that is specifically designed to protect equipment from ground faults through the use of sensors is a ____.
   a. molded-case circuit breaker
   b. dual-element fuse
   c. ground fault relay
   d. ground fault circuit interrupter

6. The maximum voltage that a piece of equipment can withstand is known as its ____.
   a. interrupting capacity
   b. basic impulse insulation level (BIL)
   c. current limit
   d. frequency

7. A transformer rated at more than 500kVA is considered a(n) ____ transformer.
   a. power
   b. control
   c. distribution
   d. isolation

8. The term *capacity* on a transformer nameplate refers to ____.
   a. its voltage rating
   b. its ability to transfer energy
   c. the voltage produced by the secondary
   d. the number of secondary windings

9. The term *class* on a transformer nameplate refers to ____.
   a. its use, such as control or power
   b. the type of cooling it uses
   c. whether it is step-up or step-down
   d. its range of operating frequencies

10. When testing switchboards, what should you do in order to ensure a stable reference point to ground?
    a. Bond all three phases together.
    b. Bond the neutral to each phase.
    c. Disconnect the neutral from the grounding conductor.
    d. Bond neutrals to one another and to the grounding conductor.

11. Of the following types of current transformers, which is normally used with circuit breakers or power transformers?
    a. Bar
    b. Bushing
    c. Oil filled
    d. Window

12. Which of the following is *not* a general classification of circuit breakers?
    a. Fuse circuit breaker
    b. Air circuit breaker
    c. Oil circuit breaker
    d. Gas circuit breaker

13. What type of diagram shows the actual wiring connections between unit assemblies or equipment with each wire labeled to indicate where to terminate it?
    a. Schematic diagram
    b. Front panel diagram
    c. Interconnection diagram
    d. Block diagram

14. The trip rating of a circuit breaker used as the main protective device in a panelboard cannot exceed ____.
    a. the total amperage of the branch breakers
    b. the amperage capacity of the busbars in the panelboard
    c. the amperage of the individual branch fuses
    d. 250V

15. According to the IEEE Identification System found in the *Appendix* of this module, if device No. 51 is indicated on an electrical print, it would be a(n) ____.
    a. circuit breaker
    b. reverse power relay
    c. field circuit breaker
    d. AC time overcurrent relay

Classroom

Have the trainees complete the Review Questions and go over the answers prior to administering the Module Examination.

Examination

Administer the Module Examination. Record the results on Craft Training Report Form 200 and submit the results to the Training Program Sponsor.

GLOSSARY

Trade Terms Introduced in This Module

Air circuit breaker: A circuit breaker in which the interruption occurs in air.

Basic impulse insulation level (BIL): The maximum impulse voltage the winding insulation can withstand without failure.

Branch circuit: A set of conductors that extends beyond the last overcurrent device in the low-voltage system of a given building.

Bus: A conductor or group of conductors that serves as a common connection for two or more circuits in a switchgear assembly.

Bushing: An insulating structure including a through conductor, or providing a passageway for such a conductor, for the purpose of insulating the conductor from the barrier and conducting from one side of the barrier to the other.

Capacity: The rated load-carrying ability, expressed in kilovolt-amperes or kilowatts, of generating equipment or other electric apparatus.

Contactor: An automatic electric power switch designed for frequent operation.

Distribution system equipment: Switchboard equipment that is downstream from the service-entrance equipment.

Distribution transformer: A transformer that is used for transferring electric energy from a primary distribution circuit to a secondary distribution circuit. Distribution transformers are usually rated between 5kVA and 500kVA.

Feeder: A set of conductors originating at a main distribution center that supply one or more secondary distribution centers, one or more branch circuit distribution centers, or any combination of these two types of load.

Metal-enclosed switchgear: Switchgear that is primarily used in indoor applications up to 600V.

Service-entrance equipment: Equipment located at the service entrance of a given building that provides overcurrent protection to the feeder and service conductors and also provides a means of disconnecting the feeders from the energized service equipment.

Switchboard: A large single panel, frame, or assembly of panels on which switches, fuses, buses, and instruments are mounted.

Switchgear: A general term covering switching or interrupting devices and any combination thereof with associated control, instrumentation, metering, protective, and regulating devices.

Instructor's Notes:

APPENDIX

IEEE Identification System

The devices in switching equipment are referred to by numbers with appropriate suffix letters when necessary, according to the functions they perform.

These numbers are based on a system adopted as standard for automatic switchgear by IEEE and incorporated in American Standard C37.2-1970. This system is used in connection diagrams, instruction books, and specifications.

Device Number	Definition and Function
1	**Master Element** – The initiating device, such as a control switch, voltage relay, float switch, that serves either directly, or through such permissive devices as protective and time-delay relays, to place equipment in or out of operation.
2	**Time-Delay Starting or Closing Relay** – A device that functions to give a desired amount of time delay before or after any point of operation in a switching sequence or protective relay system, except as specifically provided by device functions 48, 62, and 79 described later.
3	**Checking or Interlocking Relay** – A device that operates in response to the position of a number of other devices (or to a number of predetermined conditions) in equipment to allow an operating sequence to proceed, to stop, or to provide a check of the position of these devices or of these conditions for any purpose.
4	**Master Contactor** – A device, generally controlled by device No. 1 or equivalent, and the required permissive and protective devices, that serves to make and break the necessary control circuits to place equipment into operation under the desired conditions and to take it out of operation under other or abnormal conditions.
5	**Stopping Device** – A control device used primarily to shut down equipment and hold it out of operation. [This device may be manually or electrically actuated, but excludes the function of electrical lockout (see device function 86) on abnormal conditions.]
6	**Starting Circuit Breaker** – A device whose principal function is to connect a machine to its source of starting voltage.
7	**Anode Circuit Breaker** – A device used in the anode circuits of a power rectifier for the primary purpose of interrupting the rectifier circuit if an arc-back should occur.
8	**Control Power Disconnecting Device** – A disconnective device, such as a knife switch, circuit breaker, or pullout fuse block, that is used for the purpose of connecting and disconnecting the source of control power to and from the control bus or equipment. ***Note:*** Control power is considered to include auxiliary power which supplies such apparatus as small motors and heaters.
9	**Reversing Device** – A device used for the purpose of reversing a machine field or for performing any other reversing functions.
10	**Unit Sequence Switch** – A device used to change the sequence in which units may be placed in and out of service in multiple-unit equipment.
11	Reserved for future application.
12	**Over-Speed Device** – Usually a direct-connected speed switch that functions on machine over-speed.
13	**Synchronous-Speed Device** – A device such as a centrifugal-speed switch, slip-frequency relay, voltage relay, or undercurrent relay, that operates at approximately the synchronous speed of a machine.
14	**Under-Speed Device** – A device that functions when the speed of a machine falls below a predetermined value.
15	**Speed- or Frequency-Matching Device** – A device that functions to match and hold the speed or frequency of a machine or of a system equal to, or approximately equal to, that of another machine, source, or system.

306A01.EPS

Instructor's Notes:

Device Number	Definition and Function
16	Reserved for future application.
17	**Shunting or Discharge Switch** – A device that serves to open or close a shunting circuit around any piece of apparatus (except a resistor), such as machine field, machine armature, capacitor, or reactor.
	Note: This excludes devices that perform such shunting operations as may be necessary in the process of starting a machine by devices 6 or 42, or their equivalent, and also excludes the device 73 function, which serves for the switching of resistors.
18	**Accelerating or Decelerating Device** – A device used to close or to cause the closing of circuits that are used to increase or decrease the speed of a machine.
19	**Starting-to-Running Transition Contactor** – A device that operates to initiate or cause the automatic transfer of a machine from the starting to the running power connection.
20	**Electrically Operated Valve** – An electrically operated, controlled, or monitored valve in a fluid line.
	Note: The function of the valve may be indicated by the use of suffixes.
21	**Distance Relay** – A device that functions when the circuit admittance, impedance, or reactance increases or decreases beyond predetermined limits.
22	**Equalizer Circuit Breaker** – A breaker that serves to control or to make and break the equalizer or the current-balancing connections for a field, or for regulating equipment, in a multiple-unit installation.
23	**Temperature Control Device** – A device that functions to raise or lower the temperature of a machine or other apparatus, or of any medium, when its temperature falls below or rises above a predetermined value.
	Note: An example is a thermostat that switches on a space heater in a switchgear assembly when the temperature falls to a desired value as distinguished from a device that is used to provide automatic temperature regulation between close limits and would be designated as 90T.
24	Reserved for future application.
25	**Synchronizing or Synchronism-Check Device** – A device that operates when two AC circuits are within the desired limits of frequency, phase angle, or voltage to permit or to cause the paralleling of these two circuits.
26	**Apparatus Thermal Device** – A device that functions when the temperature of the shunt field or the armortisseur winding of a machine or that of a load limiting or load shifting resistor or of a liquid or other medium exceeds a predetermined value; it also functions if the temperature of the protected apparatus, such as a power rectifier, or of any medium decreases below a predetermined value.
27	**Undervoltage Relay** – A device that functions on a given value of undervoltage.
28	**Flame Detector** – A device that monitors the presence of the pilot or main flame in such apparatus as a gas turbine or steam boiler.
29	**Isolating Contactor** – A device used expressly for disconnecting one circuit from another for the purposes of emergency operation, maintenance, or testing.
30	**Annunciator Relay** – A nonautomatic reset device that gives a number of separate visual indications upon the functioning of protective devices and that may also be arranged to perform a lockout function.
31	**Separate Excitation Device** – A device that connects a circuit, such as the shunt field of a synchronous converter, to a source of separate excitation during the starting sequence or one that energizes the excitation and ignition circuits of a power rectifier.

306A02.EPS

Device Number	Definition and Function
32	**Directional Power Relay** – A device that functions on a desired value of power flow in a given direction or upon reverse power resulting from arc-back in the anode or cathode circuits of a power rectifier.
33	**Position Switch** – A device that makes or breaks its contacts when the main device or piece of apparatus that has no device function number reaches a given position.
34	**Master Sequence Device** – A device, such as a motor-operated multi-contact switch or the equivalent, or a programming device, such as a computer, that establishes or determines the operating sequence of the major devices in equipment during starting and stopping or during other sequential switching operations.
35	**Brush-Operating or Slip-Ring Short-Circuiting Device** – A device used for raising, lowering, or shifting the brushes of a machine; for short-circuiting its slip rings, or for engaging or disengaging the contacts of a mechanical rectifier.
36	**Polarity or Polarizing Voltage Device** – A device that operates or permits the operation of another device on a predetermined polarity only, or one that verifies the presence of a polarizing voltage in equipment.
37	**Undercurrent or Underpower Relay** – A device that functions when the current or power flow decreases below a predetermined value.
38	**Bearing Protective Device** – A device that functions on excessive bearing temperature or on other abnormal mechanical conditions, such as undue wear, that may eventually result in excessive bearing temperature.
39	**Mechanical Condition Monitor** – A device that functions upon the occurrence of an abnormal mechanical condition (except that associated with bearings as covered under device function 38), such as excessive vibration, eccentricity, expansion, shock, tilting, or seal failure.
40	**Field Relay** – A device that functions on a given or abnormally low value or failure of machine field current or on an excessive value of the reactive component of armature current in an AC machine indicating abnormally low field excitation.
41	**Field Circuit Breaker** – A device that functions to apply or remove the field excitation of a machine.
42	**Running Circuit Breaker** – A device whose principal function is to connect a machine to its source of running or operating voltage. This function may also be used for a device, such as a contactor, that is used in series with a circuit breaker or other fault protecting means, primarily for frequent opening and closing of the circuit.
43	**Manual Transfer or Selector Device** – A device that transfers the control circuits so as to modify the plan of operation of the switching equipment or of some of the devices.
44	**Unit Sequence Starting Relay** – A device that functions to start the next available unit in multiple-unit equipment on the failure or non-availability of the normally preceding unit.
45	**Atmospheric Condition Monitor** – A device that functions upon the occurrence of an abnormal atmospheric condition, such as damaging fumes, explosive mixtures, smoke, or fire.
46	**Reverse-Phase or Phase-Balance Current Relay** – A device that functions when the polyphase currents are of reverse-phase sequence or when the polyphase currents are unbalanced or contain negative phase-sequence components above a given amount.
47	**Phase-Sequence Voltage Relay** – A relay that functions upon a predetermined value of polyphase voltage in the desired phase sequence.
48	**Incomplete Sequence Relay** – A relay that generally returns the equipment to the normal or off position and locks it out if the normal starting, operating, or stopping sequence is not properly completed within a predetermined time. If the device is used for alarm purposes only, it should preferably be designated as 48A (alarm).

306A03.EPS

Instructor's Notes:

Device Number	Definition and Function
49	**Machine or Transformer Thermal Relay** – A relay that functions when the temperature of a machine armature, or other load-carrying winding or element of a machine or the temperature of a power rectifier or power transformer (including a power rectifier transformer) exceeds a predetermined value.
50	**Instantaneous Overcurrent or Rate-of-Rise Relay** – A relay that functions instantaneously on an excessive value of current or on an excessive rate of current rise, indicating a fault in the apparatus or circuit being protected.
51	**AC Time Overcurrent Relay** – A relay with either a definite or inverse time characteristic that functions when the current in an AC circuit exceeds a predetermined value.
52	**AC Circuit Breaker** – A device that is used to close and interrupt an AC power circuit under normal conditions or to interrupt this circuit under fault or emergency conditions.
53	**Exciter or DC Generator Relay** – A relay that forces the DC machine field excitation to build up during starting or which functions when the machine voltage has built up to a given value.
54	Reserved for future application.
55	**Power Factor Relay** – A relay that operates when the power factor in an AC circuit rises above or below a predetermined value.
56	**Field Application Relay** – A relay that automatically controls the application of the field excitation to an AC motor at some predetermined point in the slip cycle.
57	**Short-Circuiting or Grounding Device** – A primary circuit switching device that functions to short-circuit or ground a circuit in response to automatic or manual means.
58	**Rectification Failure Relay** – A device that functions if one or more anodes of a power rectifier fail to fire, to detect an arc-back, or on failure of a diode to conduct or block properly.
59	**Overvoltage Relay** – A relay that functions on a given value of overvoltage.
60	**Voltage or Current Balance Relay** – A relay that operates on a given difference in voltage or current input or output of two circuits.
61	Reserved for future application.
62	**Time-Delay Stopping or Opening Relay** – A time-delay relay that serves in conjunction with the device that initiates the shutdown, stopping, or opening operation in an automatic sequence.
63	**Pressure Switch** – A switch that operates on given values or on a given rate of change of pressure.
64	**Ground Protective Relay** – A relay that functions on failure of the insulation of a machine, transformer, or other apparatus to ground or on flashover of a DC machine to ground. ***Note:*** This function is assigned only to a relay that detects the flow of current from the frame of a machine or enclosing case or structure of a piece of apparatus to ground or one that detects a ground on a normally ungrounded winding or circuit. It is not applied to a device connected in the secondary circuit or secondary neutral of a current transformer connected in the power circuit of a normally grounded system.
65	**Governor** – The assembly of fluid, electrical, or mechanical control equipment used for regulating the flow of water, steam, or other medium to the prime mover for such purposes as starting, holding speed or load, or stopping.

306A04.EPS

Device Number	Definition and Function
66	**Notching or Jogging Device** – A device that functions to allow only a specified number of operations of a given device or equipment or a specified number of successive operations within a given time of each other. It also functions to energize a circuit periodically or for fractions of specified time intervals or that is used to permit intermittent acceleration or jogging of a machine at low speeds for mechanical positioning.
67	**AC Directional Overcurrent Relay** – A relay that functions on a desired value of AC overcurrent flowing in a predetermined direction.
68	**Blocking Relay** – A relay that initiates a pilot signal for blocking of tripping on external faults in a transmission line or in other apparatus under predetermined conditions, or a relay cooperates with other devices to block tripping or to block reclosing on an out-of-step condition or on power swings.
69	**Permissive Control Device** – Generally a two-position, manually operated switch that in one position permits the closing of a circuit breaker or the placing of equipment into operation and in the other position prevents the circuit breaker or the equipment from being operated.
70	**Rheostat** – A variable resistance device used in an electric circuit that is electrically operated or has other electrical accessories, such as auxiliary, position, or limit switches.
71	**Level Switch** – A switch that operates on given values or on a given rate of change of level.
72	**DC Circuit Breaker** – A circuit breaker used to close and interrupt a DC power circuit under normal conditions or to interrupt this circuit under fault or emergency conditions.
73	**Load-Resistor Contactor** – A contactor used to shunt or insert a step of load limiting, shifting, or indicating resistance in a power circuit, to switch a space heater in a circuit, or to switch a light or regenerative load resistor of a power rectifier or other machine in and out of a circuit.
74	**Alarm Relay** – A device other than an annunciator, as covered under device No. 30, that is used to operate or to operate in connection with a visual or audible alarm.
75	**Position Changing Mechanism** – A mechanism that is used for moving a main device from one position to another in equipment (for example, shifting a removable circuit breaker unit to and from the connected, disconnected, and test positions).
76	**DC Overcurrent Relay** – A relay that functions when the current in a DC circuit exceeds a given value.
77	**Pulse Transmitter** – A device used to generate and transmit pulses over a telemetering or pilot-wire circuit to remove the indicating or receiving device.
78	**Phase Angle Measuring or Out-of-Step Protective Relay** – A relay that functions at a predetermined phase angle between two voltages, between two currents, or between voltage and current.
79	**AC Reclosing Relay** – A relay that controls the automatic reclosing and locking out of an AC circuit interrupter.
80	**Flow Switch** – A switch that operates on given values, or a given rate of change of flow.
81	**Frequency Relay** – A relay that functions on a predetermined value of frequency, either under, over, or on normal system frequency or rate of change of frequency.
82	**DC Reclosing Relay** – A relay that controls the automatic closing and reclosing of a DC circuit interrupter, generally in response to load circuit conditions.
83	**Automatic Selective Control or Transfer Relay** – A relay that operates to select automatically between certain sources or conditions in equipment or that performs a transfer operation automatically.

306A05.EPS

Instructor's Notes:

Device Number	Definition and Function
84	**Operating Mechanism** – The complete electrical mechanism or servo-mechanism, including the operating motor, solenoids, position switches, and for a tap changer, induction regulator, or any similar piece of apparatus that has no device function number.
85	**Carrier or Pilot-Wire Receiver Relay** – A relay that is operated or restrained by a signal used in connection with carrier-current or DC pilot-wire fault directional relaying.
86	**Locking-Out Relay** – An electrically operated relay that functions to shut down and hold equipment out of service on the occurrence of abnormal conditions. It may be reset either manually or electrically.
87	**Differential Protective Relay** – A protective relay that functions on a percentage of phase angle or other quantitative difference of two currents or of some other electrical quantities.
88	**Auxiliary Motor or Motor Generator** – A device used for operating auxiliary equipment, such as pumps, blowers, exciters, and rotating magnetic amplifiers.
89	**Line Switch** – A switch used as a disconnecting load-interrupter or isolating switch in an AC or DC power circuit when this device is electrically operated or has electrical accessories, such as an auxiliary switch or magnetic lock.
90	**Regulating Device** – A device that functions to regulate a quantity, or quantities, such as voltage, current, power, speed, frequency, temperature, and load, at a certain value or between certain (generally close) limits for machines, tie lines, or other apparatus.
91	**Voltage Directional Relay** – A relay that operates when the voltage across an open circuit breaker or contactor exceeds a given value in a given direction.
92	**Voltage and Power Directional Relay** – A relay that permits or causes the connection of two circuits when the voltage difference between them exceeds a given value in a predetermined direction and causes these two circuits to be disconnected from each other when the power flowing between them exceeds a given value in the opposite direction.
93	**Field Changing Contactor** – A device that functions to increase or decrease in one step the value of field excitation on a machine.
94	**Tripping or Trip-Free Relay** – A device that functions to trip a circuit breaker, contactor, or equipment, to permit immediate tripping by other devices, or to prevent immediate reclosure of a circuit interrupter in case it should open automatically even though its closing circuit is maintained closed.
95 96 97	Used only for specific applications on individual installations where none of the assigned numbered functions from 1 to 94 is suitable.

306A06.EPS

Additional Resources

This module is intended to present thorough resources for task training. The following reference works are suggested for further study. These are optional materials for continued education rather than for task training.

American Electrician's Handbook, 1996. Terrell Croft and Wilfred I. Summers. New York, NY: McGraw-Hill.

National Electrical Code Handbook, 2002. Quincy, MA: National Fire Protection Association.

Instructor's Notes:

ANSWER KEY

Answers to Review Questions

Answer	Section
1. c	2.0.0
2. d	6.1.0
3. d	6.2.0
4. c	7.0.0
5. c	7.2.0
6. b	8.1.0
7. a	10.2.0
8. b	10.5.0
9. b	10.5.0
10. d	5.0.0
11. b	11.2.0
12. a	12.0.0
13. c	14.3.0
14. b	16.3.0
15. d	Appendix A

Section 11.3.0 *Think About It—Transformers*

An instrument transformer generally has a low current rating since its output is used for metering only (indications), while a control transformer's output is used to supply control circuit components and generally requires a higher current rating. For this reason, one device cannot serve both functions.

Section 7.0.0 *What's Wrong with This Picture?*

Other than the obvious disarray, note the uncapped pipe opening that could allow water or other contaminants to enter the enclosures.

The NCCER makes every effort to keep these textbooks up-to-date and free of technical errors. We appreciate your help in this process. If you have an idea for improving this textbook, or if you find an error, a typographical mistake, or an inaccuracy in NCCER's Contren™ textbooks, please write us, using this form or a photocopy. Be sure to include the exact module number, page number, a detailed description, and the correction, if applicable. Your input will be brought to the attention of the Technical Review Committee. Thank you for your assistance.

Instructors – If you found that additional materials were necessary in order to teach this module effectively, please let us know so that we may include them in the Equipment/Materials list in the Instructor's Guide.

Write: Curriculum Revision and Development Department
National Center for Construction Education and Research
P.O. Box 141104, Gainesville, FL 32614-1104

Fax: 352-334-0932

E-mail: curriculum@nccer.org

Craft ____________ Module Name ____________

Copyright Date ____________ Module Number ____________ Page Number(s) ____________

Description ____________

(Optional) Correction ____________

(Optional) Your Name and Address ____________

Distribution System Transformers

26307-02

MODULE OVERVIEW

This module introduces the Electrical trainee to the methods and procedures used in selecting and wiring distribution system transformers.

PREREQUISITES

Please refer to the Course Map in the Trainee Module. Prior to training with this module, it is recommended that the trainee shall have successfully completed the following:

Core Curriculum; Electrical Level One; Electrical Level Two; Electrical Level Three, Modules 26301 through 26306.

OBJECTIVES

Upon completion of this module, the trainee will be able to:

1. Describe transformer operation.
2. Explain the principle of mutual induction.
3. Describe the operating characteristics of various types of transformers.
4. Connect a multi-tap transformer for the required secondary voltage.
5. Explain NEC requirements governing the installation of transformers.
6. Compute transformer sizes for various applications.
7. Explain types and purposes of grounding transformers.
8. Connect a control transformer for a given application.
9. Size the maximum load allowed on open delta systems.
10. Describe how current transformers are used in conjunction with watt-hour meters.
11. Apply capacitors and rectifiers to practical applications.
12. Calculate the power factor of any given electrical circuit.

PERFORMANCE TASKS

Under supervision of the instructor, the trainee should be able to:

1. Sketch or physically connect a dual-voltage transformer showing a high primary and low secondary, including proper bonding.
2. Sketch or physically connect a dual-voltage transformer showing a low primary and high secondary, including proper bonding.
3. Sketch or physically connect a single-phase transformer in a delta-wye and open delta configuration, including proper bonding.

NCCER STANDARDIZED CRAFT TRAINING PROGRAMS

The National Center for Construction Education and Research (NCCER) provides a standardized national program of accredited craft training. Key features of the program include instructor certification, competency-based training, and performance testing. The program provides trainees, instructors, and companies with a standard form of recognition through a National Craft Training Registry. The program is described in full in the *Guidelines for Accreditation*, published by the NCCER. For more information on standardized craft training, contact the NCCER by writing us at P.O. Box 141104, Gainesville, FL 32614-1104, calling 352-334-0911, or e-mailing info@nccer.org. More information may be found at our Web site, www.nccer.org.

HOW TO USE THIS ANNOTATED INSTRUCTOR'S GUIDE

Each page presents two sections of information. The larger section displays each page exactly as it appears in the Trainee Module. The narrow column ties suggested trainee and instructor actions to each page and provides icons to call your attention to material, safety, audiovisual, or testing requirements. The bottom of each page includes space for your notes.

If you see the Teaching Tip icon, that means there is a teaching tip associated with this section. Also refer to any suggested teaching tips at the end of the module.

SAFETY CONSIDERATIONS

Ensure that the trainees are equipped with appropriate personal protective equipment.

PREPARATION

Before teaching this module, you should review the Module Outline, Objectives, Performance Tasks, and the Materials and Equipment List. Be sure to allow ample time to prepare your own training or lesson plan and gather all required equipment and materials.

MATERIALS AND EQUIPMENT LIST

Materials:
Transparencies
Markers/chalk
Copy of the latest edition of the *National Electrical Code®*
Samples of various types of transformers
Samples of transformer cores, including:
- Mitered
- Butt
- Wound

Iron filings
Multi-tap control transformers
Control cable
Module Examinations*
Performance Profile Sheets*

Equipment:
Overhead projector and screen
Whiteboard/chalkboard
Appropriate personal protective equipment
Volt-ohm-milliammeter (VOM)
Tools necessary for making transformer wiring connections

*Located in the Test Booklet packaged with this Annotated Instructor's Guide.

ADDITIONAL RESOURCES

This module is intended to present thorough resources for task training. The following reference works are suggested for both instructors and motivated trainees interested in further study. These are optional materials for continued education rather than for task training.

American Electrician's Handbook, Terrell Croft and Wilfred I. Summers. New York, NY: McGraw-Hill, 1996.

National Electrical Code Handbook. Quincy, MA: National Fire Protection Association, 2002.

TEACHING TIME FOR THIS MODULE

An outline for use in developing your lesson plan is presented below. Note that each Roman numeral in the outline equates to one session of instruction. Each session has a suggested time period of 2½ hours. This includes 10 minutes at the beginning of each session for administrative tasks and one 10-minute break during the session. Approximately 15 hours are suggested to cover *Distribution System Transformers*. You will need to adjust the time required for hands-on activity and testing based on your class size and resources.

Topic	Planned Time
Session I. Introduction; Transformer Basics; Transformer Construction; Transformer Taps	
A. Introduction	__________
B. Transformer Basics	__________
1. Mutual Induction	__________
2. Induction in Transformers	__________
3. Magnetic Flux in Transformers	__________
C. Transformer Construction	__________
1. Cores	__________
2. Types of Cores	__________
3. Wound Cores	__________
4. Mitered Cores	__________
5. Transformer Characteristics	__________
D. Transformer Taps	__________
1. Laboratory	__________
Have the trainees practice connecting a dual-voltage transformer showing a high primary and low secondary, and vice versa. Note the proficiency of each trainee.	
Session II. Basic Transformer Connections; Autotransformers	
A. Basic Transformer Connections	__________
1. Single-Phase Light and Power Systems	__________
2. Three-Phase Power Systems	__________
a. Delta-Wye Transformers	__________
b. Delta-Delta Transformers	__________
c. Open Delta Transformers	__________
3. Laboratory	__________
Have the trainees practice connecting a single-phase transformer in a delta-wye and open delta configuration. Note the proficiency of each trainee.	
4. Parallel Operation of Transformers	__________
a. Impedance in Parallel-Operated Transformers	__________
b. Parallel Operation of Three-Phase Transformers	__________
B. Autotransformers	__________
Session III. Dry-Type Transformer Connections; Control Transformers; Potential and Current Transformers	
A. Dry-Type Transformer Connections	__________
1. Zig-Zag Connections	__________
2. Buck-and-Boost Transformers	__________
B. Control Transformers	__________
C. Potential and Current Transformers	__________

Session IV. NEC Requirements

A. NEC Requirements ____________

1. Overcurrent Protection for Transformers (600V or Less) ____________
    a. Overcurrent Protection for Small Power Transformers ____________
2. Transformer Grounding ____________

Session V. Power Factor; Vectors; Troubleshooting

A. Power Factor ____________

1. Capacitors ____________
2. Resistors and Reactors ____________
3. Diodes and Rectifiers ____________

B. Vectors ____________

1. Practical Applications of Phasor Diagrams ____________
2. Voltage Drop ____________

C. Troubleshooting ____________

1. Double-Check Before Beginning ____________
2. Find and Correct the Basic Cause of the Trouble ____________
3. Troubleshooting Transformers ____________
    a. Open Circuit ____________
    b. Shorted Turns ____________
    c. Complete Short ____________
    d. Grounded Windings ____________

Session VI. Review; Module Examination; Performance Testing

A. Review ____________

B. Module Examination ____________

1. Trainees must score 70% or higher to receive recognition from the NCCER.
2. Record the testing results on Craft Training Report Form 200 and submit the results to the Training Program Sponsor.

C. Performance Testing ____________

1. Trainees must perform each task to the satisfaction of the instructor to receive recognition from the NCCER.
2. Record the testing results on Craft Training Report Form 200 and submit the results to the Training Program Sponsor.

Module 26307-02

Distribution System Transformers

Course Map

This course map shows all of the modules in the third level of the Electrical curriculum. The suggested training order begins at the bottom and proceeds up. Skill levels increase as you advance on the course map. The local Training Program Sponsor may adjust the training order.

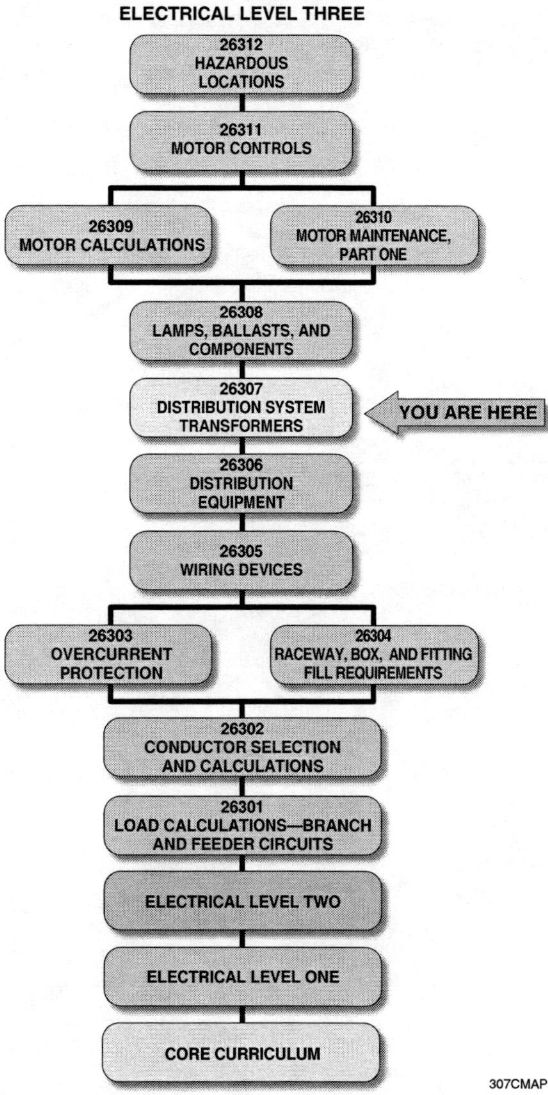

307CMAP.EPS

Homework

Assign reading of Module 26307.

MODULE 26307 CONTENTS

Figures

Instructor's Notes:

Tables

MODULE 26307

Distribution System Transformers

Objectives

When you have completed this module, you will be able to do the following:

1. Describe transformer operation.
2. Explain the principle of mutual induction.
3. Describe the operating characteristics of various types of transformers.
4. Connect a multi-tap transformer for the required secondary voltage.
5. Explain NEC requirements governing the installation of transformers.
6. Compute transformer sizes for various applications.
7. Explain types and purposes of grounding transformers.
8. Connect a control transformer for a given application.
9. Size the maximum load allowed on open delta systems.
10. Describe how current transformers are used in conjunction with watt-hour meters.
11. Apply capacitors and rectifiers to practical applications.
12. Calculate the power factor of any given electrical circuit.

Prerequisites

Before you begin this module, it is recommended that you successfully complete the following: Core Curriculum; Electrical Level One; Electrical Level Two; Electrical Level Three, Modules 26301 through 26306.

Required Trainee Materials

1. Pencil and paper
2. Appropriate personal protective equipment
3. Copy of the latest edition of the *National Electrical Code*

1.0.0 ◆ INTRODUCTION

The electric power produced by alternators in a generating station is transmitted to locations where it is utilized and distributed to users. Many different types of **transformers** play an important role in the distribution of electricity. The main purpose of a transformer is to change the output voltage. **Power transformers** are located at generating stations to step up the voltage for more economical transmission. Substations with additional power transformers and distribution equipment are installed along the transmission line. Finally, distribution transformers are used to step down the voltage to a level suitable for utilization.

Transformers are also used quite extensively in all types of control work to raise and lower AC voltage on control circuits. They are also used in 480Y/277V systems to reduce the voltage for operating 208Y/120V lighting and other electrically operated equipment. Buck-and-boost transformers are used for maintaining appropriate voltage levels in certain electrical systems.

It is important for anyone working with electricity to become familiar with all aspects of transformer operation—how they work, how they are connected into circuits, their practical applications,

Note: The designations "National Electrical Code," "NE Code," and "NEC," where used in this document, refer to the National Electrical Code®, which is a registered trademark of the National Fire Protection Association, Quincy, MA. *All National Electrical Code (NEC) references in this module refer to the 2002 edition of the NEC.*

Materials

Ensure that you have everything required to teach the course. Check the Materials and Equipment List at the front of this Instructor's Guide.

Classroom

Review the goals of the course.

Explain that terms shown in bold (blue) are defined in the Glossary at the back of this module.

Audiovisual

Show Transparency 1, Course Objectives.

Show Transparency 2, Performance Tasks.

Classroom

Discuss the role of transformers in power distribution systems.

Classroom

Describe the construction of a basic transformer. Provide an example for the trainees to examine.

Audiovisual

Show Transparency 3 (Figure 1).

Classroom

Define *mutual induction.*

Audiovisual

Show Transparency 4 (Figure 2).

and precautions to take during the installation or while working on them. This module is designed to cover these items, as well as overcurrent protection and grounding. Other subjects include correcting power factor with capacitors and the application of **rectifiers.**

2.0.0 ◆ TRANSFORMER BASICS

A very basic transformer consists of two coils, or windings, formed on a single magnetic core, as shown in *Figure 1.* Such an arrangement will allow transforming a large alternating current at a low voltage into a small alternating current at a high voltage, or vice versa.

2.1.0 Mutual Induction

The term **mutual induction** refers to the condition in which two circuits are sharing the energy of one of the circuits. It means that energy is being transferred from one circuit to the other.

Consider the diagram in *Figure 2.* Coil A is the primary circuit that obtains energy from the battery. When the switch is closed, the current starts to flow and a **magnetic field** expands out of coil A. Coil A then changes the electrical energy of the battery into the magnetic energy (**induction**) of a magnetic field. When the field of coil A is expanding, it cuts across coil B, the secondary circuit, inducing a voltage in coil B. The indicator (a galvanometer) in the secondary circuit is deflected

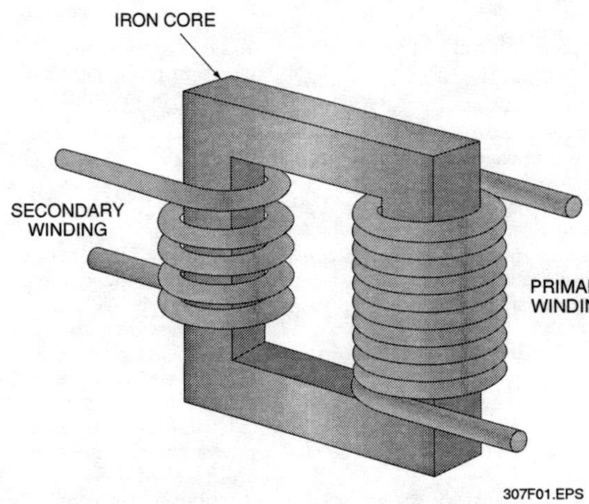

Figure 1 ◆ Basic components of a transformer.

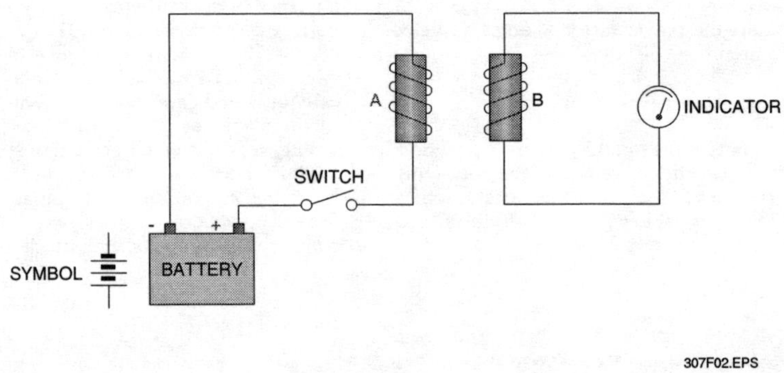

Figure 2 ◆ Mutual induction circuits.

Instructor's Notes:

and shows that a current, developed by the induced voltage, is flowing in the circuit.

The induced voltage may be generated by moving coil B through the **flux** of coil A. However, this voltage is induced without moving coil B. When the switch in the primary circuit is open, coil A has no current and no field. As soon as the switch is closed, current passes through the coil, and the magnetic field is generated. This expanding field moves or cuts across the wires of coil B, thus inducing a voltage without the movement of coil B.

The magnetic field expands to its maximum strength and remains constant as long as full current flows. Flux lines stop their cutting action across the **turns** of coil B because the expansion of the field has ceased. At this point, the indicator needle on the meter reads zero because the induced voltage no longer exists. If the switch is opened, the field collapses back to the wires of coil A. As it does so, the changing flux cuts across the wires of coil B, but in the opposite direction. The current present in the coil causes the indicator needle to deflect, showing this new direction. Therefore, the indicator shows current flow only when the field is changing, either building up or collapsing. In effect, the changing field produces an induced voltage in the same way as a magnetic field moving across a conductor. This principle of inducing voltage by holding the coils steady and forcing the field to change is used in innumerable applications. The transformer is particularly suitable for operation by mutual induction. Transformers are the ideal components for transferring and changing AC voltages as needed.

Transformers are generally composed of two coils placed close to each other but not connected. Refer once more to *Figure 1.* The coil that receives energy from the line voltage source is called the primary, and the coil that delivers energy to a load is called the secondary. Even though the coils are not physically connected, they manage to convert and transfer energy as required by a process known as mutual induction.

Transformers, therefore, enable changing or converting power from one voltage to another. For example, generators that produce moderately large alternating currents at moderately high voltages use transformers to convert the power to a very high voltage and proportionately small current in transmission lines, permitting the use of smaller cable and producing less power **loss.**

When alternating current (AC) flows through a coil, an alternating magnetic field is generated around the coil. This alternating magnetic field expands outward from the center of the coil and collapses into the coil as the AC through the coil varies from zero to a maximum and back to zero again, as discussed in an earlier module. Since the alternating magnetic field must cut through the turns of the coil, a self-inducing voltage occurs in the coil, which opposes the change in current flow.

If the alternating magnetic field generated by one coil cuts through the turns of a second coil, voltage will be generated in this second coil just as voltage is induced in a coil that is cut by its own magnetic field. The induced voltage in the second coil is called the voltage of mutual induction, and the action of generating this voltage is called transformer action. In transformer action, electrical energy is transferred from one coil (the primary) to another (the secondary) by means of a varying magnetic field.

2.2.0 Induction in Transformers

As stated previously, a simple transformer consists of two coils located very close together and electrically insulated from each other. The primary coil generates a magnetic field that cuts through the turns of the secondary coil and generates a voltage in it. The coils are magnetically coupled to each other, and consequently, a transformer transfers electrical power from one coil to another by means of an alternating magnetic field.

Assuming that all the magnetic lines of force from the primary cut through all the turns of the secondary, the voltage induced in the secondary will depend on the ratio of the number of turns in the primary to the number of turns in the secondary. For example, if there are 100 turns in the primary and only 10 turns in the secondary, the voltage in the primary will be 10 times the voltage in the secondary. Since there are more turns in the primary than there are in the secondary, the transformer is called a step-down transformer. Transformers are rated in **kilovolt-amperes (kVA)** because they are independent of power factor. *Figure 3* shows a diagram of a step-down transformer with a **turns ratio** of 100:10, or 10:1.

$$\frac{10 \text{ turns}}{100 \text{ turns}} = 0.10 = 0.10 \times 120\text{V} = 12\text{V}$$

Assuming that all the primary magnetic lines of force cut through all the turns of the secondary, the amount of induced voltage in the secondary will vary with the ratio of the number of turns in the secondary to the number of turns in the primary.

If there are more turns in the secondary winding than in the primary winding, the secondary voltage will be higher than that in the primary and by the same proportion as the number of turns in the winding. The secondary current, in turn, will be proportionately smaller than the primary current.

Classroom

Describe how a transformer operates.

Explain that transformers are rated in kVA because it is independent of power factor.

Show the trainees how to calculate the output of a transformer using various turns ratios.

Audiovisual

Show Transparency 5 (Figure 3).

Audiovisual

Show Transparency 6 (Figure 4).

Classroom

Define *leakage flux*. Explain the difference between a low-leakage flux transformer and a high-leakage flux transformer.

Audiovisual

Show Transparencies 7 and 8 (Figures 5 and 6).

Teaching Tip

Demonstrate the principle of leakage flux using iron filings and a transformer.

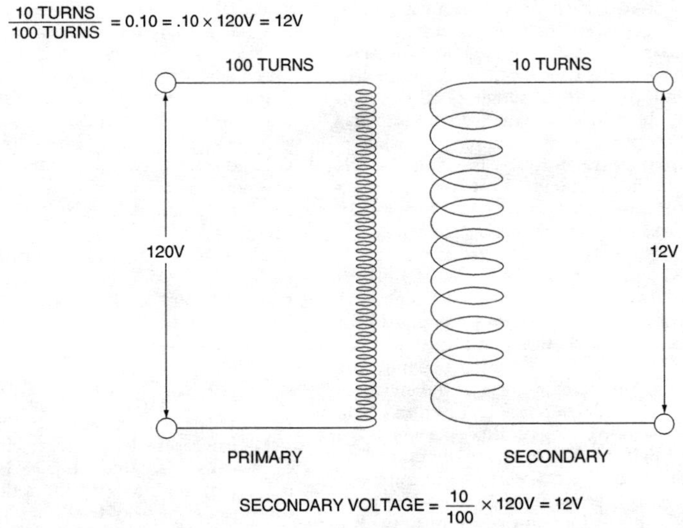

Figure 3 ◆ Step-down transformer with a 10:1 turns ratio.

307F03.EPS

With fewer turns in the secondary than in the primary, the secondary voltage will be proportionately lower than that in the primary, and the secondary current will be proportionately larger. Since alternating current continually increases and decreases in value, every change in the primary winding of the transformer produces a similar change of flux in the core. Every change of flux in the core and every corresponding movement of the magnetic field around the core produce a similarly changing voltage in the secondary winding, causing an alternating current to flow in the circuit that is connected to the secondary.

For example, if there are 100 turns in the secondary and only 10 turns in the primary, the voltage induced in the secondary will be 10 times the voltage applied to the primary. See *Figure 4*. Since there are more turns in the secondary than in the primary, the transformer is called a step-up transformer.

$$\frac{100}{10} = 10 \times 12V = 120V$$

> **Note**
>
> A transformer does not generate electric power. It simply transfers electric power from one coil to another by **magnetic induction.** Transformers are rated in either volt-amperes (VA) or kilovolt-amperes (kVA).

2.3.0 Magnetic Flux in Transformers

Figure 5 shows a cross section of what is known as a high-leakage flux transformer. In these transformers, if no load were connected to the secondary or output winding, a voltmeter would indicate a specific voltage reading across the secondary terminals. If a load were applied, the voltage would drop, and if the terminals were shorted, the voltage would drop to zero. During these circuit changes, the flux in the core of the transformer would also change; it is forced out of the transformer core and is known as leakage flux. Leaking flux can actually be demonstrated with iron filings placed close to the transformer core. As the changes take place, the filings will shift their position, clearly showing the change in the flux pattern.

What actually happens is that as the current flows in the secondary, it tries to create its own magnetic field, which is in opposition to the original flux field. This action, like a valve in a water system, restricts the flux flow, which forces the excess flux to find another path, either through the air or in adjacent structural steel, such as transformer housings or supporting clamps.

Note that the coils in *Figure 5* are wrapped on the same iron core but are separated from each other, while the transformer in *Figure 6* has its coils wrapped around each other, which results in a low-leakage transformer design.

Instructor's Notes:

$$\frac{100}{10} = 10 \times 12V = 120V$$

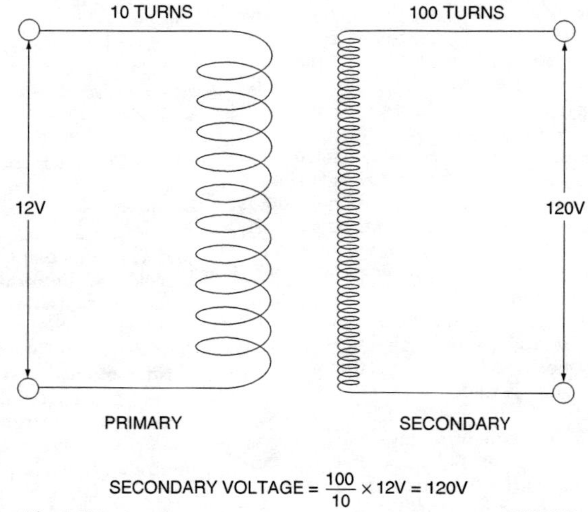

Figure 4 ◆ Step-up transformer with a 1:10 turns ratio. 307F04.EPS

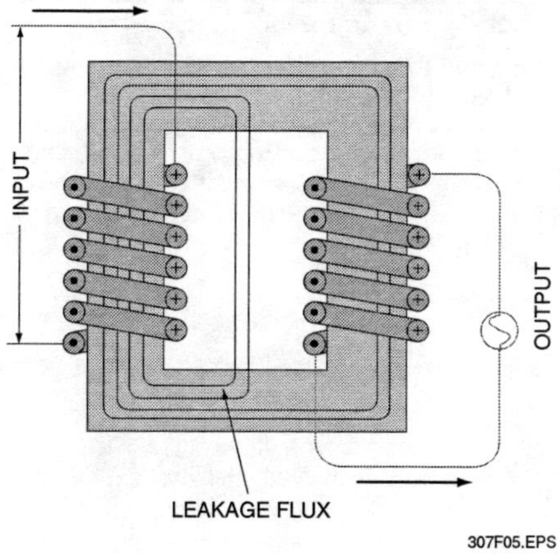

Figure 5 ◆ Transformer with high-leakage flux.

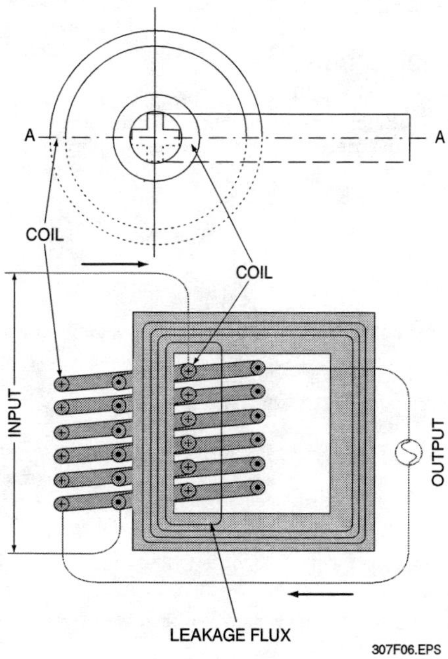

Figure 6 ◆ Low-leakage transformer.

Classroom

Explain that transformers consist of two windings on an iron core. Discuss the three types of cores used in transformers:
- **Open core**
- **Closed core**
- **Shell core**

Audiovisual

Show Transparencies 9 through 11 (Figures 7 through 9).

Classroom

Discuss the special characteristics of the steel used in transformer cores.

3.0.0 ◆ TRANSFORMER CONSTRUCTION

Transformers that are designed to operate on low frequencies have their coils, called windings, wound on iron cores. Since iron offers little resistance to magnetic lines, nearly all the magnetic field of the primary flows through the iron core and cuts the secondary.

Iron cores of transformers are constructed in three basic types: the open core, the closed core, and the shell type. See *Figure 7*. The open core is the least expensive to manufacture because the primary and secondary are wound on one cylindrical core. The magnetic path, as shown in *Figure 7*, is partially through the core and partially through the surrounding air. The air path opposes the magnetic field so that the magnetic interaction or linkage is weakened. Therefore, the open core transformer is highly inefficient.

The closed core improves the transformer efficiency by offering more iron paths and a reduced air path for the magnetic field. The shell-type core further increases the magnetic coupling, and therefore, the transformer efficiency is greater due to two parallel magnetic paths for the magnetic field, providing maximum coupling between the primary and the secondary.

3.1.0 Cores

Special core steel is used to provide a controlled path for the flow of magnetic flux generated in a transformer. In most practical applications, the transformer core is not a solid bar of steel but is constructed of many layers of thin sheet steel called laminations. Although the specifications of the core steel are primarily of interest to the transformer design engineer, the electrical worker should at least have a conversational knowledge of the materials used.

The steel used for transformer core laminations will vary with the manufacturer, but a popular size is .014" thick and is called 29-gauge steel. It is processed from silicon iron alloys containing approximately 3¼% silicon. The addition of silicon to the iron increases its ability to be magnetized and also renders it essentially non-aging.

The most important characteristic of electrical steel is core loss. It is measured in watts per pound at a specified frequency and flux density. The core loss is responsible for the heating in the transformer and also contributes to the heating of the windings. Much of the core loss is a result of eddy currents that are induced in the laminations when the core is energized. To hold this loss to a minimum, adjacent laminations are coated with an inorganic varnish.

Cores may either be of the core type, as shown in *Figure 8*, or the shell type, as shown in *Figure 9*. Of the two, the core type is favored for dry-type transformers for the following reasons:

- Only three core legs require stacking, which reduces cost.
- Steel does not encircle the two outer coils; this provides better cooling.
- The required floor space is reduced.

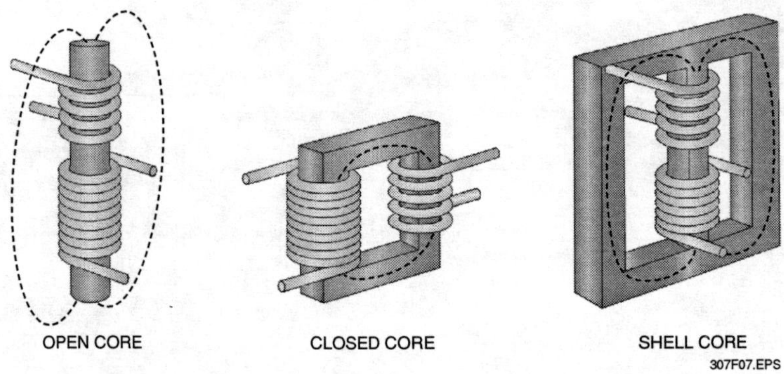

Figure 7 ◆ Three types of iron core transformers.

Instructor's Notes:

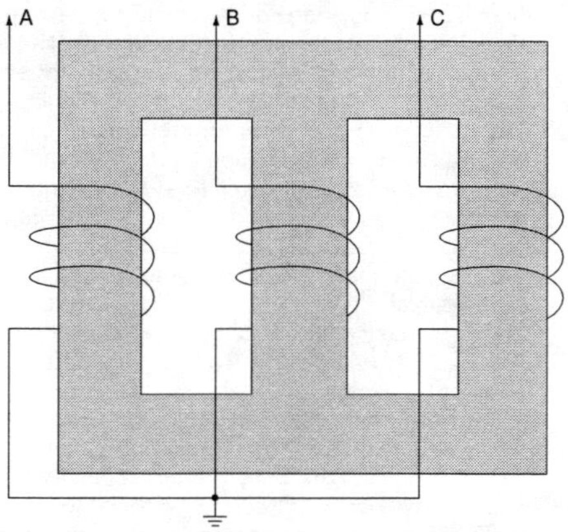

Figure 8 ◆ Core-type transformer construction.

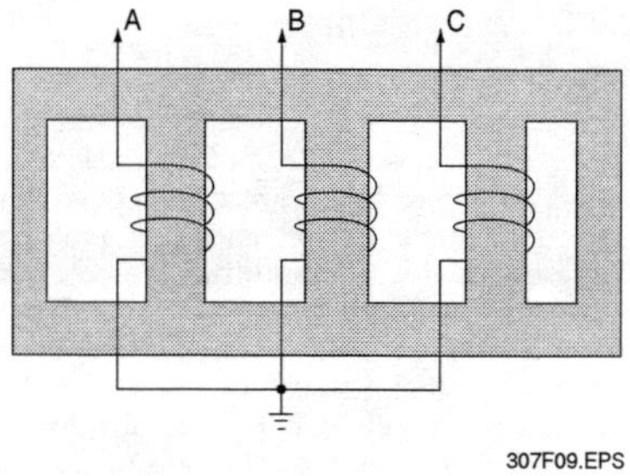

Figure 9 ◆ Shell-type transformer core.

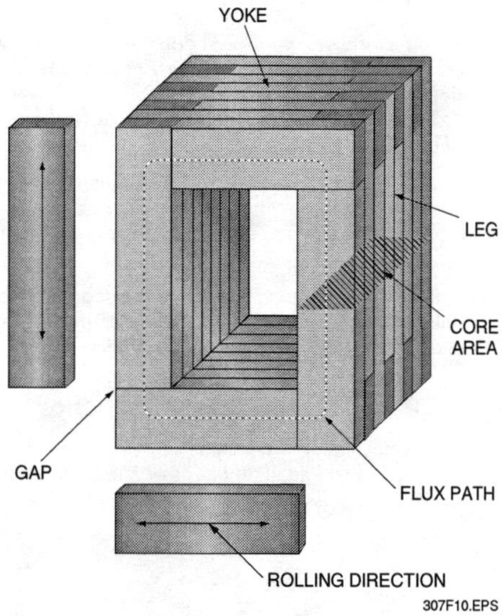

Figure 10 ◆ Butt-and-lap transformer core.

3.2.0 Types of Cores

Transformer cores are normally available in three types:

- Butt
- Wound
- Mitered cores

The butt-and-lap core is shown in *Figure 10.* Only two sizes of core steel are needed in this type of core due to the lap construction shown at the top and right side. For ease of understanding, the core strips are shown much thicker than the .014" thickness mentioned earlier. Each strip is carefully cut so that the air gap indicated in the lower left corner is as small as possible. The permeability of steel to the passage of flux is about 10,000 times as effective as air, hence, the air gap must be held to the barest minimum to reduce the **ampere turns** necessary to achieve adequate flux density. Also, the amount of sound produced by a transformer is a function of the flux density, which produces a difference between this construction and other types.

Another phenomenon in core steel is that the flux flows more easily in the direction in which the steel was rolled. This also varies between hot-rolled steel and cold-rolled steel. For example, the core loss due to flux passing at right angles to the rolling direction is almost 1½ times as great in hot-rolled steel and 2½ times as great in cold-rolled steel when compared with the core loss in the direction of rolling. The difference in exciting current is more dramatic, with ratios of two to one in hot-rolled steel and almost 40 to 1 in cold-rolled steel. These are primarily the designer's concern, but you should know that there is a difference.

Eddy currents are restricted from passage from one lamination to another due to the inorganic insulating coating. However, the magnetic lines of flux easily transfer at adjacent laminations in the lap area, but in so doing, they are forced to cross at an angle to the preferred direction.

Classroom

Point out that transformers are also classified according to the method by which the laminations are arranged in the core.

Discuss the three types of core lamination:
- **Butt-and-lap**
- **Wound**
- **Mitered**

Provide examples for the trainees to examine.

Audiovisual

Show Transparency 12 (Figure 10).

Audiovisual

Show Transparencies 13 and 14 (Figures 11 and 12).

Classroom

Point out the advantages of a mitered transformer core.

Explain that transformers are designed to minimize winding resistance, eddy current losses, and magnetic leakage.

Discuss the function of a transformer tap.

3.3.0 Wound Cores

Because of the unique characteristics of core steel, some core designs are made to take advantage of these differences. One such type is shown in *Figure 11.* The core loops are cut to predetermined lengths so that the gap locations do not coincide. These cuts permit assembling the core around a prewound coil that passes through both openings. Another design, now discontinued because of unfavorable cost, used a continuous core with no cuts. Separate coils had to be wound on each of the vertical legs of the completed core. You may encounter transformers of this type in existing installations.

3.4.0 Mitered Cores

Figure 12 shows a mitered core design. It is basically a butt-lap core with the joints made at 45-degree angles.

There are two benefits derived from this type of joint:

- It eliminates all cross grain flux, thereby improving the core loss and exciting current values.
- It reduces the flux density in the air gap, resulting in lower sound levels.

This type of core is normally used only with cold-rolled, grain-oriented steel and permits this steel to be used to its fullest capability.

3.5.0 Transformer Characteristics

In a well-designed transformer, there is very little magnetic leakage. The effect of the leakage is to cause a decrease of secondary voltage when the transformer is loaded. When a current flows through the secondary in phase with the secondary voltage, a corresponding current flows through the primary in addition to the magnetizing current. The magnetizing effects of the two currents are equal and opposite.

In a perfect transformer (one having no eddy current losses, no resistance in its windings, and no magnetic leakage), the magnetizing effects of the primary load current and the secondary current neutralize each other, leaving only the constant primary magnetizing current effective in setting up the constant flux. If supplied with a constant primary pressure, such a transformer would maintain constant secondary pressure at all loads. Obviously, the perfect transformer has yet to be built; the best transformers available today have a very small eddy current loss where the drop in pressure in the secondary windings is not more than 1% to 3%, depending on the size of the transformer.

4.0.0 ◆ TRANSFORMER TAPS

If the exact rated voltage could be delivered at every transformer location, transformer taps would be unnecessary. However, this is not possible, so taps are provided to either increase or decrease the secondary voltage.

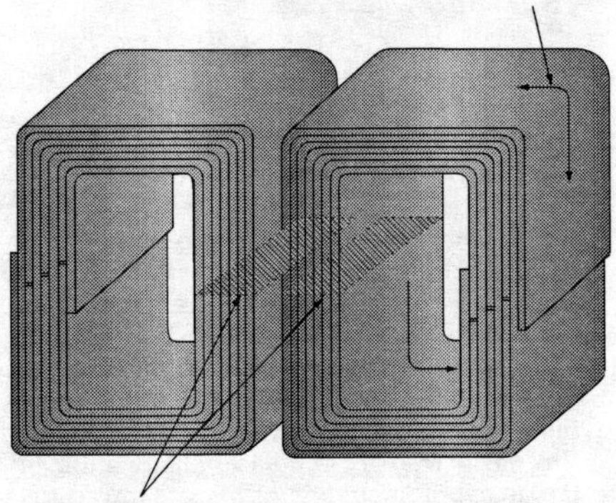

Figure 11 ◆ Wound transformer coil.

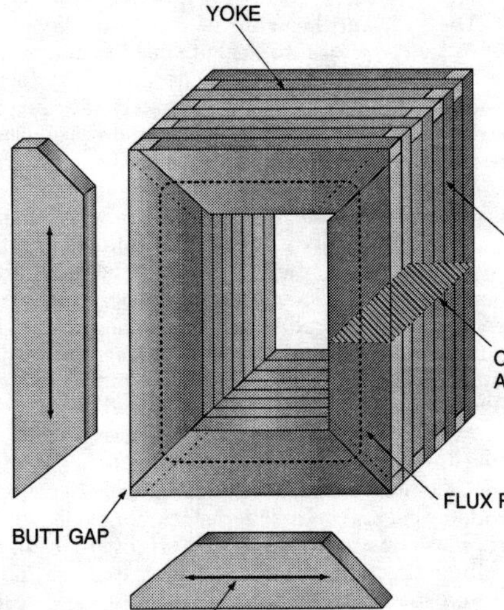

Figure 12 ◆ Mitered transformer core.

Instructor's Notes:

Generally, if a load is very close to a substation or power plant, the voltage will consistently be above normal. Near the end of the line, the voltage may be below normal.

In large transformers, it would naturally be very inconvenient to move the thick, well-insulated primary leads to different tap positions when changes in source voltage levels make this necessary. Therefore, taps are used, such as those shown in the wiring diagram in *Figure 13.* In this transformer, the permanent high-voltage leads would be connected to H_1 and H_2, and the secondary leads, in their normal fashion, to X_1 and X_2, and X_3 and X_4. Note, however, the tap arrangements available at taps 2 through 7. Until a pair of these taps is interconnected with a jumper wire, the primary circuit is not completed. If this were a typical 7,200V primary, the transformer would normally have 1,620 turns. Assume 810 of these turns are between H_1 and H_6 and another 810 between H_3 and H_2. Then, if taps 6 and 3 are connected with a flexible jumper on which lugs have already been installed, the primary circuit is completed, and we have a normal ratio transformer that could deliver 120/240V from the secondary.

Between taps 6 and either 5 or 7, 40 turns of wire exist. Similarly, between taps 3 and either 2 or 4, 40 turns are present. Changing the jumper from 3 to 6 to 3 to 7 removes 40 turns from the left half of the primary. The same condition would apply on the right half of the winding if the jumper were between taps 6 and 2. Either connection would boost secondary voltage by 2½%. Had taps 2 and 7 been connected, 80 turns would have been omitted, and a 5% boost would result. Placing the jumper between taps 6 and 4 or 3 and 5 would reduce the output voltage by 5%.

5.0.0 ◆ BASIC TRANSFORMER CONNECTIONS

Transformer connections are many, and space does not permit the description of all of them here. However, an understanding of a few connection types will give the basic requirements and make it possible to use manufacturer's data for others should the need arise.

THINK ABOUT IT

Secondary Neutrals

Why do we ground and bond secondary neutrals?

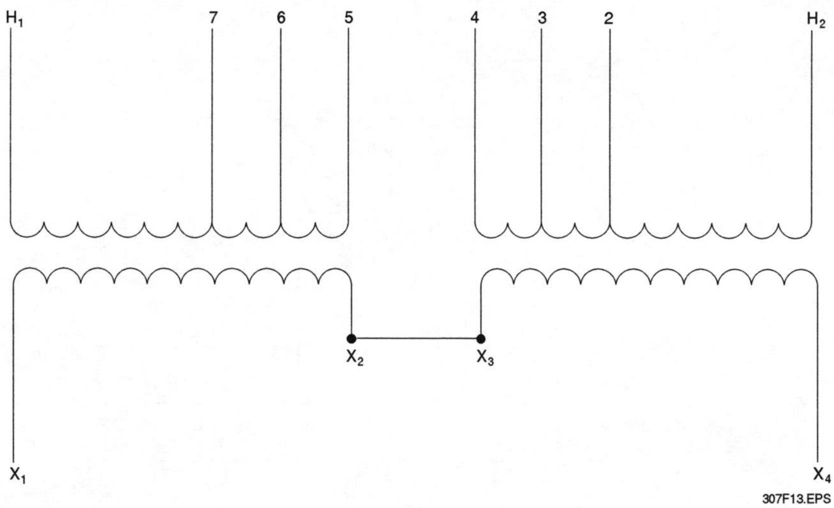

Figure 13 ◆ Transformer taps to adjust secondary voltage.

Audiovisual

Show Transparency 15 (Figure 13).

Classroom

Discuss the relationship between the position of a transformer tap on the winding and the final output voltage.

Demonstration

Demonstrate how to connect a dual-voltage transformer showing a high primary and low secondary, and vice versa. Emphasize proper bonding.

Laboratory

Have the trainees practice connecting a dual-voltage transformer showing a high primary and low secondary, and vice versa. Note the proficiency of each trainee.

Classroom

Point out that transformers can be connected in numerous ways to suit the application.

Teaching Tip

Discuss the "Think About It."

Classroom

Discuss the construction and operation of a three-wire, single-phase transformer. Point out that this is the most widely used type of transformer connection.

Audiovisual

Show Transparency 16 (Figure 14).

Classroom

Discuss the factors involved in selecting a three-phase power system.

Discuss the construction and operation of a four-wire, three-phase, delta-wye transformer. Point out that this type of system is widely used in commercial and industrial applications.

Audiovisual

Show Transparency 17 (Figure 15).

5.1.0 Single-Phase Light and Power Systems

Figure 14 is a single-phase transformer line diagram showing a connection used primarily for residential and small commercial applications. It is the most common single-phase distribution transformer in use today. It is known as a three-wire, 120/240V, single-phase system. Because of its configuration, it is easy to balance the load between the two coils. The kVA of a single-phase transformer is calculated by dividing the total VA by 1,000.

5.2.0 Three-Phase Power Systems

The following factors need to be considered when choosing a three-phase power system:

- The load(s) to be supplied
- The voltages needed for the application
- Future expansion

There are advantages and disadvantages associated with the different types of three-phase systems. The wye system can be used for both single-phase and three-phase loads. Because of its configuration, it is easy to balance the single-phase loads while still having the ability to use it for three-phase loads. Transformers installed in this type of system are sized by first dividing the total single-phase load in kVA by three, then taking this result and adding it to the total three-phase load in kVA divided by three. The result of these two loads added together identifies the required kVA rating of the transformer. The nominal voltage levels provided by this type of system are 120/208V or 277/480V. Note that 240V is not an option, and this may prove to be a disadvantage in many applications since 240V is a common operating voltage.

5.2.1 Delta-Wye Transformers

One of the most common transformer systems found in today's commercial and industrial settings is the delta primary, wye secondary transformer system (*Figure 15*). This is a three-phase, four-wire system that has the advantage of providing both three-phase power and also allowing lighting to be connected between any of the secondary phases and the neutral.

In *Figure 15,* the system is basically made up of three typical single-phase, step-down transformers with the interconnections between each individual transformer's primary and secondary coils determining the output voltage on the secondary side. However, this does not mean that any three single-phase transformers can be developed into a functional three-phase, delta-wye system for power and lighting. The coils must be rated for

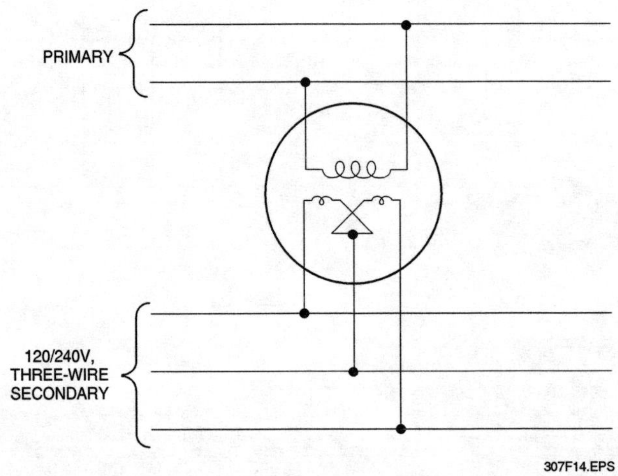

Figure 14 ◆ Single-phase transformer connection.

Instructor's Notes:

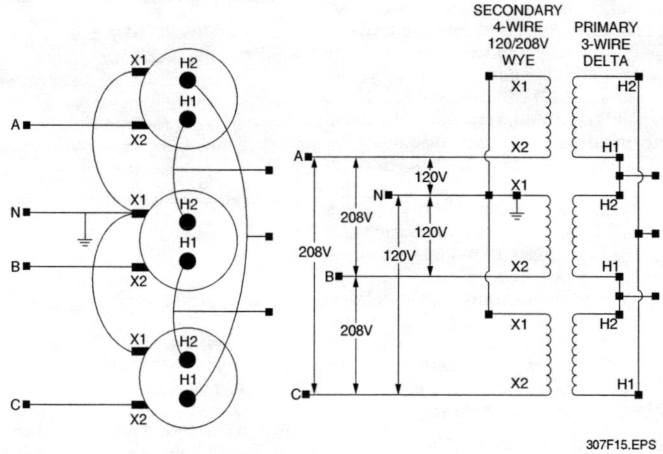

Figure 15 ◆ Delta-wye transformer system.

the loads to be served, as well as the primary voltage level to be connected. Power and light delta-wye transformers are universally found in both commercial and industrial installations as floor-mounted, dry-type transformers.

Although this transformer provides the convenience of three-phase and single-phase power, as well as each phase of the 208V system sharing the neutral to supply three legs of 120V lighting circuits, remember that the three-phase and single-phase power availability is at the 208V level and not 240V. Should a power requirement specifically call for 240V and not permit a 208V supply, this type of transformer is not the best selection.

5.2.2 Delta-Delta Transformers

The delta-delta system in *Figure 16* operates a little differently from the delta-wye system. Whereas the wye-connected system is formed by connecting one terminal from each of three equal voltage transformer windings together to make a

$120V \times 1.732 = 207.84V = 208V$

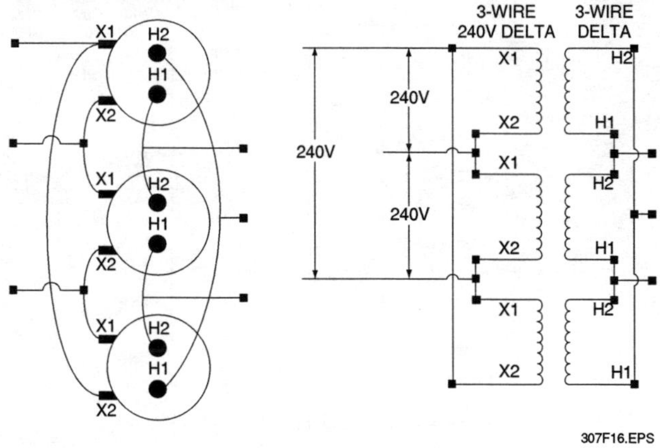

Figure 16 ◆ Delta-connected secondary.

Classroom

Discuss the construction and operation of a delta-delta transformer and compare it to the delta-wye transformer.

Audiovisual

Show Transparency 18 (Figure 16).

Classroom

Point out the NEC requirement for marking the high leg of a center-tapped, delta-connected system.

Audiovisual

Show Transparency 19 (Figure 17).

Safety

Emphasize that 120V loads should never be connected to the high leg of a center-tapped, delta-connected system.

Classroom

Discuss the construction and operation of an open-delta transformer and compare it to the other transformer types.

Audiovisual

Show Transparency 20 (Figure 18).

common terminal, the delta-connected system has its windings connected in series, forming a triangle, or the Greek symbol delta (Δ). In *Figure 17*, a center-tapped terminal is used on one winding to ground the system. A 120/240V system has 120V between the center-tapped terminal and each ungrounded terminal on either side such as phases A and C, and 240V across the full winding of each phase.

Refer to *Figure 17* and note that a high leg results at point *B*. This is also known as the wild leg. This high leg has a higher voltage to ground than the other two phases. The voltage of the high leg can be determined by multiplying the voltage to ground of either of the other two legs by the square root of 3, which we round to a value of 1.732. Therefore, if the voltage between phase A to ground is 120V, the voltage between phase B to ground may be determined as follows:

$$120V \times 1.732 = 207.84V = 208V$$

From this, it should be obvious that no single-pole breakers should be connected to the high leg of a center-tapped, four-wire, delta-connected system. In fact, ***NEC Sections 110.15 and 215.8*** require that the phase busbar or conductor having the higher voltage to ground be permanently marked by an outer finish that is orange in color. This prevents future workers from connecting 120V single-phase loads to this high leg, which would probably damage any equipment connected to the circuit. Remember the color orange; no line-to-neutral loads are to be connected to this phase.

> **WARNING!**
> Always use caution when working on a center-tapped, four-wire, delta-connected system. Phase B has a higher voltage to ground than phases A and C. Never connect 120V circuits to the high leg. Doing so will result in damage to the circuits and equipment.

5.2.3 Open Delta Transformers

Three-phase, delta-connected systems may be connected so that only two transformers are used; this arrangement is known as an open delta system, as shown in *Figure 18*. It is frequently used on a delta system when one of the three transformers becomes damaged. The damaged transformer is disconnected from the circuit, and the remaining two transformers carry the load. In doing so, the three-phase load carried by the open delta bank is only 86.6% of the combined rating of the remaining two equally sized units. It is only 57.7% of the normal full-load capability of a full bank of transformers. In an emergency, however, this capability permits single-phase and three-phase power at a location where one unit burned out and a replacement was not readily available. The total load must be curtailed to avoid another burnout.

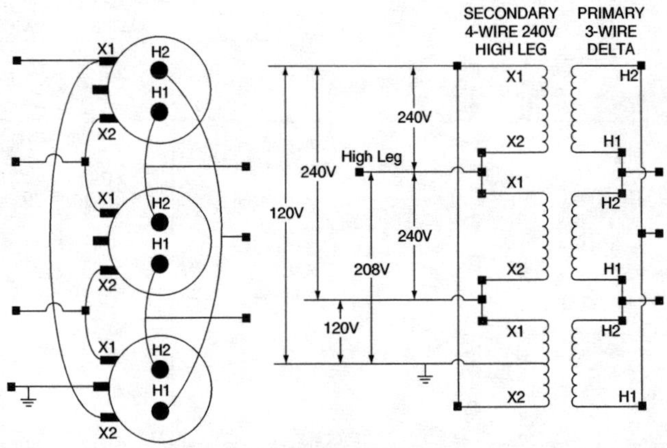

Figure 17 ◆ Characteristics of a center-tapped, delta-connected system.

Instructor's Notes:

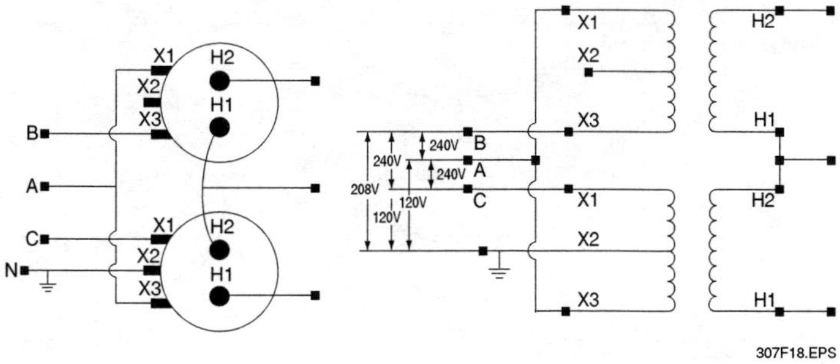

Figure 18 ◆ Open delta system.

5.3.0 Parallel Operation of Transformers

Transformers will operate satisfactorily in parallel on a single-phase, three-wire system if the terminals with the same relative polarity are connected together. However, the practice is not very economical because the individual cost and losses of the smaller transformers are greater than one larger unit giving the same output. Therefore, paralleling of smaller transformers is usually done only in an emergency. In large transformers, however, it is often practical to operate units in parallel as a regular practice. See *Figure 19*.

When connecting large transformers in parallel, especially when one of the windings is for a comparatively low voltage, the resistance of the joints and interconnecting leads must not vary significantly between the different transformers, or it will cause an unequal division of load.

Two three-phase transformers may also be connected in parallel, provided they have the same winding arrangement, are connected with the same polarity, and have the same phase rotation. If two transformers, or two banks of transformers, have the same voltage ratings, the same turns ratios, the same impedances, and the same ratios of **reactance** to resistance, they will divide the load current in proportion to their kVA ratings with no phase difference between the currents in the two transformers. However, if any of the preceding conditions are not met, then it is possible for the

INSIDE TRACK — *Transformers*

These photos show single-phase and three-phase transformers. Note the orange marking on the high leg of the three-phase transformer in accordance with ***NEC Sections 110.15 and 215.8.***

100 kVA SINGLE-PHASE TRANSFORMER
480V PRIMARY, 120V/240V SECONDARY

300 kVA THREE-PHASE CENTER-TAPPED
DELTA TRANSFORMER
480V PRIMARY, 120V/240V SECONDARY

307P0701.EPS

Demonstration

Demonstrate how to connect a single-phase transformer in a delta-wye and open delta configuration. Emphasize proper bonding.

Laboratory

Have the trainees practice connecting a single-phase transformer in a delta-wye and open delta configuration. Note the proficiency of each trainee.

Classroom

Explain that large transformers are often connected in parallel. Point out the design considerations necessary when connecting two transformers in parallel.

Audiovisual

Show Transparency 21 (Figure 19).

Classroom

Note that some three-phase transformers may not operate properly when connected in parallel. Explain how to test the polarity to determine whether the connection will operate.

Audiovisual

Show Transparency 22 (Figure 20).

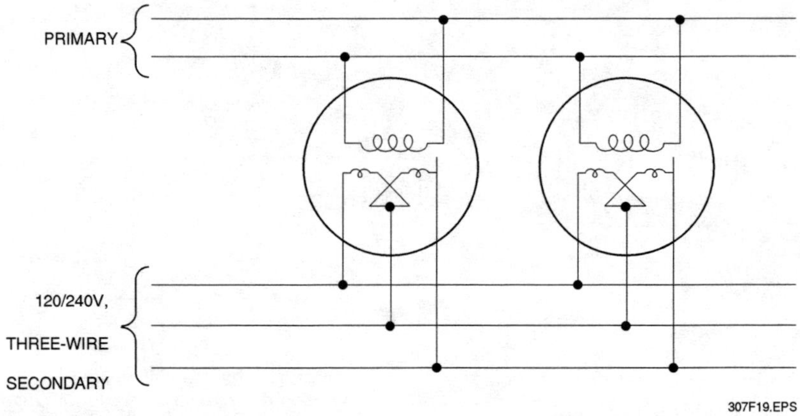

Figure 19 ◆ Parallel operation of single-phase transformers.

load current to divide between the two transformers in proportion to their kVA ratings. There may also be a phase difference between currents in the two transformers or banks of transformers.

Some three-phase transformers cannot be operated properly in parallel. For example, a transformer having both its primary and secondary windings connected in delta cannot be connected in parallel with another transformer that is connected either with a primary delta or a secondary wye. However, a transformer with a delta primary and a wye secondary can be made to parallel with transformers having their windings joined in certain ways; that is, a wye primary connection and a delta secondary connection.

To determine whether or not three-phase transformers will operate in parallel, connect them as shown in *Figure 20,* leaving two leads on one of the transformers unjoined. Test with a voltmeter across the unjoined leads. If there is no voltage between the points shown in the drawing, the polarities of the two transformers are the same,

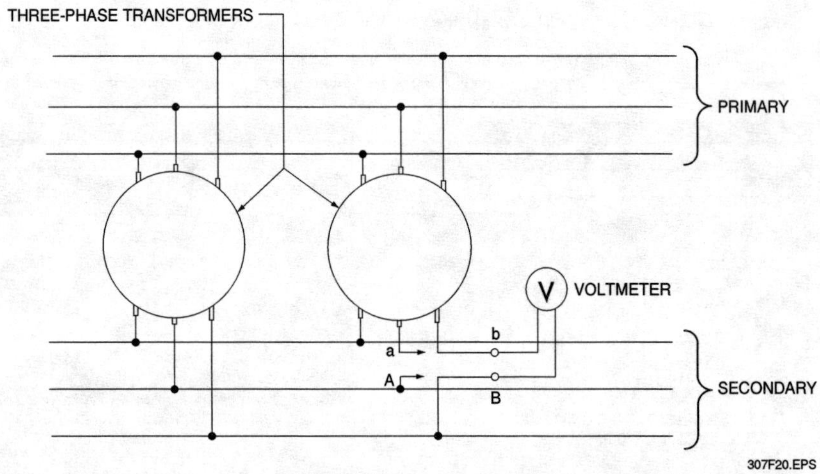

Figure 20 ◆ Testing three-phase transformers for parallel operation.

Instructor's Notes:

and the connections may then be made and the transformer put into service.

If a reading indicates a voltage between the points indicated in the drawing (either one of the two or both), the polarities of the two transformers are different. Should this occur, disconnect transformer lead A successively to mains 1, 2, and 3, as shown in *Figure 20,* and at each connection, test with a voltmeter between b and B and the legs of the main to which lead A is connected. If with any trial connection the voltmeter readings between b and B and either of the two legs is found to be zero, the transformer will operate with leads b and B connected to those two legs. If no system of connections can be discovered that will satisfy this condition, the transformer will not operate in parallel without changes to its internal connections, or it may not operate in parallel at all.

In parallel operation, the primaries of the two or more transformers involved are connected together, and the secondaries are also connected together. With the primaries so connected, the voltages in both primaries and secondaries will be in certain directions. It is necessary that the secondaries be so connected that the voltage from one secondary line to the other will be in the same direction through both transformers. Proper connections to obtain this condition for single-phase transformers of various polarities are shown in *Figure 21.* In *Figure 21(A),* both transformers A and B have additive polarity; in *Figure 21(B),* both transformers have subtractive polarity; and in *Figure 21(C),* transformer A has additive polarity and B has subtractive polarity.

Transformers, even when properly connected, will not operate satisfactorily in parallel unless their transformation ratios are very close to being equal and their impedance voltage drops are also approximately equal. A difference in transformation ratios will cause a circulating current to flow, even at no load, in each winding of both transformers. For example, in a loaded parallel bank of two transformers of equal capacities, if there is a difference in the transformation ratios, the load circuit will be superimposed on the circulating current. The result in such a case is that in one transformer, the total circulating current will be added to the load current; whereas in the other transformer, the actual current will be the difference between the load current and the circulating current. This may lead to unsatisfactory operation. Therefore, the transformation ratios of transformers for parallel operation must be definitely known.

When two transformers are connected in parallel, the circulating current caused by the difference in the ratios of the two is equal to the difference in the open circuit voltage divided by the sum of the transformer impedances. This is because the current is circulated through the windings of both transformers due to this voltage difference. To illustrate, let I represent the amount of circulating current, in percent of full-load current, and the equation will be:

$$I = \frac{\text{percent voltage difference} \times 100}{\text{sum of percent impedances}}$$

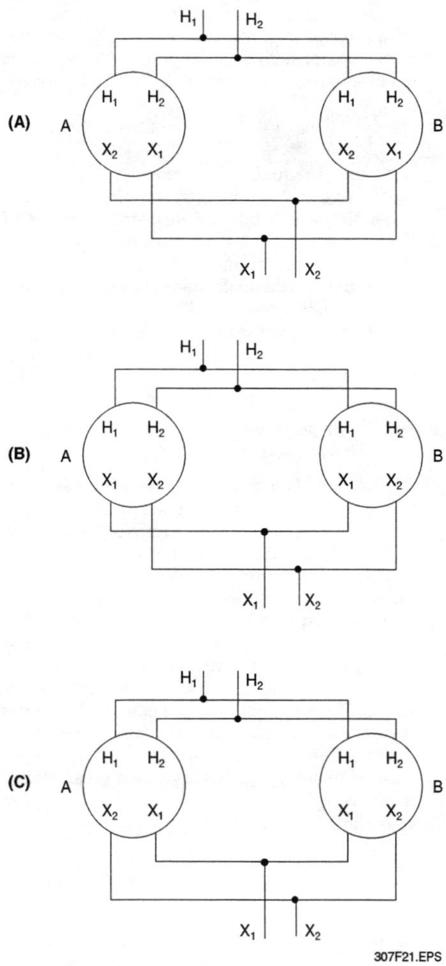

Figure 21 ◆ Transformers connected in parallel.

Classroom

Discuss acceptable polarity combinations for single-phase transformers.

Audiovisual

Show Transparency 23 (Figure 21).

Classroom

Explain how to calculate the circulating current for transformers connected in parallel.

Classroom

Point out that a high circulating current may result in poor system operation and overloading.

Explain that the voltage drop from no load to full load must be equal in all transformers connected in parallel. Show the trainees how to calculate the division of loads in parallel-connected transformer banks.

Discuss the special considerations for connecting three-phase transformers in parallel.

Describe the operation and advantages of autotransformers.

Audiovisual

Show Transparencies 24 and 25 (Figures 22 and 23).

Assume an open circuit voltage difference of 3% between two transformers connected in parallel. If each transformer has an impedance of 5%, the circulating current, in percent of full-load current, is:

$$I = \frac{3 \times 100}{5 + 5} = 30\%$$

A current equal to 30% of the full-load current therefore circulates in both the high-voltage and low-voltage windings. This current adds to the load current in the transformer having the higher induced voltage and subtracts from the load current of the other transformer. Therefore, one transformer will be overloaded, while the other may or may not be, depending on the phase angle difference between the circulating current and the load current.

5.3.1 Impedance in Parallel-Operated Transformers

Impedance plays an important role in the successful operation of transformers connected in parallel. The impedance of the transformers must be such that the voltage drop from no load to full load is the same in all transformer units in both magnitude and phase. In most applications, you will find that the total resistance drop is relatively small when compared with the reactance drop, and the total percent impedance drop can be taken as approximately equal to the percent reactance drop. If the percent impedances of the given transformers at full load are the same, they will, of course, divide the load equally.

The following equation may be used to obtain the division of loads between two transformer banks operating in parallel on single-phase systems:

$$\text{Power} = \frac{kVA_1 \div Z_1}{(kVA_1 \div Z_1) + (kVA_2 \div Z_2)} \times \text{total kVA load}$$

Where:

kVA_1 = kVA rating of transformer 1

kVA_2 = kVA rating of transformer 2

Z_1 = percent impedance of transformer 1

Z_2 = percent impedance of transformer 2

In this equation, it can be assumed that the ratio of resistance to reactance is the same in all units since the error introduced by differences in this ratio is usually so small as to be negligible.

The preceding equation may also be applied to more than two transformers operated in parallel by adding to the denominator of the fraction the kVA of each additional transformer divided by its percent impedance.

5.3.2 Parallel Operation of Three-Phase Transformers

Three-phase transformers, or banks of single-phase transformers, may be connected in parallel, provided each of the three primary leads in one three-phase transformer is connected in parallel with a corresponding primary lead of the other transformer. The secondaries are then connected in the same way. The corresponding leads are the leads that have the same potential at all times and the same polarity. Furthermore, the transformers must have the same voltage ratio and the same impedance voltage drop.

When three-phase transformer banks operate in parallel and the three units in each bank are similar, the division of the load can be determined by the same method previously described for single-phase transformers connected in parallel on a single-phase system.

In addition to the requirements of polarity, ratio, and impedance, paralleling of three-phase transformers also requires that the angular displacement between the voltages in the windings be taken into consideration when they are connected together.

Phasor diagrams of three-phase transformers that are to be paralleled greatly simplify matters. With these, all that is required is to compare the two diagrams to make sure they consist of phasors that can be made to coincide and then to connect the terminals corresponding to coinciding voltage phasors. If the phasor diagrams can be made to coincide, leads that are connected together will have the same potential at all times. This is one of the fundamental requirements for paralleling. Phasor diagrams are covered in more detail later in this module.

6.0.0 ◆ AUTOTRANSFORMERS

An **autotransformer** is a transformer whose primary and secondary circuits have part of a winding in common; therefore, the two circuits are not isolated from each other. See *Figure 22*. The application of an autotransformer is a good choice where a 480Y/277V or 208Y/120V, three-phase, four-wire distribution system is used. Some of the advantages are:

- Lower purchase price
- Lower operating cost due to lower losses
- Smaller size, easier to install
- Better voltage regulation
- Lower sound levels

For example, when the ratio of transformation from the primary to the secondary voltage is small, the most economical way of stepping down

Instructor's Notes:

> ### Autotransformers
> Autotransformers are not separately derived systems and must not be bonded on the secondary side.

the voltage is by using autotransformers, as shown in *Figure 23.* For this application, it is necessary that the neutral of the autotransformer bank be connected to the system neutral, similar to a wye connection.

7.0.0 ◆ DRY-TYPE TRANSFORMER CONNECTIONS

Electricians performing work on commercial and industrial installations will be concerned with the installation and connections of dry-type transformers. Dry-type transformers are available in both single-phase and three-phase types with a wide range of sizes from small control transformers to those rated at 500kVA or more. Such transformers have wide application in electrical systems of all types.

NEC Section 450.11 requires that each transformer must be provided with a nameplate giving the manufacturer, rated kVA, frequency, primary and secondary voltage, impedance of transformers 25kVA and larger, required clearances for transformers with ventilating openings, and the amount and type of insulating liquid (where used). In addition, the nameplate of each dry-type transformer must include the temperature class for the insulation system. See *Figure 24.*

In addition, most manufacturers include a wiring diagram and a connection chart, as shown in *Figure 25* for a 480V delta primary to 208Y/120V wye secondary. It is recommended that all transformers be connected as shown on the manufacturer's nameplate.

In general, this wiring diagram and accompanying table indicate that the 480V, three-phase, three-wire primary conductors are connected to terminals H_1, H_2, and H_3, respectively, regardless of the desired voltage on the primary. A neutral

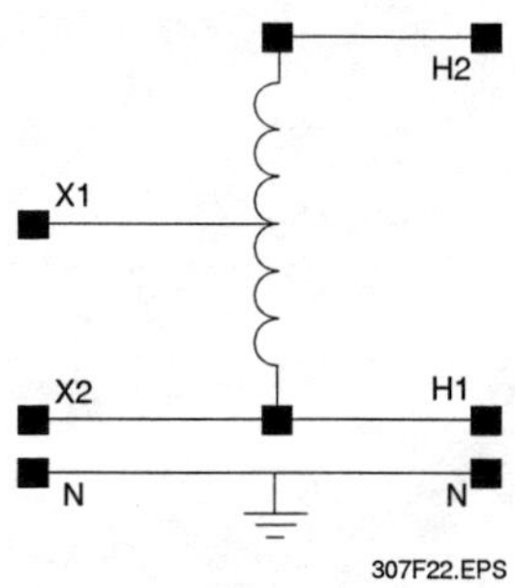

Figure 22 ◆ Step-down autotransformer.

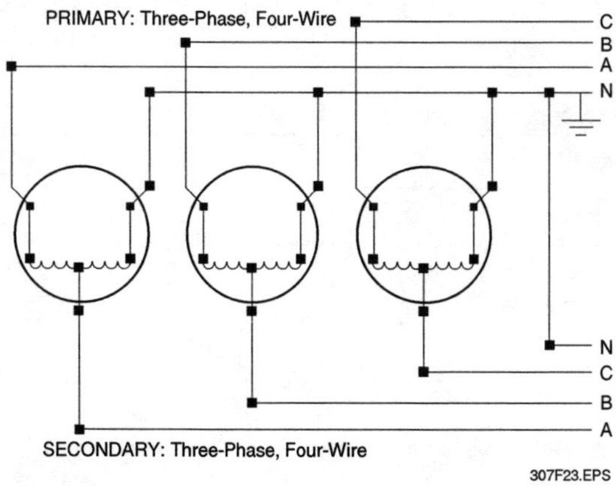

Figure 23 ◆ Autotransformers supplying power from a three-phase, four-wire system.

Classroom

Discuss the NEC requirements for transformer installations. Refer the trainees to *NEC Article 450.*

Audiovisual

Show Transparency 26 (Figure 24).

Classroom

Note that most manufacturers include a wiring diagram to help simplify transformer connections.

Audiovisual

Show Transparency 27 (Figure 25).

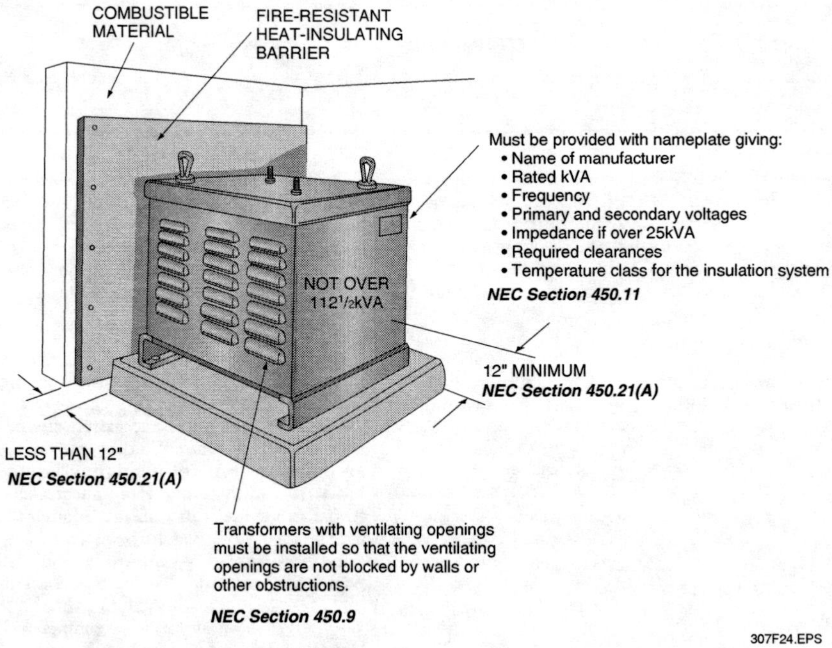

Figure 24 ◆ Dry-type transformer installed indoors.

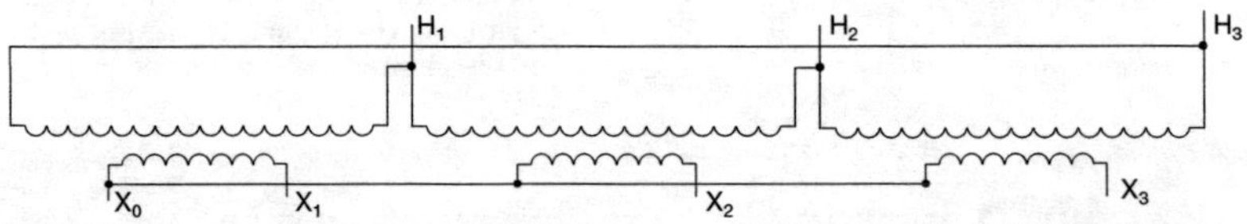

PRIMARY VOLTS	CONNECT PRIMARY LINES TO	CONNECT SECONDARY LINES TO
480V	H_1, H_2, H_3	———
SECONDARY VOLTS		
208V	———	X_1, X_2, X_3
120V SINGLE-PHASE	———	X_1 to X_0 X_2 to X_0 X_3 to X_0

307F25.EPS

Figure 25 ◆ Typical manufacturer's wiring diagram for a delta-wye transformer.

Instructor's Notes:

conductor, if required, is carried from the primary, through the transformer, to the secondary. Two variations are possible on the secondary side of this transformer: 208V, three-phase, three-wire or four-wire or 120V, single-phase, two-wire. To connect the secondary side of the transformer as a 208V, three-phase, three-wire system, the secondary conductors are connected to terminals X_1, X_2, and X_3; the neutral is carried through with conductors usually terminating at a solid neutral bus in the transformer.

Another popular dry-type transformer connection is the 480V primary to 240V delta/120V secondary. This configuration is shown in *Figure 26.* Again, the primary conductors are connected to transformer terminals H_1, H_2, and H_3. The secondary connections for the desired voltages are made as indicated in the table.

7.1.0 Zig-Zag Connections

There are many situations in which it is desirable to upgrade a building's lighting system from 120V fixtures to 277V fluorescent fixtures. Often, these buildings have a 240/480V, three-phase, four-wire delta system. One way to obtain 277V from a 240/480V system is to connect 240/480V transformers in a zig-zag fashion, as shown in *Figure 27.* In doing so, the secondary of one phase is connected in series with the primary of another phase, thus changing the phase angle.

The zig-zag connection may also be used as a grounding transformer where its function is to obtain a neutral point from an ungrounded system. With a neutral being available, the system may then be grounded. When the system is grounded through the zig-zag transformer, its sole function is to pass ground current. A zig-zag transformer is essentially six impedances connected in a zig-zag configuration.

The operation of a zig-zag transformer is slightly different from that of the conventional transformer. We will consider current rather than voltage. While a voltage rating is necessary for the connection to function, this is actually line voltage and is not transformed. It provides only exciting current for the core. The dynamic portion of the zig-zag grounding system is the fault current. To understand its function, the system must also be viewed backward; that is, the fault current will flow into the transformer through the neutral, as shown in *Figure 28.*

The zero sequence currents are all in phase in each line (they all hit the peak at the same time). In reviewing *Figure 28,* we see that the current leaves the motor, goes to ground, flows up the neutral, and splits three ways. It then flows back down the line to the motor through the fuses, which then open, shutting down the motor.

The neutral conductor will carry full fault current and must be sized accordingly. It is also time rated (0–60 seconds) and can therefore be reduced in size. This should be coordinated with the manufacturer's time/current curves for the fuse. See the Level Three module entitled *Overcurrent Protection.*

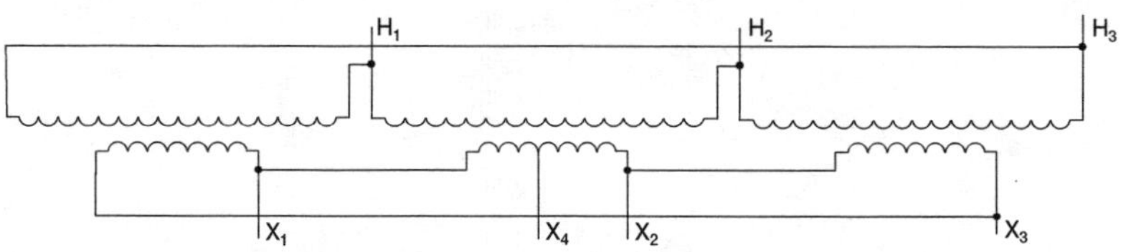

PRIMARY VOLTS	CONNECT PRIMARY LINES TO	CONNECT SECONDARY LINES TO
480V	H_1, H_2, H_3	———
SECONDARY VOLTS		
240V	———	X_1, X_2, X_3
120V	———	X_1, X_4 or X_2, X_4

307F26.EPS

Figure 26 ◆ 480V delta to 240V delta transformer connections.

Classroom

Describe the connections for a 480V delta/240V transformer.

Audiovisual

Show Transparency 28 (Figure 26).

Classroom

Discuss the operation and applications of zig-zag connections. Point out that they are often used to supply fluorescent lighting fixtures and may also be used as grounding transformers.

Audiovisual

Show Transparencies 29 and 30 (Figures 27 and 28).

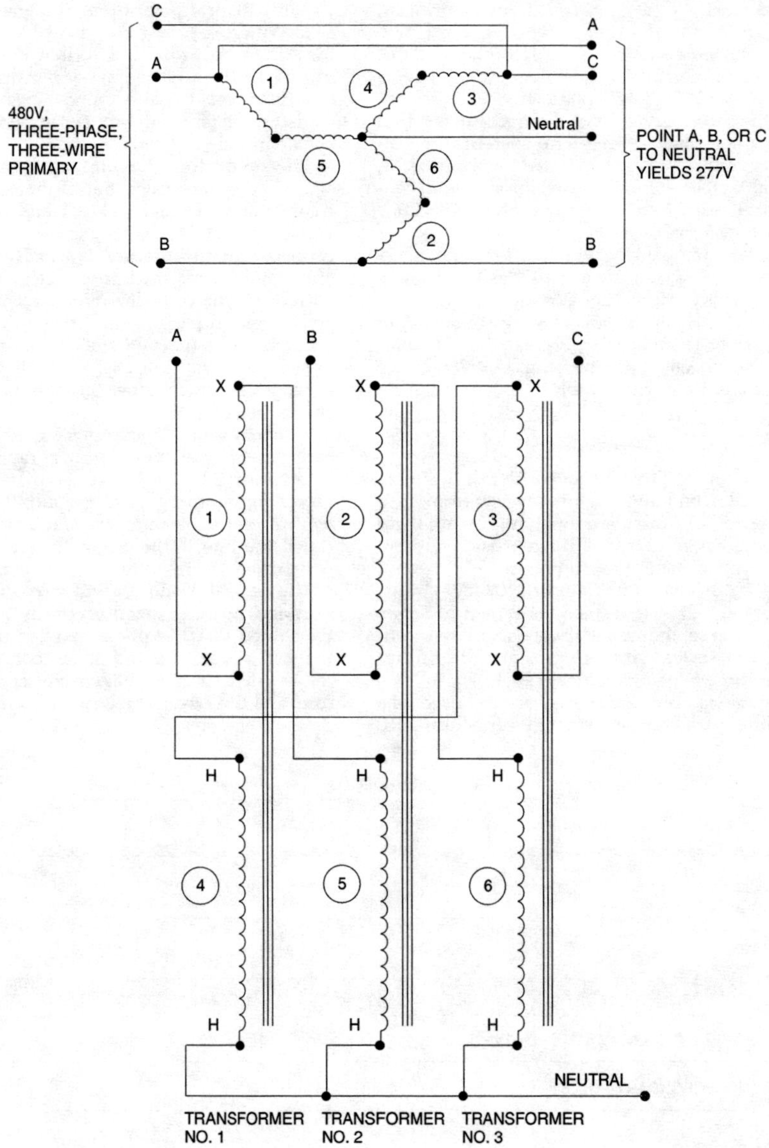

307F27.EPS

Figure 27 ◆ Zig-zag connection.

Instructor's Notes:

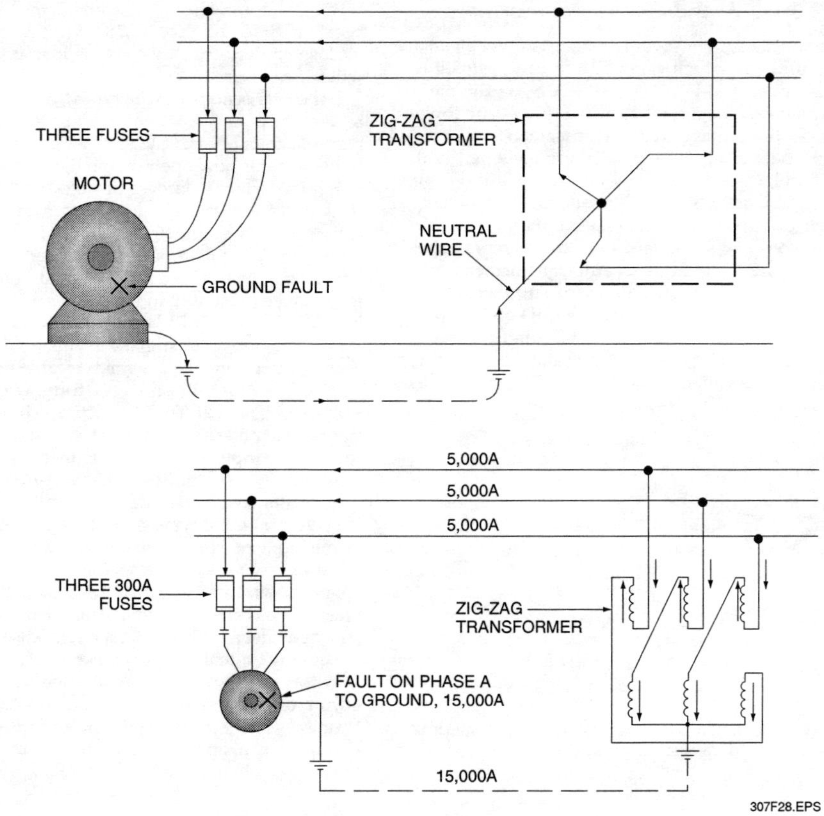

Figure 28 ◆ Fault current paths for a three-phase system.

To determine the size of a zig-zag grounding transformer, proceed as follows:

Step 1 Calculate the system line-to-ground asymmetrical fault current.

Step 2 If relaying is present, consider reducing the fault current by installing a resistor in the neutral. If fuses or circuit breakers are the protective device, you may need all the fault current to quickly open the overcurrent protective devices.

Step 3 Obtain the time/current curves of the relay and the fuses or circuit breakers.

Step 4 Select the zig-zag transformer for:

- Fault current (line-to-ground fault)
- Line-to-line voltage
- Duration of fault (determined from time/current curves)
- Impedance per phase at 100% (for any other, contact the manufacturer)

Classroom

Explain how to size a zig-zag grounding transformer.

Classroom

Discuss the operation and applications of buck-and-boost transformers.

Audiovisual

Show Transparency 31 (Figure 29).

Classroom

Describe various types of buck-and-boost connections:
- **Open delta, three-phase**
- **Wye-connected, three-phase**
- **Single-phase**

Audiovisual

Show Transparencies 32 through 34 (Figures 30 through 32).

7.2.0 Buck-and-Boost Transformers

The buck-and-boost transformer is a very versatile unit for which a multitude of applications exist. Buck-and-boost transformers, as the name implies, are designed to raise (boost) or lower (buck) the voltage in an electrical system or circuit. In their simplest form, these insulated units will deliver 12V or 24V when the primaries are energized at 120V or 240V, respectively. However, their prime use and value lies in the fact that the primaries and secondaries can be interconnected, permitting their use as an autotransformer.

Assume that an installation is supplied with a 208Y/120V service, but one piece of equipment in the installation is rated for 230V single phase. A buck-and-boost transformer may be used to increase the voltage from 208V to 230V (*Figure 29*).

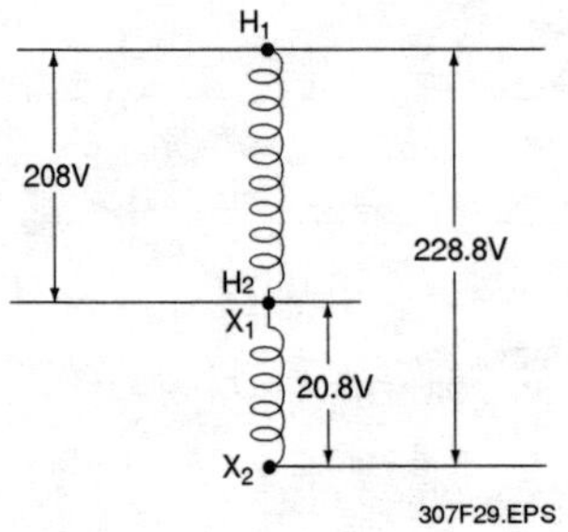

Figure 29 ◆ Buck-and-boost transformer connected to a 208V system to obtain 230V.

With this connection, the transformer is in the boost mode and delivers 228.8V at the load. This is close enough to 230V that the load equipment will function properly.

If the connections were reversed, this would also reverse the polarity of the secondary with the result being a voltage of 208V − 20.8V = 187.2V. The transformer is now operating in the buck mode.

It is important to know how to calculate sizes of buck-and-boost transformers for any given application. However, due to the amount of basic material covered in this module, advanced sizing and application techniques for buck-and-boost transformers are presented in Level Four of your training. Still, you should be familiar with the basic buck-and-boost wiring diagrams at this time. Transformer connections for typical three-phase, buck-and-boost, open delta transformers are shown in *Figure 30*. The connections shown are in the boost mode; to convert to the buck mode, reverse the input and output connections.

Another three-phase buck-and-boost transformer connection is shown in *Figure 31;* this time it is a wye-connected type. While the open delta transformers (*Figure 30*) can be converted from buck to boost or vice versa by reversing the input/output connections, this is not the case with the three-phase, wye-connected transformer. The connection shown in *Figure 31* is for the boost mode only.

Several typical single-phase, buck-and-boost transformer connections are shown in *Figure 32*. Other diagrams may be found on the transformer's nameplate or in the manufacturer's instructions supplied with each new transformer.

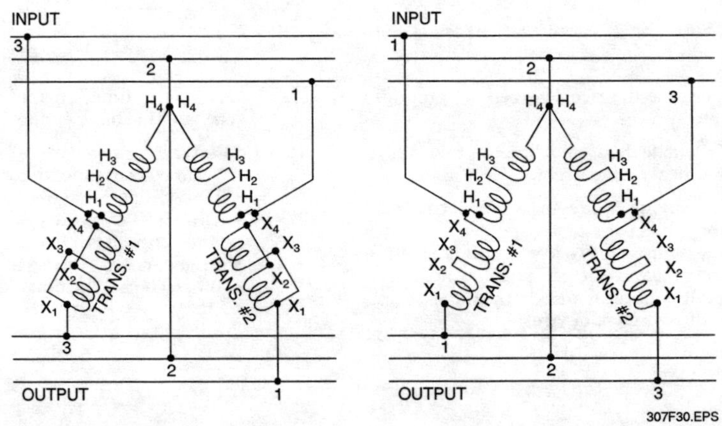

Figure 30 ◆ Open delta, three-phase, buck-and-boost transformer connections.

Instructor's Notes:

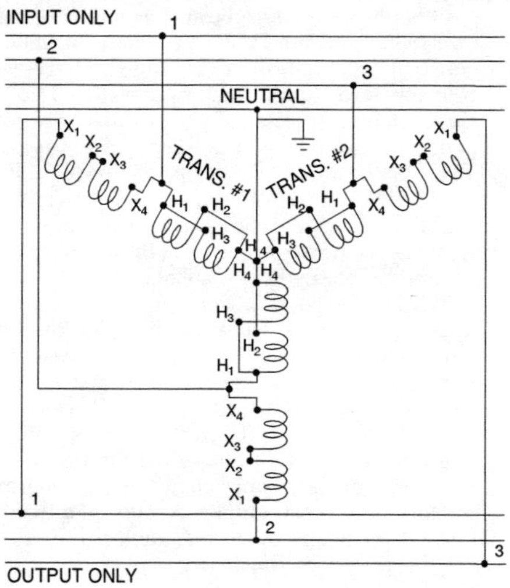

Figure 31 ◆ Three-phase, wye-connected, buck-and-boost transformer in the boost mode.

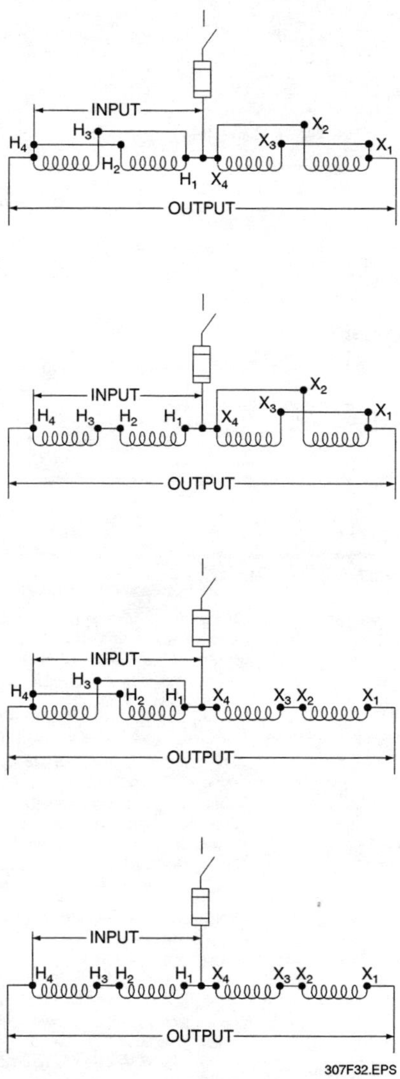

Figure 32 ◆ Typical single-phase, buck-and-boost transformer connections.

THINK ABOUT IT

Calculating Secondary Voltages

Given a turns ratio of 10:1 and an input voltage of 208V, determine what the secondary voltage should be for each of the connections shown in *Figure 32*.

8.0.0 ◆ CONTROL TRANSFORMERS

Control transformers are available in numerous types, but most are dry-type, step-down units with the secondary control circuit isolated from the primary line circuit to ensure maximum safety. See *Figure 33*. Industrial control transformers are designed to accommodate the momentary current inrush caused when electromagnetic components are energized without sacrificing secondary voltage stability beyond practical limits. Refer to *Table 1* for control circuit transformer requirements.

Teaching Tip

Discuss the "Think About It."

Classroom

Discuss the operation and applications of control transformers. Point out the NEC requirements shown in Table 1 of the Trainee Module.

Audiovisual

Show Transparency 35 (Figure 33).

Classroom

Explain the purpose of control and signal transformers. Discuss the application considerations for control transformers.

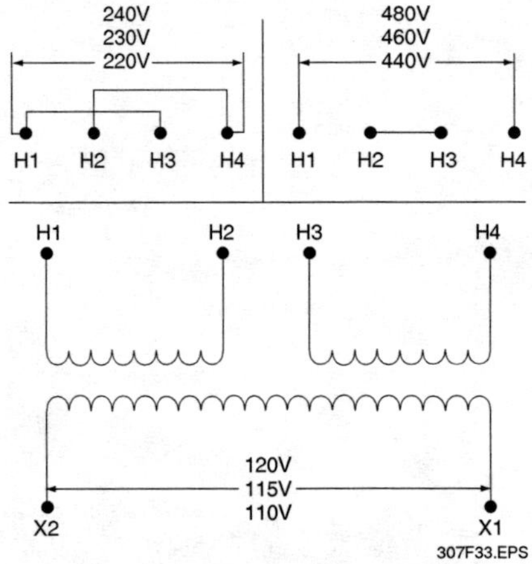

Figure 33 ◆ Typical control transformer wiring diagram.

Other types of control transformers, sometimes referred to as control and signal transformers, are constant-potential, air-cooled transformers. Their purpose is to supply the proper reduced voltage for control circuits of electrically operated switches, signal circuits, or other equipment. These transformers do not normally require the industrial regulation characteristics found in other transformers. Some are of the open type with no protective casing over the windings; others are enclosed within a metal casing.

When choosing control transformers for any application, the loads must be calculated and completely analyzed before the proper transformer selection can be made. This analysis must consider every electrically energized component in the control circuit. To select an appropriate control transformer, first determine the voltage and frequency of the supply circuit. Next, determine the total inrush volt-amperes (watts) of the control circuit. In doing so, do not neglect the current requirements of indicating lights and timing

Table 1 Summary of NEC Transformer Installation and Overcurrent Protection Requirements

Application	NEC Regulation	NEC Reference
Location	Transformers must be readily accessible to qualified personnel for maintenance and inspection.	***NEC Section 450.13***
	Dry-type transformers rated at 112½kVA or less may be located out in the open provided they are separated from combustible material by 12" or a suitable fire/heat barrier.	***NEC Section 450.21(A)***
	Dry-type transformers rated at more than 112½kVA must be installed in a transformer room of fire-resistant construction.	***NEC Section 450.21(B)***
	Dry-type transformers not exceeding 600V and 50kVA are not required to be readily accessible and are permitted in fire-resistant hollow spaces of a building under the conditions specified in the NEC.	***NEC Sections 450.13(A) and (B)***
	Dry-type transformers installed outdoors must have a weatherproof enclosure.	***NEC Section 450.22***
	Liquid-filled transformers must be installed as specified in the NEC and usually in vaults when installed indoors.	***NEC Section 450.23***
Overcurrent protection	The primary protection must be rated or set as follows: • 9A or more—125% • Less than 9A—167% • Less than 2A—300% If the primary current (line side) is 9A or more, the next higher standard size overcurrent protective device greater than 125% of the primary current is used. For example, if the primary current is 15A, 125% of 15A = 18.75A. The next standard size circuit breaker is 20A. Therefore, this size, 20A, may be used. Conductors on the secondary side of a single-phase transformer with a two-wire secondary may be protected by the primary overcurrent device under certain NEC conditions.	***NEC Table 450.3(B)***
Transformers used in motor control circuits	Special rules apply to these circuits for the various types of transformers.	***NEC Section 430.72(C)***
Over 600V	Special NEC rules apply to transformers operating at over 600V.	***NEC Table 450.3(A)***

Instructor's Notes:

devices that do not have inrush volt-amperes but are energized at the same time as the other components in the circuit. Their total volt-amperes should be added to the total inrush volt-amperes.

Again, control transformers will be covered in more detail in Level Four, as will other types of transformers. The material presented in this module is designed to introduce you to these devices; additional study of practical applications is forthcoming.

9.0.0 ◆ POTENTIAL AND CURRENT TRANSFORMERS

In general, a **potential transformer** (*Figure 34*) is used to supply voltage to devices, such as voltmeters, frequency meters, power factor meters, watt-hour meters, and protective relays. The voltage is proportional to the primary voltage, but it is small enough to be safe for the test instrument. The secondary of a potential transformer may be designed for several different voltages, but most are designed for 120V. The potential transformer is primarily a distribution transformer especially designed for voltage regulation so that the secondary voltage (under all conditions) will be as close as possible to a specified percentage of the primary voltage.

A **current transformer** is used to supply current to an instrument connected to its secondary with the current being proportional to the primary current but small enough to be safe for the instrument. The secondary of a current transformer is usually designed for a rated current of 5A.

A current transformer operates in the same way as any other transformer in that the same relationship exists between the primary current and the secondary current. A current transformer uses the circuit conductors as its primary winding. The secondary of the current transformer is connected to current devices, such as ammeters, wattmeters, watt-hour meters, power factor meters, some forms of relays, and the trip coils of some types of circuit breakers.

When no instruments or other devices are connected to the secondary of the current transformer, a short circuit device or shunt is placed across the secondary to prevent the secondary circuit from being opened while the primary winding is carrying current.

> **WARNING!**
> If the secondary circuit is open, there will be no secondary ampere turns to balance the primary ampere turns, so the total primary current becomes exciting current and magnetizes the core to a high flux density. This produces a high voltage across both the primary and secondary windings and endangers the life of anyone coming in contact with the meters or leads. This is why current transformers should never be fused. A current transformer is the only transformer that may be short-circuited on the secondary while energized.

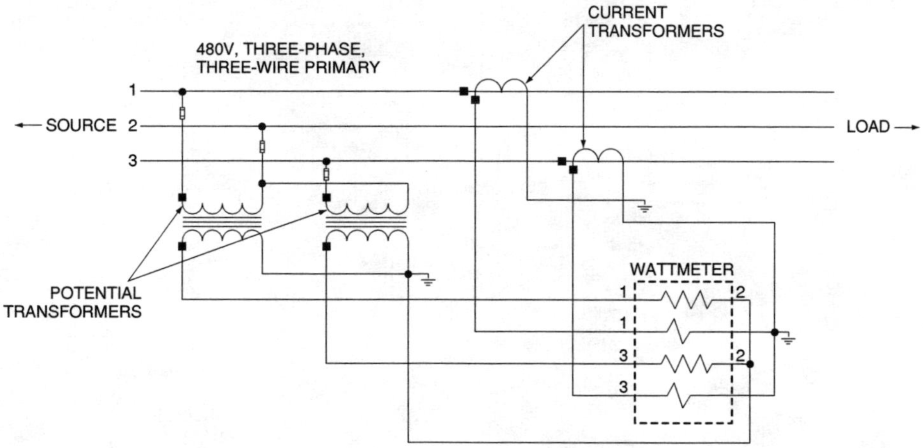

307F34.EPS

Figure 34 ◆ Current and potential transformers connected for power metering of a three-phase circuit.

Classroom

Discuss the operation and applications of potential and current transformers.

Audiovisual

Show Transparency 36 (Figure 34).

Safety

Explain why current transformers should never be fused.

![Classroom]

Point out that certain types of high-voltage transformers must be enclosed in transformer rooms or vaults. List the NEC requirements for these installations.

![Audiovisual]

Show Transparency 37 (Figure 35).

10.0.0 ◆ NEC REQUIREMENTS

Transformers must normally be accessible for inspection, except for dry-type transformers under certain specified conditions. Certain types of transformers with a high voltage or kVA rating are required to be enclosed in transformer rooms or vaults when installed indoors. The construction of these vaults is covered in ***NEC Sections 450.41 through 450.48*** and described in *Figure 35* and *Table 1*.

In general, the NEC specifies that the walls and roofs of vaults must be constructed of materials that have adequate structural strength for the conditions with a minimum fire resistance of three

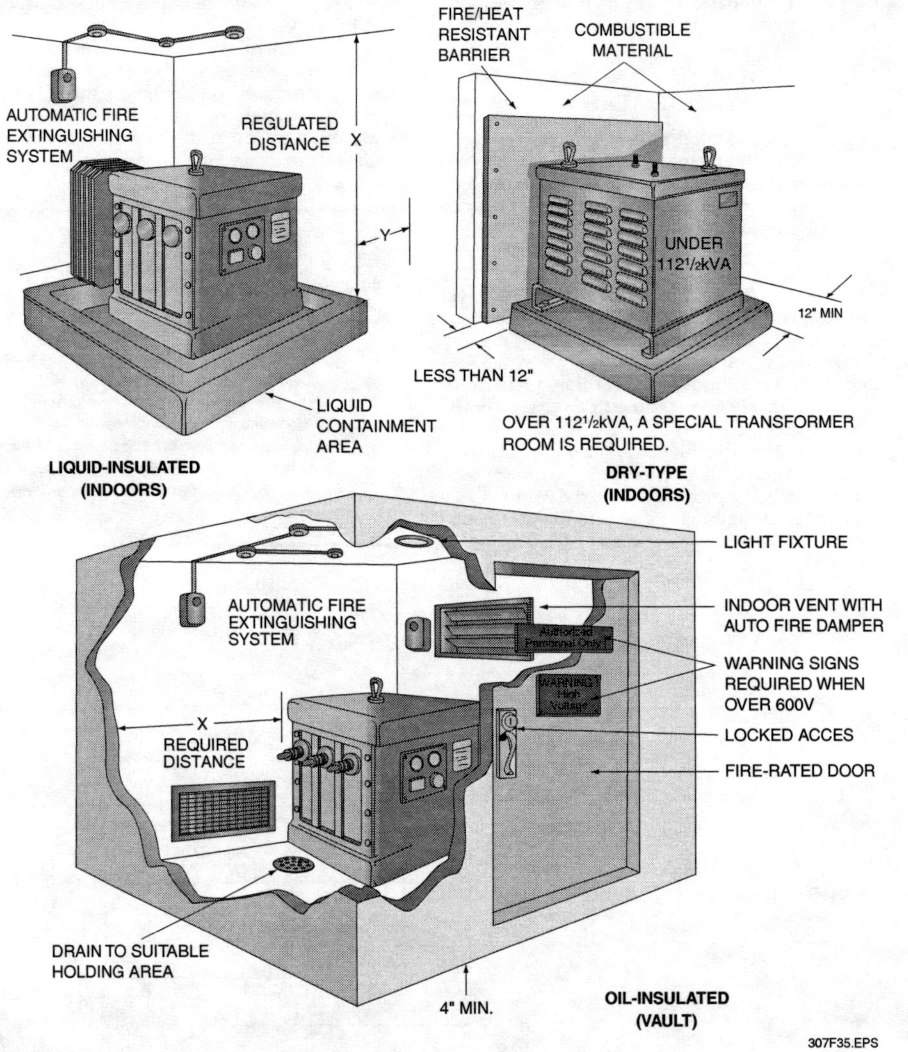

Figure 35 ◆ Transformer installation requirements.

Instructor's Notes:

hours. However, where transformers are protected with an automatic sprinkler system, water spray, carbon dioxide, or halon, the fire resistance construction may be lowered to only one hour. The floors of vaults in contact with the earth must be of concrete and not less than 4" thick. If the vault is built with a vacant space or other floors (stories) below it, the floor must have adequate structural strength for the load imposed thereon and a minimum fire resistance of three hours. Again, if the fire extinguishing facilities are provided, as outlined above, the fire resistance construction need only be one hour. The NEC does not permit the use of studs and wallboard construction for transformer vaults.

10.1.0 Overcurrent Protection for Transformers (600V or Less)

The overcurrent protection for transformers is based on their rated current, not on the load to be served. The primary circuit may be protected by a device rated or set at not more than 125% of the rated primary current of the transformer for transformers with a rated primary current of 9A or more.

Instead of individual protection on the primary side, the transformer may be protected only on the secondary side if all of the following conditions are met:

- The overcurrent device on the secondary side is rated or set at not more than 125% of the rated secondary current.
- The primary feeder overcurrent device is rated or set at not more than 250% of the rated primary current.

For example, if a 12kVA transformer has a primary voltage rating of 480V, calculate the amperage as follows:

$$\frac{12{,}000\text{VA}}{480\text{V}} = 25\text{A}$$

With a secondary voltage rated at 120V, the amperage becomes:

$$\frac{12{,}000\text{VA}}{120\text{V}} = 100\text{A}$$

The individual primary protection must be set at:

$$1.25 \times 25\text{A} = 31.25\text{A}$$

In this case, a standard 30A cartridge fuse rated at 600V could be used, as could a circuit breaker approved for use on 480V. However, if certain conditions are met, individual primary protection for the transformer is not necessary if the feeder overcurrent protective device is rated at not more than:

$$2.5 \times 25\text{A} = 62.5\text{A}$$

In addition, the protection of the secondary side must be set at not more than:

$$1.25 \times 100\text{A} = 125\text{A}$$

In this case, a standard 125A circuit breaker could be used.

> **Note**
> The example cited is for the transformer only, not the secondary conductors. The secondary conductors must be provided with overcurrent protection as outlined in ***NEC Section 210.20(B)***.

The requirements of ***NEC Section 450.3*** cover only transformer protection; in practice, other components must be considered when applying circuit overcurrent protection. Circuits with transformers must meet the requirements for conductor protection in ***NEC Articles 240 and 310***. Panelboards must meet the requirements of ***NEC Article 408***.

- *Primary fuse protection only*—If secondary fuse protection is not provided, then the primary fuses must not be sized larger than 125% of the transformer primary full-load amperes (FLA), except if the transformer primary FLA is that shown in ***NEC Table 450.3(B)***. See *Figure 36*. Individual transformer primary fuses are not necessary where the primary circuit fuse provides this protection.
- *Primary and secondary protection*—According to ***NEC Table 450.3(A)***, a transformer with a primary voltage over 600V, located in unsupervised areas, is permitted to have the primary fuse sized at a maximum of 300%. If the secondary is also over 600V, the secondary fuses can be sized at a maximum of 250% for transformers with impedances not greater than 6% and 225% for transformers with impedances greater than 6% and not more than 10%. If the secondary is 600V or below, the secondary fuses can be sized at a maximum of 125%. Where these settings do not correspond to a standard fuse size, the next higher standard size is permitted.

Classroom

Explain how to calculate the size of the overcurrent protective device for a transformer.

Point out the NEC sections applicable to transformer protection.

Audiovisual

Show Transparency 38 (Figure 36).

Audiovisual

Show Transparency 39 (Figure 37).

Classroom

Discuss the overcurrent protection requirements for small power transformers. Point out that the inrush current must be calculated to ensure that the fuse selected has sufficient inrush capacity.

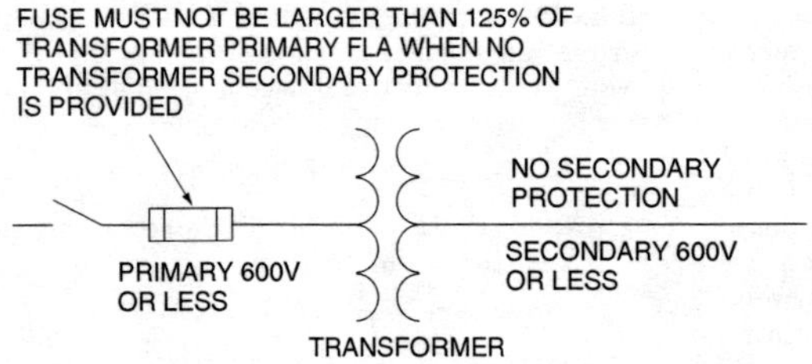

PRIMARY CURRENT	PRIMARY FUSE RATING
9A or more	125% or next higher standard rating if 125% does not correspond to a standard fuse size
2A to 9A	167% maximum
Less than 2A	300% maximum

307F36.EPS

Figure 36 ◆ Transformer circuit with primary fuse only.

In supervised locations, the maximum settings are as shown in *Figure 37,* except for secondary voltages of 600V or below, where the secondary fuses can be sized at a maximum of 250%.

- *Primary protection only*—In supervised locations, the primary fuses can be sized at a maximum of 250% or the next larger standard size if 250% does not correspond to a standard fuse size.

> **Note**
>
> The use of primary protection *only* does not remove the requirements for overcurrent compliance found in ***NEC Articles 240 and 408.*** See FPN No. 1 in ***NEC Section 450.3,*** which references ***NEC Sections 240.4, 240.21, 240.100, and 240.101,*** for proper protection of secondary conductors.

10.1.1 Overcurrent Protection for Small Power Transformers

Low-amperage, E-rated, medium-voltage fuses are general-purpose, current-limiting fuses. The E rating defines the melting time current characteristic of the fuse and permits electrical interchangeability of fuses with the same E rating. For a general-purpose fuse to have an E rating, the current responsive element shall melt in 300 seconds at an rms current within the range of 200% to 240% of the continuous current rating of the fuse, fuse refill, or link (*ANSI C37.46*).

Low-amperage, E-rated fuses are designed to provide primary protection for potential, small service, and control transformers. These fuses offer a high level of fault current interruption in a self-contained, non-venting package that can be mounted indoors or in an enclosure.

As for all current-limiting fuses, the basic application rules found in the NEC and the manufacturer's literature should be adhered to. In addition, potential transformer fuses must have sufficient inrush capacity to successfully pass through the magnetizing inrush current of the transformer. If the fuse is not sized properly, it will open before the load is energized. The maximum magnetizing inrush currents to the transformer at system voltage and the duration of this inrush current vary with the transformer design. Magnetizing inrush currents are usually denoted as a percentage of the transformer full-load current (10X, 12X, 15X, etc.). The inrush current duration is usually given in seconds. Where this information is available, an easy check can be made on the appropriate minimum melting curve to verify proper fuse selection. In lieu of transformer inrush data, the rule of thumb is to select a fuse size rated at 300% of the primary full-load current or the next larger standard size.

For example, a transformer manufacturer states that an 800VA, 2,400V, single-phase potential transformer has a magnetizing inrush current of 12X lasting for 0.1 second. Therefore:

$$I = \frac{800VA}{2{,}400V} = 0.333A$$

$$\text{Inrush current} = 12A \times 0.333A = 4A$$

Since the voltage is 2,400V, we can use either a JCW or a JCD fuse. Using the 300% rule of thumb:

$$300\% \text{ of } 0.333A = 0.999A$$

Therefore, we would choose a JCW-1E or JCD-1E fuse.

Instructor's Notes:

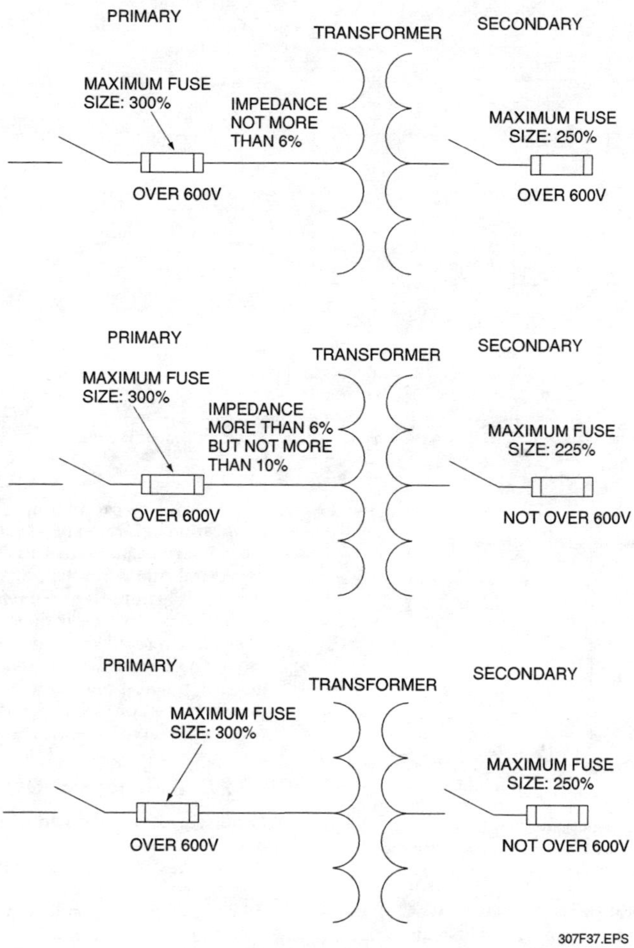

Figure 37 ◆ Minimum overcurrent protection for transformers in supervised locations.

Typical potential transformer connections can be grouped into two categories:

- Those connections that require the fuse to pass only the magnetizing inrush of one potential transformer (*Figure 38*)
- Those connections that must pass the magnetizing inrush of more than one potential transformer (*Figure 39*)

Fuses for medium-voltage transformers and feeders are E-rated, medium-voltage fuses, which are general-purpose, current-limiting fuses. The fuses carry either an E or an X rating, which defines the melting time current characteristic of the fuse. The ratings are used to allow electrical interchangeability among different manufacturers' fuses.

Classroom

Discuss the two categories of potential transformer connections:

- **Single transformer**
- **Multiple transformers**

Audiovisual

Show Transparencies 40 and 41 (Figures 38 and 39).

Classroom

List the conditions required for the application of E-rated fuses. Point out that in some situations, the use of a fuse with an X rating may be preferable.

Discuss transformer grounding requirements. Refer the trainees to *NEC Section 250.30.*

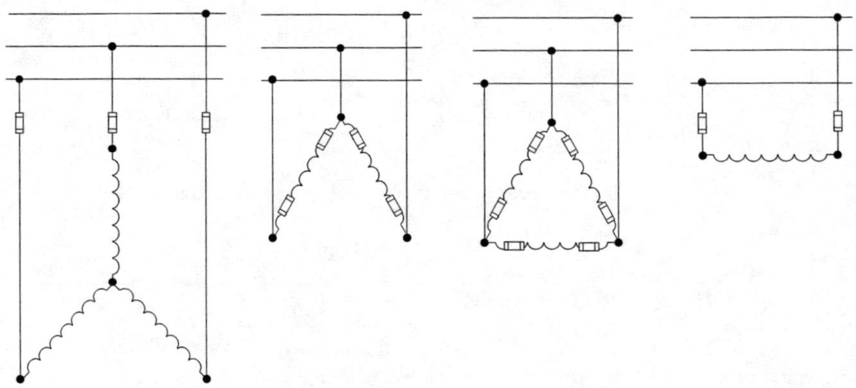

307F38.EPS

Figure 38 ◆ Connections requiring fuses to pass only the magnetizing inrush of one transformer.

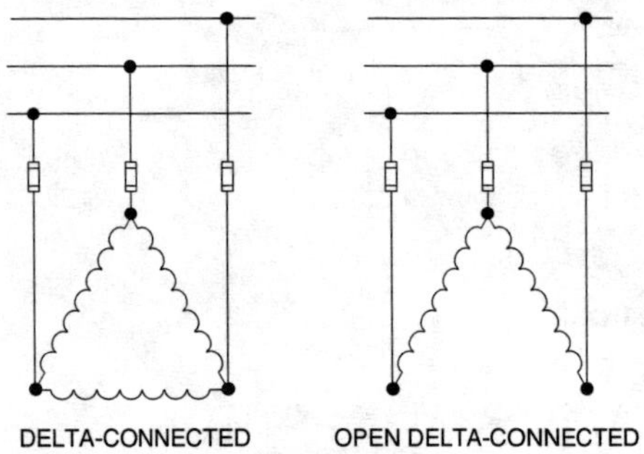

307F39.EPS

Figure 39 ◆ Connections requiring fuses to pass the magnetizing inrush of more than one transformer.

For a general-purpose fuse to have an E rating, the following conditions must be met:

- Current responsive elements with ratings 100A or below shall melt in 300 seconds at an rms current within the range of 200% to 240% of the continuous current rating of the fuse unit (*ANSI C37.46*).
- Current responsive elements with ratings above 100A shall melt in 600 seconds at an rms current within the range of 220% to 264% of the continuous current rating of the fuse unit (*ANSI C37.46*).

A fuse with an X rating does not meet the electrical interchangeability for an E-rated fuse, but it offers the user other ratings that may provide better protection for the particular application.

Transformer protection is the most popular application of E-rated fuses. The fuse is applied to the primary of the transformer and is solely used to prevent rupture of the transformer due to short circuits. It is important, therefore, to size the fuse so that it does not clear on system inrush or permissible overload currents. Magnetizing inrush must also be considered when sizing a fuse. In general, power transformers have a magnetizing inrush current of 12X, which is the full-load rating for a duration of $\frac{1}{10}$ second.

10.2.0 Transformer Grounding

Grounding is necessary to remove static electricity and also as a precautionary measure in case the transformer windings accidentally come in contact with the core or enclosure. All transformers should be grounded and bonded to meet NEC requirements and also local codes where applicable.

The case of every power transformer should be grounded to eliminate the possibility of obtaining static shocks from it or being injured by accidental grounding of the winding to the case. A grounding lug is provided on the base of most transformers for the purpose of grounding the case and fittings.

The NEC specifically states the requirements for grounding and should be followed in every respect. Furthermore, certain advisory rules recommended by manufacturers provide additional protection beyond that of the NEC. In general, the code requires that separately derived alternating current systems be grounded as stated in *NEC Section 250.30.*

Instructor's Notes:

> ### Transformers
> INSIDE TRACK
>
> Transformers are not considered separately derived systems when they have a direct electrical connection, including a solidly grounded neutral, from the primary through the secondary.

11.0.0 ◆ POWER FACTOR

Power factor was covered in the Level Two module entitled *Alternating Current,* so a brief review of the subject should suffice here. The equation for power factor is:

$$\text{power factor (pf)} = \frac{\text{kW}}{\text{kVA}}$$

Where:

kW = kilowatts
kVA = kilovolt-amperes

Calculating the power factor of an electrical system requires that the true power, inductive reactance, and capacitive reactance of the system be determined. An analogy should enhance your understanding of these terms.

Imagine a farm wagon to which three horses are hitched, as shown in *Figure 40.* The horse in the middle (#1) is pulling straight ahead; we will call this horse true power because all of the effort is in the direction that the work should be done. Horse #2 wants to nibble at the grass growing along the side of the road. This horse does not contribute an ounce of pull in the desired direction and causes a problem by pulling the wagon toward the ditch. We will call horse #2 inductance. The third horse has about the same strength as horse #2 and enjoys the grass on the opposite side of the road, which causes this horse to pull in the exact opposite direction of horse #2. Horse #3 also contributes nothing to the forward motion. This horse is called **capacitance.**

We will forget about horse #3 for the moment. If only horse #1 and horse #2 were pulling, the wagon would go in the direction of dotted line A. Notice that the length of that line is greater than the line to #1, so the horse named inductance has an effect on the final result. The direction and length of A might well be called apparent power and happens to be the hypotenuse (or diagonal) of a right angle triangle.

If the #2 horse were unhitched, then horses #1 and #3 would cause the wagon to move in the direction of line B, and the length of that line would also be apparent power. If all three horses were pulling, horses #2 and #3 would cancel one another out, and the only useful animal would be reliable horse #1, true power.

In an AC circuit, there are always three forces working in varying lengths. Inductance (horse #2)

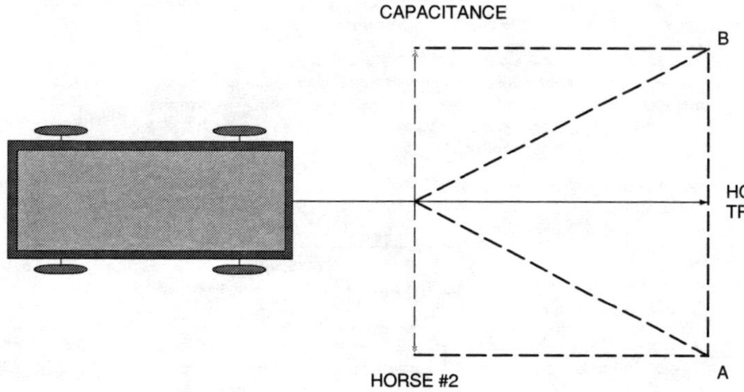

Figure 40 ◆ Depicting power factor.

Classroom

Explain how to calculate power factor.

Audiovisual

Show Transparency 42 (Figure 40).

Classroom

Discuss the installation requirements for capacitors. Refer the trainees to *NEC Article 460* and the specific rules listed in Table 2 of the Trainee Module.

is present in every magnetic circuit and always works at a 90° angle with true power. Therefore, we have a power factor of less than 100% because the wagon does not move straight ahead, but travels towards the right due to the pull of horse #2. However, we can improve the power factor by adding horse #3, capacitance, which tends to cancel out the pull of inductance, enabling the wagon to travel straight ahead.

11.1.0 Capacitors

NEC Article 460 states specific rules for the installation and protection of capacitors other than surge capacitors or capacitors that are part of another apparatus. The chief use of capacitors is to improve the power factor of an electrical installation or an individual piece of electrically operated equipment. In general, this efficiency lowers the cost of power. The NEC requirements for capacitors operating under 600V are summarized in *Table 2.*

Since capacitors may store an electrical charge and hold a voltage that is present even when a capacitor is disconnected from a circuit, capacitors must be enclosed, guarded, or located so that persons cannot accidentally contact the terminals. In most installations, capacitors are installed out of reach or are placed in an enclosure accessible only to qualified persons. The stored charge of a capacitor must be drained by a discharge circuit either permanently connected to the capacitor or automatically connected when the line voltage of the capacitor circuit is removed. The windings of a motor or a circuit consisting of resistors and reactors will serve to drain the capacitor charge.

Capacitor circuit conductors must have an ampacity of not less than 135% of the rated current

Table 2 NEC Capacitor Installation Requirements

Application	NEC Regulation	NEC Reference
Enclosing and guarding	Capacitors must be enclosed, located, or guarded so that persons cannot come into accidental contact or bring conducting materials into accidental contact with exposed energized parts, terminals, or buses associated with them. However, no additional guarding is required for enclosures accessible only to authorized and qualified persons.	***NEC Section 460.2(B)***
Stored charge	Capacitors must be provided with a means of draining the stored charge. The discharge circuit must be either permanently connected to the terminals of the capacitor or capacitor bank or provided with automatic means of connecting it to the terminals of the capacitor bank on removal of voltage from the line. Manual means of switching or connecting the discharge circuit shall not be used.	***NEC Section 460.6***
Capacitors on circuits over 600V	Special NEC regulations apply to capacitors operating at over 600V.	***NEC Articles 460, Part II and 490***
Conductor ampacity	The ampacity of capacitor circuit conductors must not be less than 135% of the rated current of the capacitor.	***NEC Section 460.8(A)***
Capacitors on motor circuits	The ampacity of conductors that connect a capacitor to the terminals of a motor or to motor circuit conductors shall not be less than one-third the ampacity of the motor circuit conductors and in no case less than 135% of the rated current of the capacitor.	***NEC Section 460.8(A)***
Overcurrent protection	Overcurrent protection is required in each ungrounded conductor unless the capacitor is connected on the load side of a motor running overcurrent device. The setting must be as low as practicable.	***NEC Section 460.8(B)***
Disconnecting means	A disconnecting means is required for a capacitor unless it is connected to the load side of a motor controller. The rating must be not less than 135% of the rated current of the capacitor.	***NEC Section 460.8(C)***
Overcurrent protection for improved pf	If the power factor is improved, the motor running overcurrent device must be selected based on the reduced current draw, not the full-load current of the motor.	***NEC Section 460.9***
Grounding	Capacitor cases must be grounded except when the system is designed to operate at other than ground potential.	***NEC Section 460.10***

Instructor's Notes:

Increasing Power Factors

INSIDE TRACK

Industrial facilities with large inductive loads that must rely on commercial power to operate pay a rate depending on power factor and consumption. The rate increases when the power factor is low because the utility must supply additional kVARs not metered by the watt-hour meter. In order to decrease the cost of electricity caused by these inductive loads, capacitors are often installed to increase the power factor.

of the capacitor. This current is determined from the VA rating of the capacitor as for any other load. For example, a 100kVA (100,000VA), three-phase capacitor operating at 480V has a rated current of:

$$I = \frac{100{,}000\text{VA}}{1.732 \times 480\text{V}} = 120.3\text{A}$$

The minimum conductor ampacity is then:

$$I = 1.35 \times 120.3\text{A} = 162.4\text{A}$$

When a capacitor is switched into a circuit, a large inrush current results to charge the capacitor to the circuit voltage. Therefore, an overcurrent protective device for the capacitor must be rated or set high enough to allow the capacitor to charge. Although the exact setting is not specified in the NEC, typical settings vary between 150% and 250% of the rated capacitor current.

In addition to overcurrent protection, a capacitor must have a disconnecting means rated at not less than 135% of the rated current of the capacitor unless the capacitor is connected to the load side of the motor running overcurrent device. In this case, the motor disconnecting means would serve to disconnect the capacitor and the motor.

A capacitor connected to a motor circuit serves to increase the power factor and reduce the total kVA required by the motor capacitor circuit. As stated earlier, the power factor (pf) is defined as the true power in kilowatts divided by the total kVA, or:

$$\text{pf} = \frac{\text{kW}}{\text{kVA}}$$

A power factor of less than one represents a lagging current for motors and inductive devices. The capacitor introduces a leading current that reduces the total kVA and raises the power factor to a value closer to unity (one). If the inductive load of the motor is completely balanced by the capacitor, a maximum power factor of unity results, and all of the input energy serves to perform useful work.

The capacitor circuit conductors for a power factor correction capacitor must have an ampacity of not less than 135% of the rated current of the capacitor. In addition, the ampacity must not be less than one-third the ampacity of the motor circuit conductors.

The connection of a capacitor reduces the current in the feeder up to the point of connection. If the capacitor is connected on the load side of the motor running overcurrent device, the current through this device is reduced, and its rating must be based on the actual current, not on the full-load current of the motor.

11.2.0 Resistors and Reactors

NEC Article 470 covers the installation of separate resistors and reactors on electric circuits. However, this article does not cover such devices that are component parts of other machines and equipment.

In general, *NEC Section 470.2* requires resistors and reactors to be installed where they will not be exposed to physical damage. Therefore, such devices are normally installed in a protective enclosure, such as a controller housing or other type of cabinet. When these enclosures are constructed of metal, they must be grounded as specified in *NEC Article 250.* Furthermore, a thermal barrier must be provided between resistors and/or reactors and any combustible material that is less than 12" away. A space of 12" or more between the devices and combustible material is considered a sufficient distance so as not to require a thermal barrier.

Insulated conductors used for connections between resistors and motor controllers must be rated at not less than 90°C (194°F) except for motor starting service. In this case, other conductor insulation is permitted, provided other sections of the NEC are not violated.

11.3.0 Diodes and Rectifiers

The diode and the rectifier are the simplest form of electronic components. The major difference between the two is their current rating. A component that is rated less than 1A is called a diode; a

Classroom

Provide general guidelines for sizing the overcurrent protection and the disconnecting means of a capacitor.

Point out that *NEC Article 470* covers the installation of separate resistors and reactors on electric circuits.

Explain the difference between a diode and a rectifier.

Classroom

Discuss the use of rectifiers in control circuits.

Audiovisual

Show Transparency 43 (Figure 41).

Classroom

Point out that vector diagrams are used to illustrate the voltages and currents in transformer windings. Explain how to read a simplified vector diagram.

Audiovisual

Show Transparency 44 (Figure 42).

similar component rated above 1A is called a rectifier. The main purpose of either device is to convert or rectify alternating current to direct current.

Diodes and rectifiers are composed of two types of semiconductor materials and are classified as either P or N types. One has free electrons; the other has a shortage of electrons. When the two types of material are bonded together, a solid-state component is produced that will allow electrons to flow in one direction and act as an insulator when the voltage is reversed.

Diodes and rectifiers are used extensively in control circuits. For example, *Figure 41* shows an AC voltage supplying a control transformer that must supply a DC electronic controller. Consequently, two rectifiers are installed on the secondary side of the transformer to change AC to DC. The rectifiers allow current to flow in one direction but will not allow the normal alternating current reversal, simulating direct current.

12.0.0 ◆ VECTORS

The theoretical study of transformers includes the use of phasor diagrams, or vectors, that graphically represent voltages and currents in transformer windings.

A vector or phasor diagram is a line with direction and length. Reading a vector diagram is like reading a road map and is not much more difficult, just a bit more refined. For example, road directions that instruct you to go east 40 miles and then south 30 miles to get to your destination are simple to understand. In other words, you had to drive 70 miles to get there. However, had you been able to drive as the crow flies, the distance would have been shorter—only 50 miles. See *Figure 42(A)*. If the miles were converted into electrical terms, this same triangle would be the classic 3/4/5 triangle or the 80% power factor relationship, as shown in *Figure 42(B)*.

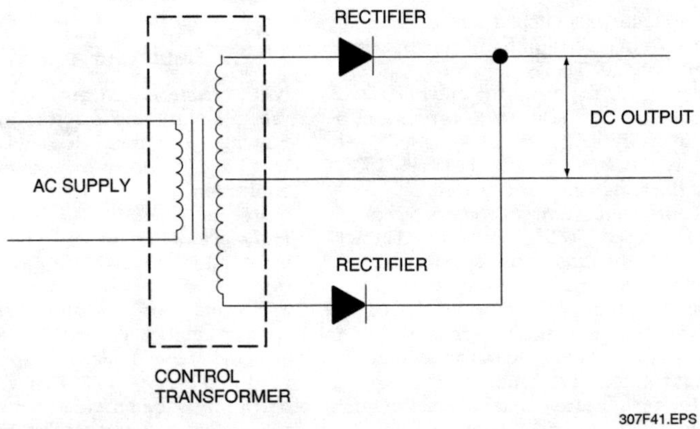

Figure 41 ◆ Rectifiers used in a control circuit to change AC to DC.

> INSIDE TRACK
>
> ### *Types N and P Diodes or Rectifiers*
>
> Type N semiconductor diodes or rectifiers are called *donor diodes* because they have an abundance of free electrons and will give away electrons. Type P semiconductor diodes or rectifiers are referred to as *acceptor diodes* because they have a shortage of free electrons and will receive electrons. This helps you remember which is which; the word donor has the letter *n* in it, while the word acceptor has the letter *p* in it.

Instructor's Notes:

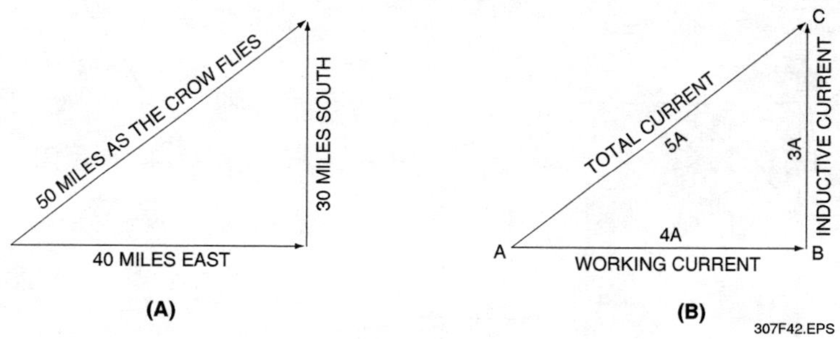

Figure 42 ◆ Typical vectors.

Referring to *Figure 42(B)*, if the working current (line A-B) is 4A and the inductive current (line B-C) is 3A, then the diagonal line A-C, or hypotenuse, will be 5A. This value may be proven by drawing lines A-B and B-C to scale and then connecting line A-C and measuring it. However, the same results may be obtained mathematically using the Pythagorean theorem, which states that the hypotenuse is equal to the square root of the sum of the squares of the other two sides:

$$AC = \sqrt{AB^2 + BC^2}$$

$$AC = \sqrt{4^2 + 3^2} = \sqrt{16 + 9} = \sqrt{25}$$

$$AC = 5$$

In this example, the working current is related to kilowatts (kW), and the total current is related to kVA. The ratio between the two is 4 ÷ 5 = 0.80. Since the power factor equals kW ÷ kVA, the power factor of this circuit is 80%.

12.1.0 Practical Applications of Phasor Diagrams

When three single-phase transformers are used as a three-phase bank, the direction of the voltage in each of the six phase windings may be represented by a voltage phasor. A voltage phasor diagram of the six voltages involved provides a convenient way to study the relative direction and amounts of the primary and secondary voltages.

The same is true for one three-phase transformer, which also has six voltages to be considered, because its phase windings on the high-voltage and low-voltage sides are connected together in the same way as the phase windings of three single-phase transformers.

To show how a phasor diagram is drawn, consider the three-phase transformer shown in *Figure 43.* Here we have a wye/delta-connected, three-phase transformer with three legs, with each leg carrying a high-voltage and a low-voltage winding. The high-voltage windings are connected in wye (with a common neutral point at N) with the leads to the high-voltage terminals designated H_1, H_2, and H_3. The three low-voltage windings are connected in delta. The junction points of the three windings serve as low-voltage terminals X_1, X_2, and X_3.

The voltages in the low-voltage windings are assumed to be equal to each other in amount but are displaced from each other by 120 degrees.

When drawing the phasor diagram for the low-voltage windings, phasor X_1X_2 is drawn first in any selected direction and to any convenient scale. The arrowhead indicates the instantaneous direction of the alternating voltage in winding X_1X_2, and the length of the phasor represents the amount of voltage in the winding. The broken lines that extend past the arrowheads represent reference lines for phase angles.

Since winding X_2X_3 is physically connected to the end of X_2 in the X_1X_2 winding, phasor X_2X_3 will be started at point X_2, 120-degrees out of phase with phasor X_1X_2 in the clockwise direction. The length of phasor X_2X_3 is equal to the length of phasor X_1X_2. Winding X_3X_1 is drawn in a similar manner.

The high-voltage phasor comes next. Since the low-voltage winding X_1X_2 is wound on the same leg as the high-voltage winding NH_1, the voltages in these two windings are in phase and are represented by parallel voltage phasors. Therefore, the phasor NH_1 is drawn from a selected point N parallel to X_1X_2. Note that all three high-voltage windings are physically connected to a common point N. Therefore, the phasors representing the voltages NH_1, NH_2, and NH_3 will all start at the common point (point N in the phasor diagram). Again, the high-voltage phasors are all of the same length but are displaced from each other by 120-degrees as shown.

Classroom

Explain how to calculate the total current in a circuit using a vector diagram.

Demonstrate how to apply a vector diagram to a three-phase transformer circuit.

Audiovisual

Show Transparency 45 (Figure 43).

Audiovisual

Show Transparency 46 (Figure 44).

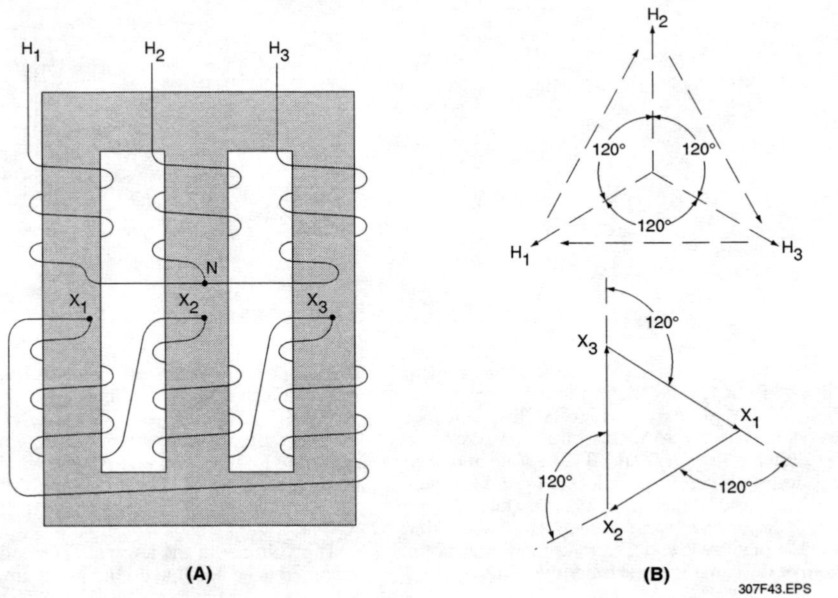

Figure 43 ◆ Windings and phasor diagrams of a three-phase transformer.

The high-voltage phasors have the same length as the low-voltage phasors. Each phase of the high voltage is, however, proportional to the low voltage in the same phase according to the turns ratio, making the line voltage 1.732 times higher than the phase voltage in any one winding. It can be geometrically proven, for example, that $H_1H_2 \times NH_1 = 1.732 \times NH_2$ and so forth for each line voltage.

When comparing three-phase transformers for possible operation in parallel, draw the voltage phasor diagram for transformer bank A, and mark the terminals as shown in *Figure 44(A)*. Next, draw the phasor diagram for bank B on drafting or other transparent paper; then draw a heavy reference line m-n as shown in *Figure 44(B)*. Cut the transparent diagram into two parts as indicated by the dashed lines in the drawing. The two parts

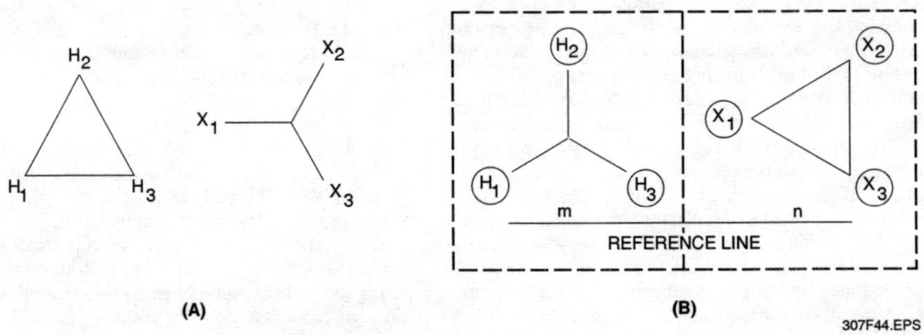

Figure 44 ◆ Phasor diagrams for three-phase transformer banks in parallel.

Instructor's Notes:

of the reference line are now marked m and n. Place diagram m on the high-voltage diagram in *Figure 44(A)* so that the terminals that are desired to be connected together coincide. Place diagram n on the low-voltage diagram in *Figure 44(A)* so that the heavy reference line of n is parallel to the heavy reference line of m. If the terminals of n can be made to coincide with the low-voltage terminals, the terminals that coincide can be connected together for parallel operation. If the low-voltage terminals cannot be made to coincide, parallel operation is not possible with the assumed high-voltage connection.

12.2.0 Voltage Drop

Figure 45 shows a simple 100% power factor circuit in which a 1Ω resistance appears between the source of power and the load. The current in this series circuit is 10A, and the voltage at the source is 120V. Because Ohm's law states that E = IR, we must have a voltage drop across the resistance equal to 10 times 1, or 10V. With the voltage and current in phase, only 110V will be available at the load because we must subtract the resistance drop from the source voltage. Voltage drop was covered more thoroughly earlier in your training.

Voltage drop, however, becomes a little more complicated when inductance is introduced into the circuit, as shown in *Figure 46*. In this case, the load voltage will be equal to the source voltage minus the voltage drops through R and X, but they cannot be added arithmetically. The vector diagram in *Figure 47* shows that the dotted line A-C is the combination of the two voltages across R and X and represents the voltage drop in the line only.

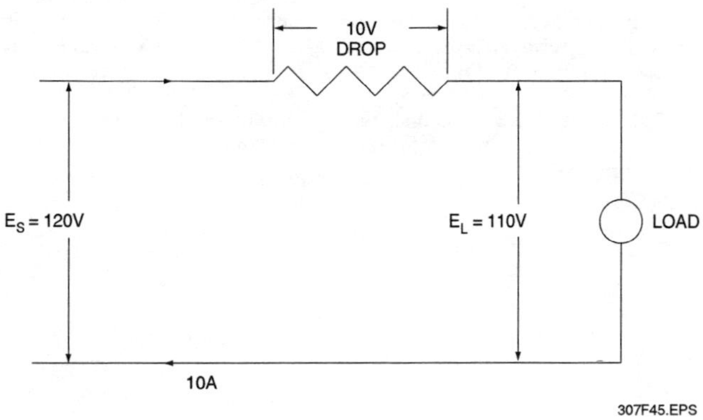

Figure 45 ◆ Circuit containing resistance only.

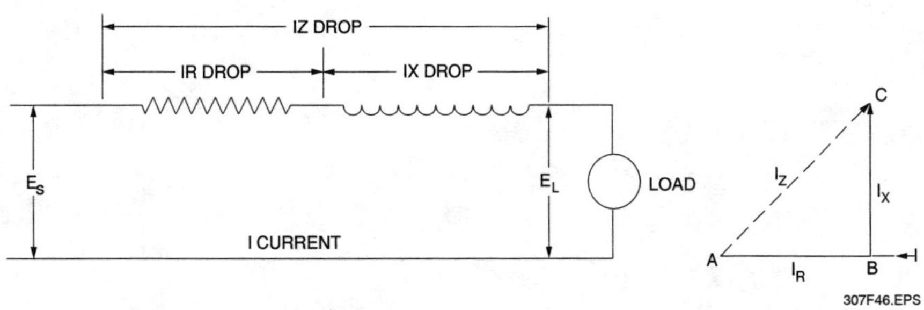

Figure 46 ◆ Electrical circuit with both resistance and inductance.

Classroom

Explain how to calculate the voltage drop in a resistance-only circuit.

Audiovisual

Show Transparency 47 (Figure 45).

Classroom

Explain how to calculate the voltage drop in a circuit containing both inductance and resistance.

Audiovisual

Show Transparency 48 (Figure 46).

Classroom

Discuss the effect of voltage drop in an AC circuit.

Audiovisual

Show Transparency 49 (Figure 47).

Classroom

Explain the relationship between voltage and current in a transformer.

Audiovisual

Show Transparency 50 (Figure 48).

Classroom

Explain that most transformer problems are easily identified and corrected.

Figure 47 shows that line G-A, which is the load voltage E_L, is less than the source voltage G-C due to the voltage drop in line A-C. Because of the effect of the reactance in the line, this voltage drop cannot be subtracted arithmetically from the source voltage to obtain the load voltage; vectors must be used.

Calculations of impedance can be simplified by using equivalent circuits. The reactance voltage drop is governed by the leakage flux, and the voltage regulation depends on the power factor of the load. To determine transformer efficiency at various loads, it is necessary to first calculate the core loss, hysteresis loss, eddy current loss, and load loss.

However, for all practical purposes in electrical construction applications, a transformer's efficiency may be considered to be 100%. Therefore, for our purposes, a transformer may be defined as a device that transfers power from its primary circuit to the secondary circuit without any significant loss.

Since apparent power (VA) equals voltage (E) times current (I), if E_PI_P represents the primary apparent power and E_SI_S represents the secondary apparent power, then $E_PI_P = E_SI_S$ (the subscript P = primary, and the subscript S = secondary). See *Figure 48*. If the primary and secondary voltages are equal, the primary and secondary currents must also be equal. Assume that E_P is twice as large as E_S. For E_PI_P to equal E_SI_S, I_P must be one-half of I_S. Therefore, a transformer that steps voltage down always steps current up. Conversely, a transformer that steps voltage up always steps current down. However, transformers are classified as step-up or step-down only in relation to their effect on voltage.

13.0.0 ◆ TROUBLESHOOTING

Since transformers are an essential part of every electrical installation, electricians must know how to test and locate problems that develop in transformers, especially in the smaller power supply or control transformers. The procedure for accomplishing this is commonly known as troubleshooting.

The term troubleshooting, as used in this module, covers the investigation, analysis, and corrective action required to eliminate faults in electrical systems, including circuits, components, and equipment. Most problems are simple and easily corrected, such as an open circuit, a ground fault or short circuit, or a change in resistance.

There are many useful troubleshooting charts available. Most charts list the complaint on the left side of the chart, then the possible cause, followed by the proper corrective action.

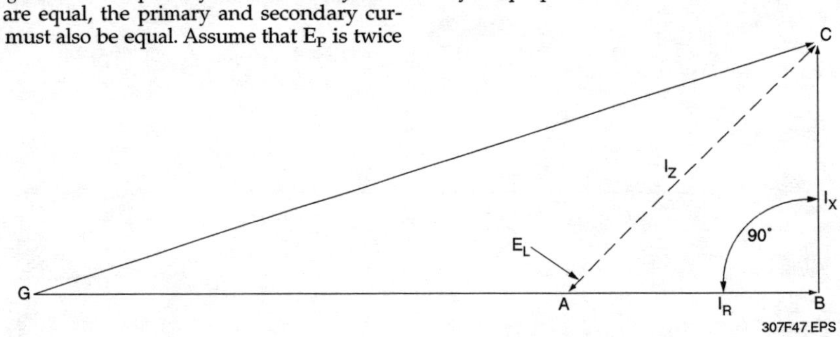

Figure 47 ◆ Effect of voltage drop in an AC circuit.

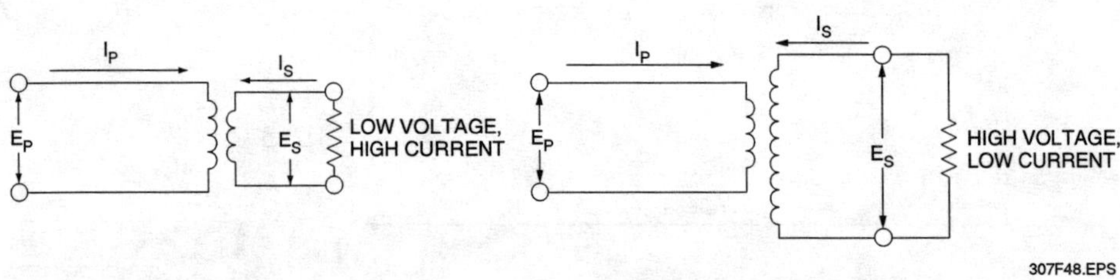

Figure 48 ◆ Voltage-current relationship in a transformer.

Instructor's Notes:

When troubleshooting, think before acting, study the problem thoroughly, then ask yourself these questions:

- What were the warning signs preceding the trouble?
- What previous repair and maintenance work has been done?
- Has similar trouble occurred before?
- If the circuit, component, or piece of equipment still operates, is it safe to continue operation before further testing?

The answers to these questions can usually be obtained by:

- Questioning the owner of the equipment
- Taking time to think the problem through
- Looking for additional symptoms
- Consulting troubleshooting charts
- Checking the simplest things first
- Referring to repair and maintenance records
- Checking with calibrated instruments
- Double-checking all conclusions before beginning any repair

> **Note**
> Always check the easiest and most obvious things first; following this simple rule will save time and trouble.

13.1.0 Double-Check Before Beginning

The source of many problems can be traced not to one part alone but to the relationship of one part with another. For instance, a tripped circuit breaker may be reset to restart a piece of equipment, but what caused the breaker to trip in the first place? It could have been caused by a vibrating hot conductor momentarily coming into contact with a ground, a loose connection, or any number of other causes.

Too often, electrically operated equipment is completely disassembled in search of the cause of a certain complaint and all evidence is destroyed during disassembly. Check again to be certain an easy solution to the problem has not been overlooked.

13.2.0 Find and Correct the Basic Cause of the Trouble

After an electrical failure has been corrected in any type of electrical circuit or piece of equipment, be sure to locate and correct the cause so the same failure will not be repeated. Further investigation may reveal other faulty components.

Also be aware that although troubleshooting charts and procedures greatly help in diagnosing malfunctions, they can never be complete. There are too many variations and solutions for any given problem.

To solve electrical problems consistently, you must first understand the basic parts of electrical circuits, how they function, and for what purpose. If you know that a particular part is not performing its job, then the cause of the malfunction must be within this part or series of parts.

13.3.0 Troubleshooting Transformers

This section discusses common transformer problems.

13.3.1 Open Circuit

Should one of the windings in a transformer develop a break or open condition, no current can flow, and therefore, the transformer will not deliver any output. The main symptom of an open circuit in a transformer is that the circuits that derive power from the transformer are de-energized or dead. Use an AC voltmeter or a volt-ohm-milliammeter (VOM) to check across the transformer output terminals, as shown in *Figure 49.* A reading of zero volts indicates an open circuit.

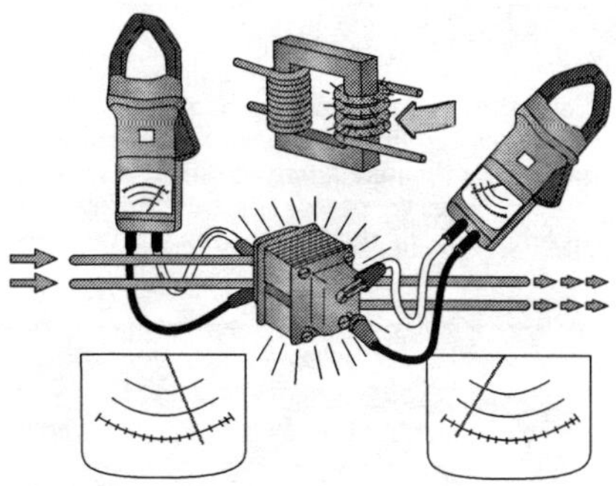

307F49.EPS

Figure 49 ◆ Checking a transformer for an open winding using a VOM.

Classroom

Point out that effective troubleshooting begins with collecting as much information as possible about the equipment failure.

Emphasize the importance of checking the simplest and most obvious problems first.

Discuss the common problems related to transformers.

Explain that shorted windings can be verified using a VOM.

Audiovisual

Show Transparency 51 (Figure 49).

Next, take a voltage reading across the input terminals. If a voltage reading is present, then the conclusion is that one of the windings in the transformer is open. However, if no voltage reading is on the input terminals either, then the conclusion is that the open is elsewhere on the line side of the circuit; perhaps a disconnect switch is open.

> **WARNING!**
> Make absolutely certain that your testing instruments are designed for the job and are calibrated for the correct voltage. Never test the primary side of any transformer over 600V unless you are qualified and have the correct high-voltage testing instruments and the test is made under the proper supervision. Always set the range selector switch on the test equipment to the highest setting, and decrease it in increments until the most accurate reading is indicated.

If voltage is present on the line or primary side and is not present on the secondary or load side, open the switch to de-energize the circuit, and place a warning tag (tagout and lock) on this switch so that it is not inadvertently closed again while someone is working on the circuit. Disconnect all of the transformer primary and secondary leads; then check each winding in the transformer for continuity, as indicated by a resistance reading taken with an ohmmeter (*Figure 50*).

Continuity is indicated by a relatively low resistance reading on control transformers; an open winding will be indicated by an infinite resistance reading on the ohmmeter. In most cases, small transformers will have to be replaced unless the break is accessible and can be repaired.

13.3.2 Shorted Turns

Sometimes a few turns in the secondary winding of a transformer will acquire a partial short, which in turn will cause a voltage drop across the secondary. The symptom of this condition is usually overheating of the transformer caused by large circulating currents flowing in the shorted windings. The most accurate way to check for this condition is with a transformer turns ratio tester (TTR). However, another way to check for this condition is with a VOM set at the proper voltage scale (*Figure 51*). Take a reading on the line or primary side of the transformer first to make certain normal voltage is present; then take a reading on the secondary side. If the transformer has a partial short or ground fault, the voltage reading should be lower than normal.

Replace the faulty transformer with a new one, and again take a reading on the secondary. If the voltage reading is now normal and the circuit operates satisfactorily, leave the replacement transformer in the circuit, and either discard or repair the original transformer.

A highly sensitive ohmmeter may also be used to test for this condition when the system is de-energized and the leads are disconnected; a lower than normal reading on the ohmmeter indicates this condition. However, the difference will usually be so slight that the average ohmmeter is not sensitive enough to detect it. Therefore, the recommended method is to use the voltmeter test.

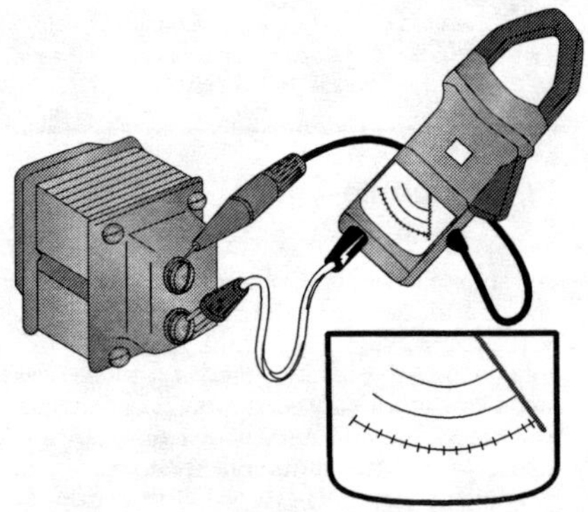

307F50.EPS

Figure 50 ◆ Checking for an open winding with a continuity test using a VOM.

> THINK ABOUT IT
>
> ***Open and Unbonded Neutrals***
>
> What is the difference between an open neutral and an unbonded neutral?

Safety

Review the safety precautions associated with the use of meters.

Classroom

Explain that open windings can be isolated by performing a continuity test.

Audiovisual

Show Transparency 52 (Figure 50).

Classroom

Explain that shorted turns can be identified by checking the voltage.

Audiovisual

Show Transparency 53 (Figure 51).

Teaching Tip

Discuss the "Think About It."

Instructor's Notes:

Testing External Circuits

INSIDE TRACK

Always retest the external circuits supplied by the transformer before energizing a replacement transformer. This is done to avoid damaging or destroying the replacement due to a shorted external circuit, which may have caused the original transformer to burn out.

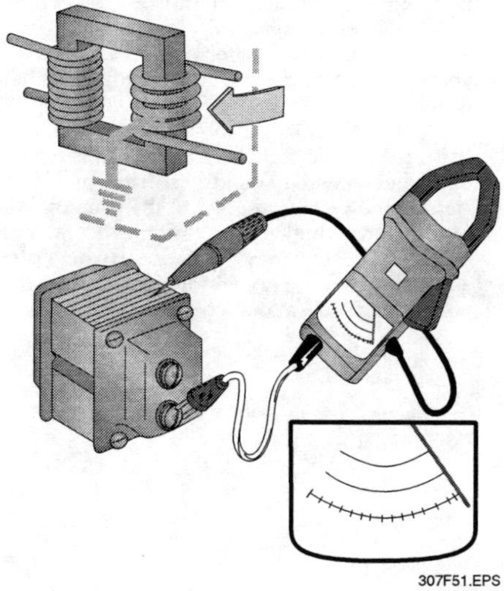

Figure 51 ◆ Testing a transformer for a ground fault using a VOM.

13.3.3 Complete Short

Occasionally, a transformer winding will become completely shorted. In most cases, this will activate the overload protective device and de-energize the circuit, but in other instances, the transformer may continue trying to operate with excessive overheating due to the very large circulating current. This heat will often melt the wax or insulation inside the transformer, which is easily detected by the odor. Also, there will be no voltage output across the shorted winding, and the circuit across the winding will be dead.

The short may be in the external secondary circuit, or it may be in the transformer's winding. To determine its location, disconnect the external secondary circuit from the winding, and take a reading with a voltmeter. If the voltage is normal with the external circuit disconnected, then the problem lies within the external circuit. However, if the voltage reading is still zero across the secondary leads, the transformer is shorted and will have to be replaced.

13.3.4 Grounded Windings

Insulation breakdown is quite common in older transformers, especially those that have been overloaded. At some point, the insulation breaks or deteriorates, and the wire becomes exposed. The exposed wire often comes into contact with the transformer housing and grounds the winding.

If a winding develops a ground and a point in the external circuit connected to this winding is also grounded, part of the winding will be shorted out. The symptoms will be overheating, which is usually detected by heat or a burning smell and a low-voltage reading, as indicated on a voltmeter (*Figure 52*). In most cases, transformers with this condition will have to be replaced.

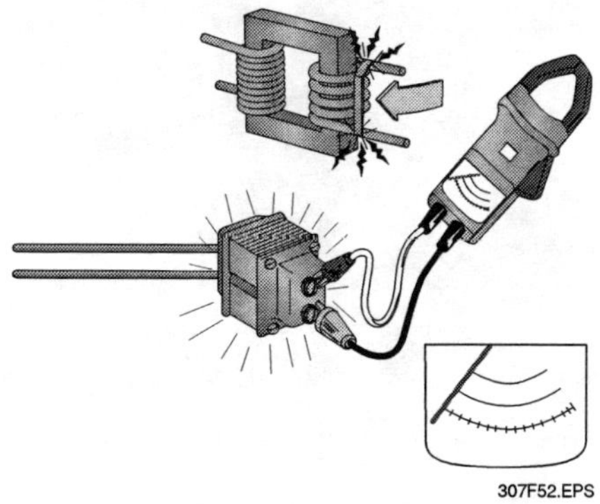

Figure 52 ◆ Transformers that overheat usually have a partial short in the windings.

Classroom

Describe the symptoms of a completely shorted winding.

Explain that grounded windings can be identified by using a megger to test the insulation resistance.

Audiovisual

Show Transparency 54 (Figure 52).

Safety

Point out that only qualified and trained individuals may use a megger.

A megohmmeter (megger) is the best test instrument to check for this condition. Disconnect the leads from both the primary and secondary windings. Tests can then be performed on either winding by connecting the megger negative test lead to an associated ground and the positive test lead to the winding to be measured.

> **WARNING!**
> Do not use a megger unless you are qualified and properly supervised.

The insulation resistance should then be measured between the windings themselves. This is accomplished by connecting one test lead to the primary and the second test lead to the secondary and ground. After that, connect the first lead to the secondary and the second test lead to the primary and ground. All such tests should be noted on a record card under the proper identifying labels.

Summary

This module covered the basic components and applications of distribution system transformers. When the AC voltage needed for an application is lower or higher than the voltage available from the source, a transformer is used. The essential components of a transformer are the primary winding, which is connected to the source, and the secondary winding, which is connected to the load. Both are wound on an iron core. The two windings are not physically connected, with the exception of autotransformers. The alternating voltage in the primary winding induces an alternating voltage in the secondary winding. The ratio of the primary and secondary voltages is equal to the ratio of the number of turns in the primary and secondary windings. Transformers may step up the voltage applied to the primary winding and have a higher voltage at the secondary terminals, or they may step down the voltage applied to the primary winding and have a lower voltage available at the secondary terminals.

Instructor's Notes:

Review Questions

1. The main purpose of a transformer is to ____.
   a. change the current
   b. improve the power factor
   c. change the output voltage
   d. enter impedance into the circuit
2. The three parts of a basic transformer are the ____.
   a. housing, lifting hooks, and base
   b. dry, oil-filled, and gas-filled chambers
   c. shell, open winding, and closed winding
   d. primary winding, secondary winding, and core
3. When AC flows through a transformer coil, a ____ field is generated around the coils.
   a. magnetic
   b. non-magnetic
   c. high-impedance
   d. rotating
4. When the field from one coil cuts through the turns of a second coil, ____.
   a. the second coil will not be affected
   b. the coils will rotate
   c. voltage will be generated
   d. no current will flow in the circuit
5. What causes voltage to be induced in a transformer?
   a. Transformer taps
   b. Mutual induction
   c. Reluctance
   d. Capacitance
6. In a transformer with a turns ratio of 5:1 (primary has five times the number of turns as the secondary), what will be the voltage on the secondary if the primary voltage is 120V?
   a. 12V
   b. 24V
   c. 48V
   d. 60V
7. The three basic types of iron core transformers are ____.
   a. dry, oil-filled, and auto
   b. closed, open, and shell
   c. control, power, and lighting
   d. metal, nonmetallic, and high-temperature
8. One effect caused by magnetic leakage in transformers is a ____.
   a. reactance voltage drop
   b. low impedance
   c. higher secondary voltage
   d. lower secondary voltage
9. A symptom of a transformer with an open circuit is ____.
   a. high voltage on the secondary
   b. excessive overheating
   c. no output on the secondary
   d. voltage drop across the secondary
10. A(n) ____ should be used to measure for insulation resistance.
    a. ammeter
    b. megger
    c. voltmeter
    d. TTR

Classroom

Have the trainees complete the Review Questions and go over the answers prior to administering the Module Examination.

Examination

Administer the Module Examination. Record the results on Craft Training Report Form 200 and submit the results to the Training Program Sponsor.

Performance Testing

Administer the Performance Test and fill out Performance Profile Sheets for each trainee. If desired, trainee proficiency noted during laboratory sessions may be used to complete the Performance Test. Be sure to record the results on Craft Training Report Form 200 and submit the results to the Training Program Sponsor.

GLOSSARY

Trade Terms Introduced in This Module

Ampere turn: The product of amperes times the number of turns in a coil.

Autotransformer: Any transformer in which primary and secondary connections are made to a single winding. The application of an autotransformer is a good choice where a 480Y/277V or 208Y/120V, three-phase, four-wire distribution system is used.

Capacitance: The storage of electricity in a capacitor or the opposition to voltage change. Capacitance is measured in farads (f) or microfarads (μf).

Current transformer: A single-phase instrument transformer connected in series in a line that carries the full-load current. The turns ratio is designed to produce a reduced current in the secondary suitable for the current coil of standard measuring instruments and in proportion to the load current.

Flux: The rate of energy flow across or through a surface. Also a substance used to promote or facilitate soldering or welding by removing surface oxides.

Induction: The production of magnetization or electrification in a body by the mere proximity of magnetized or electrified bodies, or the production of an electric current in a conductor by the variation of the magnetic field in its vicinity.

Kilovolt-amperes (kVA): 1,000 volt-amperes (VA).

Loss: The power expended without doing useful work.

Magnetic field: The area around a magnet in which the effect of the magnet can be felt.

Magnetic induction: The number of magnetic lines or the magnetic flux per unit of cross-sectional area perpendicular to the direction of the flux.

Mutual induction: The condition of voltage in a second conductor because of current in another conductor.

Potential transformer: A special transformer designed for use in measuring high voltage; normally, the secondary voltage is 120V.

Power transformer: A transformer that is designed to transfer electrical power from the primary circuit to the secondary circuit(s) to step up the secondary voltage at less current or step down the secondary voltage at more current, with the voltage-current product being constant for either the primary or secondary.

Reactance: The opposition to AC due to capacitance and/or inductance.

Rectifiers: Devices used to change alternating current to direct current.

Transformer: A static device consisting of one or more windings with a magnetic core. Transformers are used for introducing mutual coupling by induction between circuits.

Turn: The basic coil element that forms a single conducting loop comprised of one insulated conductor.

Turns ratio: The ratio between the number of turns between windings in a transformer; normally the primary to the secondary, except for current transformers, in which it is the ratio of the secondary to the primary.

Instructor's Notes:

Additional Resources

This module is intended to present thorough resources for task training. The following reference works are suggested for further study. These are optional materials for continued education rather than for task training.

American Electrician's Handbook. Terrell Croft and Wilfred I. Summers. New York, NY: McGraw-Hill, 1996.

National Electrical Code Handbook. Quincy, MA: National Fire Protection Association, 2002.

ANSWER KEY

Answers to Review Questions

Answer	Section
1. c	1.0.0
2. d	2.0.0
3. a	2.1.0
4. c	2.1.0
5. b	2.1.0
6. b	2.2.0
7. b	3.0.0
8. d	3.5.0
9. c	13.3.1
10. b	13.3.4

Section 5.0.0 ***Think About It—Secondary Neutrals***

Secondary neutrals are bonded and grounded in order to limit the voltage to ground on the secondary side and facilitate the operation of the secondary overcurrent devices in the event of a ground fault.

Section 7.2.0 ***Think About It—Calculating Secondary Voltages***

First = 228.8V
Second = 218.4V
Third = 249.6V
Fourth = 228.8V

Section 13.3.1 ***Think About It—Open and Unbonded Neutrals***

An open neutral is missing the connection between the neutral and the transformer. An unbonded neutral is not bonded or connected to ground. An open neutral causes loads requiring a neutral to operate in series with line 1 and line 2, potentially resulting in severe equipment damage. In an unbonded neutral, the voltages measured from line 1 or line 2 or neutral to ground will be erratic. This may prevent the overcurrent devices from operating correctly. In addition, the inconsistent voltage may interfere with equipment operation.

CONTREN™ LEARNING SERIES — USER UPDATES

The NCCER makes every effort to keep these textbooks up-to-date and free of technical errors. We appreciate your help in this process. If you have an idea for improving this textbook, or if you find an error, a typographical mistake, or an inaccuracy in NCCER's Contren™ textbooks, please write us, using this form or a photocopy. Be sure to include the exact module number, page number, a detailed description, and the correction, if applicable. Your input will be brought to the attention of the Technical Review Committee. Thank you for your assistance.

Instructors – If you found that additional materials were necessary in order to teach this module effectively, please let us know so that we may include them in the Equipment/Materials list in the Instructor's Guide.

Write: Curriculum Revision and Development Department
National Center for Construction Education and Research
P.O. Box 141104, Gainesville, FL 32614-1104

Fax: 352-334-0932

E-mail: curriculum@nccer.org

Craft ____________________ Module Name ____________________

Copyright Date ____________ Module Number ____________ Page Number(s) ____________

Description

____

____

____

____

(Optional) Correction

____

____

____

(Optional) Your Name and Address

____

____

____

Lamps, Ballasts, and Components

26308-02

Lamps, Ballasts, and Components
Annotated Instructor's Guide

Module 26308-02

MODULE OVERVIEW

This module is the second of three modules that cover electric lighting. It builds on the information and lighting principles previously covered in the first module entitled *Electric Lighting*. This module provides information on the operation of specific types of lamps and related lighting fixtures. Also covered are some common lighting circuit control devices and energy conservation schemes used to control lighting. Troubleshooting procedures for fluorescent and HID lamps are also covered.

PREREQUISITES

Please refer to the Course Map in the Trainee Module. Prior to training with this module, it is recommended that the trainee shall have successfully completed the following:

Core Curriculum; Electrical Level One; Electrical Level Two; Electrical Level Three, Modules 26301 through 26307.

OBJECTIVES

Upon completion of this module, the trainee will be able to:

1. Recognize incandescent, fluorescent, and high-intensity discharge (HID) lamps and describe how each type of lamp operates.
2. Recognize ballasts and describe their purpose for use in fluorescent and HID lighting fixtures.
3. Explain the relationship of Kelvin temperature to the color of light produced by a lamp.
4. Recognize basic occupancy sensors, photoelectric sensors, and timers used to control lighting circuits and describe how each device operates.
5. Use troubleshooting checklists to troubleshoot fluorescent and HID lamps and lighting fixtures.

PERFORMANCE TASKS

Under supervision of the instructor, the trainee should be able to:

1. Use troubleshooting checklists and guidelines to troubleshoot selected fluorescent and/or HID lamps and lighting fixtures.

NCCER STANDARDIZED CRAFT TRAINING PROGRAMS

The National Center for Construction Education and Research (NCCER) provides a standardized national program of accredited craft training. Key features of the program include instructor certification, competency-based training, and performance testing. The program provides trainees, instructors, and companies with a standard form of recognition through a National Craft Training Registry. The program is described in full in the *Guidelines for Accreditation*, published by the NCCER. For more information on standardized craft training, contact the NCCER by writing us at P.O. Box 141104, Gainesville, FL 32614-1104, calling 352-334-0911, or e-mailing info@nccer.org. More information may be found at our Web site, www.nccer.org.

HOW TO USE THIS ANNOTATED INSTRUCTOR'S GUIDE

Each page presents two sections of information. The larger section displays each page exactly as it appears in the Trainee Module. The narrow column ties suggested trainee and instructor actions to each page and provides icons to call your attention to material, safety, audiovisual, or testing requirements. The bottom of each page includes space for your notes.

Teaching Tip

If you see the Teaching Tip icon, that means there is a teaching tip associated with this section. Also refer to any suggested teaching tips at the end of the module.

SAFETY CONSIDERATIONS

Ensure that the trainees are equipped with appropriate personal protective equipment.

PREPARATION

Before teaching this module, you should review the Module Outline, Objectives, Performance Tasks, and the Materials and Equipment List. Be sure to allow ample time to prepare your own training or lesson plan and gather all required equipment and materials.

MATERIALS AND EQUIPMENT LIST

Materials:
Transparencies
Markers/chalk
Copy of the latest edition of the *National Electrical Code*®
Examples of manufacturers' lamp and lighting fixture catalogs
Module Examinations*
Performance Profile Sheets*

Equipment:
Overhead projector and screen
Whiteboard/chalkboard
Appropriate personal protective equipment
Assortment of electric lamps, including:
- Incandescent
- Halogen
- Fluorescent
- High-intensity discharge (HID)

Assortment of lighting fixtures, including:
- Incandescent
- Halogen
- Fluorescent
- High-intensity discharge (HID)

Assortment of fluorescent and HID ballasts
Motion and infrared-type occupancy sensors
Photosensors
Timers and time clocks
Volt-ohm-milliammeter (VOM)

*Located in the Test Booklet packaged with this Annotated Instructor's Guide.

ADDITIONAL RESOURCES

This module is intended to present thorough resources for task training. The following reference works are suggested for both instructors and motivated trainees interested in further study. These are optional materials for continued education rather than for task training.

American Electrician's Handbook, Terrell Croft and Wilfred I. Summers. New York, NY: McGraw-Hill, 1996.

Lighting Handbook. New York, NY: Illuminating Engineering Society of North America (IESNA), 2000.

National Electrical Code Handbook. Quincy, MA: National Fire Protection Association, 2002.

TEACHING TIME FOR THIS MODULE

An outline for use in developing your lesson plan is presented below. Note that each Roman numeral in the outline equates to one session of instruction. Each session has a suggested time period of 2½ hours. This includes 10 minutes at the beginning of each session for administrative tasks and one 10-minute break during the session. Approximately 5 hours are suggested to cover *Lamps, Ballasts, and Components.* You will need to adjust the time required for hands-on activity and testing based on your class size and resources.

Topic	Planned Time
Session I. Introduction; Types of Lamps; Fluorescent and HID Lighting Fixture Ballasts; Troubleshooting Lamps and Lighting Fixtures	
A. Introduction	
B. Types of Lamps	
1. Standard Incandescent Lamps	
2. Tungsten Halogen Incandescent Lamps	
3. Fluorescent Lamps	
4. High-Intensity Discharge (HID) Lamps	
5. Lamp Color Rendering and Color Temperature Characteristics	
C. Fluorescent and HID Lighting Fixture Ballasts	
1. Fluorescent Lighting Fixture Ballasts	
a. Preheat Lamp and Ballast Operation	
b. Rapid-Start Lamp and Ballast Operation	
c. Instant-Start Lamp and Ballast Operation	
d. Types of Fluorescent Ballasts	
e. Fluorescent Dimming Ballasts	
f. Emergency Lighting Ballasts	
2. HID Lighting Fixture Ballasts	
a. Linear, Nonregulating Circuit Ballasts	
b. Constant-Wattage Autotransformer Ballasts	
c. Three-Coil Ballasts	
d. Dimming HID Lamps	
D. Troubleshooting Lamps and Lighting Fixtures	
1. Laboratory	
Have the trainees use troubleshooting checklists and guidelines to troubleshoot various types of lamps and lighting fixtures. Note the proficiency of each trainee.	
Session II. Controls for Lighting; Energy Management Systems; Review; Module Examination; Performance Testing	
A. Controls for Lighting	
1. Occupancy Sensors	
2. Photosensors	
3. Timers	
B. Energy Management Systems	
C. Review	
D. Module Examination	
1. Trainees must score 70% or higher to receive recognition from the NCCER.	
2. Record the testing results on Craft Training Report Form 200 and submit the results to the Training Program Sponsor.	
E. Performance Testing	
1. Trainees must perform each task to the satisfaction of the instructor to receive recognition from the NCCER.	
2. Record the testing results on Craft Training Report Form 200 and submit the results to the Training Program Sponsor.	

Module 26308-02

Lamps, Ballasts, and Components

Course Map

This course map shows all of the modules in the third level of the Electrical curriculum. The suggested training order begins at the bottom and proceeds up. Skill levels increase as you advance on the course map. The local Training Program Sponsor may adjust the training order.

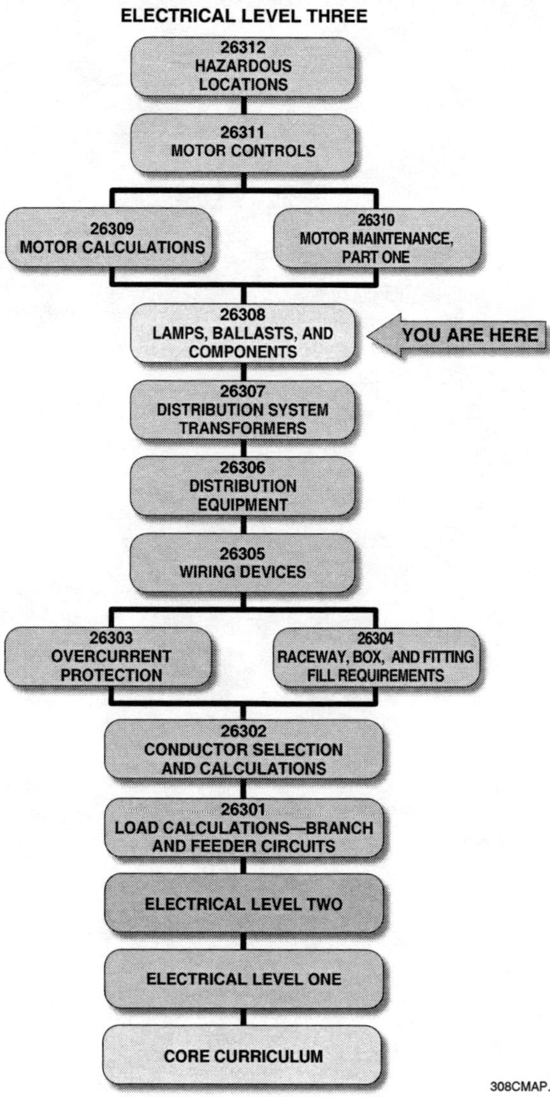

308CMAP.EPS

Homework

Assign reading of Module 26308.

MODULE 26308 CONTENTS

Figures

Instructor's Notes:

MODULE 26308

Lamps, Ballasts, and Components

Objectives

When you have completed this module, you will be able to do the following:

1. Recognize incandescent, fluorescent, and high-intensity discharge (HID) lamps and describe how each type of lamp operates.
2. Recognize ballasts and describe their purpose for use in fluorescent and HID lighting fixtures.
3. Explain the relationship of Kelvin temperature to the color of light produced by a lamp.
4. Recognize basic occupancy sensors, photoelectric sensors, and timers used to control lighting circuits and describe how each device operates.
5. Use troubleshooting checklists to troubleshoot fluorescent and HID lamps and lighting fixtures.

Prerequisites

Before you begin this module, it is recommended that you successfully complete the following modules: Core Curriculum; Electrical Level One; Electrical Level Two; Electrical Level Three, Modules 26301 through 26307.

Required Trainee Materials

1. Pencil and paper
2. Appropriate personal protective equipment
3. Copy of the latest edition of the *National Electrical Code*

1.0.0 ◆ INTRODUCTION

This is the second of three modules that cover the subject of electric lighting. This module builds on the information and lighting principles previously covered in the first module, *Electric Lighting*. The third module, presented later in your training, will provide detailed technical information on the applications for the different types of lighting **luminaires** (fixtures).

This module provides information on the operation of specific types of incandescent, fluorescent, and high-intensity discharge (HID) lamps and related lighting fixture components. Also introduced are some common lighting circuit control devices and energy conservation schemes used to control lighting. Guidelines for troubleshooting fluorescent and HID lighting fixtures are also covered.

2.0.0 ◆ STANDARD INCANDESCENT LAMPS

Incandescent lamps, also called filament lamps, are used for general lighting and typically provide a warm and natural light. Incandescent lamps were invented over a century ago. With some refinements, the basic construction of a standard incandescent lamp remains the same today. Incandescent lamps consist of a thin coiled or shaped tungsten-wire filament (*Figure 1*) supported inside an evacuated glass envelope (bulb) filled with an inert gas, typically a mix of argon and nitrogen. The inert gas helps to prevent the filament from combining with oxygen and burning out. The envelopes of most lamps are made of regular lead or soda lime (soft) glass. Envelopes of lamps that must withstand higher temperatures are typically made of borosilicate heat-resistant (hard) glass. The lamp's base supports the lamp envelope and filament and provides the electrical connection between the lamp and its power source.

Note: The designations "National Electrical Code," "NE Code," and "NEC," where used in this document, refer to the *National Electrical Code®*, which is a registered trademark of the National Fire Protection Association, Quincy, MA. *All National Electrical Code (NEC) references in this module refer to the 2002 edition of the NEC.*

Materials

Ensure that you have everything required to teach the course. Check the Materials and Equipment List at the front of this Instructor's Guide.

Classroom

Review the goals of the course.

Explain that terms shown in bold (blue) are defined in the Glossary at the back of this module.

Audiovisual

Show Transparency 1, Course Objectives.

Show Transparency 2, Performance Tasks.

Classroom

Discuss the construction and operation of incandescent lamps. Provide an example for the trainees to examine.

Audiovisual

Show Transparency 3 (Figure 1).

Classroom

Explain that incandescent lamps are the least efficient type of lamp because most of the energy they consume is given off as heat.

Teaching Tip

Discuss the "Think About It."

Classroom

Point out that incandescent lamps are available in a wide variety of sizes, shapes, and filament arrangements.

Audiovisual

Show Transparencies 4 and 5 (Figures 2 and 3).

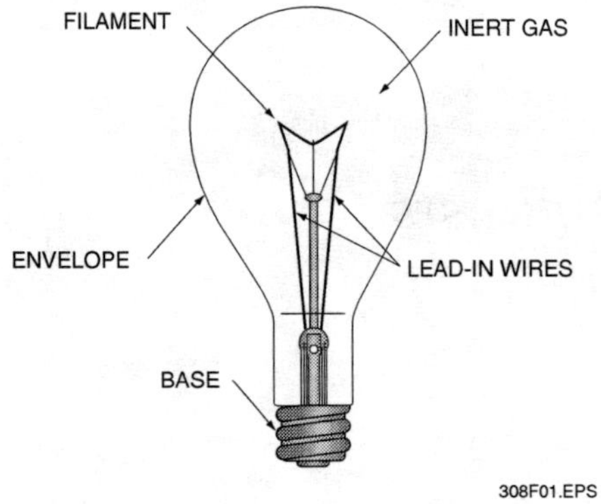

Figure 1 ◆ Components of an incandescent lamp.

> **THINK ABOUT IT**
>
> ### *Energy Consumed by Incandescent Lamps*
>
> How much of the energy consumed by incandescent lamps is dissipated as heat?
>
> a. 25%
> b. 50%
> c. 75%
> d. 90%
>
> 
>
> 308P0801.EPS

In an incandescent lamp, light is generated by passing an electric current through the filament, and its resistance causes it to heat to **incandescence.** The hotter the filament gets, the more efficient it is in converting electricity to light output. Tungsten has a positive resistance characteristic that makes its resistance at operating temperature much greater than its cold resistance (typically 12 to 16 times greater). It should be pointed out that when a filament operates hotter, its life is shortened. This makes the design of each type of lamp a trade-off between efficiency and lamp life. This is why lamps of equal wattage may have different lumen and life ratings.

Of all the lamp types, standard incandescent lamps are the most inefficient. Because they produce light by heating the filament until it glows, most of the energy they consume is given off as heat, resulting in a low efficiency (**efficacy**) of typically 5 to 22 **lumens per watt (LPW).** Incandescent lamps also have the shortest life expectancy of all lamp types, typically between 750 and 2,000 average hours, depending on the type. This is because tungsten from the filament evaporates over time and is deposited on the walls of the bulb, thus reducing the light output. Also, the filament gets thinner and thinner with use and eventually breaks, causing the lamp to fail.

Incandescent lamps are made in numerous sizes and shapes, with different filament form arrangements and mounting bases. *Figure 2* shows some examples of typical lamp sizes and shapes. Lamps are identified by a letter referring to their shape and a number that indicates the maximum diameter stated in eighths of an inch. For example, A-40 identifies a lamp with an A-shape that is 40/8 (5") in diameter. Lamps with standard, tubular, and similar envelope shapes provide lighting in all directions (omni-directional). Those with shapes designated as *R*, *ER*, and *PAR* are all reflector-type lamps. They direct their light out in front by reflecting it from their cone-shaped inside walls.

Figure 3 shows some examples of incandescent lamp filament forms used by one lamp manufacturer. The form of a filament is determined mainly by service requirements. Filament forms are identified by a letter or letters followed by an arbitrary number. Commonly used letters are C (coiled), indicating that the filament wire is wound into a helical coil; CC (coiled coil), indicating the coil itself is wound into a helical coil; and S (straight), indicating that the filament wire is uncoiled. The numbers shown in *Figure 3* indicate the arrangement of the filament on the supports.

Instructor's Notes:

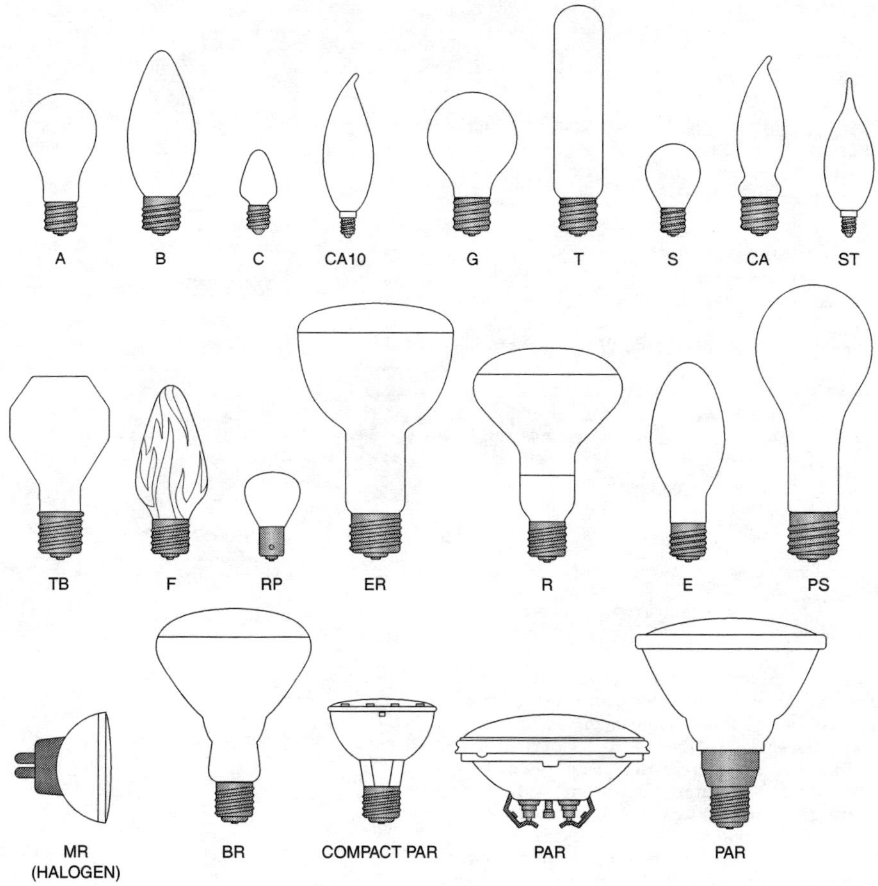

Figure 2 ◆ Incandescent lamp shapes.

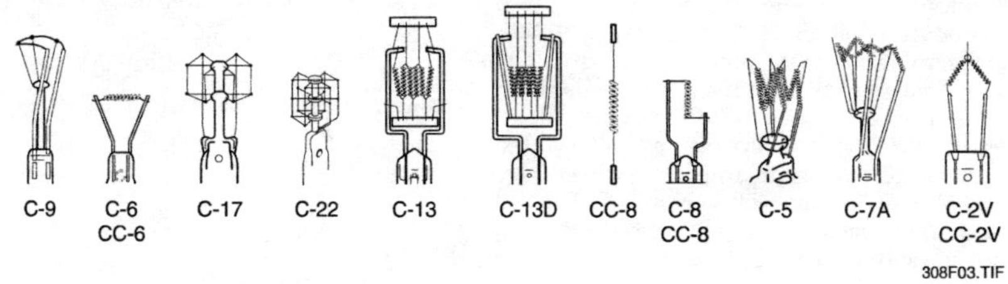

Figure 3 ◆ Examples of incandescent lamp filament forms.

Teaching Tip

Provide the trainees with various incandescent lamps and have them identify each type.

Classroom

Discuss incandescent lamp bases.

Audiovisual

Show Transparency 6 (Figure 4).

Classroom

Explain how the voltage of a lamp is related to watts, lumens, and lamp life. Note the effect of excess voltage on light output and lamp life.

Audiovisual

Show Transparency 7 (Figure 5).

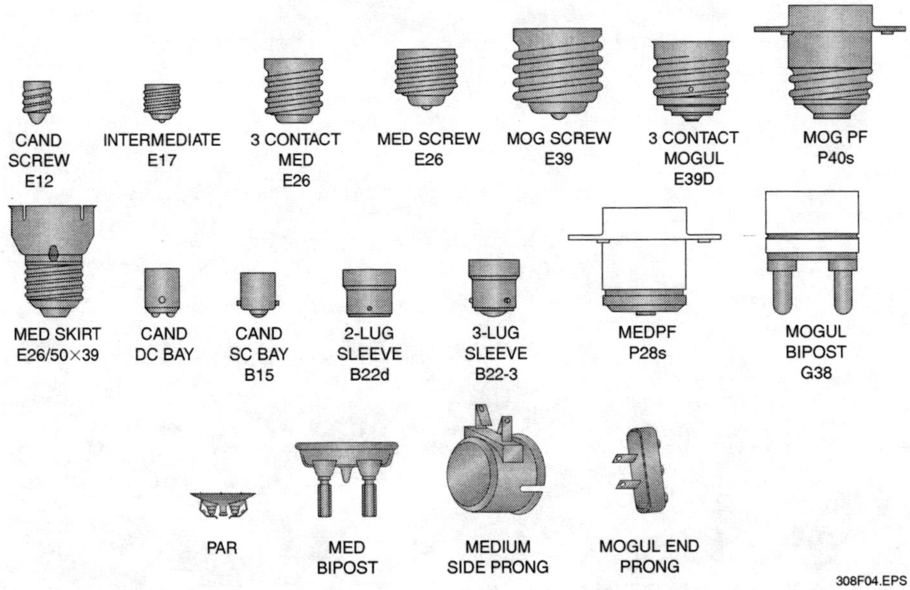

Figure 4 ◆ Examples of incandescent lamp bases.

The lamp base supports the lamp and provides the connection between the lamp and the power source. *Figure 4* shows some examples of common incandescent lamp bases.

Incandescent lamps and other types of lamps are available in many different voltage and wattage ratings. When installing incandescent lamps, it is important to select the correct voltage rating. This is because a small difference between the rating and the actual supply voltage has a great effect on the lamp life and light output. The wattage rating is an indication of the consumption of electrical energy used by a lamp to produce its rated light output. It is not a measure of the lamp's light output. For example, a standard 100W lamp may produce 1,600 lumens, while a typical 32W fluorescent lamp may produce about 2,600 lumens. *Figure 5* shows the relationships between watts, lumens, and lamp life for a typical lamp when operated at different percentages of its rated voltage. For example, reducing the supply voltage for a 120V lamp to approximately 94% (113V) will increase its life by 220%, but reduce the light output to 80% and the wattage to about 90%.

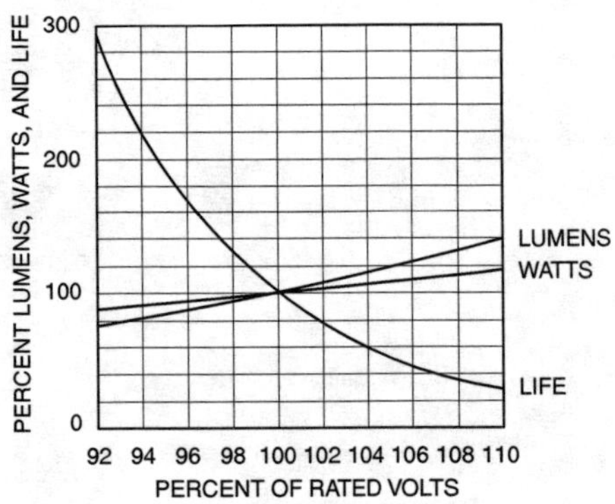

Figure 5 ◆ Relationship of rated lamp voltage to watts, lumens, and lamp life.

Instructor's Notes:

3.0.0 ◆ TUNGSTEN HALOGEN INCANDESCENT LAMPS

Tungsten halogen incandescent lamps (halogen lamps) are a refinement over the standard incandescent lamp. Like standard incandescent lamps, halogen lamps are made in many sizes, shapes, and wattages. When compared to the standard incandescent lamp, they provide greater efficacy (12 to 36 LPW), longer service life (2,000 to 5,000 average hours), and improved light quality. Their light output contains more blue and less yellow than standard incandescent lamps, making their light appear whiter and brighter.

Halogen lamps (*Figure 6*) typically have a short, thick tungsten filament encased in a capsule filled with halogen gases, such as iodine or bromine, that allow the filaments to operate at higher temperatures than a standard incandescent lamp. This increases their efficacy (LPW) by more than 20%. The use of halogen gas in the lamp accounts for the longer life and excellent **lumen maintenance.** During operation, tungsten atoms evaporated from the filament combine with halogen atoms to form a gaseous compound that circulates inside the lamp, causing the tungsten atoms to be redeposited on the hot filament, rather than on the inside surface of the lamp envelope. The halogen atoms are then released, allowing them to combine with additional tungsten atoms, thus repeating the process. This action slows down any deterioration of the filament, thereby improving lumen maintenance and extending the lamp's life.

Because of the higher filament operating temperatures, there is more ultraviolet (UV) radiation generated from a halogen lamp than from standard incandescent lamps. The amount of UV radiation emitted is determined by the lamp envelope material. Fused quartz and high-silica glass transmit most of the UV radiation emitted by the filament; special high-silica glass and aluminosilicate glasses absorb UV radiation. For general lighting applications, it is recommended that lighting fixtures for halogen lamps have a lens or cover glass which will, in addition to providing the required safety protection, filter out most of the UV radiation.

> **CAUTION**
>
> Operating halogen lamps at voltages above and below the manufacturer's recommendations can have adverse effects on the internal chemical process because the temperature will differ from the design value. Also, it is important to follow the manufacturer's instructions as to burning position, lamp handling, and lighting fixture temperatures.

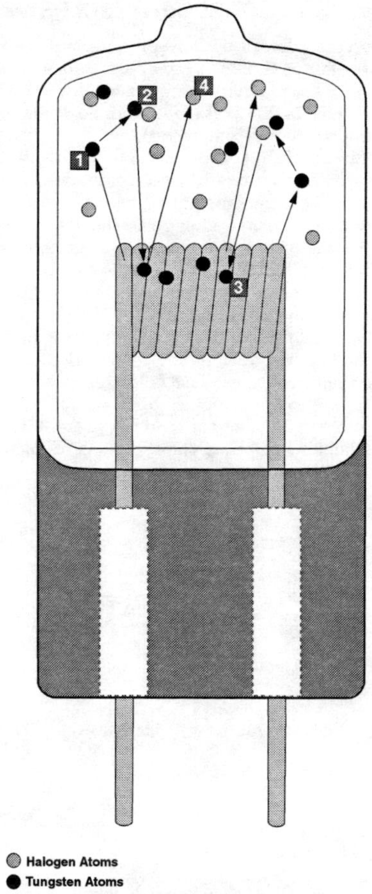

1. Tungsten atoms evaporate from filament.
2. Tungsten atoms combine with halogen atoms.
3. Gaseous compound returns to hot filament, redepositing tungsten atoms.
4. Halogen atoms are released to combine with additional tungsten atoms.

308F06.EPS

Figure 6 ◆ Basic tungsten halogen lamp.

Classroom

Discuss the construction and operation of tungsten halogen lamps. Provide an example for the trainees to examine.

Audiovisual

Show Transparency 8 (Figure 6).

Safety

Discuss the special safety precautions associated with the use of halogen lamps.

Classroom

Discuss the construction and operation of fluorescent lamps. Provide an example for the trainees to examine.

Audiovisual

Show Transparency 9 (Figure 7).

Classroom

Explain that fluorescent lamps are available in a wide variety of shapes and sizes. Refer the trainees to Figure 8 in the Trainee Module.

Explain how to interpret the T rating of a fluorescent lamp.

4.0.0 ◆ FLUORESCENT LAMPS

Fluorescent lamps are low-pressure mercury discharge lamps that are very energy efficient (75 to 100 LPW) and have a long service life (12,000 to more than 24,000 average hours). Each requires a ballast to effectively start the lamp and regulate its operation. Light is produced by passing an electric arc between two tungsten cathodes at opposite ends of a glass tube filled with a low-pressure mercury vapor and other gases (*Figure 7*). The arc excites the atoms of mercury. This generates UV radiation, which causes the phosphor coating on the inside of the tube to fluoresce and produce visible light. By using different phosphor coatings, the spectral light output of a fluorescent lamp can be made to produce warm, intermediate, or cool color temperatures. The color temperatures of lamps are covered later in this module.

Fluorescent bulbs are made in straight, U-bent, circular, and compact varieties, several of which are shown in *Figure 8*. Not only do they come in a wide variety of wattages, sizes, and bases, but they are also available in several color temperatures and color rendition capabilities. Fluorescent lamps are designated by the letter T followed by the diameter of the lamp tube expressed in eighths of an inch. They vary in diameter from T-5 (5/8") to PG (power groove)-17 (2 1/8"). In overall length, straight fluorescent lamps range from 6" to 96". Higher wattages go with longer tubes. For example, a 20W straight T-12 tube is shorter than a 40W T-12 tube.

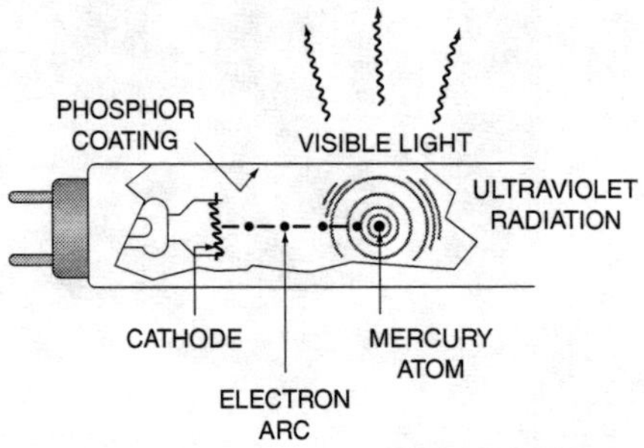

Figure 7 ◆ Basic fluorescent lamp.

INSIDE TRACK

Bi-Pin Lamps

Most bi-pin lamps, like the one shown here, have indents or index marks to aid in the proper alignment when inserting into a fixture. When properly inserted, the indents or marks will be perpendicular to the lampholders.

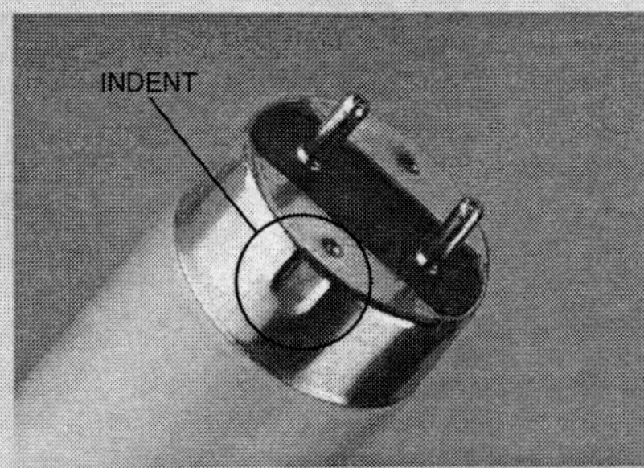

308P0802.EPS

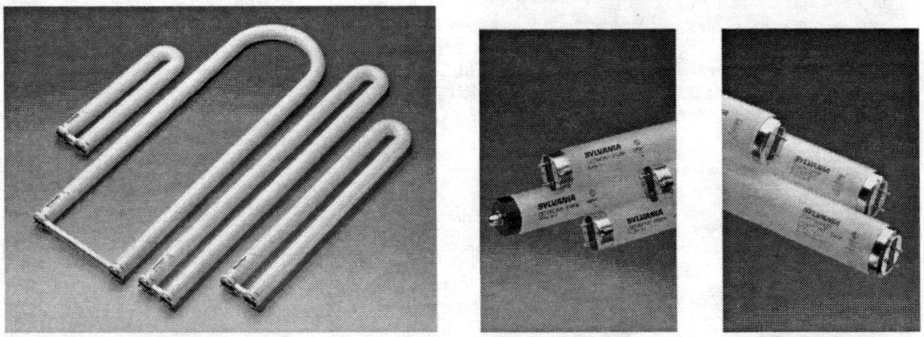

Figure 8 ◆ Typical fluorescent lamps.

Instructor's Notes:

Fluorescent lamps have two electrical requirements. To start the lamp, a high-voltage surge is needed to establish an arc in the mercury vapor. Once the lamp is started, the gas offers a decreasing amount of resistance, which means that current must be regulated to match this drop. Otherwise, the lamp would draw more and more power and rapidly burn itself out. This is why fluorescent lamps are operated in lighting fixture circuits containing a ballast that provides the required voltage surge at startup and then controls the subsequent flow of current to the lamp. Ballasts are covered in detail later in this module.

There are three electrical classes of fluorescent lamps and lighting fixtures: preheat, rapid start, and instant start. The term *preheat* refers to a lighting fixture circuit used with fluorescent lamps wherein the lamp electrodes are heated or warmed to a glow stage by a replaceable starter separate from the ballast. When power is applied to the lighting fixture, the starter functions to preheat the lamp's cathodes before the lamp is started. Note that preheat lamps and lighting fixtures are nearly obsolete and are not used in new construction. The term *rapid start* refers to a lighting fixture circuit designed to start the lamp by continuously heating or preheating the lamp electrodes by means of heater windings built into the ballast. Unlike the preheat circuit, the ballast does not require a separate starter, but the fixture must be properly grounded. During operation, standard rapid-start lamps draw about 430mA of current. The term *instant start* refers to a lighting fixture circuit used to start specially designed lamps without the aid of a starter. To strike the arc instantly, the circuit uses a higher open circuit voltage than is required for a preheat or rapid-start lamp of the same length, a voltage that is approximately three times the normal lamp operating voltage. Both preheat and rapid-start lamps have a bi-pin (2-pin) base at each end. Instant-start lamps have a single pin at each end of the lamp. Normally, lamps identified as preheat, rapid-start, or instant-start types should be used only with the corresponding type of ballast. Other terms commonly used to designate types of fluorescent lamps include:

- *Slimline lamps*—A group of instant-start lamps with single-pin bases.
- *High-output (HO) lamps*—A group of rapid-start lamps designed to operate at higher operating currents (800mA to 1,000mA) that produce higher levels of light output. Because of the higher operating currents involved, the lamps have a recessed double-contact base. HO lamps are typically used in industrial areas and retail stores with high ceilings.
- *Very high-output (VHO) lamps*—A group of rapid-start lamps designed to operate at high currents (1,500mA) in order to produce high light output levels. Because of the higher operating currents involved, the lamps have a recessed double-contact base. VHO lamps are typically used in factories, warehouses, gymnasiums, and open areas.

Note

Although the bases on HO and VHO lamps are the same, each must be matched to the correct ballast.

- *Compact lamps*—Lamps made of ½" to ⅝" single or multiple U-shaped tubes that terminate in a plastic base (*Figure 9*). The base contains the cathodes and, in some versions, a magnetic or electronic ballast. Some have replaceable tubes. A few of the newer versions are dimmable, and others look much like a standard lamp in size and shape. They are used both to replace incandescent lamps and in lighting fixtures designed for their use. They can provide up to 75% energy cost savings when compared to incandescent lamps of comparable light output. They also have a lifespan of up to 13 times longer than standard incandescent lamps.
- *T-8 lamps*—Newer and more energy-efficient lamps than older T-12 lamps. For example, a 32W, T-8 lamp uses 20% less energy to provide the same light output as a 40W, T-12 lamp. T-8 lamps use special triphosphor coatings that improve control over the lamp's color properties.

INSIDE TRACK

Fluorescent Lamps

As fluorescent lamps age, they tend to darken at the ends. This helps in identifying lamps for future replacement.

Classroom

Discuss the electrical requirements of fluorescent lamps. Explain that there are three classes of fluorescent lamps:

- **Preheat**
- **Rapid start**
- **Instant start**

List some of the other ways by which fluorescent lamps are classified. Discuss the applications and advantages of each type.

Refer the trainees to Figure 9 in the Trainee Module.

Classroom

Identify the types of lamp holders used with fluorescent lamps. Point out that high-voltage applications may require the use of a lamp holder with an integral disconnect.

Refer the trainees to Figure 10 in the Trainee Module.

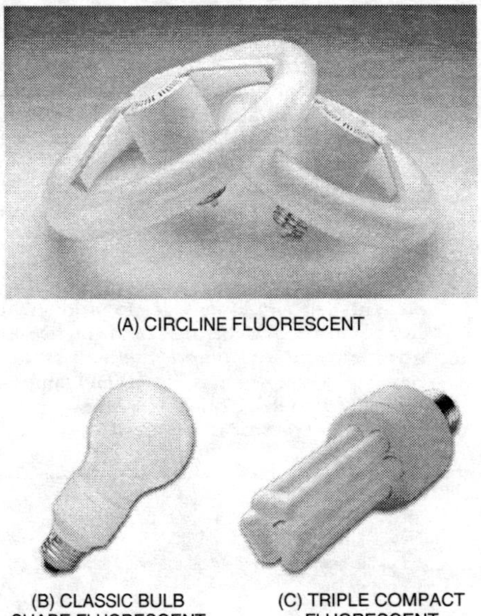

Figure 9 ◆ Compact fluorescent lamps.

> **INSIDE TRACK**
>
> ### *Cost Effectiveness of Compact Fluorescent Lamps*
>
> The 14W compact fluorescent lamp (CFL) shown here has approximately the same light output in lumens as a 60W incandescent lamp. However, the CFL has a much longer lifespan at one-fourth the operating cost.
>
> 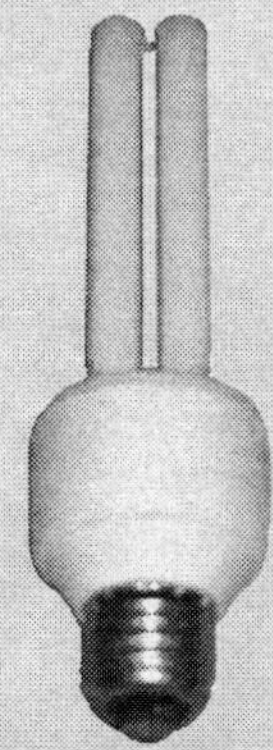
>
> 308P0804.EPS

Lamp holders (*Figure 10*) are made in several variations for each lamp base style to allow for various spacings and mounting methods in fixtures. When fluorescent lamps are used in circuits providing an open circuit voltage in excess of 330V, or in circuits that may permit a lamp to ionize and conduct current with only one end inserted in the lamp holder, electrical codes require some automatic means for deenergizing the circuit when the lamp is removed. This is usually accomplished by the lamp holder so that upon removal, the ballast primary circuit is opened. Note that the use of recessed contact bases for HO and VHO lamps has eliminated the need for this disconnect feature in the lamp holders for these lamps.

Figure 10 ◆ Typical fluorescent lamp holders.

Instructor's Notes:

5.0.0 ◆ HIGH-INTENSITY DISCHARGE (HID) LAMPS

High-intensity discharge (HID) lamps provide long life and high efficiency. They are somewhat similar to fluorescent lamps in that they produce light when electricity excites specific gases in pressurized bulbs. An arc is established between two electrodes in a gas-filled tube, which causes mercury vapor to produce radiant energy. Unlike a fluorescent lamp, a combination of factors shifts the wavelength of much of the energy to within the visible range, so light is produced without phosphors. First, the electrodes are only a few inches apart at opposite ends of a sealed arc tube (*Figure 11*), and the gases in the tube are highly pressurized. This allows the arc to generate extremely high temperatures, causing metallic elements within the gas atmosphere to vaporize and release large amounts of visible radiant energy. Like fluorescent lamps, HID lamps must be used in matched lighting fixtures with a ballast specifically designed for the lamp type and wattage. In addition, HID lamps require a warmup period to achieve full light output.

> INSIDE TRACK
>
> ### *Lamp Disposal*
>
> Fluorescent, metal halide, and mercury vapor lamps may have special disposal requirements. Always check your local environmental laws prior to disposal.

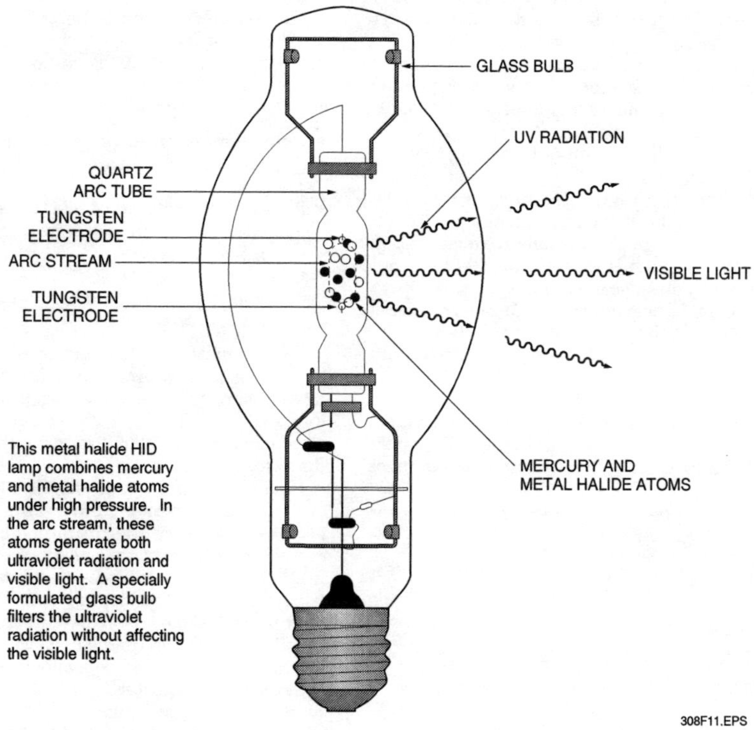

Figure 11 ◆ Basic metal halide lamp.

Classroom

Discuss the construction and operation of high-intensity discharge (HID) lamps. Provide an example for the trainees to examine.

Audiovisual

Show Transparency 10 (Figure 11).

Safety

Explain that certain lamps have special disposal requirements.

Classroom

Explain that there are three classes of HID lamps:
- **Mercury vapor**
- **Metal halide**
- **High-pressure sodium**

Discuss the operating characteristics and applications of each type.

Define *CRI.* Explain that the type of lamp selected can have a significant effect on the way colors are perceived.

Point out that fluorescent lamps offer the most color options.

There are three types of HID lamps: mercury vapor, metal halide, and high-pressure sodium. The names refer to elements that are added to the gases in the lamp, which cause each type to have somewhat different color characteristics and efficiency. Mercury vapor lamps are the oldest HID technology. They are energy efficient (50 to 60 LPW) and have a long service life (12,000 to more than 24,000 average hours). These lamps produce light energy by radiation from excited mercury vapor in both the visible and ultraviolet range. They normally have specially formulated glass outer jackets to filter the UV energy. The phosphor coatings used in some mercury vapor lamp types add additional light and improve color rendering. Today, the use of mercury vapor lamps is limited mainly to the replacement of existing lamps and landscape lighting of evergreen trees. Other HID lamps that have better efficiency and color properties are being used for new construction.

Metal halide lamps are the most energy-efficient source of white light. They have high efficacy (80 to 115 LPW), excellent color rendition, long service life (10,000 to more than 20,000 average hours), and good lumen maintenance (longevity). The metal halide HID lamp combines mercury and metal halide atoms under high pressure. In the arc stream, these atoms generate both UV radiation and visible light. A special glass bulb filters the UV radiation without affecting the visible light.

High-pressure sodium HID lamps use mercury and sodium in the arc stream contained within a tube made of a special ceramic material. They have extremely high efficacy (90 to 140 LPW) and exceptionally long service life, typically up to 24,000 hours. However, they produce light that is concentrated in the yellow/orange portion of the spectrum, which causes them to render colors poorly.

Low-pressure sodium lamps use sodium in a low-pressure arc stream and produce light that is limited to a single wavelength in the yellow portion of the spectrum. These lamps are the most efficient of any lamp type, but are used only where energy efficiency and long life are the only requirements. It should be pointed out here that technically speaking, low-pressure sodium lamps are not actually a type of HID lamp.

> INSIDE TRACK
>
> ***Lumen Output***
>
> The lumen output of an electric discharge lamp decreases significantly as the lamp ages. This should be taken into account during the lighting system design in order to ensure adequate lighting.

6.0.0 ◆ LAMP COLOR RENDERING AND COLOR TEMPERATURE CHARACTERISTICS

Colors appear differently under various light sources. The **color rendering index (CRI),** a scale from 0 to 100, is used by lamp manufacturers to indicate how normal and natural a specific lamp makes objects appear. Generally, the higher the CRI, the better it makes people and objects appear. Note that the CRI of different lamps can be compared only if the sources have approximately the same color temperature. Also, CRI differences among lamps are not usually visible to the eye unless the difference is greater than three to five points.

Lamps can create atmospheres that are warm or cool in appearance. The color temperature, expressed in kelvins (K), is one way lamp manufacturers describe the color tone (warmth or coolness) produced by a lamp. For example:

- Color temperatures of 3,000K and lower are described as warm in tone and slightly enhance reds and yellows.
- A color temperature of 3,500K is considered moderate in tone, producing a balance between warmth and coolness.
- Color temperatures of 4,100K and higher are considered cool in tone, slightly biased toward blues and greens.

Some typical color temperatures are 2,200K for high-pressure sodium lamps, 2,800K for incandescent lamps, 3,000K for halogen lamps, 3,500K for metal halide lamps, 4,100K for cool white fluorescent lamps, and 5,000K for daylight-simulating fluorescent lamps.

Fluorescent lamps have more options in terms of light quality than any other lamp type. This is because of the variations (formulations) available in the composition of the phosphor coating on the inside of the lamp tube. Early fluorescent lamps used a single halophosphor coating and could offer improved color quality with only a decrease in efficacy (LPW). With the newer lamps, triphosphor coatings are used that allow precise control over the generation of red, green, and blue (the primary colors of light). This enables the manufacture of high-LPW lamps in a variety of color temperatures that provide excellent color rendition.

Instructor's Notes:

7.0.0 ◆ FLUORESCENT AND HID LIGHTING FIXTURE BALLASTS

Ballasts are a main component of fluorescent and HID lighting fixtures. They are used to start and properly control the flow of power to the fluorescent or HID lamps.

7.1.0 Fluorescent Lighting Fixture Ballasts

In fluorescent lighting fixtures, the ballasts perform the following functions:

- Provide the proper voltage to establish an arc between two electrodes
- Regulate the electric current flowing through the lamp to stabilize the light output
- Supply the correct voltage required for proper lamp operation and compensate for voltage variations in the electrical current
- Provide continuous voltage to maintain heat in the lamp electrodes while the lamp operates (rapid-start circuits)

Fluorescent ballasts (*Figure 12*) are made to operate in the three basic fluorescent lighting fixture operating circuits: preheat, rapid start, and instant start.

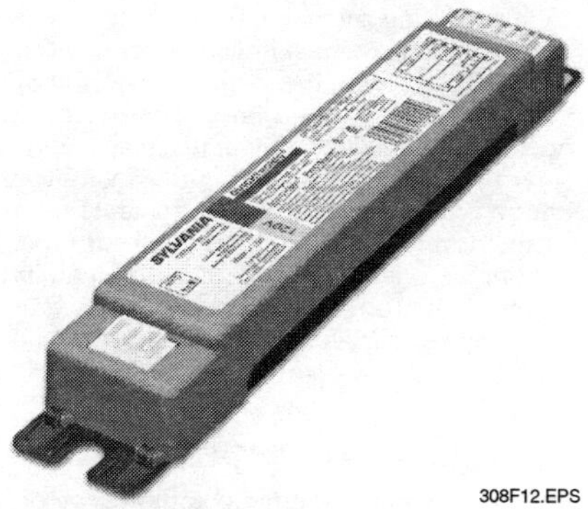

Figure 12 ◆ Fluorescent ballast.

7.1.1 Preheat Lamp and Ballast Operation

In preheat circuits, such as the one shown in *Figure 13(A)*, the lamp electrodes (cathodes) are heated before application of the high voltage across the lamp(s). The preheating requires a few seconds, and the necessary delay is provided by an automatic switch called a starter. When power is first applied to the lamp circuit, the starter places the lamp's electrodes in series across the

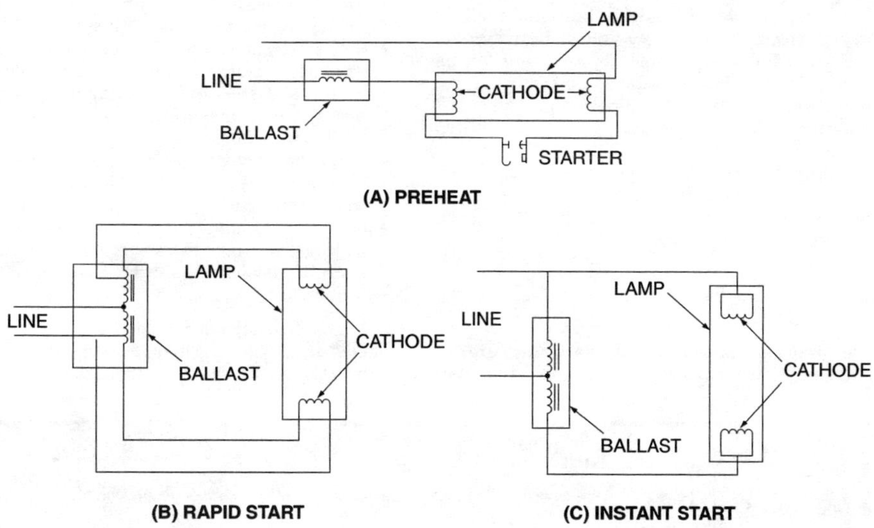

Figure 13 ◆ Basic fluorescent lighting fixture circuits.

Classroom

Discuss the operation and applications of the three classes of fluorescent ballasts:

- **Preheat ballasts**
- **Rapid-start ballasts**
- **Instant-start ballasts**

Provide examples for the trainees to examine.

Refer the trainees to Figure 12 in the Trainee Module.

Audiovisual

Show Transparency 11 (Figure 13).

Classroom

List the common categories of ballasts. Point out the advantages of rapid-start ballasts.

Teaching Tip

Discuss the "Think About It."

ballast, causing current to flow through both electrodes, heating them. After the electrodes are sufficiently preheated, the switch opens and applies the voltage across the lamp. Because the switch opens under load, a transient voltage (inductive spike) is developed in the circuit, which aids in the ignition of the lamp. Note that the first fluorescent lamps developed were of the preheat type. This type of lamp is now obsolete and is seldom used except in smaller sizes, such as those used in desk lamps and similar luminaires.

7.1.2 Rapid-Start Lamp and Ballast Operation

This is probably the most common type of lamp and luminaire used today. Lamps designed for rapid-start operation, such as the one shown in *Figure 13(B),* normally have low-resistance electrodes. These remain energized by low voltage applied from the ballast while the lamps are in operation. They usually start in one second, the time required to bring the electrodes up to proper temperature. The standard rapid-start circuit operates with a typical lamp current of about 430mA. Rapid-start circuits used with high-output (HO) and very high-output (VHO) lamps draw currents of about 800mA and 1500mA, respectively. In some energy-saving circuits, the electrode voltage is reduced or disconnected after the starting of the lamps. Heating is accomplished through low-voltage windings built into the ballast or through separate low-voltage transformers designed for this purpose. Fluorescent lamps used with rapid-start ballasts are bi-pin lamps. Rapid-start lamps can be dimmed using special dimming ballasts. These are covered later in this module.

Another version of a rapid-start ballast is the trigger-start ballast. Trigger-start ballasts are used with preheat fluorescent lamps up to 32W without the need for a starter.

7.1.3 Instant-Start Lamp and Ballast Operation

The lamp electrodes in instant-start lamps are not preheated. The ballasts provide a high voltage (100V to 1,000V) across the electrodes that causes electrons to be emitted from the electrodes, as shown in *Figure 13(C).* These electrons flow through the tube, ionizing the gas and initiating an arc discharge. Thereafter, the arc current provides electrode heating. Because no preheating of electrodes is required, instant-start lamps need only a single contact at each end. Thus, a single pin is used on most instant-start lamps, commonly called slimline lamps. However, T-8 bi-pin lamps can be operated with either rapid-start or instant-start electronic ballasts. When used with an instant-start ballast, the terminals in the lamp holders must be connected together. New fixtures come with the lamp holders wired in this way.

7.1.4 Types of Fluorescent Ballasts

Each fluorescent lamp must be operated by a ballast that is specifically designed to provide the proper starting and operating voltage required by the particular lamp. Lamp and lighting fixture manufacturers make a wide variety of ballasts designed for use in lighting fixtures that operate in all three of the lamp starting modes previously described. Ballasts are made for single-lamp, two-lamp, three-lamp, and four-lamp operation. The names used to identify the different types of ballasts can vary by manufacturer. Some common categories of ballasts are:

- *Standard ballast*—Lowest priced, least efficient, and highest wattage ballast. Their use is obsolete.
- *High-efficiency ballast*—Lower wattage, better efficiency, and longer life than a standard ballast.
- *Hybrid ballast*—Lower wattage, higher efficiency, and longer life than a standard high-efficiency ballast.
- *Electronic rapid-start ballast*—Low wattage and longest life available with various ballast factors and wattage packages for T-12 and T-8 lamps.
- *Electronic instant-start ballast*—Lowest system wattage and highest system efficiency. The lamp life is slightly shorter than with rapid-start ballasts.

High-efficiency ballasts can be of the magnetic or electronic type. The magnetic type usually contains coils, capacitors, transformers, and a thermal

THINK ABOUT IT

Rapid-Start Lamps

Why do some rapid-start lamps fail to ignite at low ambient temperatures?

Instructor's Notes:

INSIDE TRACK

Electronic Fixture Ballasts

Modern electronic fixture ballasts offer superior application flexibility. These ballasts can be rated for two, three, or four lamps. In addition, electronic ballasts may also be used to operate less than the maximum number of lamps listed. For example, a ballast rated for four lamps may be used to operate a two- or three-lamp fixture. These ballasts also provide for the ability to use lamps of different lengths.

Replacing Ballasts

When replacing ballasts, be sure to use an exact equivalent. Nuisance tripping may occur if a non-Class P ballast is replaced with a Class P ballast.

Power Factor

The power factor of a ballast is not a measurement of the ballast's ability to supply light through the ballast but is an indication of power consumption. The power factor is based on a value of one. The higher the power factor, the lower the power consumption. Some ballasts offer power factors of >0.90, which means that they are very efficient devices. Always look for a high power factor when selecting devices for a new installation.

protector installed in a metal case. The coils and transformers are generally made with steel laminations and copper wire. Once assembled, the ballast components are encapsulated in the ballast case with a potting compound to improve the heat dissipation and reduce ballast noise.

If the ballasts are of the electronic type, they may be made with discrete electronic components and/or with integrated circuits. Some of those with integrated circuits are compatible for use with dimming systems, occupancy sensors, and daylight sensors. Electronic ballasts are quieter, more efficient, and weigh less, but are also more expensive. With electronic ballasts, the power input is 50Hz to 60Hz, but the ballast operates the lamps at 20kHz to 50kHz, with resulting improvements in ballast and lamp efficacy. The operating frequency is selected so that it is high enough to increase the lamp's efficacy and to shift the ballast noise to the inaudible range, but not so high as to cause electromagnetic interference (EMI) problems. Note that many electronic ballasts do not come equipped with wire leads. These ballasts are connected into the circuit using wire harnesses specifically designed for that purpose.

NEC Section 410.73(E) requires that all fluorescent fixture ballasts used indoors, including any replacement ballasts, have a thermal protection device integral within the ballast. All such ballasts are marked *Class P.* The exceptions to this requirement include simple reactance ballasts used with fluorescent fixtures using straight tubular lamps, ballasts used with exit fixtures and so identified, and egress lighting energized only during an emergency. Class P ballasts with thermal protection operate to open the circuit at a predetermined temperature in order to prevent abnormal heat buildup caused by a fault in one or more of the ballast components, or by some lamp holder or wiring fault.

High-efficiency, energy-saving ballasts have a high power factor rating. The power factor is the ratio of watts to volt-amperes. To be classified as a high power factor ballast, a ballast must have a power factor of at least 90%. Anything less is considered a normal or low power factor. A ballast's power factor rating is marked on the ballast nameplate. Energy-saving high power factor ballasts cost more than low power factor ballasts, but over time, the savings in energy consumption far exceed the higher initial cost. In any application where there are to be a large number of ballasts, it is best to install ballasts with a high power factor. When compared to magnetic ballasts, electronic ballasts are more energy efficient.

Ballasts can emit a hum, especially the magnetic types. This is caused by magnetic vibration in the ballast core. Ballast manufacturers give their ballasts a sound rating ranging from A to F, with A being the quietest. The need for quiet ballast operation is determined mainly by the desired ambient noise level of the location where the ballast is to be installed. For example, a ballast with an A rating might be used in a doctor's office, while one with an F rating might be suitable for a factory application.

Classroom

Discuss the advantages of electronic ballasts.

Note that the NEC requires that all fluorescent ballasts used indoors contain a thermal protection device. Emphasize the importance of replacing any ballast with an exact equivalent.

Point out the advantages of selecting a ballast with a high power factor.

Teaching Tip

Discuss the "Case History." Emphasize the life-saving importance of the ground connection.

Classroom

Discuss the disposal requirements for fluorescent ballasts.

Describe the operation of electronic and magnetic fluorescent dimming ballasts. Point out the advantages of the electronic type.

Audiovisual

Show Transparency 12 (Figure 14).

CASE HISTORY

The Importance of Connecting Grounds

A painter in an industrial facility was standing on a 10' wooden ladder and painting steel I beams. He received a fatal shock while leaning across one of many suspended fluorescent light fixtures, while touching a metal pipe with his other arm. Later investigation revealed that the ground wire in the fixture was disconnected. It is presumed that the ground wire had not been reconnected when the ballast was last replaced. Numerous burn marks were noted within the light fixture at the points where the conductors were connected to the ballast.

The Bottom Line: Don't forget to connect the green (ground) wire when replacing any electrical device.

When installing a replacement ballast, make sure to dispose of the old ballast in a proper manner. Unless you see a label stamped *No PCBs* on the failed ballast you are disposing of, you must assume that it contains toxic PCBs and must be disposed of in accordance with the prevailing EPA and local requirements. Failure to do so can expose you and your employer to potential liability for cleanup in the event of PCB leakage.

7.1.5 Fluorescent Dimming Ballasts

Special dimming ballasts and dimmer switches are required to dim fluorescent lamps. The dimmer control allows the dimming ballast to maintain a voltage to the lamp's electrodes that will maintain the electrodes' proper operating temperature. It also allows the dimming ballast to vary the current flowing in the arc. This in turn varies the intensity of light coming from the lamp. Dimming fluorescent lamps differs from dimming incandescent lamps in two main ways: first, fluorescent dimmers do not provide dimming to zero light as do incandescent dimmers; and second, when dimming fluorescent lamps, the color temperature does not vary much over the dimming range. This is unlike incandescent lamps, which tend to turn yellower when dimmed.

Most fluorescent dimming ballasts are of the electronic type. However, older autotransformer magnetic types are also available. Electronic dimming ballasts are normally more efficient and less bulky than magnetic ballasts. *Figure 14* shows a wiring diagram for an electronic dimmer used with a dimming ballast for rapid-start lamps. It is important to point out that the performance of a dimming system may not be satisfactory if the lamp is not correctly matched with the dimming ballast and the controller. Also, when connecting such dimmer circuits, always check the dimmer and ballast manufacturer's wiring diagrams to determine the proper connections.

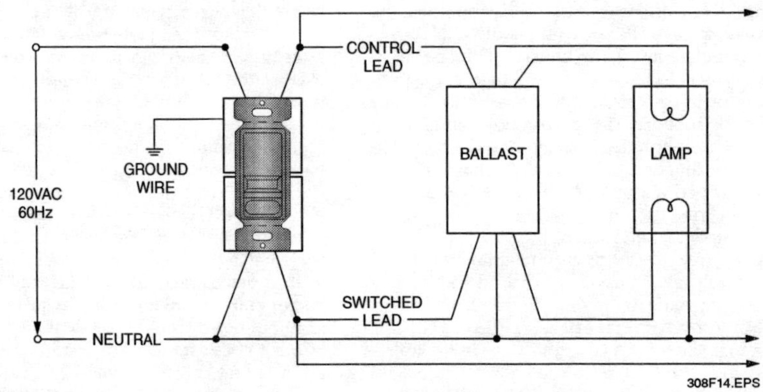

Figure 14 ◆ Dimmer circuit using dimming ballast for rapid-start lamp.

Instructor's Notes:

INSIDE TRACK

Ballast Wiring

Manufacturers color-code ballast wiring for ease of installation. Always follow the manufacturer's color code when wiring ballasts.

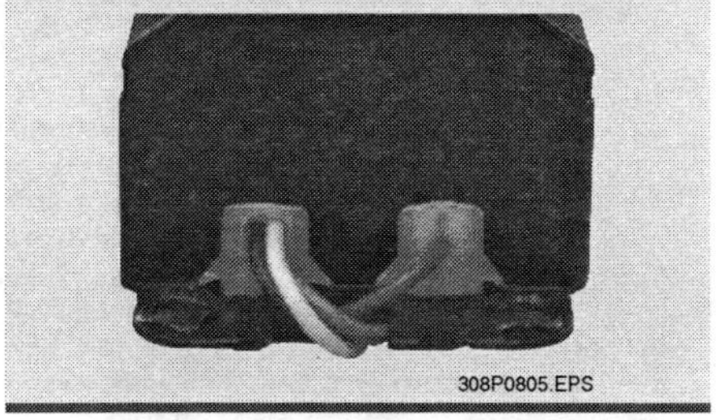

308P0805.EPS

7.1.6 Emergency Lighting Ballasts

Special emergency ballasts with self-contained battery-operated power packs are made for use in some fluorescent lighting fixtures. Upon the loss of input power to the lighting fixture, these ballasts typically function to operate one 8' lamp at emergency lighting levels for about 90 minutes, or one 4' lamp for about 120 minutes.

7.2.0 HID Lighting Fixture Ballasts

HID lamps require the use of a ballast to provide enough voltage to strike the arc in the lamp. This function may be accomplished by the ballast itself or in conjunction with a separate electronic ignitor circuit. An ignitor (*Figure 15*) is an electronic device used in the circuitry for high-pressure sodium and some metal halide HID lamps. It provides a pulse of at least 2,500V peak root mean square (rms) to initiate the lamp arc. When the system is energized, the ignitor provides the required pulse until the lamp is completely lit and then automatically stops pulsing.

The HID lamp ballast also acts to control the arc wattage during warmup and normal operation. In addition, some ballasts may also provide a line voltage matching transformer function, enhance lamp wattage regulation with respect to changes in line voltage and/or lamp voltage, and dimming or other control interface functions.

Physically, there are numerous types and shapes of ballasts used with HID lamps (*Figure 16*). The same is true electrically. Some ballasts are made with primary leads that allow the ballast to be connected to different supply voltages, such as 120V, 208V, 240V, or 277V. Such ballasts are called

308F16.EPS

Figure 16 ◆ HID ballast and lamp.

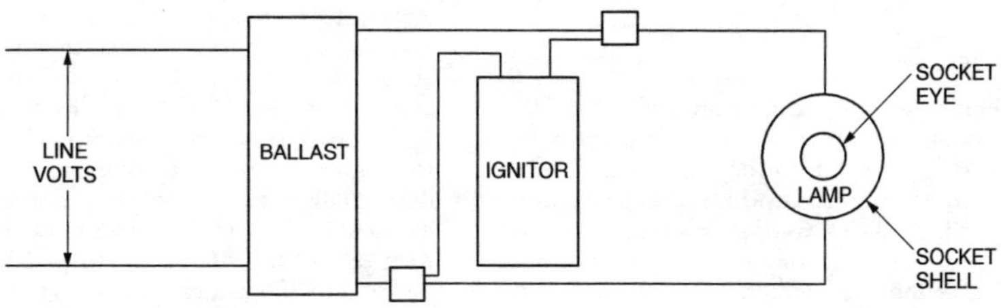

308F15.EPS

Figure 15 ◆ Simplified HID lamp ignitor circuit.

Classroom

Explain that emergency lighting ballasts contain battery packs to provide power for up to two hours.

Explain the purpose of the ignitor in an HID lamp.

Audiovisual

Show Transparency 13 (Figure 15).

Classroom

Discuss the various functions performed by HID ballasts. Refer the trainees to Figure 16 in the Trainee Module.

Classroom

Discuss the operation and applications of the three classes of HID ballasts:
- **Linear, nonregulating circuit ballasts**
- **Constant-wattage autotransformer ballasts**
- **Three-coil ballasts**

Provide examples for the trainees to examine.

Audiovisual

Show Transparency 14 (Figure 17).

Classroom

Discuss the types of dimmers used with HID lamps.

INSIDE TRACK

Unused Wire Taps

The unused wire taps in multi-tap ballasts must be electrically insulated in order to avoid damage to the fixture.

multi-tap ballasts. It is extremely important that only the proper voltage lead be connected to the supply voltage. The types of ballasts used by a major HID lighting fixture manufacturer (Hubbell) are described here. The ballasts used by other manufacturers are similar. HID ballasts can be grouped into three basic categories:

- Linear, nonregulating circuit ballasts
- Constant wattage autotransformer ballasts
- Three-coil ballasts

7.2.1 Linear, Nonregulating Circuit Ballasts

Linear, nonregulating circuits include reactor ballasts and auto-lag ballasts. These provide for the basic operation of some mercury and high-pressure sodium HID lamps. With the exception of the self-ballasted lamp, the reactor ballast is the most basic form of ballast. See *Figure 17(A)*. It consists of a single coil wound on an iron core placed in series with the lamp. The only function performed by the reactor is to limit the current delivered to the lamp. This type of ballast can only be used when the line voltage applied to the lamp is within the required starting voltage range of the lamp. The power factor for a reactor ballast is typically in the 40% to 50% range. However, a capacitor is normally added to the circuit to improve the power factor to better than 90%. Because this type of ballast provides for no line voltage regulation, outages due to line dips and brownouts are typical. The auto-lag ballast, as shown in *Figure 17(B)*, is a reactor ballast combined with a step-up or step-down autotransformer, which provides for some input voltage regulation.

7.2.2 Constant-Wattage Autotransformer Ballasts

The constant-wattage autotransformer (CWA) ballast, shown in *Figure 17(C)*, is a ballast circuit that uses magnetic saturation to maintain better lamp wattage regulation and improved **dip tolerance.** CWA ballasts are used mainly with mercury and high-pressure sodium lamps. A variation of the CWA, called the peaked lead auto-regulator (PLA), is used with metal halide lamps. Another variation, called the constant-wattage isolated (CWI) ballast, shown in *Figure 17(D)*, is an isolated winding version of the PLA. It is typically used with mercury lamps.

The term *dip tolerance* relates to the dips in line voltage experienced by all power systems as loads are switched in and out, or as other transitory conditions occur. A well-regulated distribution system will seldom experience voltage dips of more than 10%, but on some circuits, dips of 20% or more may occur. If a ballast is not capable of riding through the voltage dip and sustaining the lamp, it will extinguish and have to cool down before reignition. Lamp dropout due to line voltage dips generally increases with lamp age. The use of ballasts with improved dip tolerance may delay the onset of such lamp dropout problems.

7.2.3 Three-Coil Ballasts

Three-coil ballasts are isolated winding ballasts in which the input and lamp windings are separated (isolated) by a third winding, which helps to eliminate drastic changes in the demands of the lamp on the supply system and maintains lamp stability during supply system variations. See *Figure 17(E)*. This type of circuit provides the highest degree of lamp operating stability and waveform control, with the lowest harmonics and best performance consistency through the life of the lamp. A magnetic regulator version is used with high-pressure sodium lamps and an electro-regulator version is used with metal halide lamps.

7.2.4 Dimming HID Lamps

HID lamps can be controlled using equipment similar to that used for dimming fluorescent lamps. However, the long warmup and restrike times associated with HID lamps may limit their applications. Multi-level ballasts are made for HID lamps that allow their light to be reduced. This type of ballast is typically used in lighting fixtures for warehouses, parking garages, tunnels, and daytime lighting applications. Some equipment is made that allows HID lamps to be dimmed to less than 20% of their full light output; however, most lamp manufacturers will not guarantee full life expectancy if their lamps are operated below 50%. Also, color shifts in the light output of the lamps may limit their use in some applications.

Instructor's Notes:

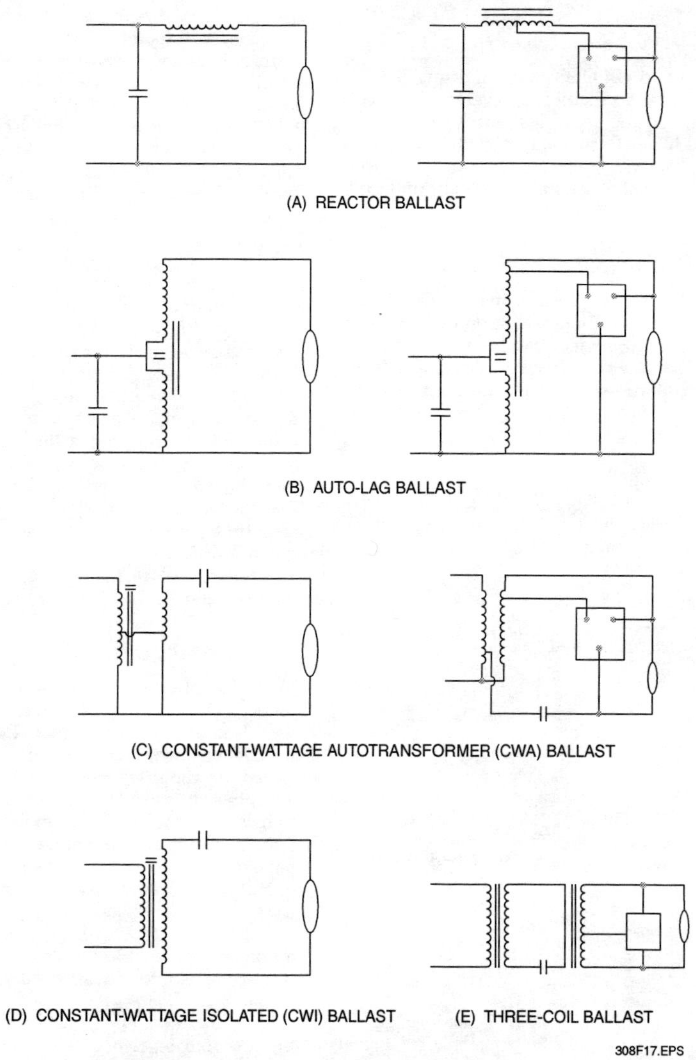

Figure 17 ◆ Simplified HID ballast circuits.

THINK ABOUT IT

HID Ballasts

What would happen if the output leads of an HID ballast were shorted together?

Teaching Tip

Discuss the "Think About It."

Demonstration

Demonstrate how to use troubleshooting checklists and guidelines to troubleshoot various types of lamps and lighting fixtures. Refer the trainees to the *Appendix* in the Trainee Module.

Laboratory

Have the trainees use troubleshooting checklists and guidelines to troubleshoot various types of lamps and lighting fixtures. Note the proficiency of each trainee.

Classroom

Point out that several options are available to provide automatic lighting control. Refer the trainees to Figures 18 and 19 in the Trainee Module.

Discuss the operation and applications of occupancy sensors and photosensors. Provide examples for the trainees to examine.

8.0.0 ◆ TROUBLESHOOTING LAMPS AND LIGHTING FIXTURES

Fluorescent and HID lamps and lighting fixtures are normally very reliable; however, problems do occur that require troubleshooting the lamps and/or lighting fixtures. *Appendix A* contains guidelines and checklists to aid you when troubleshooting problems with fluorescent and HID lighting systems.

9.0.0 ◆ CONTROLS FOR LIGHTING

Because of high energy costs, the use of lighting controls to manage the application of lighting is common. Several devices can be used to control lighting circuits in order to conserve energy. These include occupancy sensors, photosensors, timers, and similar devices.

9.1.0 Occupancy Sensors

Occupancy sensors (*Figure 18*) are devices that can be used to turn lights on and off automatically in an individual space such as a private office, restroom, or storage area. Occupancy sensors can be motion detecting (ultrasonic), heat sensing (infrared), or sound sensing. They can be recessed or surface-mounted on a wall or ceiling, they can replace wall switches, or they can plug into receptacles. The sensor turns the lights on when it senses someone coming into the room or area and turns the lights off some time after no longer sensing anyone present. Units come either with fixed, preset time delays and sensitivity levels, or with adjustable ones.

Ultrasonic sensors transmit ultrasound and receive a reflected signal to sense the presence of occupants in a space. They typically operate at a frequency between 25kHz and 40kHz. Passive infrared sensors detect the changes in infrared patterns across their segmented detection regions. The type of sensor used must be compatible with the application. For example, a motion detector or sound detector may not be the right choice if occupants of the space sit very quietly at desks. People in such situations have been known to complain that they must deliberately move or make noise from time to time to prevent the sensor from turning off the lights. On the other hand, infrared sensors must be placed so that no obstruction blocks their sensing field.

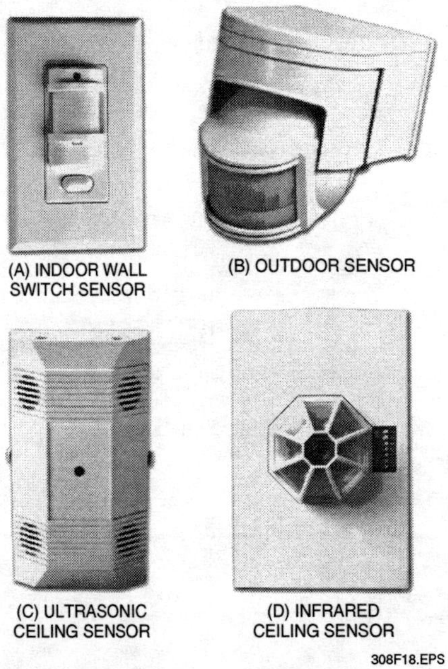

Figure 18 ◆ Examples of infrared occupancy sensors.

9.2.0 Photosensors

Electronic photosensors sense the level of visible light in the surrounding area and convert this level into an electrical signal. This signal can be used in one of two ways, depending on the type of system. In the first type, the signal can be used to activate a simple on/off switch or relay that functions to control the power to the related lighting fixtures. In the second type, a variable output signal is produced that can be sent to a controller that operates to continuously adjust the output of electric lighting in the area. The device shown in *Figure 19* includes a photoelectric sensor to automatically enable the device at dusk plus an infrared sensor to activate the lamps when motion is detected.

Photosensors can be an integral part of a lighting fixture, remote from the lighting fixture, or may control a circuit relay that operates several lighting fixtures. It is important that the area controlled by one photosensor have the same daylight illumination conditions (amount and direction) and that the area be contiguous with no high walls or partitions to divide it. When photosensors are used outdoors, the sensor should be aimed due north.

Instructor's Notes:

308F19.EPS

Figure 19 ◆ Combination photoelectric/infrared sensor.

9.3.0 Timers

Timers are used to turn lighting on or off in response to known or scheduled sequences of events. Timers can be very simple clock-like mechanisms, or they can be microprocessors that can program a sequence of events for years at a time. With a simple timer, the load is switched on and held energized for a preset period of time. Timer limits can range from a few minutes to 12 hours. Some models have a hold position for continuous service.

An electromechanical time clock/timer (*Figure 20*) is driven by an electric motor, with contactors actuated by mechanical stops or arms attached to the clock face. Time clocks have periods ranging from 24 hours to seven days. They can initiate many operations. Some can actuate a momentary contact switch to provide on and off pulses for actuating relays or contactors. Electronic time clocks/timers provide programmable selection of many switching operations and can typically be controlled to the nearest minute over a seven-day period. Most can control multiple channels and have time-of-day scheduling, holiday programming, daylight savings time adjustment, leap year correction, annual override, and battery carryover for protection against power outages.

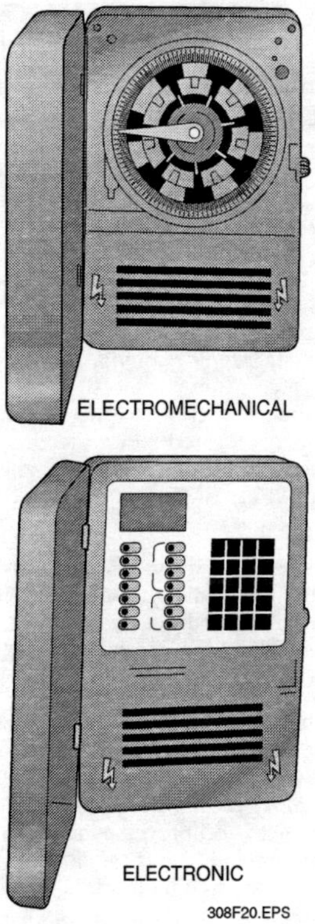

308F20.EPS

Figure 20 ◆ Typical timer controls.

Time clocks are often used in conjunction with photosensors in order to turn off lighting when there is no longer a need for it. For example, an industrial building may use a photosensor signal to activate outdoor lighting at dusk, then a time clock to turn off that lighting after the last worker has left the facility. Another example is when a photosensor is used to signal when to turn the lights on, then a time clock is used to initiate dimming the lights when a high level of lighting is no longer needed.

Classroom

Discuss the operation and applications of timer controls. Provide examples for the trainees to examine.

Audiovisual

Show Transparency 15 (Figure 20).

Classroom

Provide an overview of the functions performed by energy management systems.

Audiovisual

Show Transparency 16 (Figure 21).

Classroom

Explain how relays can be used to provide various levels of lighting control.

Audiovisual

Show Transparency 17 (Figure 22).

10.0.0 ◆ ENERGY MANAGEMENT SYSTEMS

Modern buildings normally use some form of an energy management system (EMS) to economically control the amount of energy consumed by the building's lighting circuits and HVAC equipment. An EMS can be a fairly simple stand-alone unit connected to one or more pieces of equipment (e.g., room lighting, a heat pump or rooftop HVAC unit, etc.), or it can be more extensive and control all the lights and equipment throughout an entire large facility.

Whether large or small, an EMS typically consists of a computer or control processor, energy management and scheduling software, sensors and controls located where needed, and, in large systems, a communications network. When programmed, an EMS can automatically control lighting to:

- Turn off lights in unoccupied areas
- Maintain partial lighting before and after working hours or public-use hours
- Schedule lighting operation by hour of day and time of year

You will recognize the tasks performed by an EMS as being similar to those performed by the sensors and timers that were discussed earlier. The EMS simply receives signal inputs applied from these devices and processes them so as to perform tasks more reliably and precisely than can be performed by manual methods. Lighting control can be implemented in a building by a local approach, a central system, or both. The method used is determined both by the size of the controlled areas and by how the control inputs are integrated into the system. A local lighting system is divided into small, independently controllable zones based on size, environment, etc., or according to functional need. The inputs from the sensors located in the zone are wired directly to a control that is also located in the zone (*Figure 21*). In central systems, the sensor inputs from the individual zone sensors are all wired to a central control.

An EMS commonly turns lights on and off and/or initiates partial lighting in an area via relays that are activated or deactivated by the EMS control unit. *Figure 22* shows a typical lighting switching scheme involving the use of split-wired, multi-ballasted lighting fixtures. By split-wiring three-lamp and four-lamp lighting fixtures, multiple light intensities can be provided in a single zone. The relay-based control system provides full lighting for specific times of the day, while allowing a reduced lighting level and reduced power for those times when less lighting is needed.

Figure 22(A) shows a relay-controlled split-wiring system connected to two three-lamp lighting fixtures. This arrangement allows for four levels of lighting: 0%, 33⅓%, 66⅔%, and 100%. As shown, the relay to the inboard ballasts is closed, allowing two of the six lamps to be turned on, thus providing light at the 33⅓% level.

Figure 22(B) shows a similar relay-controlled split-wiring system connected to two four-lamp lighting fixtures. This arrangement allows for three levels of lighting: 0%, 50%, and 100%. As shown, the relay to the inboard ballasts is closed, allowing four of the eight lamps to be turned on, thus providing light at the 50% level.

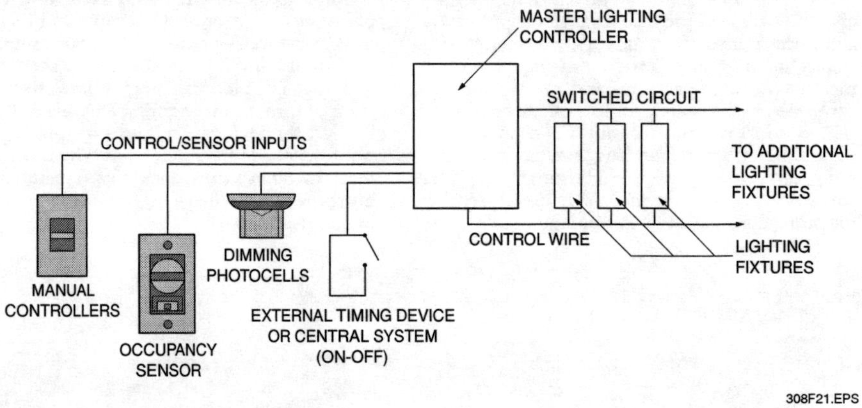

Figure 21 ◆ Simplified zone lighting control circuit.

Instructor's Notes:

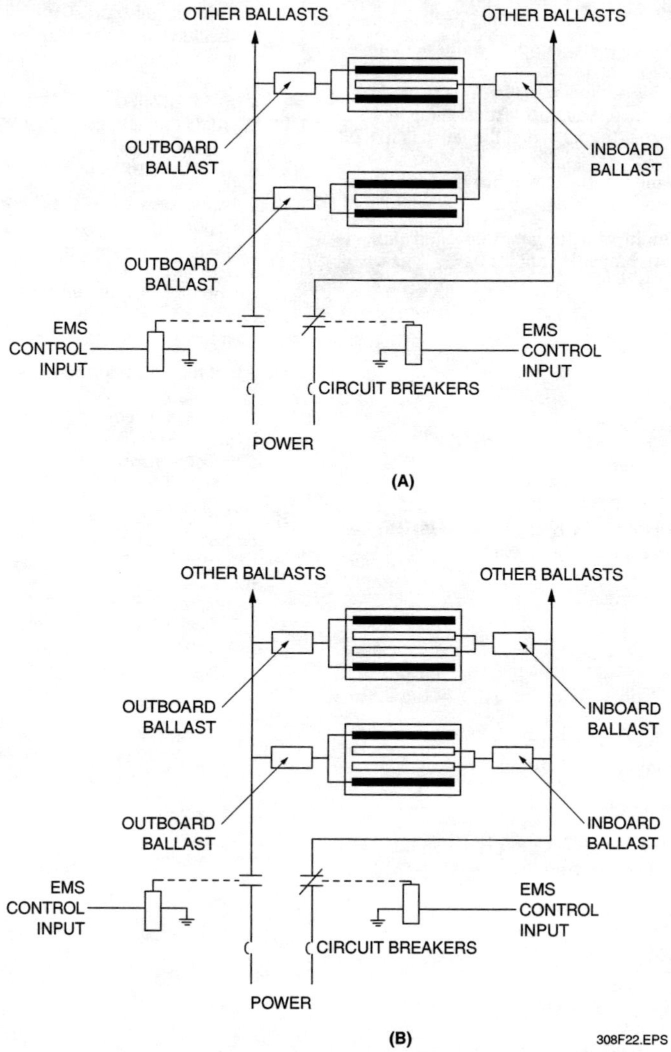

Figure 22 ◆ Simplified wiring diagrams of relay-controlled lighting fixture circuits.

Summary

This module covered the construction and theory of operation for incandescent, fluorescent, and high-intensity discharge (HID) lamps and lighting fixtures. It also introduced some common types of equipment and methods widely used for controlling the use of electric lighting in order to conserve energy. Understanding the construction and characteristics of the different kinds of lamps and lighting fixtures and how to control them for increased efficiency and economy will enable you to better install and/or maintain lighting systems so that they will provide a maximum of comfort and a minimum of eyestrain and fatigue.

Classroom

Have the trainees complete the Review Questions and go over the answers prior to administering the Module Examination.

Examination

Administer the Module Examination. Record the results on Craft Training Report Form 200 and submit the results to the Training Program Sponsor.

Performance Testing

Administer the Performance Test and fill out Performance Profile Sheets for each trainee. If desired, trainee proficiency noted during laboratory sessions may be used to complete the Performance Test. Be sure to record the results on Craft Training Report Form 200 and submit the results to the Training Program Sponsor.

Review Questions

1. In an incandescent lamp, the filament resistance is ____.
   a. much higher when the lamp is turned on
   b. lower when the lamp is turned on
   c. somewhat higher when the lamp is turned on
   d. the same whether the lamp is turned on or off

2. The filament of a tungsten halogen lamp is encased in a capsule containing ____ gas.
   a. nitrogen
   b. argon
   c. halogen
   d. sodium

3. The diameter of a 4', T-12 fluorescent lamp is ____.
   a. 1"
   b. 1⅛"
   c. 1¼"
   d. 1½"

4. During operation, a high-output (HO) fluorescent lamp typically has a current draw of ____.
   a. 430mA
   b. 800mA
   c. 1,100mA
   d. 1,500mA

5. The type of high-intensity discharge (HID) lamp that uses the oldest HID technology is the ____ lamp.
   a. low-pressure sodium
   b. high-pressure sodium
   c. mercury vapor
   d. metal halide

6. Lamps with color temperatures of ____ and lower produce a light that is considered warm in tone.
   a. 3,000K
   b. 3,500K
   c. 4,100K
   d. 5,000K

7. An electronic ballast typically operates the connected fluorescent lamps at a frequency of ____ to improve both ballast efficiency and lamp efficacy.
   a. 50Hz to 60Hz
   b. 10kHz to 20kHz
   c. 20kHz to 50kHz
   d. 60kHz to 100kHz

8. A fluorescent lamp is blinking off and on in very short cycles. The most probable cause is ____.
   a. the lamp itself
   b. incorrect supply voltage
   c. low ambient temperature
   d. a loose socket connection

9. All of the following sensors are normally used as occupancy sensors *except* a(n) ____.
   a. motion detector
   b. infrared sensor
   c. photosensor
   d. sound sensor

10. If the contacts of both relays are closed in the circuit shown in *Figure 22(A)*, what percentage of lighting will be produced by the lighting fixtures?
    a. 0%
    b. 33⅓%
    c. 66⅔%
    d. 100%

Instructor's Notes:

GLOSSARY

Trade Terms Introduced in This Module

Color rendering index (CRI): A scale from 0 to 100 used by lamp manufacturers to indicate how normal and natural a light source makes objects appear in full sunlight or in incandescent light. Generally, the higher the CRI, the better it makes people and objects appear.

Dip tolerance: The ability of an HID lamp or lighting fixture circuit to ride through voltage variations without the lamp extinguishing and cooling down.

Efficacy: The light output of a light source divided by the total power input to that source. It is expressed in *lumens per watt (LPW).*

Incandescence: The self-emission of radiant energy in the visible light spectrum resulting from thermal excitation of atoms or molecules such as occurs when an electric current is passed through the filament in an incandescent lamp.

Lumen maintenance: A measure of how a lamp maintains its light output over time. It may be expressed either numerically or as a graph of light output versus time.

Lumens per watt (LPW): A measure of the efficiency, or, more properly, the *efficacy* of a light source. The efficacy is calculated by taking the lumen output of a lamp and dividing by the lamp wattage. For example, a 100W lamp producing 1,750 lumens has an efficacy of 17.5 lumens per watt.

Luminaire: A complete lighting unit including the lamp(s), wiring, support, protective covering, and ballast (where applicable).

APPENDIX

Troubleshooting Guidelines and Checklists

Troubleshooting Guide

HID

This guide is prepared to assist electricians with normal routine maintenance and to help them understand the operation of different lamp sources and electrical systems.

Caution: High voltages, currents, and temperatures are required to operate gas discharge lamps. Shock and burn hazards exist, and testing or evaluating fixtures or components should be done by qualified individuals only.

Light Sources:

Fluorescent

Mercury Vapor (MV)

Metal Halide (MH)

High Pressure Sodium (HPS)

Lamps

Fluorescent lamps come in many sizes, shapes and colors. This lamp source can provide numerous advantages, depending on the needs and requirements of end users. These lamps are generally used when low mounting heights and quiet operation are required. Note that many new, more efficient lamps and electronic ballasts offer the customer many energy saving opportunities.

Mercury Vapor lamps were the first to be developed in the family of high intensity discharge (HID) lamps. Their advantage is that they offer long life. Their disadvantages are poor color rendition, poor lumen maintenance, and low efficiency.

MH lamps are similar to MV lamps in design and operation. Their lumen output is double that of MV lamps of the same wattage. MH lamps are used in installations that require high efficiency, and white light. Lamp life historically has been less than MV and HPS, however new MH systems are now changing this by providing excellent life, lumen maintenance and color control.

A. HPS lamps offer long life and more lumens per watt than MV or MH sources. HPS lamps emit a pale amber color compared to other whiter light sources. Energy savings is the main HPS advantage. New high-xenon HPS lamps offer even higher LPW, longer life and improved lumen maintenance.

Ballasts

All HID light sources require some form of ballast because:

Most require a starting voltage that is higher than the line voltage.

They all have a characteristic known as negative resistance. This means that once the arc is initiated, the lamp's resistance continually decreases as current increases. For all practical purposes, the lamp becomes a short circuit. The ballast limits and controls the current wave form through the lamp.

They provide voltage transformation to allow the use of many line voltages.

There are many different types of ballasts used with different lamps. Within a lamp type, ballasts vary in lamp operating wattage and regulation caused by changes in line voltage and lamp voltage. Better regulation and control normally results in higher initial cost, but may greatly improve operating characteristics and overall performance.

Capacitors

Capacitors are used for power factor correction or as current regulation devices which provide the control necessary to ensure proper lamp and ballast operation. Different wattages, voltages, and ballast types require a variety of different capacitors. The ballast ID label specifies the microfarad and voltage rating required for proper operation. If the capacitor is incorrectly wired, improper operation of the fixture as well as other component failures could result.

Ignitors/Starters

These devices are utilized to provide the proper voltage and energy to start the lamp. They are predominantly used with HPS lamps as well as some Metal Halide and Fluorescent systems.

Instructor's Notes:

Electrical Testing Procedures

Caution: High voltages, currents, and temperature are required to operate lamps. Therefore, shock and burn hazards exist, and testing or evaluating fixtures or components should be done only by qualified individuals.

A. Testing Lamps

The easiest method of troubleshooting a fixture is to try a **known** good lamp in the inoperative fixture. If the lamp being replaced exhibits any of the following conditions, replace with a new lamp.

1. **Sodium Leaker Lamp -** will have a brown/golden coating on the inside of the lamp envelope other than at the base of the lamp.
2. **Amalgam Leaker Lamp -** the lamp envelope will have a smoked bronze appearance on the inside of the envelope.
3. **Faulty Base to Lamp Envelope Seal -** a white powdery substance will appear at the base of the lamp where oxygen has leaked inside the lamp.
4. **End of Lamp Life -** the arc tube will be black on both ends or the entire length of the arc tube will be black.
5. **Broken Welds or Arc Tube Support Brackets -** mechanical breaks occasionally occur due to rough handling or internal thermal stresses. Broken welds in evacuated (HPS) lamps can also create a problem known as vacuum switching. Extremely high voltage surges occur in the lamp circuit if the weld opens while the lamp is operating. Secondary coil burnout, ignitor arcing and socket arcing can occur. Rewelding may occur and the lamp may **appear** to be satisfactory; however, if left in operation, failure of the ballast and/or ignitor is likely.

B. Ignitors/Starters

The starter provides the necessary voltage and energy required to initiate the arc in the lamp. The easiest way to check the ignitor on 35W to 150W HPS units is to install a 120V incandescent lamp in the fixture. If the incandescent lamp operates but a known good HPS lamp will not ignite,

308A01.TIF

replace starter. In 200W to 1000W HPS fixtures, install a mercury lamp of similar wattage. If the mercury lamp lights and the HPS lamp will not, replace the starter.

Do not operate Incandescents or mercury lamps used to check the starter for extended periods of time (more than ½ hour).

C. Capacitors

Testing Capacitors may be accomplished by:

1. **Visual Inspection for swollen capacitors.** If the capacitor is swollen or bulged on the sides or top where the terminals are located, remove and replace with a new one.
2. **Verify the correct microfarad rating** as specified on the ballast I.D. label.
3. **Using an ohmmeter to check capacitors:**

discharge capacitor by shorting between the terminals

disconnect capacitor from circuit

remove bleed resistor

4. **Set ohmmeter to highest resistance scale** and connect leads to capacitor terminals.
   - if resistance starts low and gradually increases, the capacitor is good.
   - if resistance starts low and doesn't increase, the capacitor is shorted and should be replaced.
   - if resistance is high and remains relatively the same, the capacitor is open and should be replaced.

D. Ballast

Visual inspection of the coil for burned or charred windings is the easiest method for checking the ballast. If lamps, ignitors and capacitors test good, replace the ballast. Testing the voltage at the socket is another method of checking the ballast. However, to use this procedure you must know specific ballast/lamp voltage and amperage requirements. The starting aid (if present) should be disconnected prior to testing the voltage at the socket. Failure to remove the starting aid could damage the test equipment

Fluorescent Service Checklist

(Identify problem and test for cause in numerical sequence.)

PROBLEM	End of Lp. Life	Incorr. Lamp	Def. Blst.	Incorr. Blst.	Incorrect Supply Voltage	No Supply Voltage	Incomp. Lamp Seating	(B) Dirty Lamps	Incorr. Fixt. Wiring	Loose Socket Connect.	(C) Low Amb Temp.	High Amb. Temp.
Failure to Start (A)	1	5	4	10	6	7	8	2	9	11	3	
Slow Starting		3			4			2			1	
Blinking Off and On - Very Short Duration Cycles	1				2					4	3	
Long Duration Cycles - Several Minutes to Hours(D)		3	2		4				5			1
Short Lamp Life		2	1	3	4		5		6		7	
Lamp End Blackening - One End Only (A)			4				1		2	3		
Lamp End Blackening - Both Ends(A)	1	2	3	6	7		4		5			
Lamp Ends Only Lighted		1	2	3					4			

A. Generally indicates lamp cathode heat is missing. Problem source must be identified and corrected.

B. In humid weather or when air conditioning systems are operated only during working hours, dust on the lamps may gather condensation and prevent reliable starting. Lamps and fixtures must be cleaned and the lamp waxed with a good silicone wax.

C. Energy saving lamp, ballast combinations will not reliably start near or below 60°F; all others 50°F unless low temperature systems are used.

D. Indicates ballast thermal protector is cycling. May be the result of high ambient temperature, fixture misapplication or restricted air circulation. With recessed fixtures, check to be sure insulation has not been placed directly on the recessed fixture body. With surface mounted fixtures mounted against insulated ceiling, make sure they are rated for such applications.

All service checklists detail the more probable problem causes. It is important to keep in mind that other elements may be involved or more than one cause present. All electrical service work must begin with a careful and detailed evaluation of the problem and thorough inspection of all components, paying special attention to all wiring connections.

The ability to field test ballasts is very limited. The normal verification is to substitute a known good ballast in the problem fixture.

On rapid start systems only, cathode heater voltage may be verified by a voltage reading between **the contacts on each socket.** (A voltage of 3.5 to 5 is considered normal.)

Caution - Voltages in excess of 500V may be present between the lamp socket parts and ground.

308A02.TIF

Instructor's Notes:

Mercury Vapor And Metal Halide Service Checklist

Identify problem and test for cause in numerical sequence.

PROBLEM	A End of Lamp Life	B Incorrect Lamp	C Open Cap.	D Def. Starter (if app.)	E Incorrect Supply Voltage	F Photo-Cont.	G New Lamp Repl.	H Line Voltage Dips	I Shorted. Cap.	J Burned Ballast Windings	K Incorrect Ballast
FAILURE TO OPERATE	1	5	3	4	2						6
LAMP CYCLING		1			2	3					
COLOR SHIFT							1	2			
LOW LIGHT OUTPUT		2			1						
TRIPPED BREAKER OR BLOWN FUSE									1	2	3

A. End of Lamp Life

At lamp's end of life, the voltage requirements of the lamp exceeds the output ability of the ballast. The usual failure mode for mercury is low light output followed by failure to operate, and for Metal Halide low light levels, color shifts, and lamp operating instability (cycling). Replace end of life lamp as dictated by lamp testing procedure.

B. Incorrect Lamp

Lamp wattage, voltage, burning position, and type must be checked against the fixture label to be sure the proper lamp has been installed.

C. Open Capacitor

This is the usual result of electrical failure or mechanical damage. Very often the capacitor can will be bulged or distorted. Where the capacitor is used in series with the lamp, the lamp will not operate. See tests for capacitors or replace with a known good capacitor.

D. Defective Starter

The function of the starter is to provide a high voltage pulse to ignite the lamp. To test, replace with a known good lamp. (Also see Ignitor/Starter testing procedure)

E. Incorrect Supply Voltage

When investigating problems of low light output or cycling, voltage readings first must be taken at the fixture to properly identify power distribution problems. In the case of multiple supply type ballasts, verification that the supply voltage is connected to the appropriate input lead is advised.

F. Photocontrol

Problems may result from electrical failure, incorrect wiring, or from an incorrect amount of light reaching the cell. First cover the eye of the cell with electrical tape to verify fixture and cell operation. If the fixture fails to operate, the cell must be bypassed electrically to identify the problem source. Problems of incorrect amount of light usually can be resolved by repositioning the fixture or by using cell caps to regulate the light level.

G. New Lamp Replacement

New lamps, when installed, go through a period of burn-in or seasoning which may extend for a period of 100 hours or more. The usual result is noticeable color variation between lamps. While metal-halide lamps are noted for this, the system will stabilize as the burn-in period ends. It is important to understand that some variation in color may be noted between lamp manufacturers, or between old and new lamps.

H. Line Voltage Dips

When investigating voltage dips, it is important to identify distribution system loading. The usual cause is the starting of large motors or the use of electric welding equipment. Line voltage recorders will usually identify the problem. It is important to understand that mild dips will cause a color shift, while severe dips will cause the lamp to go out. Dip tolerance depends on lamp type, lamp age and ballast type.

I. Shorted Capacitor

This is the direct result of electrical failure or mechanical damage. The most common result of a shorted capacitor is ballast failure. In all cases of shorted capacitors, both capacitor and transformer should be replaced. See tests for capacitors or replace with a known good capacitor.

J. Burned Ballast Coils

1. Burned Primary Coils

This is the usual result of fixture connection to incorrect supply voltage. Repeated failure often in conjunction with capacitor failure may indicate short duration high voltage spikes on the distribution system. The use of a scope along with power company assistance is generally required to identify these spikes. Equally important as a cause of failure is a shorted capacitor (See "I" above - Shorted Capacitor). See tests for ballasts on preceding page and replace if necessary.

2. Burned Secondary Coils

This may be caused by a short circuit in the lamp circuit wiring or by mechanical failure in the lamp. Carefully inspect all lamp circuit wiring. See tests for ballasts on preceding page and replace it if necessary. **In all cases, replace the lamp.**

K. Incorrect Ballast

The requires checking only a new fixture or if a recurring problem is encountered. Carefully compare details on the transformer to the fixture label and lamp used in the circuit. Change components as required.

308A03.TIF

High-Pressure Sodium Service Checklist

Identify problem and test for cause in numerical sequence.

	A	B	C	D	E	F	G	H	I	J
PROBLEM	End of Lamp Life	Incorrect Supp. Voltage	Incorrect Lamp	Shorted Capacitor	Photo-Control	Line Volt. Dip	Defective Lamp	Defective Starter	Open Capacitor	Burned Ballast Windings
LAMP CYCLING SINGLE FIXTURE	1	3	2		4					
LAMP CYCLING GROUP OF FIXTURES					2	1				
FAILURE TO START		2	3				4	1	5	6
LOW LIGHT OUTPUT		2	3	1			4			

A. End of Lamp Life

At a lamp's end of life, the operating voltage requirements of the lamp exceed the output ability of the ballast. This results in the lamp cycling off and on. It is important to understand that in the early stages of failure, the lamp may operate for several hours before cycling off. As the lamp nears total failure, the on time will decrease until the lamp fails to ignite at all. Cycling lamps should immediately be replaced to avoid starting aid damage. See lamp tests or replace with known good lamp.

B. Incorrect Supply Voltage

When investigating problems of low light output, cycling, or failure to start, voltage readings must be taken *at the fixture* to properly identify distribution system problems. For multiple supply type ballasts, proper lead connection must be verified.

C. Incorrect Lamp

Lamp wattage, voltage, burning position, and type must be checked against fixture label to be sure the proper lamp has been installed in the fixture.

D. Shorted Capacitor

This is the direct result of electrical failure or mechanical damage. The most common result is low light output; cycling may also occur.

E. Photocontrol

Problems may result from electrical failure or from an incorrect amount of light reaching the cell. First cover the eye of the cell with electrical tape, to verify fixture and cell operation. If the fixture fails to operate, the cell must be bypassed electrically to identify the problem source. Problems of incorrect amount of light usually can be resolved by repositioning the fixture or by using cell caps to regulate the light level. Occasionally, the cell will see light from the fixture reflected off a nearby object and turn itself off. Repositioning the cell or reflecting object may be required.

F. Line Voltage Dip

When investigation voltage dips, it is important to identify distribution system loading. The usual cause is the starting of large motors or the use of electric welding equipment. Line voltage recorders will usually identify the problem. It is important to understand that lamps nearing end of life will be more susceptible to voltage dips than new lamps, and lamp operating on reactor ballasts are more sensitive to voltage dips than those on regulating ballasts.

G. Defective Lamp

This is normally the result of some mechanical failure in the lamp. This can often be determined by brown or silver coating on the lamp outer jacket, or by deposits at the base of the lamp. See lamp testing procedures or replace with a known good lamp.

H. Defective Starter

The function of the starter is to provide a high voltage pulse to ignite the lamp. Failure to operate generally results from electrical failure in the starter. See tests for starters or replace with a known good starter.

I. Open Capacitor

This generally results from electrical failure or mechanical damage. Very often the capacitor can will be bulged or distorted. When the capacitor is used in the secondary (lamp) circuit, the fixture will not operate. See tests for capacitors or replace with a known good capacitor.

J. Burned Ballast Windings

This is often the result of fixture being connected to incorrect supply voltage. Repeated primary winding failures often in conjunction with capacitor failures may indicate short duration high voltage spikes on the distribution system. The use of a scope in conjunction with power company assistance is generally required to identify these spikes. See tests for ballasts and replace if necessary.

308A04.TIF

Instructor's Notes:

Light Output Service Checklist

(Investigate In numerical sequence)

A	B	C	D	E	F	G	H	I	J
Voltage	Socket Position	Lamps	Reflector	Reflectance	Obstruct.	Light Meters	Dirt	Line Current Harmonics	Spacing & Mount. Height
2	6	3	7	4	8	1	9	10	5

Voltage
Measurements must be checked a) at the fixture b) at the end of the distribution line. Confirm correct ballast voltage tap connection to supply voltage.

Socket Position
If adjustable, check fixture instructions and confirm correct position has been selected.

Lamps
Check for (a) correct wattage (b) correct burn position (c) high output vs. standard (d) does color appear to be correct. (after burn in)

Reflector
Has the correct reflector been installed? If adjustable, has the proper mounting position been selected? Verify whether reflector should be open or enclosed.

Reflectance
Recheck original calculations and confirm correct reflectance was used for walls, ceiling and floor. Ratio of incident light to reflected light is a measure of reflectance.

Obstructions
Note obstructions in the air and at the floor level that would restrict normal light distribution. In the air this would include piping, heating, crane rails, steel structure, fog, or other airborne contaminants. At floor level work in progress, racks, cabinets, machinery and partitions. In addition consideration must be given to guards, visors and other fixtures.

Light Meters
Use a second meter to confirm out of spec readings. Be sure the meter is not shadowed, not receiving reflected light, and held in the correct plane, and calibrated.

Dirt
Be sure the reflector lamp and lens (if involved) are clean and free of construction dust.

Current/Wave Form
Check for hot panels, conduit and feeder wiring indicating harmonics and overloading.

Spacing & Mounting Height
Confirm the installation is per the original design for fixture spacing mounting height and aiming (if involved). Check for pole spacing, pole heights, and setbacks.

308A05.TIF

Additional Resources

This module is intended to present thorough resources for task training. The following reference works are suggested for further study. These are optional materials for continued education rather than for task training.

American Electrician's Handbook. Terrell Croft and Wilfred I. Summers. New York, NY: McGraw-Hill, 1996.

Lighting Handbook. New York, NY: Illuminating Engineering Society of North America (IESNA), 2000.

National Electrical Code Handbook. Quincy, MA: National Fire Protection Association, 2002.

Instructor's Notes:

ANSWER KEY

Answers to Review Questions

Answer	Section
1. a	2.0.0
2. c	3.0.0
3. d	4.0.0
4. b	4.0.0
5. c	5.0.0
6. a	6.0.0
7. c	7.1.4
8. a	8.0.0; Appendix A
9. c	9.1.0
10. d	10.0.0

Section 2.0.0 *Think About It—Energy Consumed by Incandescent Lamps*

d. 90%

Section 7.1.2 *Think About It—Rapid-Start Lamps*

Rapid-start lamps may not ignite at low ambient temperatures because the electrodes must reach a preset temperature before they will emit.

Section 7.2.4 *Think About It—HID Ballasts*

Nothing will happen if the output leads of an HID ballast are shorted together. The ballast is current limiting by design and would only allow the passage of the rated current. In fact, this is the typical method used to test the output current of the ballast.

CONTREN™ LEARNING SERIES — USER UPDATES

The NCCER makes every effort to keep these textbooks up-to-date and free of technical errors. We appreciate your help in this process. If you have an idea for improving this textbook, or if you find an error, a typographical mistake, or an inaccuracy in NCCER's Contren™ textbooks, please write us, using this form or a photocopy. Be sure to include the exact module number, page number, a detailed description, and the correction, if applicable. Your input will be brought to the attention of the Technical Review Committee. Thank you for your assistance.

Instructors – If you found that additional materials were necessary in order to teach this module effectively, please let us know so that we may include them in the Equipment/Materials list in the Instructor's Guide.

Write: Curriculum Revision and Development Department
National Center for Construction Education and Research
P.O. Box 141104, Gainesville, FL 32614-1104

Fax: 352-334-0932

E-mail: curriculum@nccer.org

Craft ____________________ Module Name ____________________

Copyright Date __________ Module Number __________ Page Number(s) __________

Description

____

____

____

____

(Optional) Correction

____

____

____

(Optional) Your Name and Address

____

____

____

Motor Calculations

26309-02

Motor Calculations
Annotated Instructor's Guide

Module 26309-02

MODULE OVERVIEW

This module introduces the Electrical trainee to the calculations used to size motor branch and feeder circuits, overcurrent protection, and disconnects.

PREREQUISITES

Please refer to the Course Map in the Trainee Module. Prior to training with this module, it is recommended that the trainee shall have successfully completed the following:

Core Curriculum; Electrical Level One; Electrical Level Two; Electrical Level Three, Modules 26301 through 26308.

OBJECTIVES

Upon completion of this module, the trainee will be able to:

1. Size branch circuits and feeders for electric motors.
2. Size and select overcurrent protective devices for motors.
3. Size and select overload relays for electric motors.
4. Size and select devices to improve the power factor at motor locations.
5. Size motor short circuit protectors.
6. Size multi-motor branch circuits.
7. Size motor disconnects.

PERFORMANCE TASKS

This is a knowledge-based module; there is no performance examination.

NCCER STANDARDIZED CRAFT TRAINING PROGRAMS

The National Center for Construction Education and Research (NCCER) provides a standardized national program of accredited craft training. Key features of the program include instructor certification, competency-based training, and performance testing. The program provides trainees, instructors, and companies with a standard form of recognition through a National Craft Training Registry. The program is described in full in the *Guidelines for Accreditation*, published by the NCCER. For more information on standardized craft training, contact the NCCER by writing us at P.O. Box 141104, Gainesville, FL 32614-1104, calling 352-334-0911, or e-mailing info@nccer.org. More information may be found at our Web site, www.nccer.org.

HOW TO USE THIS ANNOTATED INSTRUCTOR'S GUIDE

Each page presents two sections of information. The larger section displays each page exactly as it appears in the Trainee Module. The narrow column ties suggested trainee and instructor actions to each page and provides icons to call your attention to material, safety, audiovisual, or testing requirements. The bottom of each page includes space for your notes.

Teaching Tip

If you see the Teaching Tip icon, that means there is a teaching tip associated with this section. Also refer to any suggested teaching tips at the end of the module.

SAFETY CONSIDERATIONS

Ensure that the trainees are equipped with appropriate personal protective equipment.

PREPARATION

Before teaching this module, you should review the Module Outline, Objectives, and the Materials and Equipment List. Be sure to allow ample time to prepare your own training or lesson plan and gather all required equipment and materials.

MATERIALS AND EQUIPMENT LIST

Materials:
Transparencies
Markers/chalk
Various types of fuses, including:
- Nontime-delay
- Dual-element, time-delay fuses

Copy of the latest edition of the *National Electrical Code®*
Module Examinations*

Equipment:
Overhead projector and screen
Whiteboard/chalkboard
Appropriate personal protective equipment
Various types of disassembled motors, including:
- Squirrel cage
- Wound-rotor
- Synchronous

Various types of circuit breakers
Motor short circuit protector

*Located in the Test Booklet packaged with this Annotated Instructor's Guide.

ADDITIONAL RESOURCES

This module is intended to present thorough resources for task training. The following reference works are suggested for both instructors and motivated trainees interested in further study. These are optional materials for continued education rather than for task training.

American Electrician's Handbook, Terrell Croft and Wilfred I. Summers. New York, NY: McGraw-Hill, 1996.

National Electrical Code Handbook. Quincy, MA: National Fire Protection Association, 2002.

TEACHING TIME FOR THIS MODULE

An outline for use in developing your lesson plan is presented below. Note that each Roman numeral in the outline equates to one session of instruction. Each session has a suggested time period of 2½ hours. This includes 10 minutes at the beginning of each session for administrative tasks and one 10-minute break during the session. Approximately 12½ hours are suggested to cover *Motor Calculations.* You will need to adjust the time required for hands-on activity and testing based on your class size and resources.

Topic	Planned Time
Session I. Introduction; Motor Basics	
A. Introduction	________
B. Motor Basics	________
1. Synchronous Speed	________
2. Stator Windings	________
a. Principles of Dual-Voltage Connections	________
3. Special Connections	________
Session II. Calculating Motor Circuit Conductors	
A. Calculating Motor Circuit Conductors	________
1. Wound-Rotor Motors	________
2. Conductors for DC Motors	________
3. Conductors for Miscellaneous Motor Applications	________
Session III. Motor Protective Devices; Circuit Breakers	
A. Motor Protective Devices	________
1. Practical Applications	________
2. Motor Overload Protection	________
B. Circuit Breakers	________
1. Application of Magnetic-Only Circuit Breakers	________
2. Motor Short Circuit Protectors	________

Session IV. Multi-Motor Branch Circuits; Power Factor Correction at Motor Terminals

A. Multi-Motor Branch Circuits ____________

B. Power Factor Correction at Motor Terminals ____________

Session V. Review; Module Examination

A. Review ____________

B. Module Examination ____________

1. Trainees must score 70% or higher to receive recognition from the NCCER.
2. Record the testing results on Craft Training Report Form 200 and submit the results to the Training Program Sponsor.

Module 26309-02

Motor Calculations

Course Map

This course map shows all of the modules in the third level of the Electrical curriculum. The suggested training order begins at the bottom and proceeds up. Skill levels increase as you advance on the course map. The local Training Program Sponsor may adjust the training order.

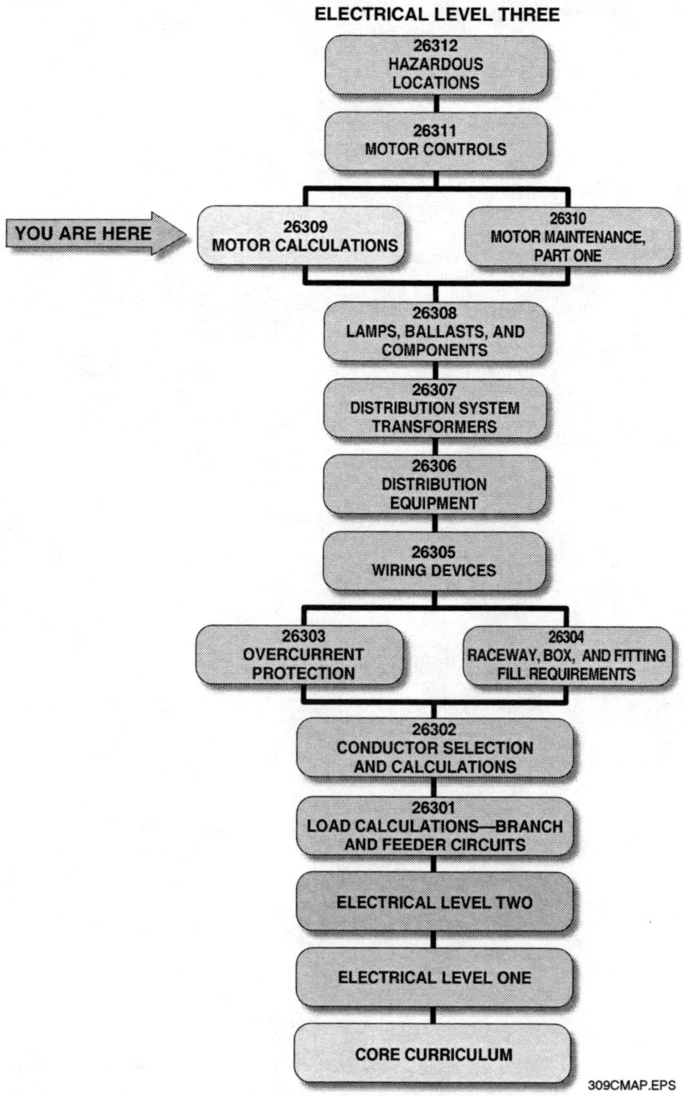

309CMAP.EPS

Homework

Assign reading of Module 26309-02.

MODULE 26309 CONTENTS

Figures

Table

Instructor's Notes:

MODULE 26309

Motor Calculations

Objectives

When you have completed this module, you will be able to do the following:

1. Size branch circuits and feeders for electric motors.
2. Size and select overcurrent protective devices for motors.
3. Size and select overload relays for electric motors.
4. Size and select devices to improve the power factor at motor locations.
5. Size motor short circuit protectors.
6. Size multi-motor branch circuits.
7. Size motor disconnects.

Prerequisites

Before you begin this module, it is recommended that you successfully complete the following: Core Curriculum, Electrical Level One; Electrical Level Two; Electrical Level Three, Modules 26301 through 26308.

Required Trainee Materials

1. Paper and pencil
2. Appropriate personal protective equipment
3. Copy of the latest edition of the *National Electrical Code*

1.0.0 ◆ INTRODUCTION

Electric motors are used in almost every type of installation imaginable, from residential appliances to heavy industrial machines. Many types of motors are available, from small shaded-pole motors (used mostly in household fans) to huge synchronous motors for use in large industrial installations. There are numerous types in between to fill every conceivable niche. None, however, have the wide application possibilities of the three-phase motor. This is the type of motor that electricians encounter most frequently. Therefore, the majority of the material in this module will deal with three-phase motors.

There are three basic types of three-phase motors:

- Squirrel cage induction motor
- Wound-rotor induction motor
- Synchronous motor

The type of three-phase motor is determined by the rotor or rotating member (*Figure 1*). The stator winding is basically the same for all three motor types.

The principle of operation for all three-phase motors is the rotating magnetic field. There are three factors that cause the magnetic field to rotate:

- The voltages of a three-phase electrical system are 120° out of phase with each other.
- The three voltages change polarity at regular intervals.
- The stator windings around the inside of the motor are arranged in a specific manner to induce rotation.

The NEC plays an important role in the installation of electric motors. ***NEC Article 430*** covers the application and installation of motor circuits and motor control connections, including conductors, short circuit and ground fault protection, controllers, disconnects, and overload protection.

Note: The designations "National Electrical Code," "NE Code," and "NEC," where used in this document, refer to the National Electrical Code®, which is a registered trademark of the National Fire Protection Association, Quincy, MA. *All National Electrical Code (NEC) references in this module refer to the 2002 edition of the NEC.*

Materials

Ensure that you have everything required to teach the course. Check the Materials and Equipment List at the front of this Instructor's Guide.

Classroom

Review the goals of the course.

Explain that terms shown in bold (blue) are defined in the Glossary at the back of this module.

Audiovisual

Show Transparency 1, Course Objectives.

Classroom

Explain that motors are categorized according to the type of rotor used. Refer the trainees to Figure 1 in the Trainee Module.

List the three types of three-phase motors. Provide examples for the trainees to examine.

Explain how three-phase motors operate.

Point out that *NEC Article 430* covers motor circuits and motor control connections.

Classroom

Explain that *NEC Article 440* covers motor-driven air conditioning and refrigeration equipment.

Point out the applicable motor requirements in *NEC Articles 422* and *424.*

Discuss the construction and operation of a squirrel cage rotor.

Audiovisual

Show Transparency 2 (Figure 2).

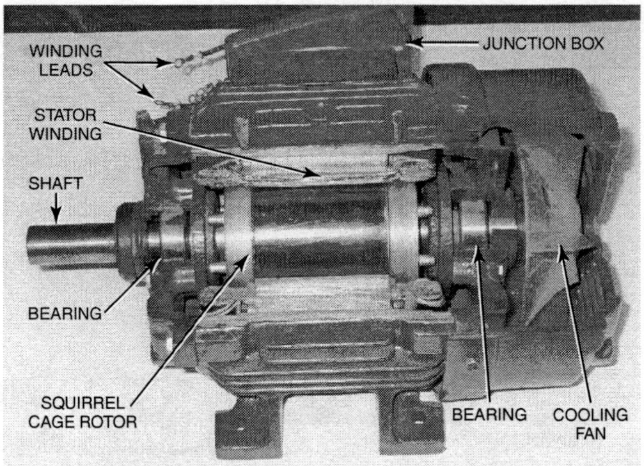

309F01.TIF

Figure 1 ◆ Basic parts of a three-phase motor.

NEC Article 440 contains provisions for motor-driven air conditioning and refrigerating equipment, including the branch circuits and controllers for the equipment. It also takes into account the special considerations involved with sealed (hermetic) motor compressors, in which the motor operates under the cooling effect of the refrigeration. When referring to *NEC Article 440,* be aware that the rules in this article are in addition to, or are amendments to, the rules given in *NEC Article 430.* Motors are also covered to some degree in *NEC Articles 422 and 424.*

2.0.0 ◆ MOTOR BASICS

The rotor of an AC squirrel cage induction motor (*Figure 2*) consists of a structure of steel laminations mounted on a shaft. Embedded in the rotor is the rotor winding, which is a series of copper or aluminum bars, short circuited at each end by a metallic end ring. The stator consists of steel laminations mounted in a frame. Slots in the stator hold stator windings that can be either copper or aluminum coils or bars. These are connected to form a circuit.

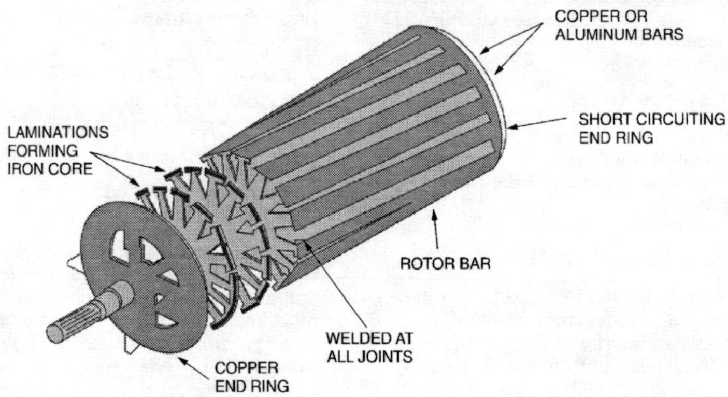

309F02.EPS

Figure 2 ◆ Squirrel cage rotor.

Instructor's Notes:

Energizing the stator coils with an AC supply voltage causes current to flow in the coils. The current produces an electromagnetic field that causes magnetic poles to be created in the stator iron. The strength and polarity of these poles vary as the AC current flows in one direction, then the other. This change causes the poles around the stator to alternate between being south and north poles, thus producing a rotating magnetic field.

The rotating magnetic field cuts through the rotor, inducing a current in the rotor bars. This induced current only circulates in the rotor, which in turn causes a rotor magnetic field. As with two conventional bar magnets, the north pole of the rotor field attempts to line up with the south pole of the stator magnetic field, and the south pole attempts to line up with the north pole. However, because the stator magnetic field is rotating, the rotor chases the stator field. The rotor field never quite catches up due to the need to furnish **torque** to the mechanical load.

2.1.0 Synchronous Speed

The speed at which the magnetic field rotates is known as the synchronous speed. The synchronous speed of a three-phase motor is determined by two factors:

- Number of stator poles
- Frequency of the AC line in hertz (Hz)

The synchronous speeds for various 60Hz motors are as follows:

- Two poles–3,600 rpm
- Four poles–1,800 rpm
- Six poles–1,200 rpm
- Eight poles–900 rpm

These speeds illustrate that the rpm of a three-phase motor decreases as the number of poles increases.

2.2.0 Stator Windings

The stator windings of three-phase motors are connected in either a wye or a delta configuration (*Figure 3*). Some motor stators are designed to operate both ways; that is, they are started as a wye-connected motor to help reduce starting current, and then changed to a delta configuration for running.

Many three-phase motors have dual-voltage stators. These stators are designed to be connected to either 240V or 480V. The leads of a dual-voltage stator use a standard numbering system. *Figure 4* shows a dual-voltage, wye-connected stator.

Note that the nine motor leads are numbered in a spiral. The leads are connected in series for use on the higher voltage and in parallel for use on the lower voltage. Therefore, for the higher voltage, leads 4 and 7, 5 and 8, and 6 and 9 are connected together. For the lower voltage, leads 4, 5, and 6 are connected together; further connections are 1 and 7, 2 and 8, and 3 and 9, which are then connected to the three-phase power source. *Figure 5* shows the equivalent parallel circuit when the motor is connected for use on the lower voltage.

The same standard numbering system is used for delta-connected motors, and many delta-wound motors also have nine leads, as shown in *Figure 6*. However, there are only three circuits of three leads each. The high-voltage and low-voltage connections for a three-phase, delta-wound, dual-voltage motor are shown in *Figure 7*.

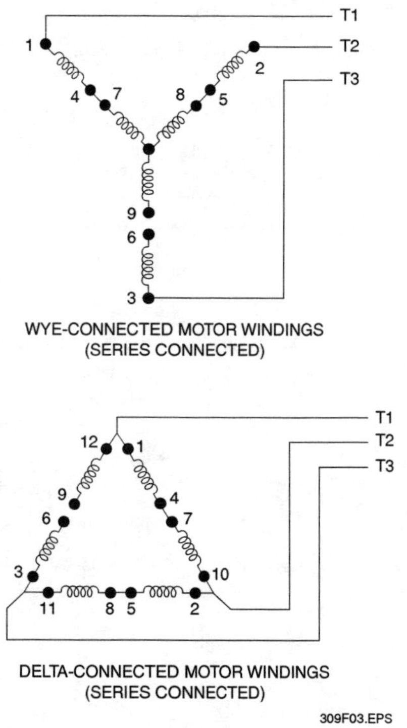

Figure 3 ◆ Types of windings found in three-phase motors.

Classroom

Define *synchronous speed*. List the synchronous speeds for various motors.

Discuss the function and operation of the stator windings in a three-phase motor. Point out that they can be constructed in either a delta or a wye configuration.

Audiovisual

Show Transparencies 3 and 4 (Figures 3 and 4).

Classroom

Explain how the motor leads are arranged in various winding configurations.

Audiovisual

Show Transparencies 5 through 7 (Figures 5 through 7).

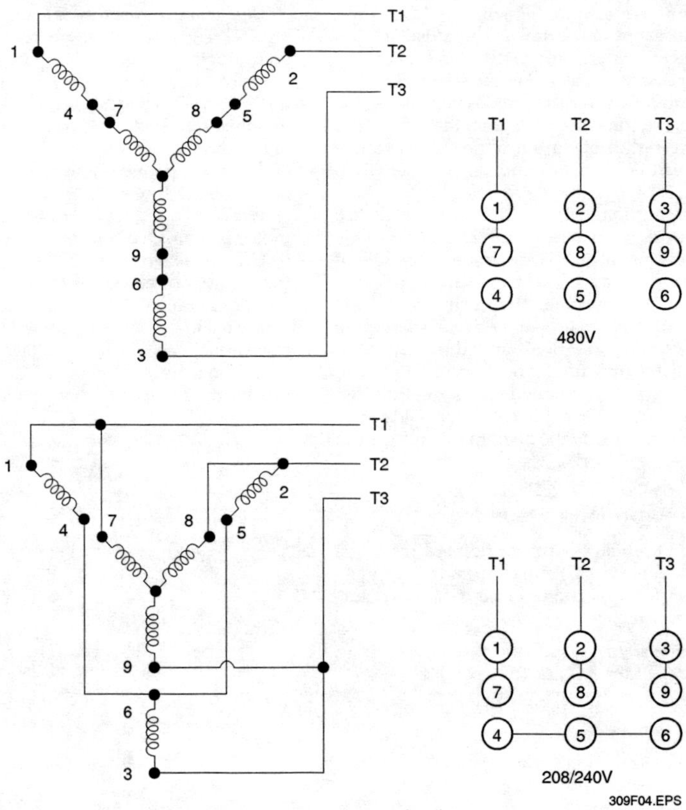

Figure 4 ◆ Dual-voltage, wye-connected, three-phase motors.

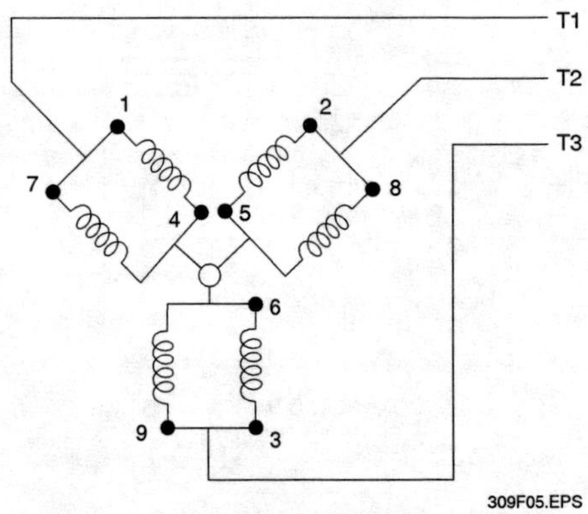

Figure 5 ◆ Equivalent parallel circuit.

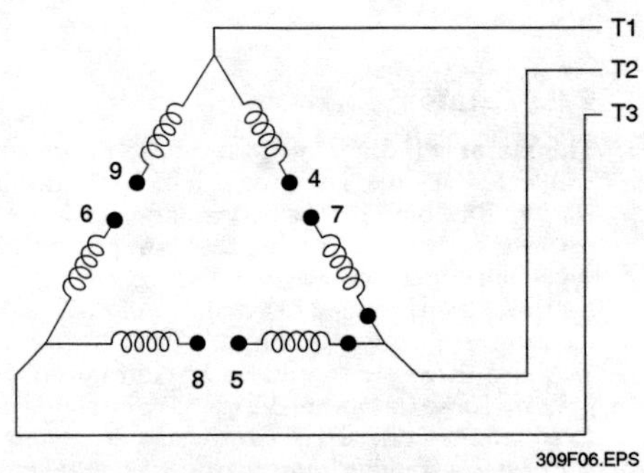

Figure 6 ◆ Arrangement of leads in a nine-lead, delta-wound, dual-voltage motor.

Instructor's Notes:

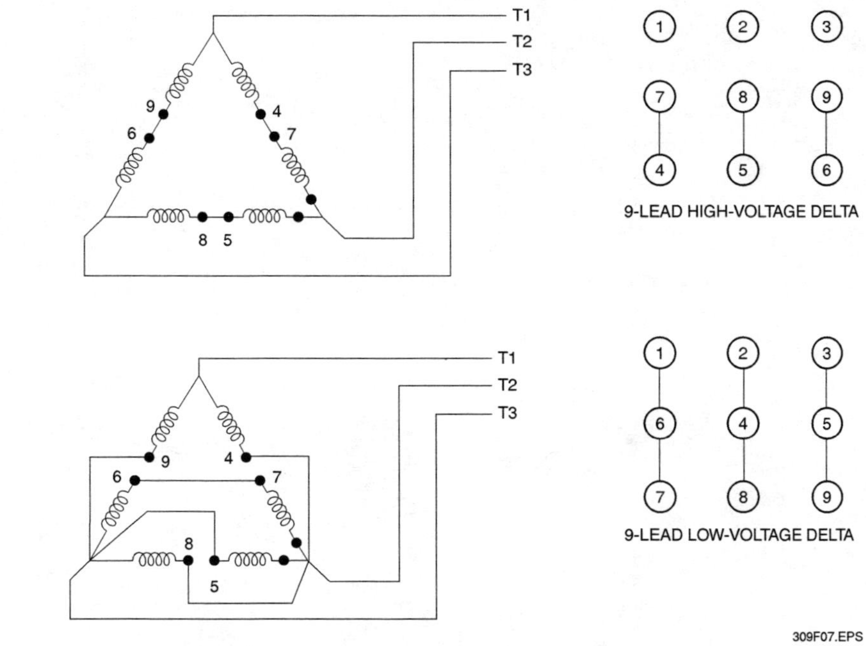

Figure 7 ◆ Lead connections for a three-phase, dual-voltage, delta-wound motor.

In some instances, a dual-voltage motor connected in a delta configuration will have 12 leads instead of nine. *Figure 8* shows the high-voltage and low-voltage connections for a dual-voltage, 12-lead, delta-wound motor.

2.2.1 Principles of Dual-Voltage Connections

When a motor is operated at 240V, the current draw of the motor is double the current draw of a 480V connection. For example, if a motor draws 10A of current when connected to 240V, it will draw only 5A when connected to 480V. The reason for this is the difference of impedance in the windings between a 240V connection and a 480V connection. Remember that the low-voltage windings are always connected in parallel, while the high-voltage windings are connected in series.

For instance, assume that the stator windings of a motor (R1 and R2) both have a resistance of 48Ω.

If the stator windings are connected in parallel, the total resistance (R_t) may be found as follows:

$$R_t = \frac{R1 \times R2}{R1 + R2}$$

$$R_t = \frac{48\Omega \times 48\Omega}{48\Omega + 48\Omega}$$

$$R_t = \frac{2{,}304\Omega}{96\Omega}$$

$$R_t = 24\Omega$$

Therefore, the total resistance (R) of the motor winding connected in parallel is 24Ω, and if a voltage (E) of 240V is applied to this connection, the following current (I) will flow:

$$I = \frac{E}{R}$$

$$I = \frac{240V}{24\Omega}$$

$$I = 10A$$

Audiovisual

Show Transparency 8 (Figure 8).

Classroom

Discuss the relationship between motor input voltage and current draw.

Explain how to calculate the total resistance and current draw of two stator windings connected in parallel.

Classroom

Explain how to calculate the total resistance and current draw of two stator windings connected in series. Point out that the current flow is half that of an equivalent parallel connection.

Note that some motors designed for operation at high voltages may have 15 or 18 leads.

Audiovisual

Show Transparency 9 (Figure 9).

Classroom

Review the basic components of a motor circuit.

Audiovisual

Show Transparency 10 (Figure 10).

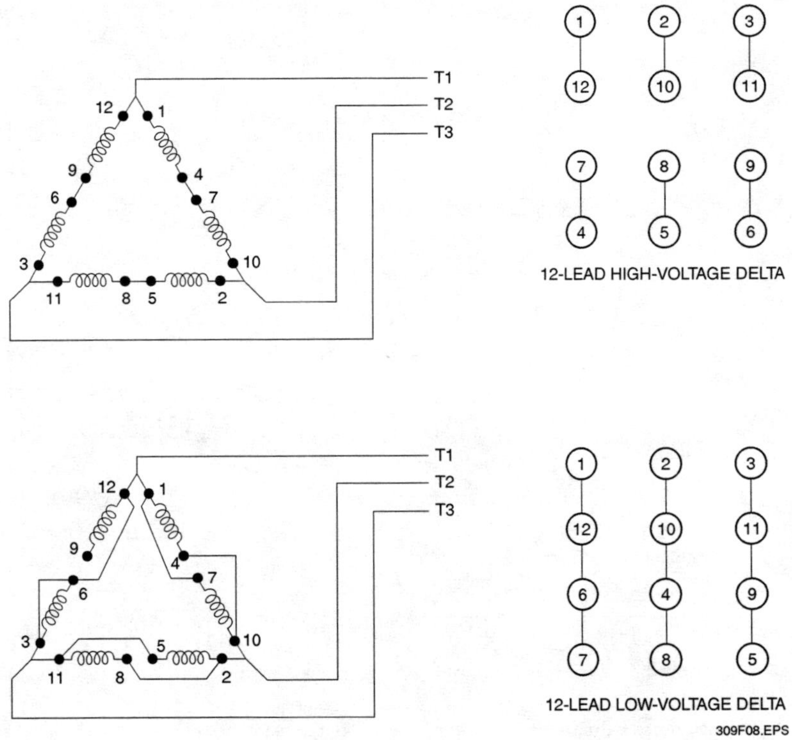

Figure 8 ◆ Lead connections for a 12-lead, dual-voltage, delta-wound motor.

If the windings are connected in series for operation on 480V, the total resistance of the winding is:

$$R_t = R1 + R2$$
$$R_t = 48\Omega + 48\Omega$$
$$R_t = 96\Omega$$

Consequently, if 480V is applied to this winding, the following current will flow:

$$I = \frac{E}{R}$$
$$I = \frac{480V}{96\Omega}$$
$$I = 5A$$

It is obvious that twice the voltage means half the current flow, or vice versa.

2.3.0 Special Connections

Some three-phase motors designed for operation on voltages higher than 600V may have more than 12 leads. Motors with 15 or 18 leads are common in high-voltage installations. A 15-lead motor has three coils per phase, as shown in *Figure 9.* Notice that the leads are numbered in the same spiral sequence as a nine-lead, wye-wound motor.

3.0.0 ◆ CALCULATING MOTOR CIRCUIT CONDUCTORS

The basic elements of a motor circuit are shown in *Figure 10.* Although these elements are shown separately in this illustration, there are certain cases in which the NEC permits a single device to serve more than one function. For example, in some

Instructor's Notes:

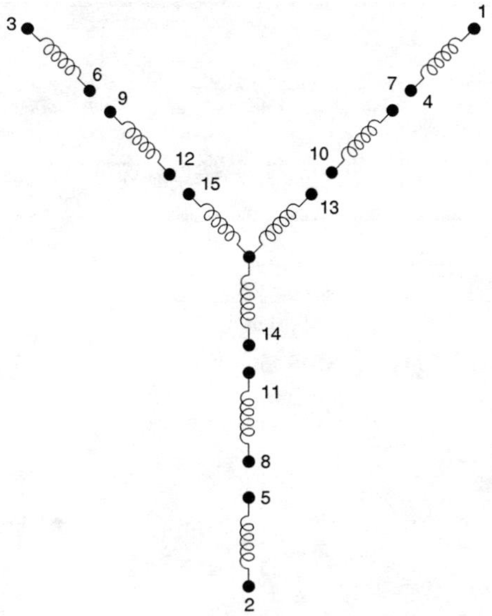

Figure 9 ◆ Fifteen-lead motor. 309F09.EPS

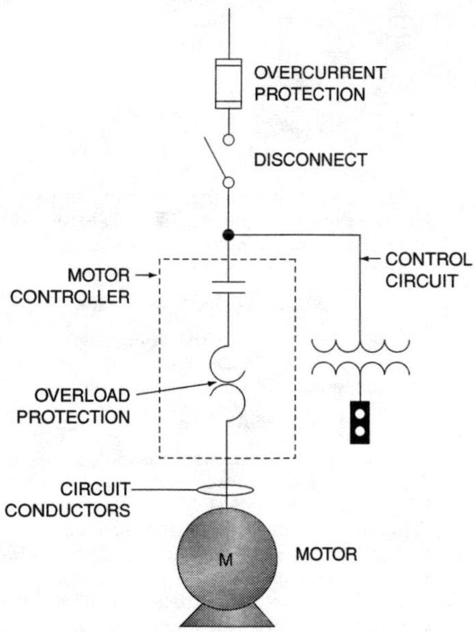

Figure 10 ◆ Basic elements of any motor circuit. 309F10.EPS

cases, one switch can serve as both the disconnecting means and the controller. In other cases, short circuit protection and overload protection can be combined in a single circuit breaker or set of fuses.

Note

NEC Section 430.22(A) states that when sizing conductors supplying a single motor used for continuous duty, the conductors must have a current-carrying capacity of not less than 125% of the motor full-load current rating. Conductors on the line side of the controller supplying multi-speed motors must be based on the highest of the full-load current ratings shown on the motor nameplate. Conductors between the controller and the motor must have a current-carrying rating based on the current rating for the speed of the motor being fed by each set of conductors.

A typical motor control center and branch circuits feeding four different motors are shown in *Figure 11*. We will see how the feeder and branch circuit conductors are sized for these motors.

Step 1 Refer to ***NEC Table 430.150*** for the full-load current of each motor.

Step 2 Determine the full-load current of the largest motor in the group.

Step 3 Calculate the sum of the full-load current **ratings** for the remaining motors in the group.

Step 4 Multiply the full-load current of the largest motor by 1.25 (125%) and then add the sum of the remaining motors to the result (***NEC Section 430.24***). The combined total will give the minimum feeder size.

THINK ABOUT IT

NEC Motor Sizes

How is the largest motor in a group determined by the NEC? Is it by frame size, horsepower, weight, or the motor's full-load amps (FLA)?

Classroom

Explain how to calculate the feeder and branch circuit conductors for a typical motor circuit.

Audiovisual

Show Transparency 11 (Figure 11).

Teaching Tip

Discuss the "Think About It."

Classroom

Work through the example feeder calculation provided in the Trainee Module.

INSIDE TRACK

Voltage Ratings

The motor voltage ratings used in the NEC can be somewhat confusing. Motors are built and tested at various voltages. For example, the 460V referenced in ***NEC Tables 430.148 and 430.150*** is based on rated motor voltages; however, in order to standardize calculations and maintain minimum safety standards, the NEC uses a higher value of 480V for motor calculations. See ***NEC Section 220.2(A).***

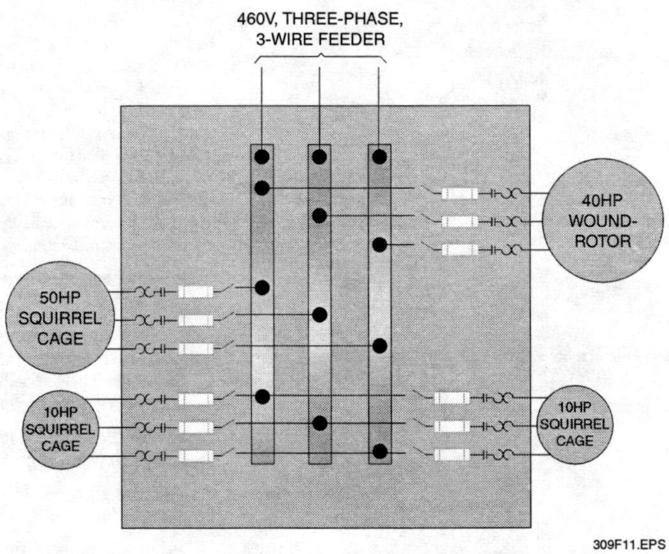

Figure 11 ◆ Typical motor control center.

When sizing feeder conductors for motors, be aware that the procedure previously described will give the minimum conductor rating based on the NEC minimum only. Consequently, it is often necessary to increase the size of conductors to compensate for voltage drop and power loss in the circuit.

Now we will complete the conductor calculations for the motor circuits in *Figure 11.*

Referring to ***NEC Table 430.150,*** the motor horsepower is shown in the far left-hand column. Follow across the appropriate row until you come to the column titled *460V,* which is the voltage of the motor circuits in *Figure 11.* We find that the ampere ratings for the motors in question are as follows:

- 50hp = 65A
- 40hp = 52A
- 10hp = 14A

The largest motor in this group is the 50hp squirrel cage motor, which has a full-load current of 65A.

The sum of the remaining motors is:

$$52A + 14A + 14A = 80A$$

Now, multiply the full-load current of the largest motor by 125% (1.25) and then add the total amperage of the remaining motors:

$$(1.25 \times 65A) + 80A = 161.25A$$

Therefore, the feeders for the 460V, three-phase, three-wire motor control center will have a minimum ampacity of 161.25A. Referring to ***NEC Table 310.16*** under the column headed 75°C, the closest conductor size is 2/0 copper (rated at 175A) or 4/0 aluminum (rated at 180A).

The branch circuit conductors feeding the individual motors are calculated somewhat differently. ***NEC Section 430.22(A)*** requires that the

Instructor's Notes:

ampacity of branch circuit conductors supplying a single continuous-duty motor must not be less than 125% of the motor full-load current rating. Therefore, the current-carrying capacity of the branch circuit conductors feeding the four motors in question are calculated as follows:

50hp motor = 65A × 1.25 = 81.25A
40hp motor = 52A × 1.25 = 65A
10hp motor = 14A × 1.25 = 17.5A

Referring to *NEC Table 310.16*, the closest size 75°C THWN copper conductors that will be permitted to be used on these various branch circuits are as follows:

- A 50hp motor at 81.25A requires No. 4 AWG THWN conductors.
- A 40hp motor at 65A requires No. 6 AWG THWN conductors.
- A 10hp motor at 17.5A requires No. 12 AWG THWN conductors [see *NEC Section 240.4(D)*].

Refer to *Figure 12* for a summary of the conductors used to feed our example motor control center, along with the branch circuits supplying the individual motors.

If voltage drop and/or power loss must be taken into consideration, please refer to the Electrical Level Three Module, *Conductor Selection and Calculations.*

For motors with other voltages (up to 2,300V) or for synchronous motors, refer to *NEC Table 430.150.*

In accordance with *NEC Section 430.22(E)*, branch circuit conductors serving motors used for short-time, intermittent, or other varying duty must have an ampacity not less than the percentage of the motor nameplate current rating shown in *NEC Table 430.22(E)*. However, to qualify as a short-time, intermittent motor, the nature of the apparatus that the motor drives must be arranged so that the motor cannot operate continuously with a load under any condition of use. Otherwise, the motor must be considered continuous duty. Consequently, the majority of motors encountered in the electrical trade must be rated for continuous duty, and the branch circuit conductors sized accordingly.

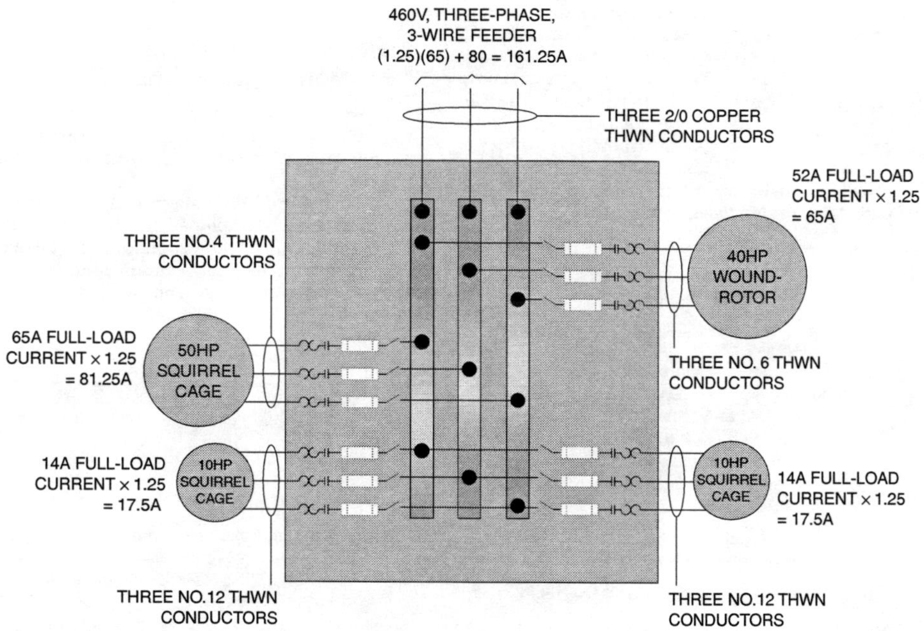

Figure 12 ◆ Sizing motor branch circuits.

Classroom

Work through the example branch circuit calculation provided in the Trainee Module and select the conductors to supply each motor. Refer the trainees to *NEC Table 310.16.*

Audiovisual

Show Transparency 12 (Figure 12).

Classroom

Explain that the full-load currents of various types of motors are listed in *NEC Table 430.150.*

Discuss intermittent and continuous-duty ratings. Point out that most motors are rated for continuous duty.

Teaching Tip

Discuss the "Think About It."

Classroom

Explain that the full-load current ratings for wound-rotor motors are the same as those for squirrel cage motors. Refer the trainees to the applicable sections in the NEC.

Discuss the NEC requirements for DC motors.

Point out the NEC sections applicable to specific types of motors.

THINK ABOUT IT

Feeder Size

If the 40hp wound-rotor motor in *Figure 12* was replaced with another 50hp squirrel cage motor, the #6 THWN conductors feeding that motor must be replaced with #4 THWN conductors. What size THWN, 75°C feeders are needed to accommodate the change?

INSIDE TRACK

Duty Ratings

Not all motors are rated for continuous duty. Some are rated for 5-, 15-, 30-, or 60-minute operation. The NEC makes special allowances for these duty ratings. Refer to ***NEC Section 430.22(E).***

3.1.0 Wound-Rotor Motors

The primary full-load current ratings for wound-rotor motors are listed in *NEC Table 430.150* and are the same as those for squirrel cage motors. Conductors connecting the secondary leads of wound-rotor induction motors to their controllers must have a current-carrying capacity at least equal to 125% of the motor full-load secondary current if the motor is used for continuous duty. If the motor is used for less than continuous duty, the conductors must have a current-carrying capacity of not less than the percentage of the full-load secondary nameplate current given in *NEC Table 430.22(E).* Conductors from the controller of a wound-rotor induction motor to its starting resistors must have an ampacity in accordance with *NEC Table 430.23(C).*

Note

NEC Section 430.6(A)(1) specifies that for general motor applications (excluding applications of torque motors and sealed hermetic-type refrigeration compressor motors), the values given in ***NEC Tables 430.147, 430.148, 430.149, and 430.150*** should be used instead of the actual current rating marked on the motor nameplate when sizing conductors, switches, and overcurrent protection. Overload protection, however, is based on the marked motor nameplate.

3.2.0 Conductors for DC Motors

NEC Sections 430.22(A), Exception 1 and 430.29 cover the rules governing the sizing of conductors from a power source to a DC motor controller and from the controller to separate resistors for power accelerating and dynamic braking. *NEC Section 430.29,* with its table of conductor ampacity percentages, assures proper application of DC constant-potential motor controls and power resistors. However, when selecting overload protection, the actual motor nameplate current rating must be used.

3.3.0 Conductors for Miscellaneous Motor Applications

NEC Section 430.6 should be referred to for torque motors, shaded-pole motors, permanent split capacitor motors, and AC adjustable-voltage motors.

NEC Section 430.6(B) specifically states that the motor's nameplate full-load current rating is used to size ground fault protection for a torque motor. However, both the branch circuit conductors and the overcurrent protection are sized by the provisions listed in *NEC Article 430, Part II* and *NEC Section 430.52.*

INSIDE TRACK

Breakdown Torque

Certain listed appliances are now rated by the breakdown torque (that is, the force it takes to stop the motor from rotating at full speed). This allows the use of the FLA directly from the nameplate of the appliance rather than having to refer to ***NEC Tables 430.148 and 430.150.***

Instructor's Notes:

For sealed (hermetic) refrigeration compressor motors, the actual nameplate full-load running current of the motor must be used in determining the current rating of the disconnecting means, controller, branch circuit conductor, overcurrent protective devices, and motor overload protection.

4.0.0 ◆ MOTOR PROTECTIVE DEVICES

NEC Sections 430.51 through 430.58 require that the branch circuit protection for motor controls protect the circuit conductors, control apparatus, and the motor itself against overcurrent due to short circuits or ground faults.

Motors and motor circuits have unique operating characteristics and circuit components. Therefore, these circuits must be dealt with differently from other types of loads. Generally, two levels of overcurrent protection are required for motor branch circuits:

- *Overload protection*—Motor running overload protection is intended to protect the system components and motor from damaging overload currents.
- *Short circuit protection (includes ground fault protection)*—Short circuit protection is intended to protect the motor circuit components such as the conductors, switches, controllers, overload relays, motor, etc., against short circuit currents or grounds. This level of protection is commonly referred to as motor branch circuit protection. Dual-element fuses are designed to give this protection, provided they are sized correctly.

There are a variety of ways to protect a motor circuit, depending upon the application. The ampere rating of a fuse selected for motor protection depends on whether the fuse is of the dual-element, time-delay type or the nontime-delay type.

In general, *NEC Table 430.52* specifies that short circuit/ground fault protection nontime-delay fuses can be sized at 300% of the motor full-load current for ordinary motors, while those for wound-rotor or direct current motors may be sized at 150% of the motor full-load current. The sizes of nontime-delay fuses for the four motors previously mentioned are listed in *Figure 13*. Because none of these sizes are standard, *NEC Section 430.52(C)(1), Exception No. 1* permits the size of the fuses to be increased to a standard size. Also, where absolutely necessary to permit motor starting, the size of the overcurrent device may be further increased, but must never be more than 400% of the full-load current [*NEC Section 430.52(C)(1), Exception No. 2(a)*]. In actual practice, most electricians would use a 200A nontime-delay fuse for the 50hp motor, a 175A fuse for the 40hp motor, and 45A fuses for the 10hp motors. If any of these fuses do not allow the motor to start without blowing, the fuses for the 50hp motor may be increased to a maximum of 400% of the full-load currents, which are 360A for the 50hp motor, 208A for the 40hp motor, and 56A for the 10hp motors.

Per *NEC Table 430.152*, dual-element, time-delay fuses are able to withstand normal motor starting current and can be sized closer to the actual motor rating than nontime-delay fuses. If necessary for proper motor operation, dual-element, time-delay fuses may be sized up to 175% of the motor's full-load current for all standard motors with the exception of wound-rotor and direct current motors. These motors must not have fuses sized for more than 150% of the motor's full-load current rating. Where absolutely necessary for proper operation,

THINK ABOUT IT

Overload Protection

In a 230V/460V motor such as the one shown here, which applied voltage requires the larger overload protection device?

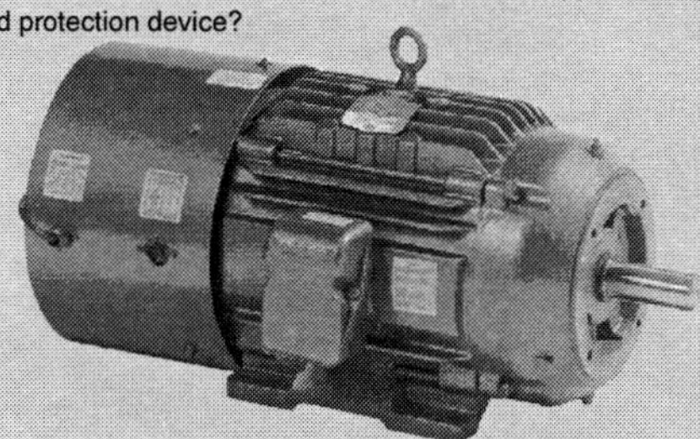

309P0901.EPS

Classroom

Explain that motors require two types of overcurrent protection:
- **Overload protection**
- **Short circuit protection**

Point out the applicable sections of the NEC.

Audiovisual

Show Transparency 13 (Figure 13).

Teaching Tip

Discuss the "Think About It."

Classroom

Explain how to size the fuses for an example motor circuit.

Discuss the application guidelines for motor branch circuits. Refer the trainees to *NEC Article 430, Part III.*

Audiovisual

Show Transparency 14 (Figure 14).

Classroom

Explain how to select the overload protection for oversized motors.

Audiovisual

Show Transparency 15 (Figure 15).

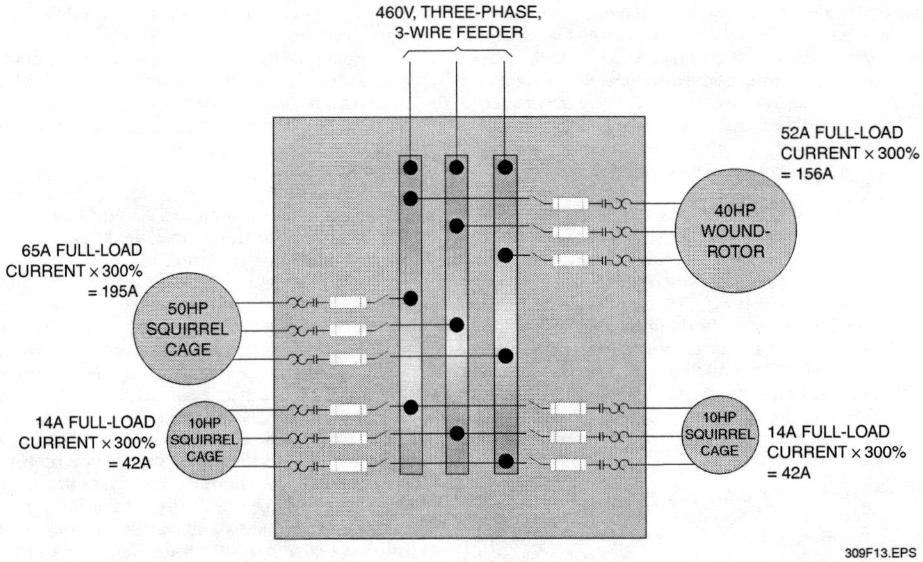

Figure 13 ◆ Ratings of nontime-delay fuses for typical motor circuits.

the rating of dual-element, time-delay fuses may be increased, but must never be more than 225% of the motor full-load current rating [*NEC Section 430.52(C)(1), Exception No. 2(b)*]. To size dual-element fuses at 175% for the four motors in *Figure 13,* proceed as follows:

50hp motor = 65A × 175% = 113.75A

40hp motor = 52A × 175% = 91A

10hp motors = 14A × 175% = 24.5A

Figure 14 gives general fuse application guidelines for motor branch circuits (*NEC Article 430, Part III*). Bear in mind that in many cases, the maximum fuse size depends on the motor design letter, motor type, and starting method.

4.1.0 Practical Applications

For various reasons, motors are often oversized. For instance, a 5hp motor may be installed when the load demand is only 3hp. In these cases, a much higher degree of overload protection can be obtained by sizing the overload relay elements and/or dual-element, time-delay fuses based on the actual full-load current draw. In existing installations, the procedure for providing the maximum overcurrent protection for oversized motors is as follows:

Step 1 With a clamp-on ammeter, determine the running rms current when the motor is at normal full-load, as shown in *Figure 15.* Be sure this current does not exceed the nameplate current rating. The advantage of this method is realized when a lightly loaded motor (especially those over 50hp) experiences a single-phase condition. Even though the relays and fuses may be sized correctly based on the motor nameplate, circulating currents within the motor may cause damage. If unable to meter the motor current, take the current rating off the motor nameplate.

Step 2 Size the overload relay elements and/or overcurrent protection based on this current.

Step 3 Use a labeling system to mark the type and ampere rating of the fuse that should be in the fuse clips. This simple system makes it easy to run spot checks for proper fuse replacements.

Instructor's Notes:

Type of Motor	Dual-Element, Time-Delay Fuses			Nontime-Delay Fuses
	Motor Overload and Short Circuit	**Backup Overload and Short Circuit**	**Short Circuit Only (Based on *NEC Tables 430.147 through 430.150* current ratings)**	**Short Circuit Only (Based on *NEC Tables 430.147 through 430.150* current ratings)**
Service Factor 1.15 or Greater or 40°C Temp. Rise or Less	125% or less of motor nameplate current	125% or next standard size (not to exceed 140% of motor nameplate current)	150% to 175%	150% to 300%
Service Factor Less Than 1.15 or Greater Than 40°C Temp. Rise	115% or less of motor nameplate current	115% or next standard size (not to exceed 130% of motor nameplate current)	150% to 175%	150% to 300%

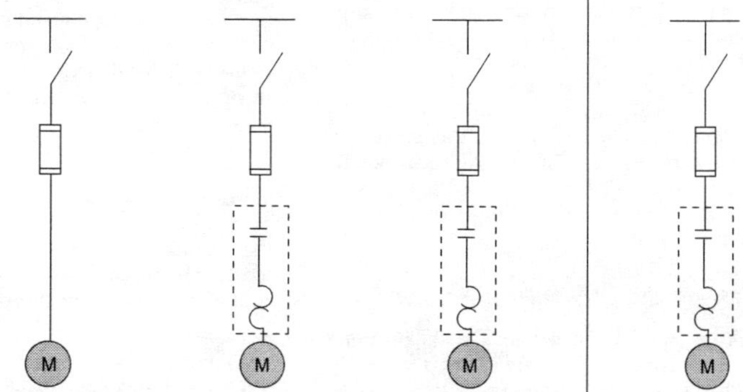

309F14.EPS

Figure 14 ◆ Fuse application guidelines for motor branch circuits.

> **Note**
> When installing the proper fuses in the switch to give the desired level of protection, it is often advisable to leave spare fuses on top of the disconnect or starter enclosure, or in a cabinet adjacent to the motor control center. This way, if the fuses open, the proper fuses can be readily reinstalled.

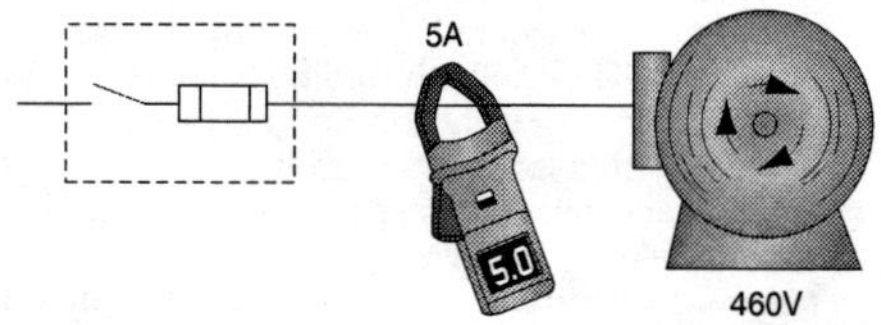

309F15.EPS

Figure 15 ◆ Determining running current with an ammeter.

Classroom

Provide example motor calculations for the trainees to solve.

Classroom

List the conditions that may require the use of a larger fuse than that sized.

Emphasize that motor overload protection is essential to prolonging equipment life. Point out that overcurrent devices must be sized to overcome the inrush current.

Audiovisual

Show Transparency 16 (Figure 16).

Classroom

Discuss the types of devices used to provide motor overload protection. Provide examples for the trainees to examine.

Audiovisual

Show Transparency 17 (Figure 17).

INSIDE TRACK

Fuses and Relay Elements

Make sure fuses are fully seated before energizing the device. If a fuse is not fully seated it may arc, causing damage to the fuseholder and creating a potential safety hazard. Also, make sure all relay elements of a three-phase motor are identical in their ratings before installing them.

Individual motor disconnect switches must have an ampere rating of at least 115% of the motor full-load ampere rating [*NEC Section 430.110(A)*] or sized as specified in *NEC Section 430.109*. The next larger size switches with fuse reducers may sometimes be required.

Some installations may require larger dual-element fuses when:

- The motor uses dual-element fuses in high ambient temperature environments.
- The motor is started frequently or rapidly reversed.
- The motor is directly connected to a machine that cannot be brought up to full speed quickly (e.g., centrifugal machines such as extractors and pulverizers, machines having large fly wheels such as large punch presses, etc.).
- The motor has a Design Letter E with full-voltage start.

4.2.0 Motor Overload Protection

A high-quality electric motor that is properly cooled and protected against overloads can be expected to have a long life. The goal of proper motor protection is to prolong motor life and postpone the failure that ultimately takes place. Good electrical protection consists of providing both proper overload protection and current-limiting short circuit protection. AC motors and other types of high inrush loads require protective devices with special characteristics. Normal, full-load running currents of motors are substantially less than the currents that result when motors start or are subjected to temporary mechanical overloads. This is illustrated by the typical motor starting current curve shown in *Figure 16*.

At the moment an AC motor circuit is energized, the starting current rapidly rises to many times the normal running current and the rotor begins to rotate. As the rotor accelerates and reaches running speed, the current declines to the normal running current. Thus, for a period of time, the overcurrent protective devices in the motor circuit must be able to tolerate the rather substantial temporary overload. Motor starting currents can vary substantially, depending on the motor type, load type, starting methods, and other factors. For the first half cycle, the momentary transient rms current may be 11 times the normal current, or even higher. After this first half cycle, the starting current subsides to 4 to 8 times (typically 6 times) the normal current and remains there for several seconds. This is called the locked-rotor current. When the motor reaches running speed, the current then subsides to its normal running level.

Motor overload protective devices must withstand the temporary overload caused by motor starting currents, and, at the same time, protect the motor from continuous or damaging overloads. The main types of devices used to provide motor overload protection include:

- Overload relays
- Fuses
- Circuit breakers

There are numerous causes of overloads, but if the overload protective devices are properly responsive, such overloads can be removed before damage occurs. To ensure this protection, the motor running protective devices should have time-current characteristics similar to motor damage curves but should be slightly faster. This is illustrated in *Figure 17*.

Note

Heaters (thermal overload relays) are discussed later in this module.

Instructor's Notes:

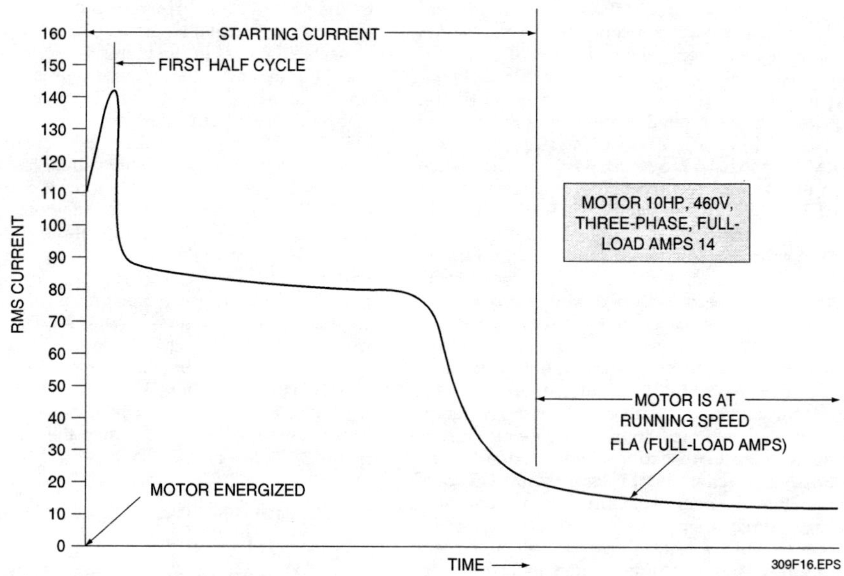

Figure 16 ◆ Motor starting current characteristics.

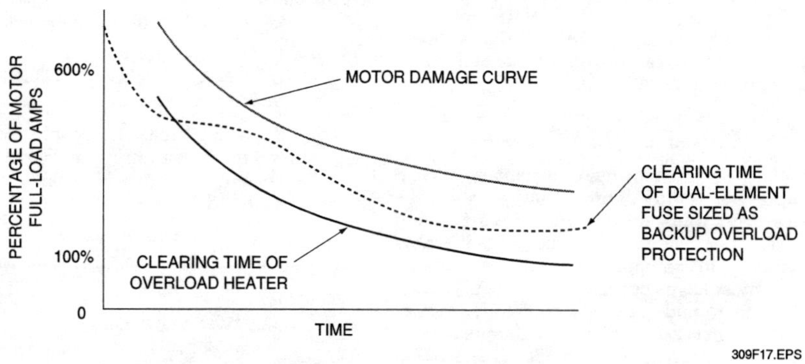

Figure 17 ◆ Time-current characteristics of dual-element fuses and overload heaters.

Audiovisual

Show Transparency 18 (Figure 18).

Classroom

Explain how to size the overcurrent protection for a motor circuit.

Discuss the NEC requirements for circuit breakers on motor branch circuits. Demonstrate the selection procedure for a magnetic-only circuit breaker.

Audiovisual

Show Transparencies 19 and 20 (Figures 19 and 20).

For example, we will take a 10hp motor and determine the proper circuit components that should be employed (refer to *Figure 18*).

To begin, select the proper size overload relays. Typically, the overload relay is rated to trip at about 115% of the rated current (in this case, 1.15 × 14A = 16.1A). The correct starter size (using NEMA standards) is a NEMA Type 1. The switch size that should be used is 30A. Switch sizes are based on NEC requirements; dual-element, time-delay fuses allow the use of smaller switches.

For short circuit protection on large motors with currents in excess of 600A, low-peak time-delay fuses are recommended. Most motors of this size will have reduced voltage starters, and the inrush currents are not as rigorous. Low-peak fuses should be sized at approximately 150% to 175% of the motor full-load current.

Motor controllers with overload relays commonly used on motor circuits provide motor running overload protection. The overload relay setting or selection must comply with *NEC Section 430.32*. On overload conditions, the overload relays should operate to protect the motor. For motor backup protection, size dual-element fuses at the next ampere rating greater than the overload relay trip setting. This can typically be achieved by sizing dual-element fuses at 125% for 1.15 **service factor** motors and 115% for 1.0 service factor motors. The service factor is the number by which the horsepower rating is multiplied to determine the maximum safe load that a motor may be expected to carry continuously at its rated voltage and frequency.

5.0.0 ◆ CIRCUIT BREAKERS

The NEC recognizes the use of instantaneous trip circuit breakers (without time delay) for short circuit protection of motor branch circuits. Such breakers are acceptable only if they are adjustable and are used in combination motor starters. Such starters must have coordinated overload, short-circuit, and ground-fault protection for each conductor and must be approved for the purpose in accordance with *NEC Section 430.52(C)(3)*. This permits the use of smaller circuit breakers than would be allowed if a standard thermal-magnetic circuit breaker was used. In this case, smaller circuit breakers offer faster operation for greater protection against grounds and short circuits. *Figure 19* shows a schematic diagram of magnetic-only circuit breakers used in a combination motor starter.

The use of magnetic-only circuit breakers in motor branch circuits requires careful consideration due to the absence of overload protection up to the short circuit trip rating that is normally available in thermal elements in circuit breakers. However, heaters in the motor starter protect the entire circuit and all equipment against overloads up to and including locked-rotor current. Heaters (thermal overload relays) are commonly set at 115% to 125% of the motor full-load current.

In dealing with such circuits, an adjustable circuit breaker can be set to take over the interrupting task at currents above locked-rotor current and up to the short circuit duty of the supply system at the point of the installation. The magnetic trip in such breakers can typically be adjusted from 3 to 13 times the breaker current rating. For example, a 100A circuit breaker can be adjusted to trip anywhere between 300A and 1,300A. Consequently, the circuit breaker may serve as motor short circuit protection. However, instantaneous trip circuit breakers used in these installations cannot be adjusted to more than the value specified in *NEC Table 430.52*.

5.1.0 Application of Magnetic-Only Circuit Breakers

We will compare the use of both thermal-magnetic and magnetic-only circuit breakers in the motor circuit shown in *Figure 20*. In doing so, our job is to select a circuit breaker that will provide short circuit protection and also qualify as the motor circuit disconnecting means.

Step 1 Determine the motor full-load current from *NEC Table 430.150*. This is found to be 80A.

Step 2 A circuit breaker suitable for use as a motor disconnecting means must have a current rating of at least 115% of the motor full-load current. Therefore:

1.15 × 80A = 92A

> **Note**
>
> ***NEC Table 430.52*** permits the use of an inverse-time circuit breaker rated at not more than 250% of the motor full-load current. However, a circuit breaker could be rated as high as 400% of the motor full-load current if necessary to hold the motor starting current without opening, according to ***NEC Section 430.52(C)(1), Exception 2(c)***.

Step 3 Assuming that a circuit breaker rated at 250% of the motor full-load current will be used, perform the following calculation:

2.5 × 80A = 200A

Instructor's Notes:

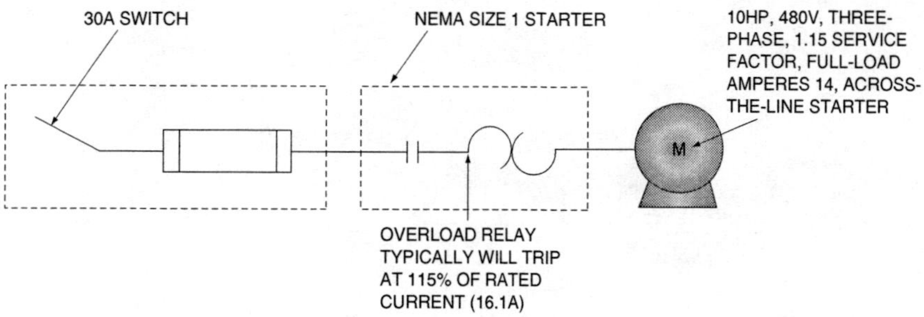

Figure 18 ◆ Circuit components of a typical 10-horsepower motor.

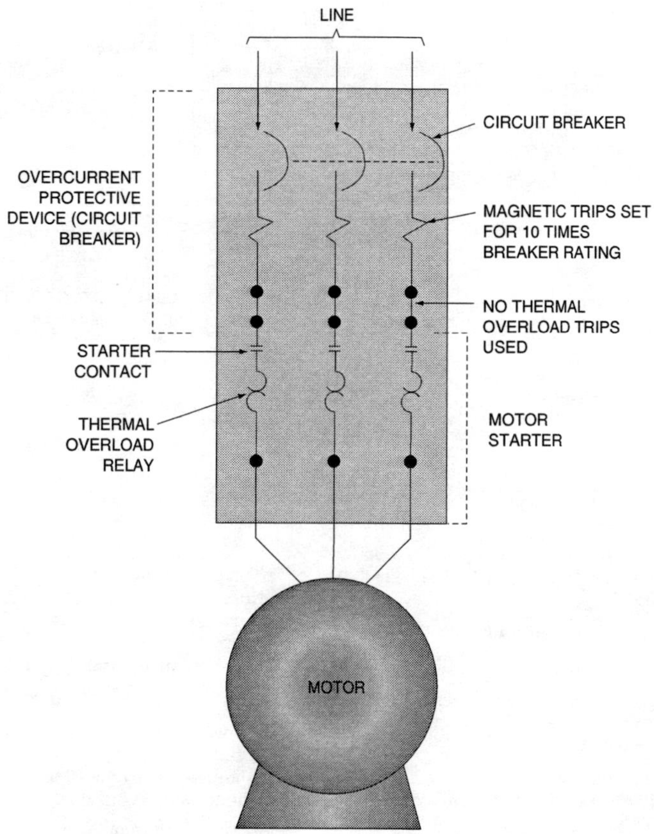

Figure 19 ◆ Combination motor starter with magnetic-only circuit breakers.

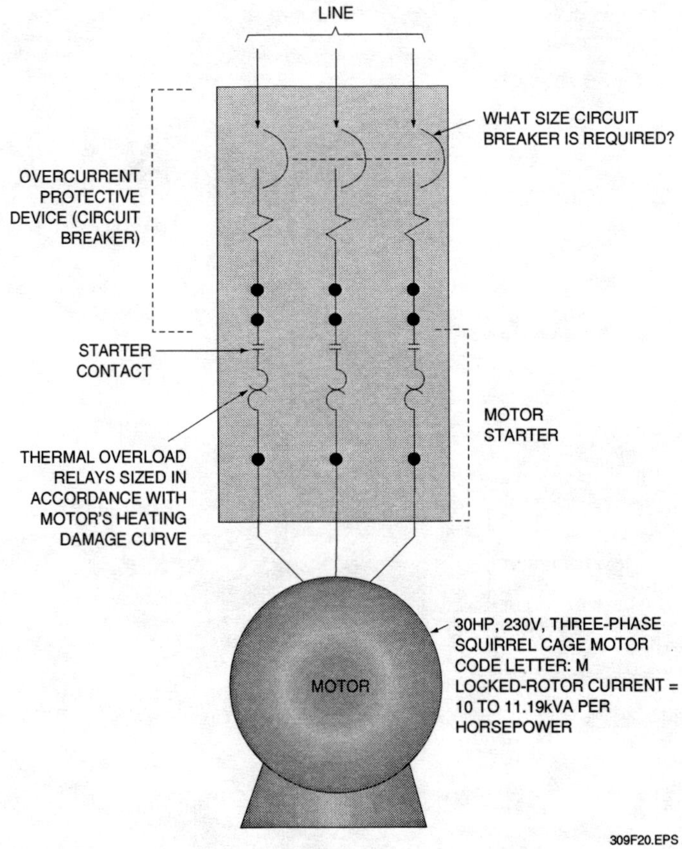

Figure 20 ◆ Typical 30-horsepower motor circuit.

Step 4 Select a regular thermal-magnetic circuit breaker with a 225A frame that is set to trip at 200A.

Step 5 Refer to *Figure 20* and note that this is a NEMA Design Letter E motor. Refer to ***NEC Table 430.52*** and note that an instantaneous breaker for a Design Letter E motor should be 1,100% of the motor full-load current.

Step 6 Determine the circuit breaker rating by multiplying the full-load current by 1,100%:

$$80\text{A} \times 1{,}100\%\ (11.00) = 880\text{A}$$

The thermal-magnetic circuit breaker selected in Step 4 will provide protection for grounds and short circuits without interfering with motor overload protection. Note, however, that the instantaneous trip setting of a 200A circuit breaker will be about 10 times the current rating, or:

$$200\text{A} \times 10 = 2{,}000\text{A}$$

Now consider the use of a 100A circuit breaker with thermal and adjustable magnetic trips. The instantaneous trip setting at 10 times the normal current rating would be:

$$100\text{A} \times 10 = 1{,}000\text{A}$$

Instructor's Notes:

Although this 1,000A instantaneous trip setting is above the 880A locked-rotor current of the 30hp motor in question, the starting current would probably trip the thermal element and open the circuit breaker.

This problem can be solved by removing the circuit breaker's thermal element and leaving only the magnetic element in the circuit breaker. Then the conditions of overload can be cleared by the overload devices (heaters) in the motor starter. If the setting of the instantaneous trip circuit breaker will not hold under the starting load in *NEC Section 430.52(C)(3)*, then *Exception No. 1* will, under engineering evaluation, permit increasing the trip setting up to, but not exceeding, 1,300% (1,700% for NEMA Design Letter E).

Therefore, since it has been determined that the 30hp motor in question has a full-load ampere rating of 80A, the maximum trip must not be set higher than:

$$80\text{A} \times 17 = 1{,}360\text{A or about } 1{,}300\text{A}$$

This circuit breaker would qualify as the circuit disconnect because it has a rating higher than 115% of the motor full-load current (80A × 1.15 = 92A). However, the use of a magnetic-only circuit breaker does not protect against low-level grounds and short circuits in the branch circuit conductors on the line side of the motor starter overload relays—such an application must be made only where the circuit breaker and motor starter are installed as a combination motor starter in a single enclosure.

5.2.0 Motor Short Circuit Protectors

Motor short circuit protectors (MSCPs) are fuse-like devices designed for use only in a special type of fusible-switch combination motor starter. The combination offers short circuit protection, overload protection, disconnecting means, and motor control, all with assured coordination between the short **circuit interrupter** and the overload devices.

The NEC recognizes MSCPs in *NEC Section 430.40 and 430.52(C)(7)*, provided the combination is identified for the purpose (i.e., a combination motor starter equipped with an MSCP and listed by Underwriters Laboratories or another nationally recognized third-party testing lab as a package called an *MSCP starter*).

6.0.0 ◆ MULTI-MOTOR BRANCH CIRCUITS

NEC Sections 430.53(A), (B), and (C) permit the use of more than one motor on a branch circuit, provided the following conditions are met:

- Two or more motors, each rated at not more than 1hp, and each drawing a full-load current not exceeding 6A, may be used on a branch circuit protected at not more than 20A at 120V or less, or 15A at 600V or less. The rating of the branch circuit protective device marked on any of the controllers must not be exceeded. Individual overload protection is necessary in such circuits unless the motor is not permanently installed, or is manually started and is within sight of the controller location, or has sufficient winding impedance to prevent overheating due to locked-rotor current, or is part of an approved assembly which does not subject the motor to overloads and which incorporates protection for the motor against locked-rotor current, or the motor cannot operate continuously under load.
- Two or more motors of any rating, each having individual overload protection, may be connected to a single branch circuit that is protected by a short circuit protective device (MSCP). The protective device must be selected in accordance with the maximum rating or setting that could protect an individual circuit to the motor of the smallest rating. This may be done only where it can be determined that the branch circuit device so selected will not open under the most severe normal conditions of service that might be encountered. This NEC section offers wide application of more than one motor on a single circuit, particularly in the use of small integral-horsepower motors installed on 208V, 240V, and 480V, three-phase industrial and commercial systems. Only such three-phase motors have full-load operating currents low enough to permit more than one motor on circuits fed from 15A protective devices.

Classroom

Discuss motor short circuit protectors and provide an example for the trainees to examine. Point out that their use is limited to special applications.

List the NEC requirements for multi-motor branch circuits.

Classroom

Show the trainees how to calculate the overcurrent protection and conductor size for multi-motor circuits.

Audiovisual

Show Transparencies 21 and 22 (Figures 21 and 22).

Using these NEC rules, we will take a typical branch circuit (*Figure 21*) with more than one motor connected and see how the calculations are made.

The full-load current of each motor is taken from *NEC Table 430.150* as required by *NEC Section 430.6(A).* A circuit breaker must be chosen that does not exceed the maximum value of short circuit protection (250%) required by *NEC Section 430.52* and *NEC Table 430.52* for the smallest motor in the group (in this case, 1.5hp). Since the listed full-load current for the smallest motor (1.5hp) is 3A, the calculation is made as follows:

$$3A \times 2.5\ (250\%) = 7.5A$$

> **Note**
>
> ***NEC Section 430.52, Exception No. 1*** allows the next higher size rating or setting for a standard circuit breaker. Since a 15A circuit breaker is the smallest standard rating recognized by ***NEC Section 240.6,*** a 15A, three-pole circuit breaker may be used.

The total load of the motor currents must be calculated as follows:

$$4.8A + 3.4A + 3.0A = 11.2A$$

The total full-load current for the three motors (11.2A) is well within the 15A circuit breaker rating, which has a sufficient time delay in its operation to permit starting of any one of these motors with the other two already operating. The torque characteristics of the loads on starting are not high. Therefore, the circuit breaker will not open under the most severe normal service.

Make certain that each motor is provided with the properly rated individual overload protection in the motor starter.

Branch circuit conductors are sized in accordance with *NEC Section 430.24.* In this case:

$$4.8A + 3.4A + 3.0A + [25\% \text{ of the largest motor } (4.8A \times 0.25 = 1.2A)] = 12.4A$$

No. 14 AWG conductors rated at 75°C will fully satisfy this application, as long as the overcurrent protection device does not exceed 15A, according to *NEC Section 240.4(D).*

Another multi-motor situation is shown in *Figure 22.* In this case, smaller motors are used. In general, *NEC Section 430.53(B)* requires branch circuit protection to be no greater than the maximum amperes permitted by *NEC Section 430.52* for the lowest rated motor of the group, which in this case is 1.1A for the ½hp motors. With this information in mind, we will size the circuit components for this application.

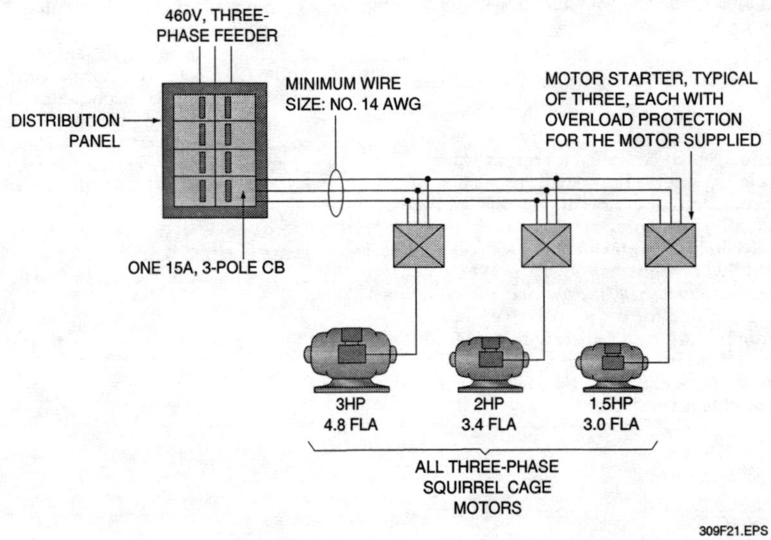

Figure 21 ◆ Several motors on one branch circuit.

Instructor's Notes:

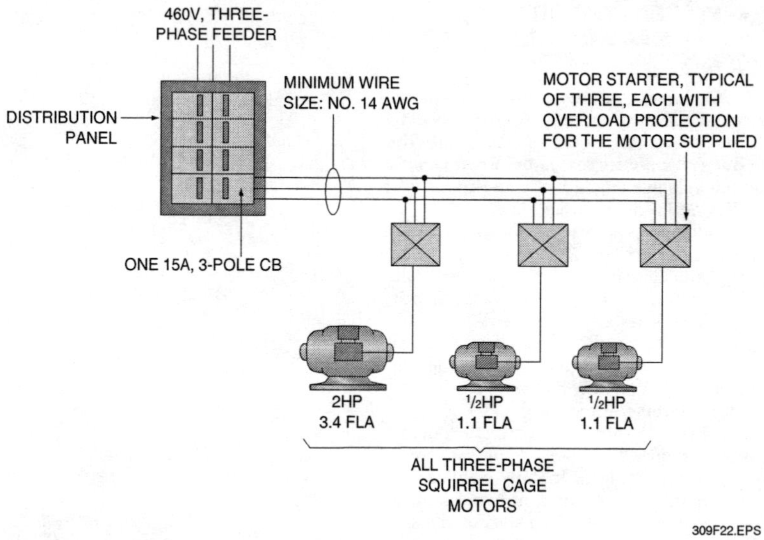

Figure 22 ◆ Several smaller motors supplied by one branch circuit.

> **THINK ABOUT IT**
>
> ### *Motor Branch Circuit Requirements*
>
> If the motors in *Figure 22* were replaced with 5hp, 3hp, and 2hp squirrel cage motors, what size feeder and circuit breaker would be needed to accommodate the changes?

From ***NEC Section 430.52*** and ***NEC Table 430.52,*** the maximum protection rating for a circuit breaker is 250% of the lowest rated motor. Since this rating is 1.1A, the calculation is performed as follows:

$$2.5 \times 1.1A = 2.75A$$

> **Note**
>
> Since 2.75A is not a standard rating for a circuit breaker (***NEC Section 240.6***), ***NEC Section 430.52(C)(1), Exception No. 1*** permits the use of the next higher rating. Because 15A is the lowest standard rating of circuit breakers, it is the next higher device rating above 2.75A and satisfies NEC rules governing the rating of the branch circuit protection.

These two previous applications permit the use of several motors up to the circuit capacity, based on ***NEC Sections 430.24 and 430.53(B)*** and on starting torque characteristics, operating duty cycles of the motors and their loads, and the time delay of the circuit breaker. Such applications greatly reduce the number of circuit breakers and panels and the amount of wire used in the total system. One limitation, however, is placed on this practice in ***NEC Section 430.52(C)(2),*** which specifies that where maximum branch circuit short circuit and ground fault protective device ratings are shown in the manufacturer's overload relay table for use with a motor controller or are otherwise marked on the equipment, they shall not be exceeded even if higher values are allowed, as shown in the preceding examples.

Teaching Tip

Discuss the "Think About It."

Classroom

Discuss the limitations imposed by *NEC Section 430.52(C)(2).*

Classroom

Explain how to make corrections for power factor. Refer the trainees to Table 1 in the Trainee Module.

Audiovisual

Show Transparency 23 (Figure 23).

7.0.0 ◆ POWER FACTOR CORRECTION AT MOTOR TERMINALS

Generally, the most effective method of power factor correction is the installation of capacitors at the cause of the poor power factor—the induction motor. This not only increases the power factor, but also releases system capacity, improves voltage stability, and reduces power losses.

When power factor correction capacitors are used, the total corrective kVAR on the load side of the motor controller should not exceed the value required to raise the no-load power factor to unity. Corrective kVAR in excess of this value may cause over-excitation that results in high transient voltages, currents, and torques that can increase safety hazards to personnel and possibly damage the motor or driven equipment.

Do not connect power factor correction capacitors at motor **terminals** on elevator motors; multi-speed motors; plugging or jogging applications; or open transition, wye-delta, autotransformer starting, and some part-winding start motors.

If possible, capacitors should be located at position No. 2, as shown in *Figure 23.* This does not change the current flowing through the motor overload protectors.

The connection of capacitors at position No. 3 requires a change of overload protectors. Capacitors should be located at position No. 1 for any of the following applications:

- Elevator motors
- Multi-speed motors
- Plugging or jogging applications
- Open transition, wye-delta, autotransformer starting motors
- Some part-winding motors

Note

To avoid over-excitation, make sure that the bus power factor is not increased above 95% under all loading conditions.

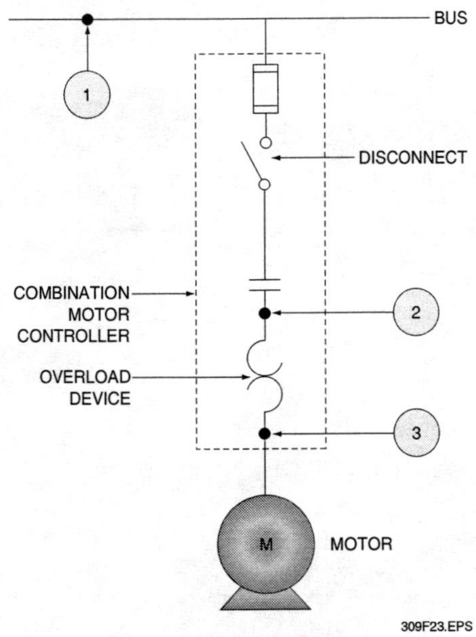

Figure 23 ◆ Placement of capacitors in motor circuit.

Table 1 allows the determination of corrective kVAR required where capacitors are individually connected at motor leads. These values should be considered the maximum capacitor rating when the motor and capacitor are switched as a unit. The figures given are for three-phase, 60Hz, NEMA Class B motors to raise the full-load power factor to 95%.

Instructor's Notes:

Table 1 Motor Power Factor Correction Table

Induction Motor Horsepower Rating	3600 Capacitor Rating kVAR	3600 Line Current Reduction %	1800 Capacitor Rating kVAR	1800 Line Current Reduction %	1200 Capacitor Rating kVAR	1200 Line Current Reduction %	900 Capacitor Rating kVAR	900 Line Current Reduction %	720 Capacitor Rating kVAR	720 Line Current Reduction %	600 Capacitor Rating kVAR	600 Line Current Reduction %
3	1.5	14	1.5	15	1.5	20	2	27	2.5	35	3.5	41
5	2	12	2	13	2	17	3	25	4	32	4.5	37
7½	2.5	11	2.5	12	3	15	4	22	5.5	30	6	34
10	3	10	3	11	3.5	14	5	21	6.5	27	7.5	31
15	4	9	4	10	5	13	6.5	18	8	23	9.5	27
20	5	9	5	10	6.5	12	7.5	16	9	21	12	25
25	6	9	6	10	7.5	11	9	15	11	20	14	23
30	7	8	7	9	9	11	10	14	12	18	16	22
40	9	8	9	9	11	10	12	13	15	16	20	20
50	12	8	11	9	13	10	15	12	19	15	24	19
60	14	8	14	8	15	10	18	11	22	15	27	19
75	17	8	16	8	18	10	21	10	26	14	32.5	18
100	22	8	21	8	25	9	27	10	32.5	13	40	17
125	27	8	26	8	30	9	32.5	10	40	13	47.5	16
150	32.5	8	30	8	35	9	37.5	10	47.5	12	52.5	15
200	40	8	37.5	8	42.5	9	47.5	10	60	12	65	14
250	50	8	45	7	52.5	8	57.5	9	70	11	77.5	13

Classroom

Have the trainees complete the Review Questions and go over the answers prior to administering the Module Examination.

Examination

Administer the Module Examination. Record the results on Craft Training Report Form 200 and submit the results to the Training Program Sponsor.

Summary

The NEC plays an important role in the selection and application of motors, including branch circuit conductors, disconnects, controllers, overcurrent protection, and overload protection. For example, *NEC Article 430* covers the application and installation of motor circuits and motor control connections, including conductors, short circuit and ground fault protection, controllers, disconnects, and overload protection.

NEC Article 440 contains provisions for motor-driven air conditioning and refrigerating equipment, including the branch circuits and controllers for the equipment. It also takes into account the special considerations involved with sealed (hermetic) motor compressors, in which the motor operates under the cooling effect of the refrigeration. When referring to *NEC Article 440,* be aware that the rules in this article are in addition to, or are amendments to, the rules given in *NEC Article 430.*

Review Questions

1. Stator windings can best be described as ____.
   a. a structure of copper or aluminum wire coils
   b. insulating fibers mounted on a shaft
   c. copper bars mounted on a spindle
   d. steel wires mounted on a shaft

2. The synchronous speed of a 60Hz, three-phase induction motor with two poles is ____.
   a. 3,600 rpm
   b. 1,800 rpm
   c. 1,200 rpm
   d. 900 rpm

3. The two most common three-phase motor configurations are ____.
   a. synchronous and rms
   b. box and star
   c. wye and delta
   d. star and wye

4. The most common number of motor leads found on a three-phase, wye-wound motor is ____.
   a. 3
   b. 6
   c. 9
   d. 12

5. Doubling the voltage on a dual-voltage motor ____.
   a. doubles the synchronous speed
   b. doubles the full-load current
   c. halves the full-load current
   d. halves the synchronous speed

6. The total resistance of the stator windings in a three-phase motor, if the windings are connected in parallel and the resistance of each winding is 96Ω, is ____.
   a. 48Ω
   b. 96Ω
   c. 192Ω
   d. 220Ω

7. If a squirrel cage induction motor draws 2A of current at 240V, the amperage will be ____ if connected for use on 480V.
   a. 1A
   b. 2A
   c. 3A
   d. 4A

8. Of the following motors, the one most likely to have 15 or 18 motor leads is the ____.
   a. single-phase capacitor-start motor
   b. 120V shaded-pole motor
   c. 480V three-phase squirrel cage motor
   d. 2,400V three-phase motor

9. The main purpose of motor overload protection is to protect the motor ____.
   a. against short circuits
   b. against ground faults
   c. from damaging overload currents
   d. against locked-rotor current

10. A circuit breaker suitable for use as a motor disconnecting means must have a current rating of at least ____% of the motor full-load current.
   a. 250
   b. 175
   c. 300
   d. 115

Instructor's Notes:

GLOSSARY

Trade Terms Introduced in This Module

Circuit interrupter: A non-automatic, manually operated device designed to open a current-carrying circuit without injury to itself.

Rating: A designated limit of operating characteristics based on definite conditions. Such operating characteristics as load, voltage, frequency, etc., may be given in the rating.

Service factor: The number by which the horsepower rating is multiplied to determine the maximum safe load that a motor may be expected to carry continuously at its rated voltage and frequency.

Terminal: A point at which an electrical component may be connected to another electrical component.

Torque: A force that produces or tends to produce rotation. Common units of measurement of torque are foot-pounds and inch-pounds.

Additional Resources

This module is intended to present thorough resources for task training. The following reference works are suggested for further study. These are optional materials for continued education rather than for task training.

American Electrician's Handbook. Terrell Croft and Wilfred I. Summers. New York, NY: McGraw-Hill, 1996.

National Electrical Code Handbook. Quincy, MA: National Fire Protection Association, 2002.

Instructor's Notes:

ANSWER KEY

Answers to Review Questions

Answer	Section
1. a	2.0.0
2. a	2.1.0
3. c	2.2.0
4. c	2.2.0
5. c	2.2.1
6. a	2.2.1
7. a	2.2.1
8. d	2.3.0
9. c	4.0.0
10. d	5.1.0

Section 3.0.0 — *Think About It—NEC Motor Sizes*

The motor's FLA is used based on *NEC Tables 430.148* through ***430.150.***

Section 3.0.0 — *Think About It—Feeder Size*

The 2/0 THWN feeders are still adequate because they are rated for 175A and changing the 40hp motor to a 50hp motor results in a conductor ampacity of 174.25A.

$$(1.25 \times 65\text{A}) + 65\text{A} + 14\text{A} + 14\text{A} = 174.25\text{A}$$

Section 4.0.0 — *Think About It—Overload Protection*

A larger overload protection device will be required for the 208V application due to the corresponding increase in FLA. Remember that the current increases as the voltage decreases per Ohm's law.

Section 6.0.0 — *Think About It—Motor Branch Circuit Requirements*

FLAs:

5hp = 7.6A

3hp = 4.8A

2hp = 3.4A

Total 15.8A

The circuit breaker selected cannot exceed the maximum value of short circuit protection (250%) for the smallest motor FLA. Therefore:

$$3.4\text{A} \times 2.50 = 8.5\text{A}$$

Since 8.5A is not a standard size, a 15A circuit breaker may be installed, but a 20A could not be used since it would represent an increase of two sizes.

Feeders:

$$(1.25 \times 7.6\text{A}) + 4.8\text{A} + 3.4\text{A} = 17.7\text{A}$$

NEC Table 310.16 indicates that No. 14 THWN would be the minimum feeder size, as long as the circuit breaker did not exceed 15A, which it does not in this case.

CONTREN™ LEARNING SERIES — USER UPDATES

The NCCER makes every effort to keep these textbooks up-to-date and free of technical errors. We appreciate your help in this process. If you have an idea for improving this textbook, or if you find an error, a typographical mistake, or an inaccuracy in NCCER's Contren™ textbooks, please write us, using this form or a photocopy. Be sure to include the exact module number, page number, a detailed description, and the correction, if applicable. Your input will be brought to the attention of the Technical Review Committee. Thank you for your assistance.

Instructors – If you found that additional materials were necessary in order to teach this module effectively, please let us know so that we may include them in the Equipment/Materials list in the Instructor's Guide.

Write: Curriculum Revision and Development Department
National Center for Construction Education and Research
P.O. Box 141104, Gainesville, FL 32614-1104

Fax: 352-334-0932

E-mail: curriculum@nccer.org

Craft ______ Module Name ______

Copyright Date ______ Module Number ______ Page Number(s) ______

Description

______

______

______

(Optional) Correction

______

______

(Optional) Your Name and Address

______

______

Motor Maintenance, Part One

26310-02

Motor Maintenance, Part One
Annotated Instructor's Guide

Module 26310-02

MODULE OVERVIEW

This module introduces the Electrical trainee to the methods and procedures used in motor maintenance.

PREREQUISITES

Please refer to the Course Map in the Trainee Module. Prior to training with this module, it is recommended that the trainee shall have successfully completed the following:

Core Curriculum; Electrical Level One; Electrical Level Two; Electrical Level Three, Modules 26301 through 26309.

OBJECTIVES

Upon completion of this module, the trainee will be able to:

1. Properly store motors and generators.
2. Test motors and generators.
3. Make connections for specific types of motors and generators.
4. Clean open-frame motors.
5. Lubricate motors that require this type of maintenance.
6. Collect and record motor data.
7. Select tools for motor maintenance.
8. Select instruments for motor testing.

PERFORMANCE TASKS

Under supervision of the instructor, the trainee should be able to:

1. Using an untagged three-phase motor without a nameplate, determine as many motor characteristics as possible.
2. Sketch and/or physically connect the motor leads in the above motor for operation on its lower voltage.
3. Sketch and/or physically connect the motor leads of the same motor for operation on its higher voltage.

NCCER STANDARDIZED CRAFT TRAINING PROGRAMS

The National Center for Construction Education and Research (NCCER) provides a standardized national program of accredited craft training. Key features of the program include instructor certification, competency-based training, and performance testing. The program provides trainees, instructors, and companies with a standard form of recognition through a National Craft Training Registry. The program is described in full in the *Guidelines for Accreditation*, published by the NCCER. For more information on standardized craft training, contact the NCCER by writing us at P.O. Box 141104, Gainesville, FL 32614-1104, calling 352-334-0911, or e-mailing info@nccer.org. More information may be found at our Web site, www.nccer.org.

HOW TO USE THIS ANNOTATED INSTRUCTOR'S GUIDE

Each page presents two sections of information. The larger section displays each page exactly as it appears in the Trainee Module. The narrow column ties suggested trainee and instructor actions to each page and provides icons to call your attention to material, safety, audiovisual, or testing requirements. The bottom of each page includes space for your notes.

Teaching Tip

If you see the Teaching Tip icon, that means there is a teaching tip associated with this section. Also refer to any suggested teaching tips at the end of the module.

SAFETY CONSIDERATIONS

Ensure that the trainees are equipped with appropriate personal protective equipment.

PREPARATION

Before teaching this module, you should review the Module Outline, Objectives, Performance Tasks, and the Materials and Equipment List. Be sure to allow ample time to prepare your own training or lesson plan and gather all required equipment and materials.

MATERIALS AND EQUIPMENT LIST

Materials:
Transparencies
Markers/chalk
Copy of the latest edition of the *National Electrical Code*®
Bearing grease
Small compass
Module Examinations*
Performance Profile Sheets*

Equipment:
Overhead projector and screen
Whiteboard/chalkboard
Appropriate personal protective equipment
Motor end bells with lubrication filler and drain plugs
Grease gun
Various types of motor bearings
Sample motors for disassembly
Defective motors for troubleshooting
Motor testing instruments
Untagged three-phase motor

*Located in the Test Booklet packaged with this Annotated Instructor's Guide.

ADDITIONAL RESOURCES

This module is intended to present thorough resources for task training. The following reference works are suggested for both instructors and motivated trainees interested in further study. These are optional materials for continued education rather than for task training.

American Electrician's Handbook, Terrell Croft and Wilfred I. Summers. New York, NY: McGraw-Hill, 1996.

National Electrical Code Handbook. Quincy, MA: National Fire Protection Association, 2002.

TEACHING TIME FOR THIS MODULE

An outline for use in developing your lesson plan is presented below. Note that each Roman numeral in the outline equates to one session of instruction. Each session has a suggested time period of 2½ hours. This includes 10 minutes at the beginning of each session for administrative tasks and one 10-minute break during the session. Approximately 12½ hours are suggested to cover *Motor Maintenance, Part One.* You will need to adjust the time required for hands-on activity and testing based on your class size and resources.

Topic	Planned Time
Session I. Introduction; Practical Maintenance Techniques; Motor Bearing Maintenance	
A. Introduction	________
B. Practical Maintenance Techniques	________
C. Motor Bearing Maintenance	________
1. Types of Bearings	________
2. Frequency of Lubrication	________
a. Lubrication Procedure	________
3. Testing Bearings	________

Session II. Troubleshooting Motors

A. Troubleshooting Motors __________

1. Tools for Troubleshooting __________
2. Grounded Coils __________
3. Shorted Coils __________
4. Open Coils __________
5. Reversed Connections __________
6. Reversed Phase __________

Session III. Troubleshooting Split-Phase Motors; Storing Motors; Identifying Motors

A. Troubleshooting Split-Phase Motors __________

B. Storing Motors __________

C. Identifying Motors __________

1. Three-Phase, Delta-Wound Motors __________
2. Recordkeeping __________

Session IV. Laboratory

A. Laboratory __________

Under instructor supervision, have the trainees identify as many characteristics as possible on an untagged motor, then sketch and/or physically connect the motor for operation at both voltages. Note the proficiency of each trainee.

Session V. Review; Module Examination; Performance Testing

A. Review __________

B. Module Examination __________

1. Trainees must score 70% or higher to receive recognition from the NCCER.
2. Record the testing results on Craft Training Report Form 200 and submit the results to the Training Program Sponsor.

C. Performance Testing __________

1. Trainees must perform each task to the satisfaction of the instructor to receive recognition from the NCCER.
2. Record the testing results on Craft Training Report Form 200 and submit the results to the Training Program Sponsor.

Module 26310-02

Motor Maintenance, Part One

Course Map

This course map shows all of the modules in the third level of the Electrical curriculum. The suggested training order begins at the bottom and proceeds up. Skill levels increase as you advance on the course map. The local Training Program Sponsor may adjust the training order.

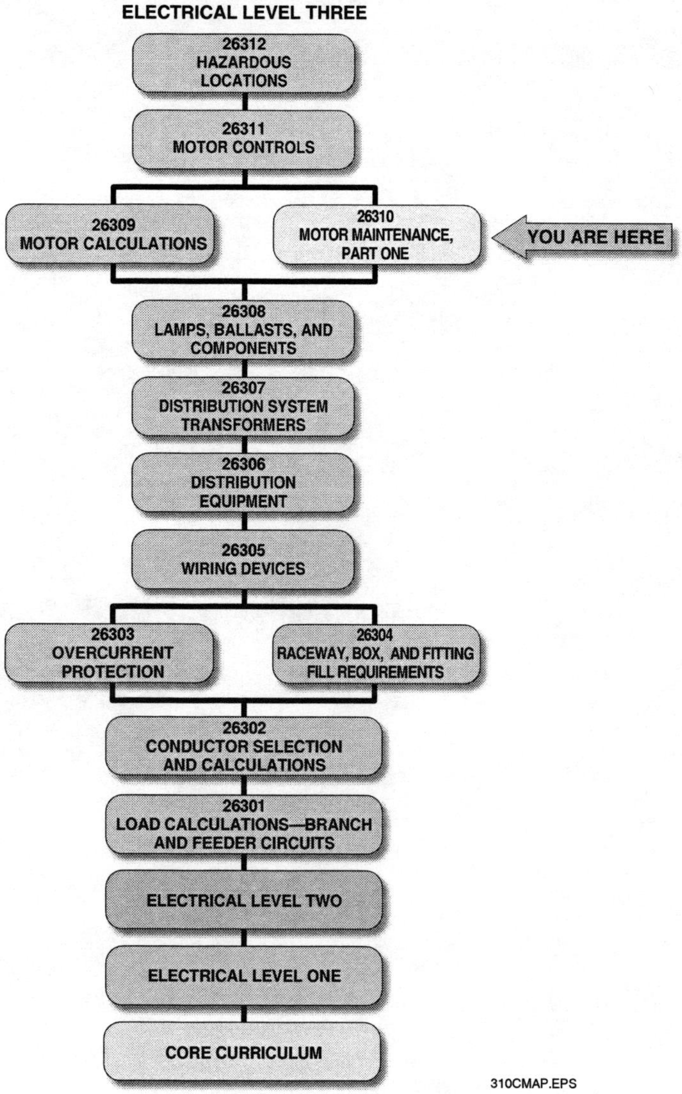

310CMAP.EPS

Homework

Assign reading of Module 26310.

MODULE 26310 CONTENTS

Figures

Tables

Instructor's Notes:

MODULE 26310

Motor Maintenance, Part One

Objectives

When you have completed this module, you will be able to do the following:

1. Properly store motors and generators.
2. Test motors and generators.
3. Make connections for specific types of motors and generators.
4. Clean open-frame motors.
5. Lubricate motors that require this type of maintenance.
6. Collect and record motor data.
7. Select tools for motor maintenance.
8. Select instruments for motor testing.

Prerequisites

Before you begin this module, it is recommended that you successfully complete the following: Core Curriculum; Electrical Level One; Electrical Level Two; Electrical Level Three, Modules 26301 through 26309.

Required Trainee Materials

1. Pencil and paper
2. Appropriate personal protective equipment
3. Copy of the latest edition of the *National Electrical Code*

1.0.0 ◆ INTRODUCTION

AC motor failure accounts for a high percentage of electrical repair work. The care given to an electric motor while it is being stored and operated affects the life and usefulness of the motor. A motor that receives good maintenance will outlast a poorly treated motor many times over. Actually, if a motor is initially installed correctly and has been properly selected for the job, very little maintenance is necessary—provided it does receive a little care at regular intervals. The basic care consists of cleaning and lubrication.

Cleaning–The frequency for cleaning an AC motor depends on the type of environment in which it is used. In general, keep both the interior and exterior of the motor free from dirt, water, oil, and grease. Motors operating in dirty areas should be periodically disassembled and thoroughly cleaned.

If the motor is totally enclosed (fan-cooled or nonventilated), such as the one shown in *Figure 1*, and is equipped with automatic drain plugs, they should be free of oil, grease, paint, grit, and dirt so they do not clog up.

Figure 1 ◆ Totally enclosed, fan-cooled motor.

Note: The designations "National Electrical Code," "NE Code," and "NEC," where used in this document, refer to the National Electrical Code®, which is a registered trademark of the National Fire Protection Association, Quincy, MA. *All National Electrical Code (NEC) references in this module refer to the 2002 edition of the NEC.*

Materials

Ensure that you have everything required to teach the course. Check the Materials and Equipment List at the front of this Instructor's Guide.

Classroom

Review the goals of the course.

Explain that terms shown in bold (blue) are defined in the Glossary at the back of this module.

Audiovisual

Show Transparency 1, Course Objectives.

Show Transparency 2, Performance Tasks.

Classroom

Explain that motor failures can often be prevented by good maintenance. Note that the two basic maintenance tasks are cleaning and lubrication.

Point out that the plugs on enclosed motors must be kept clean to avoid clogging. Refer trainees to Figure 1 in the Trainee Module.

Classroom

Explain that motors that have been in storage for six months or more should be relubricated before they are put into service.

Provide an overview of the procedure for lubricating bearings.

Audiovisual

Show Transparency 3 (Figure 2).

Classroom

Point out that motors also require periodic relubrication while in service. Refer the trainees to Tables 1 through 3 in the Trainee Module.

Lubrication–Most motors are properly lubricated at the time of manufacture, and it is not necessary to lubricate them at the time of installation. However, if a motor has been in storage for a period of six months or longer, it should be relubricated before starting.

To lubricate conventional motors with ball bearings:

Step 1 Stop the motor.

Step 2 Wipe clean all grease fittings (filler and drain).

Step 3 Remove the filler and drain plugs (A and B in *Figure 2*).

Step 4 Free the drain hole of any hard grease (use a piece of wire if necessary).

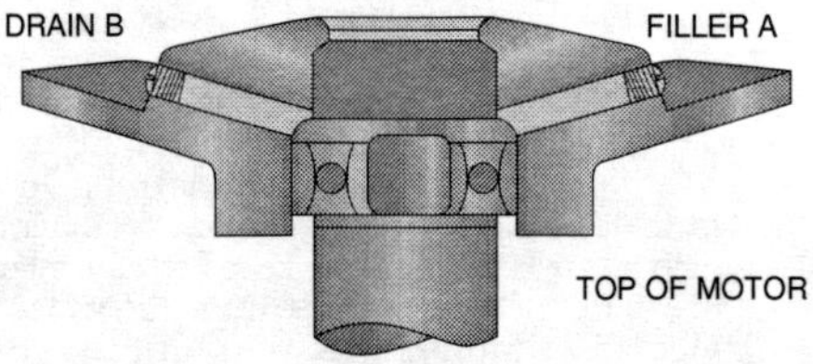

LOCATION OF FILLER AND DRAIN PLUGS IN END-BELL BEARINGS OF STANDARD INDUCTION MOTOR

310F02.EPS

Figure 2 ◆ Location of motor filler and drain plugs.

Step 5 Add grease using a low-pressure grease gun.

Step 6 Start the motor and let it run for approximately 30 minutes.

Step 7 Stop the motor, wipe off any drained grease, and replace the filler and drain plugs.

Step 8 The motor is now ready for operation.

Every four years (every year in the case of severe duty), motors with open bearings should be thoroughly cleaned, washed, and repacked with grease. *Table 1* shows the relubrication periods for various motors. Standard conditions mean operation of 8 hrs./day, normal or light loading, and at 100°F maximum ambient temperature. Severe conditions represent operation of 24 hrs./day, shock loadings, vibration, dirty or dusty areas, or areas at 100°F to 150°F ambient temperature. Extreme conditions are defined as heavy shock or vibration, large amounts of dirt or dust, or high ambient temperatures (above 150°F).

Bearings provide minimum resistance and align the motor rotor during operation. Therefore, the quantity of grease is important for proper bearing operation. The grease cavity should be filled one-third to one-half full. Always remember that too much grease is as detrimental as insufficient grease. *Table 2* shows the amount of grease required and *Table 3* gives the recommended grease for a Class B or F motor. However, always check with the motor manufacturers for their recommendations and specifications for greasing motors.

Table 1 Relubrication Periods for Various Sizes and Types of Motors

Frame Size	Relubrication Period		
900 rpm, 1,200 rpm, and Variable Speed	**Standard Conditions**	**Severe Conditions**	**Extreme Conditions**
140–180	4.5 years	18 months	9 months
210–280	4.0 years	16 months	8 months
320–400	3.5 years	14 months	7 months
440–508	3.0 years	12 months	6 months
510	2.5 years	11.5 months	6 months
1,800 rpm	**Standard Conditions**	**Severe Conditions**	**Extreme Conditions**
140–180	3.0 years	1 year	6 months
210–280	2.5 years	10.5 months	5.5 months
320–400	2.0 years	9 months	4.5 months
440–508	1.5 years	8 months	4 months
510	1 year	6 months	3.5 months
Over 1,800 rpm	**6 months**	**3 months**	**3 months**

Instructor's Notes:

CAUTION!

The amount of grease added to motor bearings is very important. Only enough grease should be added to replace the grease used by the bearings. Too much grease can be as harmful as insufficient grease.

Table 2 Typical Amount of Grease Required When Regreasing Electric Motors

Bearing Number	Amount in Cubic Inches	Approximate Equivalent Teaspoons
203	.15	.5
205	.27	.9
206	.34	1.1
207	.43	1.4
208	.52	1.7
209	.61	2.0
210	.72	2.4
212	.95	3.1
213	1.07	3.6
216	1.49	4.9
219	2.8	7.2
222	3.0	10.0
307	.53	1.8
308	.66	2.2
309	.81	2.7
310	.97	3.2
311	1.14	3.8
312	1.33	4.4
313	1.54	5.1
314	1.76	5.9
316	2.24	7.4
318	2.78	9.2

2.0.0 ◆ PRACTICAL MAINTENANCE TECHNIQUES

Once the motor has been sized and installed properly, the key to long, trouble-free motor life is proper maintenance. Maintaining a motor in good operating condition requires periodic inspection to determine if any faults exist, and then promptly correcting these faults. The frequency and thoroughness of these inspections depend on such factors as:

- Number of hours and days the motor operates
- Importance of the motor in the production scheme
- Nature of service
- Environmental conditions

Each week, every motor in operation should be inspected to see if the windings are exposed to any dripping water, acid, or alcohol fumes as well as excessive dust, chips, or lint on or about the motor. Make certain that objects that will cause problems with the motor's ventilation system are not placed too near the motor and do not come into direct contact with the motor's moving parts.

In sleeve-bearing motors, check the oil level frequently (at least once a week) and fill the oil cups to the specified line with the recommended lubricant. If the journal or motor shaft diameter is less than 2", always stop the motor before checking the oil level. For special lubricating systems such as forced, flood-and-disc, and wool-packed lubrication, follow the manufacturer's recommendations. Oil should be added to the bearing housing only when the motor is stopped, and then a check should be made to ensure that no oil creeps along the shaft toward the windings where it may harm the insulation.

Table 3 Recommended Grease for Motor Lubrication

Insulation Class Shown on Nameplate	Grease Designation	Grease Supplier
B or F	Chevron SRI-2	Standard Oil of California or Equivalent

INSIDE TRACK

Ventilation and Temperature Control

Adequate ventilation and temperature control are essential for proper motor operation. Inspect the fan and auxiliary cooling system as part of the regular maintenance program. Many larger motors include resistance temperature detectors (RTDs) embedded in the windings or in the bearing housings. This provides for continuous monitoring of the winding and bearing temperatures to warn maintenance personnel of potential motor failures.

Safety

Caution the trainees that improper lubrication can lead to motor failure.

Classroom

Explain that motor maintenance requirements are determined by the type and length of service and the environmental conditions in which the motor operates.

Describe the weekly maintenance procedures to be performed on motors.

Safety

Emphasize that the equipment must be locked out and tagged prior to performing any maintenance procedures.

Classroom

Explain that low-pressure compressed air may be used to blow motor windings clean.

Safety

Discuss the safety precautions necessary when working with compressed air.

Classroom

Describe the monthly maintenance procedures to be performed on motors.

Always be alert to any unusual noise, which may be caused by metal-to-metal contact (bad bearings, etc.), and also learn to detect any abnormal odor, which might indicate scorching insulation varnish.

Feel the bearing housing each week for evidence of excess heat and vibration. Listen for any unusual noise. Also inspect the bearing housing for the possibility of creeping grease on the inside of the motor, which might harm the insulation.

> **WARNING!**
> Before performing any motor maintenance procedures, other than external visual inspection or noise monitoring, always lock out and tag out the equipment according to approved procedures.

Commutators and **brushes** should be checked for sparking and should be observed through several cycles if the motor is on cycle duty. A stable copper oxide carbon film—as distinguished from a pure copper surface—on the commutator is an essential requirement for good commutation. Such a film, however, may vary in color from copper to straw or from chocolate brown to black. The commutator should be clean and smooth and have a high polish. All brushes should be checked for wear and connections should be checked for looseness. The commutator surface may be cleaned using a piece of dry canvas or other durable, lint-free material that is tightly wound and securely fastened to a wooden dowel and held against the commutator while manually rotating it.

The air gap on sleeve-bearing motors should be checked frequently, especially if the motor has recently been rewound or otherwise repaired. After new bearings have been installed, for example, make sure that the average reading is within specified tolerances. Check the air passages through punchings and make sure they are free of all foreign matter.

Low-pressure compressed air (with working pressures within safety standards) may be used to blow motor windings clean. Industrial-grade vacuum cleaners may also be used to remove dirt and other debris from the motor windings. The windings may be wiped clean with a dry, lint-free cloth while checking for moisture in the bottom of the motor frame. Also check to see if any oil or grease has worked its way up to the rotor or armature windings. If so, clean away the oil or grease with an approved cleaning solution.

> **WARNING!**
> When using compressed air, exercise caution by wearing appropriate personal protective equipment and using an airflow tip or pulse nozzle. Excess air pressure can result in personnel injury and cause damage to the motor windings.

As mentioned previously, always disconnect the power, and lock out/tag out the equipment per proper procedures prior to performing any motor maintenance tasks. Check other motor parts and accessories such as belts, gears, couplings, chains, and sprockets for excessive wear or misalignment. Inspect the motor starter and verify that the motor reaches proper speed each time it is started.

At least once a month, check the shunt, series, and commutator field windings for tightness in their mountings, as drying may sometimes cause these windings to loosen. Check for tightness by attempting to move the field spools on the poles. If movement is present, the motor should be removed from service and repaired immediately. Also check the motor cable connections for looseness and tighten as needed.

Also on a monthly or more frequent basis, inspect the brushes in their holders for proper fit and free play, as well as brush spring pressure. If required, tighten the brush studs in the holders to remove slack caused by washers drying out, making sure the studs are not displaced. Look carefully for chipped or cracked brushes during the inspection and replace as necessary.

During this monthly inspection, also examine the commutator surface for high bars, high mica, and evidence of scratches or roughness. See that the risers are clean and have not been damaged in any way.

Where motors having ball or roller bearings are exposed to extreme conditions or constant usage, bearings should be serviced at least monthly by purging out the old grease through the drain hole and applying new grease. After changing the grease in these bearings, inspect the bearing housing for leaking grease. If present, correct the leakage prior to starting the motor.

On motors with sleeve bearings, check the sleeve bearings for wear at least once every two months. Clean out the oil wells if there is evidence of dirt or sludge. Flush with lighter oil before refilling with the specified bearing lubricant.

On motors with gears, open the drain plug and check the oil for the presence of metal scale, sand,

Instructor's Notes:

grit, or water. If any of these conditions are present, drain, flush, and refill as recommended by the motor manufacturer. Check the rotor for slack or backlash by carefully rocking the rotor.

Loads being driven by motors have a tendency to change from time to time due to wear on the machine or the product being processed through the machine. Therefore, all loads should be checked from time to time for a changed condition, bad adjustment, and poor handling or control.

During the monthly inspection, check the tightness of all belts and adjust as necessary. Check the belts to verify that they are running steadily and near the inside edge of the pulley. On chain-driven machines, check the chain for evidence of wear and stretch, and clean the chain thoroughly. Check the chain lubricating system and note the incline of the slanting base to make sure it does not cause oil rings to rub on the housing.

Once or twice each year, all motors in operation should be given a thorough inspection consisting of the following:

- *Windings*–Check the insulation resistance using the proper instruments and techniques, which are described in detail later in this module. The windings should also be given a visual inspection; look for cracks and other evidence of insulation deterioration or damage. Clean all surfaces thoroughly, especially ventilating passages. Examine the frame for evidence of moisture or the presence of water in the bottom of the frame. The presence of either moisture or water may require the windings to be dried, varnished, and baked.
- *Air gap and bearings*–Check the air gap to make sure that average readings are within the tolerances specified by the manufacturer. All bearings (ball, roller, and sleeve) should be thoroughly checked and defective ones replaced. Waste-packed and wick-oiled bearings should have waste or wicks renewed if they have become glazed or filled with metal, grit, or dirt, making sure that the new material bears well against the shaft.
- *Squirrel cage rotors*–Check for broken parts or loose bars as well as evidence of local heating. If the fan blades are not cast in place, check for loose blades. Also look for marks on the rotor surface, which may indicate the presence of foreign matter in the air gap or a worn bearing or bearings.
- *Wound rotors*–Wound rotors should be cleaned thoroughly, especially around collector rings, washers, and connections. Tighten all connections. If the ring surfaces appear rough in finish, spotted, or out-of-round, they should be refinished by qualified personnel familiar with machine shop equipment. Make certain all components that make up the rotor are firmly secured in place; tighten any component that appears loose.
- **Armatures**–Clean all armature air passages thoroughly. Look for oil or grease seeping along the shaft and running back to the bearing. Check the commutator surface condition, looking for high bars, high mica, or eccentricity (out-of-round) conditions. If necessary, the commutator should be turned by a qualified machinist to secure a smooth, fresh surface, as illustrated in *Figure 3*.

The following procedures should only be performed by qualified personnel. For armatures with drilled center holes, a securing device called a lathe dog is installed on the shaft opposite the

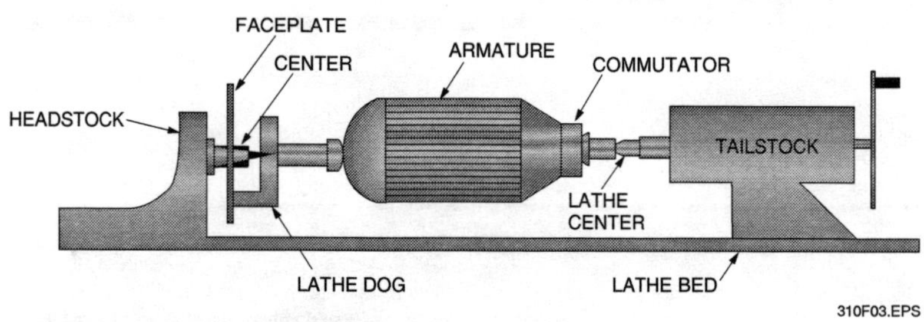

Figure 3 ◆ Motor armature secured in lathe between centers.

Classroom

Describe the annual maintenance procedures to be performed on motors.

Note that the commutator can be resurfaced by a qualified machinist if necessary.

Audiovisual

Show Transparency 4 (Figure 3).

Classroom

Describe the procedure for turning a commutator without a drilled center hole. Note that this procedure may only be performed by a qualified machinist.

Audiovisual

Show Transparency 5 (Figure 4).

Classroom

Emphasize that proper bearing lubrication is essential to ensuring long equipment life.

commutator and tightened. If it is necessary to put the lathe dog on a bearing surface, a thin piece of copper sheeting is first placed around the shaft to protect it. A faceplate is installed on the spindle end of the lathe, along with centering tools in both the headstock and tailstock. White lead or oil is applied on the tailstock centering tool and the commutator is placed between the centering tools as the tailstock is tightened firmly, but not so tight that it distorts the end of the shaft.

A sharp-pointed, lathe cutting tool is used in the tool holder to turn the commutator down, with the lathe speed set at medium speed (approximately 700 rpm). The process is finished using a fine file, followed by polishing paper.

Armatures without drilled center holes necessitate the use of chucks. This is accomplished by securing the armature shaft end opposite the commutator in a three-jaw universal chuck in the headstock. A bearing of proper shaft size is then installed in a drill-type chuck and secured in the tailstock of the lathe. The bearing is oiled and the commutator end of the shaft is placed in the bearing (see *Figure 4*). The commutator is then turned in the same manner as described previously.

When turning down any armature, a pointed tool with a sharp, smooth edge should always be used to obtain the cleanest cut possible, taking only a fine cut each time to prevent tearing the commutator. The job is finished by first smoothing down the commutator surface with a fine file while the armature is turning in the lathe, then finally polishing it with a fine polishing abrasive paper.

After the armature has been turned, clean between the bars if necessary and test the armature for shorts using a growler or other appropriate test instruments. The vibration of a hacksaw blade on any coil means that the coil is shorted at the leads or commutator. Clean between the **commutator poles** and test again. As soon as the armature tests okay, it is ready to put back into service.

Motor loads should be re-evaluated from time to time, as they can vary for several reasons. Using an ammeter, take an ampere reading on the motor, first with no load, second at full load, and finally through a full cycle of no load to full load. This test should provide information concerning the mechanical condition of the driven machine (load).

Without proper maintenance, no motor can be expected to perform well for any length of time or to remain in service as long as it should. Although motor maintenance is costly, it is far less expensive than continually replacing or overhauling motors.

3.0.0 ◆ MOTOR BEARING MAINTENANCE

AC motors account for a large percentage of industrial maintenance and repair, with many motor failures caused by faulty bearings. Consequently, most industrial facilities place a great amount of emphasis on proper care of motor bearings. Electric motors last much longer and perform better when a carefully planned motor lubrication schedule is followed.

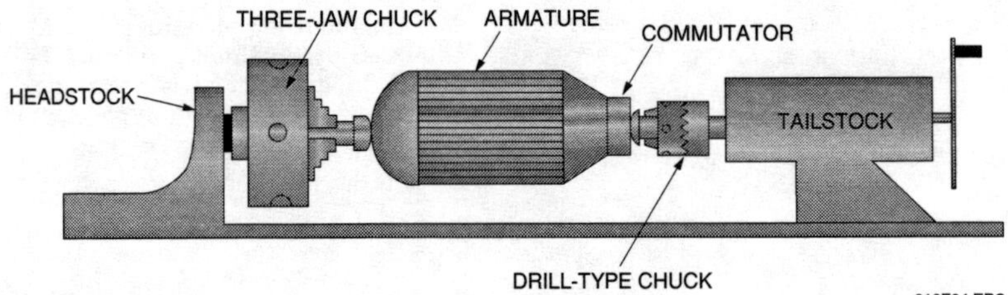

310F04.EPS

Figure 4 ◆ Motor armature secured in lathe with lathe chucks.

INSIDE TRACK

Installed Spare Motors

Installed spare motors should be operated on a regularly established time schedule. This prevents the spare motor from deteriorating while the primary motor is in use.

Instructor's Notes:

If an AC motor failure does occur, the first step is to find out why the motor failed. There are various causes of motor failures, including excessive load, binding or misalignment of motor drives, wet or dirty environments, and bearing failures. Bearing failures can occur in newer motors with high-quality bearings as often as in older motors with less reliable bearings if the bearings are not maintained properly. A notable exception in bearing failures may be found in motors equipped with sealed bearings, which are much less prone to failure.

3.1.0 Types of Bearings

Ball bearings are the most common type of bearings used in the construction of electric motors. This type of bearing is found on various sizes of motors, with bearing design types including:

- Open
- Single-shielded
- Double-shielded
- Sealed
- Double-row and other special types

Open bearings, as the name implies, are open construction and must be installed in a sealed housing. These bearings are less apt to cause churning of grease and are therefore used mostly on large motors.

The single-shield bearing has a shield on one side to keep grease from the motor windings. Double-shielded bearings have a shield on both sides of the bearing. This type of bearing is less susceptible to contamination and, because of its design, reduces the possibility of over-greasing. Sealed bearings have a double shield on each side of the bearing, which forms an excellent seal. This bearing requires no maintenance, affords protection from contamination at all times, and does not require regreasing. It is normally used on small or medium motor sizes.

Many large motors are furnished with oil-ring sleeve bearings, while some of the smaller fractional-horsepower motors are equipped with simple sleeve bearings without oil rings.

Each bearing type has characteristics that are suited for a particular application. Replacement of bearings should be made using the same type of bearing originally installed in the motor or equipment. *Figure 5* shows several types of bearings

Classroom

List the common types of bearings and their applications.

Audiovisual

Show Transparency 6 (Figure 5).

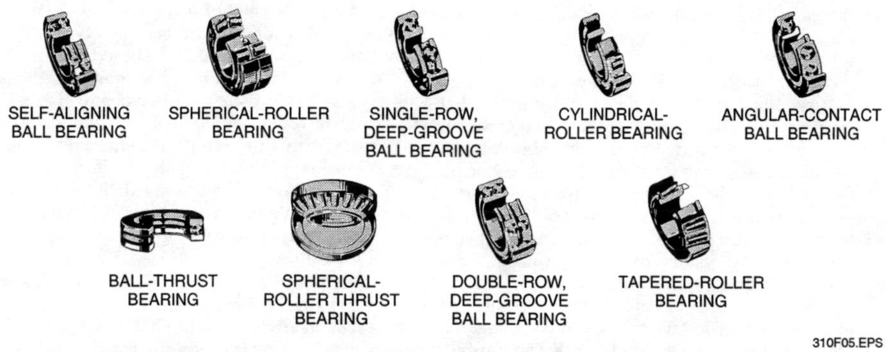

Figure 5 ◆ Various types of bearings.

INSIDE TRACK

Establishing a Rotor's Magnetic Center

Before large-frame motors are put into service, the motor should be run uncoupled in order to locate the electrical (magnetic) center of the rotor. While it is running, mark or etch the established reference on the shaft next to the bearing. This will ensure efficient motor operation when the load is mechanically coupled to the motor.

Classroom

Discuss the operation of various types of bearings. Provide examples for the trainees to examine.

used in electric motors. The following list provides a basic overview of bearing applications and a guide to analyzing bearing failures.

- *Self-aligning ball bearing*–The self-aligning ball bearing has two rows of balls rolling on the spherical surface of the outer ring, and this design compensates for angular misalignment due to errors in mounting, shaft deflection, or distortion of the foundation. This design also prevents any exertion of bending influence on the motor shaft—a most important consideration in applications requiring extreme accuracy at high speeds. Self-aligning ball bearings are used for radial loads and moderate thrust loads in either direction.
- *Single-row, deep-groove ball bearing*–The single-row, deep-groove ball bearing will sustain, in addition to radial load, a substantial thrust load in either direction, even at very high speeds. The ability to sustain these loads is made possible by the close-tolerance contact that exists between the roller balls and the continuous groove in each ring. Accurate alignment between the motor shaft and housing is essential in the application of this type of bearing. The single-row, deep-groove bearing is also available with seals and shield, which provide protection from contamination and retain lubricant.
- *Angular-contact ball bearings*–The angular-contact ball bearing can support a substantial thrust load in one direction, combined with a moderate radial load. A steep contact angle assures the highest thrust capacity and axial rigidity. This characteristic is obtained through the addition of a thrust-supporting shoulder on the inner ring, with a similar high shoulder on the opposite side of the ring. These bearings can be mounted singly, or in special applications in tandem, to allow constant thrust in one direction. They can also be mounted in pairs if the sides of the bearings have been ground to a flush finish. This installation provides for a combined load, either face-to-face or back-to-back.
- *Double-row, deep-groove ball bearing*–The double-row, deep-groove ball bearing embodies the same principle of design as the single-row bearing. However, this bearing has a lower axial displacement than occurs in the single-row design, substantial thrust capacity in either direction, and high radial capacity due to the two rows of balls.
- *Spherical-roller bearing*–The exceptional capacity of the spherical-roller bearing can be attributed to the number, size, and shape of the rollers, as well as the accuracy by which they are guided. Since the bearing is inherently self-aligning, angular misalignment between the shaft and housing has no detrimental effect on the application of the bearing. The design and proportion of the bearing are such that both thrust loads and radial loads may be carried in either direction.
- *Cylindrical-roller bearing*–This type of bearing has a high radial capacity, which provides accurate guiding of the rollers, and a close approach to true rolling. The low friction permits operation at high speeds. Cylindrical-roller bearings having flanges on one ring also allow a limited free axial movement of the shaft in relation to the housing, and are easy to dismount even when both rings are mounted within a close tolerance to one another. The double-row type is particularly suitable for machine-tool spindles.
- *Ball-thrust bearing*–The ball-thrust bearing is designed for thrust load in one direction only. The load line through the balls in parallel to the axis of the shaft results in high thrust capacity and minimum axial deflection. Flat bearing seats are essential for heavy loads or for close axial positioning of the shaft.
- *Spherical-roller thrust bearing*–The spherical-roller thrust bearing is designed to carry heavy thrust loads or combined loads that are predominantly thrust. This bearing has a single row of rollers, which roll on a spherical outer race with full self-alignment. The cage, centered by an inner ring sleeve, is constructed so that lubricant is pumped directly against the inner ring's unusually high guide flange. This bearing operates best with relatively heavy oil lubrication.
- *Tapered-roller bearings*–Since the axes of the rollers and raceways of a tapered-roller bearing form an angle with the shaft angle, the tapered-roller bearing is especially suitable for carrying coordinated radial and axial loads. This type of bearing is typically installed adjacent to another bearing capable of carrying thrust loads in the opposite direction. Tapered roller bearings are designed so that their cone (inner ring) and roller/cup assembly (outer ring) are mounted separately.

Instructor's Notes:

Recommendations for ball bearing assembly, maintenance, inspection, and lubrication are shown in *Table 4.* Refer to this list often when working with electric motors.

3.2.0 Frequency of Lubrication

The frequency of motor lubrication depends not only on the type of bearing, but also on the motor application. Small to medium-size motors equipped with ball bearings (except sealed bearings) should be greased every three to six years if the motor duty is normal. On severe applications (high temperature, wet or dirty locations, or corrosive atmospheres), lubrication may be required more often. In severe applications, past experience and condition of the grease are the best guides to the frequency of lubrication.

The lubrication in sleeve bearings should be changed at least once a year or more often when the motor duty is severe or the oil appears dirty.

3.2.1 Lubrication Procedure

Before lubricating a ball bearing motor, the bearing housing, grease gun, and fittings should be cleaned. Care must be exercised in keeping out dirt and debris during lubrication. The relief plug should be removed from the bottom of the bearing housing. This is done to prevent excessive pressure from building up inside the bearing housing during lubrication. If possible, run the motor and add grease until it begins to flow from the relief hole. The motor should be allowed to run for approximately 5 or 10 minutes to expel any excess grease. The relief plug should then be reinstalled and the bearing housing cleaned.

It is important to avoid over-lubrication. When excessive grease is forced into a bearing, a churning of the grease may occur, resulting in high bearing temperatures and eventual bearing failure.

On motors that do not have a relief plug, grease should be applied slowly and sparingly. If possible, disassemble the motor and repack the bearing

Table 4 Ball Bearing Assembly, Maintenance, and Lubrication Recommendations

Do	Do Not
DO work with clean tools in clean surroundings.	DO NOT work with poor tools, a cluttered workbench, or dirty surroundings.
DO remove all outside dirt from the housing before exposing the bearing.	DO NOT handle bearings with dirty or moist hands.
DO treat a used bearing as carefully as a new one.	DO NOT spin uncleaned bearings.
DO use clean solvents and flushing oils.	DO NOT spin any bearings using compressed air.
DO lay bearings out on clean paper or cloth.	DO NOT use the same container to clean and rinse the bearings.
DO protect disassembled bearings from dirt and moisture.	DO NOT scratch or nick the bearing surfaces.
DO use clean, lint-free rags to wipe bearings.	DO NOT remove grease or oil from new bearings.
DO keep bearings wrapped in oil-proof paper when not in use.	DO NOT use the incorrect type or amount of lubricant.
DO clean the outside of the housing before replacing the bearings.	DO NOT use a bearing as a measuring tool to check the housing bore or shaft fit.
DO keep bearing lubricants clean when applying and cover containers when not in use.	DO NOT install a bearing on a shaft that shows excessive wear.
DO be sure the shaft size is within the specified tolerances recommended for the bearing.	DO NOT open the bearing carton until the bearing is ready to be installed.
DO store bearings in their original unopened cartons in a dry place.	DO NOT determine the condition of a bearing until it has been properly cleaned.
DO use a clean, short-bristle brush with firmly embedded bristles to remove dirt, scale, or chips.	DO NOT tap directly on a bearing or ring during installation.
DO be certain that, when installed, the bearing is square with and held firmly against the shaft shoulder.	DO NOT overfill during lubrication. Excess oil or grease may enter the motor housing. Too much lubricant will also cause overheating, particularly where bearings operate at high speeds.
DO follow lubricating instructions supplied with the machinery. Use only grease where grease is specified; use only oil where oil is specified. Be sure to use the exact kind of lubricant called for.	DO NOT allow motors to remain idle for long periods of time without rotating their shafts periodically.
DO handle grease with clean paddles or grease guns. Store grease in clean containers. Keep grease containers covered.	

Classroom

Discuss the lubrication guidelines for bearings. Refer the trainees to Table 4 in the Trainee Module.

Demonstration

Demonstrate the procedure for lubricating a motor equipped with ball bearings.

Classroom

Explain how bearings are tested.

List the types of tests performed during motor troubleshooting.

Safety

Emphasize that the equipment must be locked out and tagged prior to performing any troubleshooting procedures.

with the proper amount of grease. Always maintain cleanliness during lubrication.

When lubricating sleeve-type bearings, use only the recommended type and amount of lubrication as recommended by the manufacturer of the motor or bearing.

3.3.0 Testing Bearings

Two simple methods commonly used to check bearings during motor operation are touching and listening. If the bearing housing feels unusually hot, or if a growling or grinding sound is being emitted from the area of the bearings, one of the bearings is probably nearing failure. Special stethoscopes are also available for listening to bearings while the motor is running.

> **Note**
> Keep in mind that some bearings may safely operate in a higher temperature range than other bearings, even in ranges exceeding 85°C. Check the motor manufacturer's specifications.

Using a feeler gauge, periodically check the air gap on sleeve-bearing motors for bearing wear. Four separate measurements should be taken approximately 90° apart around the diameter of the rotor. These measurements should be recorded and compared with previous measurements to determine if any deviations are present, which may indicate bearing wear since the last measurements were recorded.

Motors should also be checked for end play, which is the backward and forward movement in the shaft. Ball bearing motors typically will have $\frac{1}{32}$" to $\frac{1}{16}$" of end movement. Sleeve-bearing motors may have up to $\frac{1}{2}$" of end movement.

On larger sleeve bearings, the oil level should be checked periodically and the oil visually inspected for contamination. If safely possible, the oil rings should be checked while the motor is running.

Other inspections may include periodically checking for misalignment or bent shafts, and for excessive belt tension.

4.0.0 ◆ TROUBLESHOOTING MOTORS

To detect defects in electric motors, the windings are typically tested for ground faults, opens, shorts, and reverses. The methods used in performing these tests may depend on the type of motor being tested.

> **WARNING!**
> Before testing or troubleshooting a motor beyond a simple visual/auditory inspection, disconnect the power and follow the proper lockout/tagout procedures.

INSIDE TRACK

Three-Phase Motor Windings

The windings of motor A are good. Motor B has been run with one open phase, which caused overheating in the other two windings.

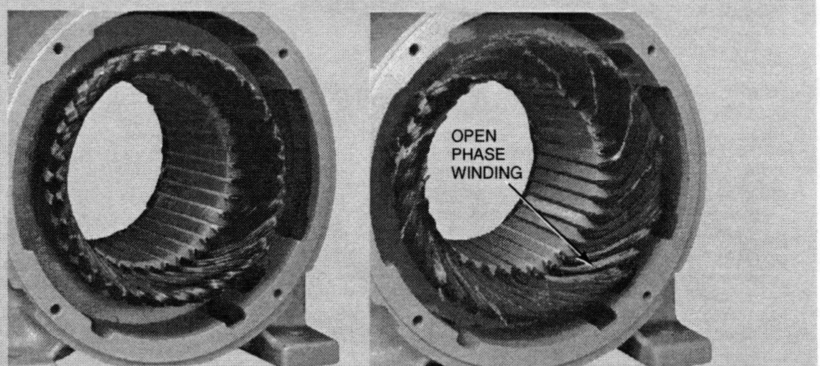

(A) (B)

310P1001.EPS

Instructor's Notes:

What's Wrong With This Picture?

310P1002.EPS

Before we can begin our study of troubleshooting, it is important to clarify some basic terms. These include the following:

- *Grounded winding*–A winding becomes grounded when it makes electrical contact with the metal frame of the motor. Causes may include end plate bolts contacting conductors or windings, winding wires pressing against the laminations at the corners of slots damaged during rewinding, or conductive parts of the centrifugal switch contacting the end plate.
- *Open circuits*–Loose connections or broken wires can cause an open circuit in an electric motor.
- *Shorts*–Two or more windings of separate phases that contact each other at a point creating a path other than the one intended cause a short circuit. This condition may develop in a relatively new winding if the winding is wound too tight at the factory and excessive force was used to place the wires in position. Shorts may also be caused by excessive heat caused by overloads, which can degrade the insulation.
- *Shorted turns*–Loose or broken turns within a winding, which contact each other in a manner that effectively causes the turns within the winding to be reduced can cause high current in the winding. This can cause excessive heat, causing degradation of insulation.

4.1.0 Tools for Troubleshooting

This section describes some of the tools used in troubleshooting electric motors.

In addition to typical equipment such as voltmeters, ammeters, brush-spring tension testers, and sound amplifiers used for checking motor bearings, motor maintenance equipment should also include an insulation resistance tester such as a 500V megger, an oil dielectric tester, and a portable oil filtering unit.

Some of these tools have been covered in previous modules, and only those not covered will be presented in this subsection.

- *Transistorized stethoscope*–A transistorized stethoscope is equipped with a transistor-type amplifier and is used to determine the condition of motor bearings. A little practice in interpreting its usage may be required, but in general, it is relatively simple to use. When the stethoscope is applied to a motor bearing housing, a smooth purring sound should be heard if the bearing is operating normally. However, if a thumping, grinding, or growling sound is detected, it could be an indication of a failing bearing.
- *Insulation resistance tests*–High-voltage cables rated at 2,300V or above should be tested according to specified standards. Conductors and equipment carrying 480V should have their insulation tested at least annually. This includes transformers, motors, motor starters, **generators,** and switches.
- *Winding resistance and surge (resonance) tests*–A winding resistance or surge test provides baseline values for comparison to other phases to detect shorts or opens between the windings.

Before performing an insulation resistance test, first perform a safety check to ensure that the circuits and equipment to be tested are rated at or above the output voltage of the megger, and that all equipment scheduled for testing is disconnected and locked out according to procedure.

Also, check the megger and other testing equipment for proper operation. A megger may be checked for an infinity reading by cranking the megger handle with the test leads disconnected from the meter. The megger may also be zeroed by cranking the megger handle with the test leads connected to the meter and to each other. A reading of zero should be indicated, and if not, the zero adjustment should be adjusted until 0Ω is indicated on the meter scale while cranking the handle with the test leads shorted together.

Once the megger has been checked and zeroed, insulation resistance readings should be taken by testing between a conductor and ground, or between two conductors, or both. Insulation readings from conductor to ground are obtained by connecting the line test lead to the conductor and

Teaching Tip

Discuss the "What's Wrong with This Picture?"

Classroom

List the types of failures that occur in motor windings.

Discuss the types of tests used to diagnose motor failures.

Explain how to use a megger to measure insulation resistance.

Classroom

List the tools used for motor maintenance and troubleshooting. Refer the trainees to Table 5 in the Trainee Module.

Describe the causes and symptoms of a grounded coil. Point out that grounded coils can be isolated using a continuity tester.

the earth test lead to ground. For the test between conductors, the test leads are connected to the two conductors to be tested.

After the proper connections have been made to the test leads, the megger handle should be turned at a uniform rate for approximately one minute, at which time the insulation resistance value should be recorded. This is considered a short time or spot megger test.

Because temperature and humidity have profound effects on insulation resistance readings, the temperature and humidity at the apparatus should be recorded immediately after the test. In addition, pertinent information such as the condition of the environment and the running or idle time of the equipment should be noted.

After the readings have been recorded, they should be corrected for temperature using a temperature chart (supplied with most meggers). Typically, maintenance departments follow a standard referred to as the One Megohm Rule of Thumb, which states that an insulation resistance rated at 600V is acceptable if the corrected resistance value is one megohm or greater. Insulation resistance values that are within these tolerances, yet show a downward trend, usually indicate insulation deterioration and demand close monitoring.

Table 5 provides a list of practical tools and equipment for effective electrical maintenance, both for motors and other electrical apparatus.

Elaborate motor insulation tests such as time resistance and step voltage tests use various megger reading ratios and indexes (i.e., absorption ratio and polarization index) to determine the effects of temperature and humidity on insulation breakdown.

4.2.0 Grounded Coils

The usual effect of one grounded coil in a winding is the repeated blowing of a fuse or tripping of the circuit breaker when the line switch is closed, provided that the machine frame and the line are both grounded. Two or more grounds will give the same result and will also short out part of the winding in that phase in which the grounds occur. A quick and simple test to determine whether or not a ground exists in the winding can be made with a conventional continuity tester. Before testing with this instrument, first make certain that the line switch is open and locked out, causing the motor leads to be de-energized. Place one test lead on the frame of the motor and the other in turn on each of the line wires leading from the motor. If there is a grounded coil at any point in the winding, the lamp of the continuity tester will light, or in the case of a meter, the reading will show low or zero resistance.

Table 5 Tools for Electrical Maintenance

Tools or Equipment	Application
Multimeters, voltmeters, ohmmeters, clamp-on ammeters, wattmeters, clamp-on power factor meter	Measure circuit voltage, resistance, current, and power. Useful for circuit tracing and troubleshooting.
Potential and current transformers, meter shunts	Increase range of test instruments to permit the reading of high-voltage and high-current circuits.
Tachometer	Checks rotating machinery speeds.
Recording meters	Provide a permanent record of voltage, current, power, temperature, etc., on charts for analytic study.
Insulation resistance tester, thermometer, psychrometer	Test and monitor insulation resistance; use a thermometer and psychrometer for temperature and humidity correction.
Portable oil dielectric tester, portable oil filter	Test OCB, transformer oil, or other insulating oils. Recondition used oil.
Transistorized stethoscope	Detects faulty rotating machinery bearings and leaky valves.
Air gap feeler gauges	Check motor or generator air gap between rotor and stator.
Cleaning solvent	Removes grease or dirt from motor windings or other electrical parts.
Hand stones (rough, medium, fine), grinding rig, canvas strip	Used for grinding, smoothing, and finishing commutators or slip rings.
Spring tension scale	Checks brush pressure on DC motor commutators or on AC motor skip rings; tests electrical contact pressure on relays, starters, or contacts.
Surge comparison tester or bridge resistance tester	Checks detailed resistance of windings/coils.

Instructor's Notes:

The One Megohm Rule of Thumb

INSIDE TRACK

The minimal acceptable reading is one megohm per kilovolt rating (with a minimum of one megohm). This only works at 68°F. To correct for the ambient temperature, the reading must be adjusted up for temperatures above 68°F and down for temperatures below 68°F. The reasoning is that equipment insulation resistance decreases at higher temperatures and increases at lower temperatures. The readings are adjusted as follows:

- For every 17° above 68°F, double the megohm reading.
- For every 17° below 68°F, halve the megohm reading.

For example, a 230V, 600hp motor has a megohm reading of 1.5 megohms. The motor heaters are on to keep the motor dry, and the conductor/motor temperature (ambient) is 105°F. The minimum acceptable reading is 2.3 megohms at 68°F. The temperature-corrected reading is as follows:

$$105°F - 68°F = 37°F$$

$$37 \div 17 = 2 \text{ (rounded off)}$$

Therefore, the reading must be doubled twice, as follows:

$$1.5 \times 2 = 3$$

$$3 \times 2 = 6$$

The temperature-corrected reading is 6 megohms. It is acceptable because it is greater than the 2.3 megohm minimum requirement.

To locate the phase that is grounded, test each phase separately. In a three-phase winding, it will be necessary to disconnect the star or delta connections, if accessible. After the grounded phase is located, the pole group connections in that phase can be disconnected and each group tested separately. When the test leads are connected (one on the frame and the other on the grounded coil group), a glowing lamp will indicate a ground in this group when using a lamp-indicating continuity tester, or near-zero ohms when using a meter. The stub connections between the coils and this group may then be disconnected and each coil tested separately until the exact coil that is grounded is located.

Sometimes moisture in old and deteriorating insulation around the coils will cause a high-resistance leakage to ground that is difficult to detect with a test lamp continuity tester. A megger can be used to detect such faults, but in many cases, a megger may not be available. If not, a homemade tester may be constructed using a headphone set (telephone receiver) and several dry cell batteries connected in series, as shown in *Figure 6*. This test set is capable of detecting grounds of relatively high resistance by producing an audible clicking sound through the receiver when one is detected, and often is more effective than an ordinary test lamp in locating grounds or leakage.

Armature windings and the commutator of a motor may be tested for grounds in a similar manner. On some motors, the **brush holders** are grounded to the end plate. Consequently, before the armature is tested for grounds, the brushes must be lifted away from the commutator.

When a grounded coil is located, it should be either removed and reinsulated or cut out of the circuit. At times, however, it may be inconvenient to stop a motor long enough for a complete rewinding or permanent repairs. In such cases, when trouble develops, it is often necessary to make a temporary repair until a later time when the motor may be taken out of service.

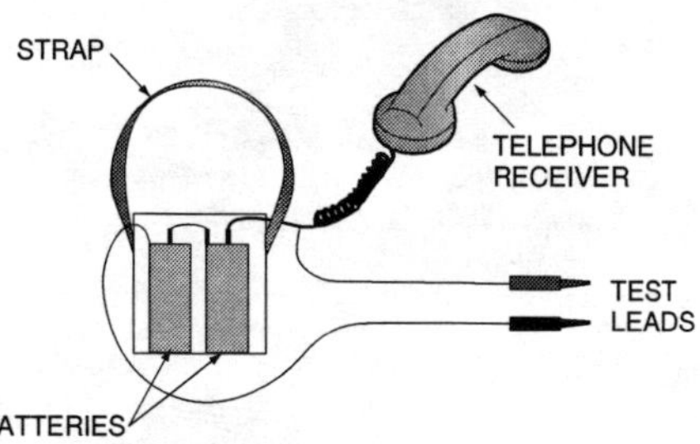

Figure 6 ◆ When used for testing, a clicking sound indicates a fault.

Classroom

Explain how to apply the one megohm rule of thumb to insulation resistance readings.

Describe how to construct a headphone set continuity tester.

Audiovisual

Show Transparency 7 (Figure 6).

Classroom

Point out that grounds in armature windings and commutator brushes may be tested in the same way as motor windings.

Classroom

Explain how to perform a temporary repair on a defective coil.

Describe the causes and symptoms of a shorted coil. Point out that shorted coils can be isolated by using a growler.

Audiovisual

Show Transparency 8 (Figure 7).

Classroom

Explain how to use a growler to test a shorted coil.

Describe the causes and symptoms of an open coil. Point out that open coils can be isolated by using a continuity tester.

Explain that opens due to loose connections can easily be repaired by resoldering the splices.

To temporarily repair a defective coil, a jumper wire of the same size as that used in the coil is connected to the bottom lead of the coil immediately adjacent to the defective coil and run across to the top lead of the coil on the other side of the defective coil, leaving the defective coil entirely out of the circuit. The defective coil should then be cut at the back of the winding and the leads taped so that they cannot function when the motor is started again. If the defective coil is grounded, it should also be disconnected from the other coils.

4.3.0 Shorted Coils

Shorted turns within coils are usually the result of failure of the insulation on the wires. This is frequently caused by the wires being crossed and having excessive pressure applied on the crossed conductors when the coils are being inserted in the slot. Quite often it is caused by using too much force when driving the coils down in the slots. In the case of windings that have been in service for several years, failure of the insulation may be caused by oil or moisture, as well as other factors. If a shorted coil is left in a winding, it will usually burn out in a short time, and if it is not located and repaired promptly, it will probably cause a ground and the burning out of a number of other coils.

One inexpensive way of locating a shorted coil when the motor is in the shop is by the use of a growler and a thin piece of steel. *Figure 7* shows a growler being used on a stator. Note that the poles are shaped to fit the curvature of the teeth inside the stator core. The growler should be placed in the core as shown, and the thin piece of steel should be placed the distance of one coil span away from the center of the growler. Then, by moving the growler around the bore of the stator and always keeping the steel strip the same distance away from it, all of the coils can be tested.

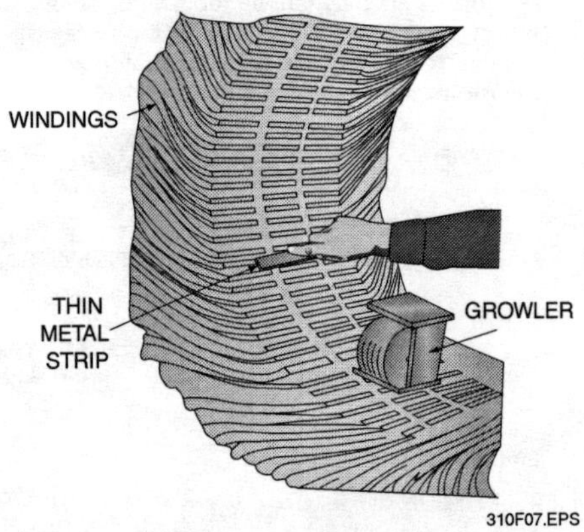

Figure 7 ◆ Growler used to test a stator of an AC motor.

If any of the coils has one or more shorted turns, the piece of steel will vibrate very rapidly and cause a loud humming noise. By locating the two slots over which the steel vibrates, both sides of the shorted coil can be found. If more than two slots cause the steel to vibrate, they should all be marked, and all shorted coils should be removed and replaced with new ones or cut out of the circuit, as previously described.

Sometimes one coil or a complete coil group becomes short circuited at the end connections. The test for this fault is the same as that for a shorted coil. If all the coils in one group are shorted, it will generally be indicated by the vibration of the steel strip over several consecutive slots, corresponding to the number of coils in the group.

The end connections should be carefully examined, and those that appear to have poor insulation should be moved during the time that the test is being made. It will often be found that when the shorted end connections are moved during the test, the vibration of the steel will stop. If these ends are reinsulated, the trouble should be eliminated.

4.4.0 Open Coils

When one or more coils become open due to a break in the turns or a poor connection at the end, they can be tested with a continuity tester, as previously explained. If this test is made at the ends of each winding, an open can be detected by the lamp failing to light. The insulation should be removed from the pole group connections, and each group should be tested separately.

An open circuit in the **starting winding** may be difficult to locate, since the problem may be in the centrifugal switch or in the winding itself. (The starting winding is only in the circuit when the motor is starting.) In fact, the centrifugal switch is probably more apt to cause trouble than the winding since parts become worn, defective, and most likely dirty. Insufficient pressure of the rotating part of centrifugal switches against the stationary part will prevent the contacts from closing and thereby produce an open circuit.

If the trouble is a loose connection at the coil ends, it can be repaired by resoldering the splices, but if it is within the coil, the coil should either be replaced or a jumper should be connected around it until a better repair can be made.

Instructor's Notes:

4.5.0 Reversed Connections

Reversed coils cause the current to flow through them in the wrong direction. This fault usually manifests itself, as do most irregularities in winding connections, by a disturbance of the magnetic circuit, which results in excessive noise and vibration. The fault can be located by the use of a magnetic compass and some source of low-voltage direct current. This voltage should be adjusted so it will send about one-fourth to one-sixth of the full-load current through the winding, and the DC leads should be placed on the start and finish of one phase. If the winding is three-phase, star-connected, this would be at the start of one phase and the star point. If the winding is delta-connected, the delta must be disconnected and each phase tested separately.

Place a compass on the inside of the stator and test each of the coil groups in that phase. If the phase is connected correctly, the needle of the compass will reverse as it is moved from one coil group to another. However, if any one of the coils is reversed, the reversed coil will build up a field in the direction opposite to the others, thus causing a neutralizing effect, which will be indicated by the compass needle refusing to point definitely to that group. If there are only two coils per group, there will be no indication if one of them is reversed, as that group will be completely neutralized.

When an entire coil group is reversed, it causes the current to flow in the wrong direction in the whole group. The test for this fault is the same as that for reversed coils. The winding should be magnetized with direct current, and when the compass needle is passed around the coil groups, they should indicate alternately north/south, north/south, etc. If one of the groups is reversed, three consecutive groups will be of the same polarity. The remedy for either reversed coil groups or reversed coils is to make a visual check of the connections at that part of the winding, locate the wrong connection, and reconnect it properly.

When the wrong number of coils are connected in two or more groups, the trouble can be located by counting the number of ends on each group. If any mistakes are found, they should be remedied by reconnecting properly.

4.6.0 Reversed Phase

Sometimes in a three-phase winding, a complete phase is reversed, caused by either taking the starts from the wrong coils or by connecting one of the windings incorrectly in relation to the others when making a star or delta connection. If the winding is connected in a delta configuration, disconnect any one of the points where the phases are connected together and pass current through the three windings in series. Place a compass on the inside of the stator and test each coil group by slowly moving the compass one complete revolution around the stator.

The reversals of the needle in moving the compass one revolution around the stator should be three times the number of poles in the winding.

When testing a star-connected or wye-connected winding, connect the three starts together and place them on one DC lead. Then connect the other DC lead to the star point, causing the current to pass through all three windings in parallel. Test with a compass as explained for the delta winding. The result should then be the same, or the reversals of the needle in making one revolution around the stator should again be three times the number of poles in the winding.

These tests for reversed phases apply to full-pitch windings only. If the winding is fractional pitch, a careful visual check should be made to determine whether there is a reversed phase or mistake in connecting the star or delta connections.

Table 6 is a troubleshooting chart for AC motors that may be used by qualified personnel who have access to the proper tools and test equipment. This table does not cover all details and variations associated with troubleshooting motors, nor does it provide a solution for every possible condition that may be encountered. Always refer to the manufacturer's instructions before testing any motor.

> THINK ABOUT IT
>
> ***Verifying Rotation***
>
> Which end of the motor should be observed when verifying rotation?

Classroom

Describe the symptoms of a reversed coil.

Explain that a compass can be used to check for a reversed coil or phase reversal.

Demonstration

Demonstrate how to troubleshoot one or more failed motors. Refer the trainees to the general troubleshooting information listed in Table 6 of the Trainee Module.

Teaching Tip

Discuss the "Think About It."

Teaching Tip

Take the trainees on a field trip to a motor repair shop to view various types of failed motors.

Table 6 General Troubleshooting Chart for AC Motors

Malfunction	Probable Cause	Corrective Action
Slow to accelerate	Excess loading	Reduce load
	Poor circuit	Check for high resistance
	Defective squirrel-cage rotor	Replace
	Applied voltage too low	Get power company to increase voltage tap
Wrong rotation	Wrong sequence of phases	Reverse connections at motor or at switchboard
Motor overheats	Overloaded motor	Reduce load
	Clogged blowers or air shields	Clean to restore proper ventilation of motor
	Motor may have one phase open	Check to make sure that all leads are well connected
	Grounded coil	Locate and repair
	Unbalanced terminal voltage	Check to make sure that all leads are well connected
	Unbalanced terminal voltage	Check for faulty leads
	Shorted stator coil	Repair and then check wattmeter reading
	Faulty connection	Indicated by high resistance; locate and repair
	High voltage	Check terminals of motor with voltmeter
	Low voltage	Same as above
Motor stalls	Wrong application	Change type or size (consult manufacturer)
	Overloaded motor	Reduce load
	Low motor voltage	See that nameplate voltage is maintained
	Open circuit	Fuses blown
	Incorrect control resistance of wound rotor	Check control sequence; replace broken resistors; repair open circuits
Motor does not start	One phase open	See that no phase is open; reduce load
	Defective rotor	Look for broken bars or rings; repair or replace
	Poor stator coil connection	Remove end bells
Motor runs, then quits	Power failure	Check for loose connections to line, fuses, and control
Slow running speed	Not applied properly	Consult supplier for proper type
	Voltage too low at motor terminals because of line drop	Use higher voltage on transformer terminals or reduce load
	If wound rotor, improper control operation of secondary	Correct secondary control
	Starting load too high	Check load that the motor is supposed to carry upon starting
	Low pull-in torque of synchronous motor	Change rotor starting resistance or change rotor design
	Brushes riding on rings	Check secondary connections; leave no leads poorly connected
	Broken rotor bars	Look for cracks near the rings; a new rotor may be required
Motor vibrates	Motor misaligned	Realign
	Weak foundation	Strengthen base
	Coupling out of balance	Balance coupling
	Driven equipment unbalanced	Rebalance driven equipment
	Defective ball bearing	Replace bearing
	Bearing not in line	Line up properly
	Balancing weights shifted	Rebalance rotor
	Wound rotor coils replaced	Rebalance rotor
	Polyphase motor running single phase	Check for open circuit
	Excessive end play	Adjust bearing or add washer
Unbalanced line current	Unequal terminal volts	Check leads and connections
	Single-phase operation	Check for open circuit
	Poor rotor contacts in wound rotor control	Check control devices
	Brushes not in proper position in wound rotor	See that brushes are properly seated and shunts are in good condition
	Fan rubbing air shield	Remove interference
	Fan striking insulation	Clear fan
	Loose on bedplate	Tighten holding bolts
Magnetic noise	Air gap not uniform	Check and correct bracket fits or bearing

Instructor's Notes:

5.0.0 ◆ TROUBLESHOOTING SPLIT-PHASE MOTORS

If a split-phase motor fails to start, the trouble may be due to one or more of the following faults:

- Tight or frozen bearings
- Worn bearings, allowing the rotor to drag on the stator
- Bent rotor shaft
- One or both bearings out of alignment
- Open circuit in either the starting or running windings
- Defective centrifugal switch
- Improper connections in either winding
- Grounds in either winding or both
- Shorts between the two windings
- Shorted or open starting capacitor

Tight or worn bearings–Tight or worn bearings may be caused by a failing bearing lubricating system. In the case of new bearings, they sometimes fail if the motor shaft is not kept properly lubricated.

The rotor may not start if the bearings are worn to such an extent that they allow the rotor to drag on the stator. When this condition exists, it can generally be detected by noticeable bright spots on the inside of the stator laminations where they have been rubbed by the dragging rotor.

Bent shaft and bearings out of line–A bent rotor shaft will usually cause the rotor to bind in a certain position and then run freely until it returns to that position. An accurate test for a bent shaft can be made by placing the rotor between centers on a lathe and turning the rotor slowly while a tool or marker is held in the tool post close to the surface of the rotor. If the rotor wobbles, it is an indication of a bent shaft.

Bearings getting out of alignment can be caused by uneven tightening of the end shield plates. When placing end shields or brackets on a motor, the bolts should be tightened alternately, first drawing up two bolts that are directly opposite one another. These two should be drawn up only a few turns and then the others tightened an equal amount all the way around. When the end shields are drawn up as far as possible with the bolts, they should be tapped tightly against the frame with a mallet and the bolts tightened again. Many motor manufacturers specify a bolt-tightening sequence to be applied when reassembling a motor.

Open circuits and defective centrifugal switches–Open circuits in either the starting or running winding will prevent the motor from starting. This fault can be detected by testing the windings using a test lamp or an ohmmeter.

A defective centrifugal switch may cause symptoms that are difficult to identify unless previously encountered while diagnosing motor failures caused by centrifugal switches. If the switch fails to close when the rotor stops, the motor will not start when the line switch is closed. Failure of the centrifugal switch to close can be caused by dirt, grit, or some other foreign matter getting into the switch, as well as a defective or broken switch. The switch should be thoroughly cleaned with a degreasing solution such as AWA 1,1,1 and then inspected for weak or broken springs.

If the winding is on the rotor, the brushes sometimes stick in the holders and fail to make good contact with the **slip rings.** This causes sparking at the brushes. Likewise, there may be a certain spot where the rotor will not start until it is moved far enough for the brush to make contact on the ring. The brush holders should be cleaned and the brushes carefully fitted so they move more freely with a minimum of friction between the brush and the holders. If a centrifugal switch fails to open when the motor is started, the motor may growl and continue to run slowly, causing the starting winding to heat up and possibly burn out if it is not promptly disconnected from the line. In most cases, however, the heaters in the motor control will take care of this before any serious damage occurs.

Reversed connections and grounds–Reversed connections are typically caused by incorrectly connecting a coil or a group of coils. Reversed connections can be identified by applying a DC voltage with a compass to indicate direction of flow, as previously described for three-phase motors. The starting and running windings should be tested separately, exciting only one winding at a time with the DC voltage. The compass should indicate alternate poles around the winding.

The symptoms demonstrated by a motor that has a grounded winding depend on the location of the ground. If the frame is grounded, a grounded winding will typically blow a line fuse or trip the overcurrent protective device.

Identifying grounded windings can be accomplished using either a test lamp or ohmmeter. One test lead should be placed on the motor frame, while the other test lead is touched to each of the motor leads. If a grounded winding is present, the test lamp will glow, or the meter will indicate a ground by reading approximately zero ohms.

Short circuits–Short circuits between the starting winding and the running windings can be detected by using a test lamp or continuity tester, as with testing for grounds. One of the test leads should be

Classroom

List common types of three-phase motor failures. For each type of failure, identify common causes and symptoms.

Classroom

Explain that motors are often kept in storage while construction is being completed or as backups for essential processes.

Safety

Explain that motors must never be lifted by the shaft because the keyways in motor shafts can cause severe cuts. In addition, improper handling may cause rotor misalignment.

Classroom

Discuss the maintenance of motors in storage and the procedures required before they are put back into service.

Safety

Emphasize that a wet or damp motor must be thoroughly dried before starting.

placed on the wire of the starting winding and the other test lead should be placed on the wire of one of the running windings. If these windings are properly insulated from each other, the lamp will not light or the meter will not indicate continuity. However, if the lamp does light or continuity is indicated, a short probably exists between the two windings. A short between the windings will usually cause part of the starting winding to burn out. The starting winding is normally wound over the running winding, and can be replaced without disturbing the running winding.

6.0.0 ◆ STORING MOTORS

Reasons for storing motors include:

- The project on which they are to be used is not complete.
- Spare motors are often kept as backups on most industrial installations.

The first consideration when storing motors for any length of time is the location in which they are to be stored. A dry location (one that does not undergo severe changes in temperature over a 24-hour period) should be selected whenever possible. When ambient temperature changes frequently, condensation is likely to form on the stored motor or motors. Moisture in motor insulation can cause motors to fail upon startup; therefore, guarding against moisture is vital when storing motors of any type.

A means for transporting the motor from the place of storage to the place where it will be used, or else shifted around in the storage area, is also important. Motors should not be lifted by their rotating shafts. Doing so can damage the alignment of the rotor in relationship to the stator. Even picking up the smaller fractional horsepower motors by the shaft is not recommended. Many workers have received bad cuts from the sharp keyways on motor shafts when picked up with bare hands.

> **WARNING!**
> Never handle a motor by its shaft without proper hand protection. Motor shaft keyways have sharp edges that can cause severe cuts.

When an electric motor is received on a job site, always follow the manufacturer's recommendations for unloading, uncrating, and installing the motor. Failure to follow these recommendations can cause injury to personnel and possible damage to the motor.

Once the motor has been uncrated, check for damage that might have occurred during shipment. Check the motor shaft to verify that it turns freely. This is also an appropriate time to clean the motor of any debris, dust, moisture, or any foreign matter that might have accumulated during shipment.

> **Note**
> Motors in storage should have their shafts turned by hand at least once a month to redistribute the grease in the bearings.

Clean the motor of any debris, dust, moisture, or any foreign matter before putting it into service.

> **WARNING!**
> Never start a wet or damp motor.

Eyebolts on motors are intended for lifting the motor and any factory motor-mounted accessory. These lifting devices should never be used when lifting or handling the motor when the motor is attached to other equipment.

The eyebolt lifting capacity is based on a lifting alignment that corresponds to the eyebolt centerline. The eyebolt capacity lessens as deviation from this alignment increases.

The following is a list of procedures that should be followed when storing motors for any length of time:

- Make sure motors are kept clean.
- Make sure motors are kept dry.
- Supply supplemental heating in the storage area, if necessary.
- Motors should be stored in an orderly fashion (i.e., grouped by horsepower, etc.).
- Motor shafts should be rotated periodically.
- Lubrication should be checked periodically.
- Protect shafts and keyways during storage and also while transporting motors from one location to another.
- Test motor winding resistance upon receiving; test again after placing in storage.

Instructor's Notes:

INSIDE TRACK

Oil Reservoirs

Empty the motor oil reservoirs to prevent oil from being spilled into the windings during shipping or when moving a motor from storage to the point of use.

Stored Motor Maintenance

To prevent shaft bowing, large horizontal motors in storage should have their rotors rotated 180° at least once a month.

Motor Storage

Motors should be stored in a humidity-controlled environment. Moisture is a major cause of insulation deterioration in motor windings.

7.0.0 ◆ IDENTIFYING MOTORS

Electricians will sometimes encounter motors with no identification (nameplate or motor lead tags) that must be put back into service or repaired. There are methods that may be applied to help identify the motor leads on motors without tags.

The NEMA standard method of motor identification can be applied to motors without tags by drawing the coils to form a wye connection. First identify one outside coil end with the number 1, then draw a decreasing spiral, numbering each coil end in sequence, as shown in *Figure 8.*

By using an ohmmeter or other continuity tester, the individual circuits can be located as follows:

Step 1 Connect one probe of the tester to any lead, and check for continuity to each of the other eight leads. A reading from only one other lead indicates one of the two-wire circuits. A reading to two other leads indicates the three-wire circuit that makes up the internal wye connection.

Step 2 Continue checking and isolating leads until all four circuits have been located. Tag the wires of the three lead circuits T-7, T-8, and T-9 in any order. The other leads should be temporarily marked T-1 and T-4 for the circuit, T-2 and T-5 for the second circuit, and T-3 and T-6 for the third and final circuit.

Note

The following test voltages are for the most common dual voltage range of 230/460V. For other motor ranges, the voltages listed should be changed in proportion to the motor rating.

As all the coils are physically mounted in slots on the same motor frame, the coils will act almost like the primary and secondary coils of a transformer. *Figure 9* shows a simplified electrical arrangement of the coils. Depending on which coil group power is applied to, the resulting voltage readings will be additive, subtractive, balanced, or unbalanced, depending on their physical location with regard to the coils themselves.

Step 3 The motor may be started on 230V by connecting leads T-7, T-8, and T-9 to the three-phase source. If the motor is too large to be connected directly to the line, the voltage should be reduced by using a reduced voltage starter or other suitable means.

Step 4 Start the motor with no load connected and bring it up to normal speed.

Step 5 With the motor running, a voltage will be induced in each of the open two-wire circuits that were tagged T-1 and T-4, T-2 and T-5, and T-3 and T-6. With a voltmeter, check the voltage reading of each circuit. The voltage should be approximately 125V to 130V and should be the same on each circuit.

Note

The voltages referred to in this section are for reference only and will vary greatly from motor to motor, depending on size, design, and manufacturer. If the test calls for equal voltages of 125V to 130V and the reading is only 80V to 90V, that is acceptable as long as the voltage readings are nearly equal.

Demonstration

Demonstrate how to identify the leads on a dual-voltage, wye-wound motor.

Audiovisual

Show Transparencies 9 and 10 (Figures 8 and 9).

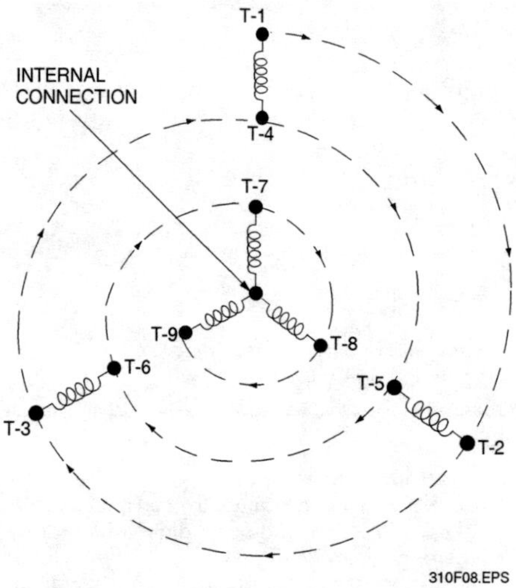

Figure 8 ◆ Identify one outside coil and then draw a decreasing spiral and number each coil.

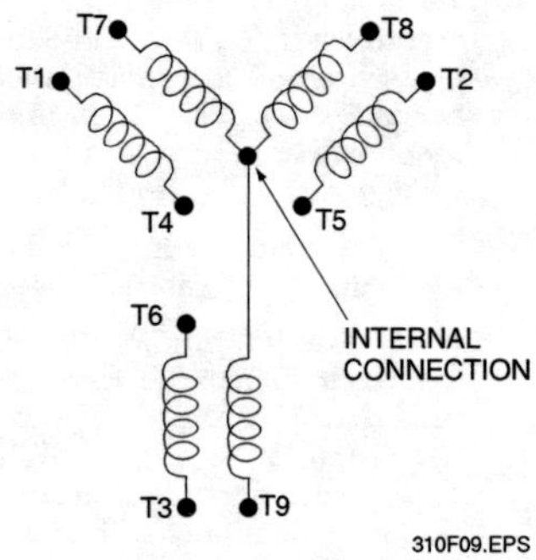

Figure 9 ◆ Simplified electrical arrangement of wye-wound motor coils.

Step 6 With the motor still running, carefully connect the lead that was temporarily marked T-4 with the T-7 and line lead. Read the voltage between T-1 and T-8 and also between T-1 and T-9. If both readings are of the same value and are approximately 330V to 340V, leads T-1 and T-4 may be disconnected and permanently marked T-1 and T-4.

Step 7 If the two voltage readings are of the same value and are approximately 125V to 130V, disconnect and interchange leads T-1 and T-4 and mark them permanently (original T-1 changed to T-4 and original T-4 changed to T-1).

Step 8 If the readings between T-1 and T-8 and also between T-1 and T-9 are of unequal values, disconnect T-4 from T-7 and reconnect T-4 to the junction of T-8 and the line.

Step 9 Measure the voltage between T-1 and T-7 and also between T-1 and T-9. If the voltages are equal and approximately 330V to 340V, tag T-1 is permanently marked T-2 and T-4 is marked T-5 and disconnected. If the readings taken are equal but are approximately 125V to 130V, leads T-1 and T-4 are disconnected, interchanged, and marked T-2 and T-5 (T-1 changed to T-5, and T-4 changed to T-2). If both voltage readings are different, the T-4 lead is disconnected from T-8 and moved to T-9. Voltage readings are taken again (between T-1 and T-7, and T-1 and T-8) and the leads permanently marked T-3 and T-6 when equal readings of approximately 330V to 340V are obtained.

Step 10 The same procedure is followed for the other two circuits that were temporarily marked T-2 and T-5, and T-3 and T-6 until a position is found where both voltage readings are equal and approximately 330V to 340V and the tags change to correspond to the standard lead markings, as shown in *Figure 10*.

Step 11 Once all leads have been properly and permanently tagged, leads T-4, T-5, and T-6 are connected together and voltage readings are taken between T-1, T-2, and T-3. The voltages should be equal and approximately 230V.

Step 12 As an additional check, the motor is shut down and leads T-7, T-8, and T-9 are disconnected, and leads T-1, T-2, and T-3 are connected to the line. Connect T-1 to the line lead T-7 was connected to, T-2 to the same line that T-8 was connected to, and T-3 to the same lead to which T-9 was connected. With T-4, T-5, and T-6 still connected together to form a wye connection, the motor can again be started without a load. If all lead markings are correct, the motor rotation with leads T-1, T-2, and T-3 connected will be the same as when T-7, T-8, and T-9 were connected.

Instructor's Notes:

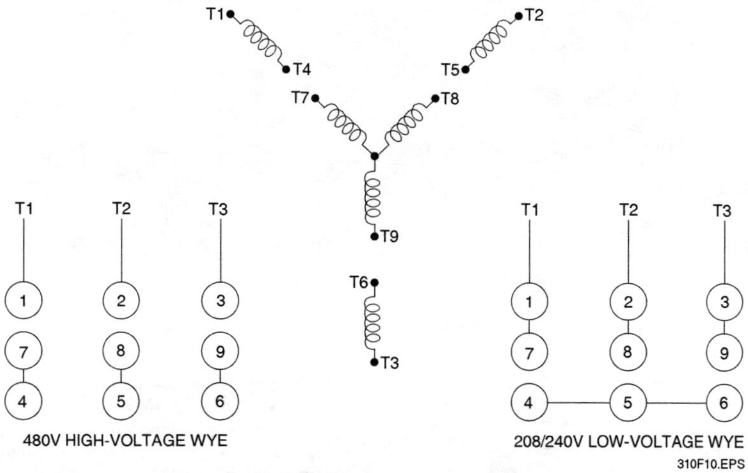

Figure 10 ◆ NEMA standard lead markings for dual-voltage, wye-wound motors.

The motor is now ready for service and is connected in series for high voltage or parallel for low, as indicated by the NEMA connections shown in *Figure 10.*

> **Note**
> This procedure may not work on some wye-wound motors with concentric coils.

7.1.0 Three-Phase, Delta-Wound Motors

Most dual-voltage, delta-wound motors have nine leads, as illustrated in *Figure 11,* with three circuits of three leads each.

Continuity tests are used to find the three coil groups as was done for the wye-wound motor. Once the coil groups are located and isolated, further resistance checks must be made to locate the common wire in each coil group. As the resistance of some delta wound motors is very low, a digital ohmmeter, Wheatstone bridge, or other sensitive device may be needed.

Each coil group consists of two coils tied together with three leads brought out to the motor junction or terminal box. Reading the resistances carefully between each of the three leads shows that the readings from one of the leads to each of the other two leads will be the same (equal), but the resistance reading between those two leads will be double the previous readings. *Figure 12* may help to clarify this technique.

The common lead found in the first coil group is permanently marked T-1, and the other two leads are temporarily marked T-4 and T-9. The common lead of the next coil group is found and permanently marked T-2, and the other leads are temporarily marked T-5 and T-7. The common lead of the last coil group is located and marked T-3, with the other leads temporarily marked T-6 and T-8.

After the leads have been marked, the motor may be connected to a 230V, three-phase line using leads T-1, T-4, and T-9. Lead T-7 is connected to the line and T-4, and the motor is started with no load connected. Voltage readings are taken between T-1 and T-2. If the voltage is approximately 460V, the markings are correct and may be permanently marked.

If the voltage reading is 400V or less, interchange T-5 and T-7 or interchange T-4 and T-9 and read the voltage again. If the voltage is approximately 230V, interchange both T-5 with T-7 and T-4 with T-9. The readings should now be approximately 460V between leads T-1 and T-2. The leads connected together now are actually T-4 and T-7 and are marked permanently. The remaining lead in each group can now be marked T-9 and T-5.

Connect one of the leads of the last coil group (not T-3) to T-9. If the reading is approximately

Demonstration

Demonstrate how to make the high-voltage and low-voltage connections for a dual-voltage motor.

Audiovisual

Show Transparency 11 (Figure 10).

Demonstration

Demonstrate how to identify the leads on a dual-voltage, delta-wound motor.

Audiovisual

Show Transparencies 12 and 13 (Figures 11 and 12).

Classroom

Emphasize the importance of accurate recordkeeping.

Laboratory

Under instructor supervision, have the trainees identify as many characteristics as possible on an untagged motor, then sketch and/or physically connect the motor for operation at both voltages. Note the proficiency of each trainee.

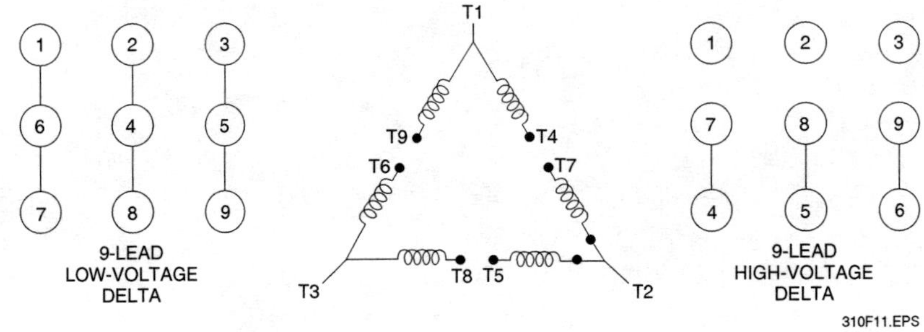

Figure 11 ◆ NEMA standard lead markings for dual-voltage, delta-wound motors.

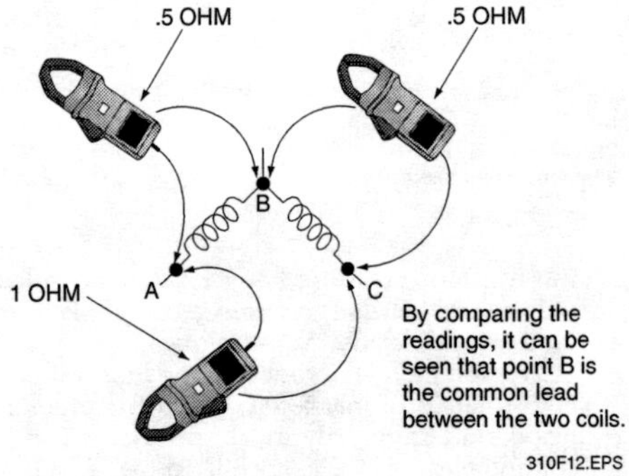

Figure 12 ◆ Using an ohmmeter to test motor leads.

460V between T-1 and T-3, the lead may be permanently marked T-6. If the reading is 400V or less, interchange T-6 and T-8. A reading of 460V should exist between T-1 and T-3. T-6 is changed to T-8 and marked permanently and temporary T-8 is changed to T-6.

If all leads are now correctly marked, equal readings of approximately 460V can be obtained between leads T-1, T-2, and T-3.

To double-check the markings, the motor is shut off and reconnected using T-2, T-5, and T-7. T-2 is connected to the same line lead as T-1, lead T-5 is connected where T-4 was, and T-7 is hooked where T-9 was previously connected. When started, the motor should rotate in the same direction as before.

Stop the motor and connect leads T-3, T-6, and T-8 to the line leads previously connected to T-2, T-5, and T-7, respectively, and when the motor is started, it should still rotate in the same direction.

The motor is now ready for service and is connected in series for high or parallel for low voltage, as indicated by the NEMA standard connections.

7.2.0 Recordkeeping

One of the first steps in establishing a reliable maintenance program is preparing accurate records. As a minimum, records on each motor should include:

- A complete description, including age and nameplate data
- Location and application, keeping such notations up-to-date if motors are transferred to different areas or used for different purposes
- Notations of scheduled preventive maintenance and previous repair work performed
- Location of duplicate or interchangeable motors
- An estimate of the motor's importance in the production process to which it relates

Summary

This module covered various motor maintenance techniques, including lubrication, storage, and elementary troubleshooting.

In determining which motors are likely to fail first, it is important to remember that motor failures are generally caused by loading, age, vibration, contamination, or commutation problems.

Advanced motor maintenance techniques will be presented in your Level Four training.

Instructor's Notes:

Review Questions

1. Which of the following provides minimum resistance and aligns the motor rotor while turning?
   a. Compensator
   b. Brushes
   c. End bells
   d. Bearings
2. Which of the following is true concerning lubricating motor bearings?
   a. Always add a little more grease than is needed.
   b. Too much grease can be as harmful as insufficient grease.
   c. No grease is better than too much grease.
   d. Too much grease is better than insufficient grease.
3. Which of the following may be eliminated as a concern in good motor maintenance?
   a. Number of hours and days the motor operates
   b. Manufacturer of the motor
   c. Environmental conditions
   d. Importance of the motor in the production scheme
4. Which of the following is usually an indication of a bad motor bearing?
   a. Hot bearing housing
   b. Low current draw
   c. Low pull-in torque
   d. Sparking at the brushes
5. When using compressed air to clean motors, which of the following precautions should be taken?
   a. Make sure the air is warmer than the ambient motor temperature.
   b. Use air pressure within safety standards.
   c. Avoid using a pulse nozzle.
   d. Use only a high-velocity nozzle.
6. Which of the following parts of a wound rotor motor requires special attention to proper cleaning?
   a. Lifting eyebolt
   b. End bells
   c. Collector rings
   d. Motor terminal enclosure
7. The ____ is/are used only during the brief period when the motor is starting.
   a. motor leads
   b. drum armature
   c. starting winding
   d. shunt-field winding
8. If a split-phase motor fails to start, which of the following may be *eliminated* as the cause?
   a. Tight or frozen bearings
   b. Bent rotor shaft
   c. Defective centrifugal switch
   d. Improper air gap
9. Which of the following statements is true concerning motors in storage?
   a. Rotors should not be moved or turned until the motor is put in use.
   b. Rotors should be turned once a month to distribute bearing grease.
   c. Rotors should be turned once a year to distribute bearing grease.
   d. Rotors should be turned once every three years to keep them from rusting.
10. The best attachment point for lifting heavy motors is/are the _____.
    a. eyebolt
    b. shaft
    c. base
    d. end bells

Classroom

Have the trainees complete the Review Questions and go over the answers prior to administering the Module Examination.

Examination

Administer the Module Examination. Record the results on Craft Training Report Form 200 and submit the results to the Training Program Sponsor.

Performance Testing

Administer the Performance Test and fill out Performance Profile Sheets for each trainee. If desired, trainee proficiency noted during laboratory sessions may be used to complete the Performance Test. Be sure to record the results on Craft Training Report Form 200 and submit the results to the Training Program Sponsor.

GLOSSARY

Trade Terms Introduced in This Module

Armature: 1. Rotating machine: the member in which alternating voltage is generated. 2. Electromagnetic: the member that is moved by magnetic force.

Brush: A conductor between the stationary and rotating parts of a machine; usually made of carbon.

Brush holders: Adjustable arms for holding the commutator brushes of a generator against the *commutator,* feeding them forward to maintain proper contact as they wear, and permitting them to be lifted from the contact when necessary.

Commutator: A device used on electric motors or generators to maintain a unidirectional current.

Commutator pole: An electromagnetic bar inserted between the pole pieces of a generator to offset the cross-magnetization of the *armature* currents.

Generator: 1. A rotating machine that is used to convert mechanical energy to electrical energy. 2. General apparatus, equipment, etc., that is used to convert or change energy from one form to another.

Slip rings: The means by which the current is conducted to a revolving electrical circuit.

Starting winding: A winding in an electric motor used only during the brief period when the motor is starting.

Instructor's Notes:

Additional Resources

This module is intended to present thorough resources for task training. The following reference works are suggested for further study. These are optional materials for continued education rather than for task training.

American Electrician's Handbook. Terrell Croft and Wilfred I. Summers. New York, NY: McGraw-Hill, 1996.

National Electrical Code Handbook. Quincy, MA: National Fire Protection Association, 2002.

Answers to Review Questions

Answer	Section
1. d	1.0.0
2. b	1.0.0
3. b	2.0.0
4. a	2.0.0
5. b	2.0.0
6. c	2.0.0
7. c	4.4.0
8. d	5.0.0
9. b	6.0.0
10. a	6.0.0

Section 4.0.0 *What's Wrong with This Picture?*

This motor has a shorted winding.

Section 4.6.0 *Think About It—Verifying Rotation*

To verify rotation, look at the bell end (opposite the load).

CONTREN™ LEARNING SERIES — USER UPDATES

The NCCER makes every effort to keep these textbooks up-to-date and free of technical errors. We appreciate your help in this process. If you have an idea for improving this textbook, or if you find an error, a typographical mistake, or an inaccuracy in NCCER's Contren™ textbooks, please write us, using this form or a photocopy. Be sure to include the exact module number, page number, a detailed description, and the correction, if applicable. Your input will be brought to the attention of the Technical Review Committee. Thank you for your assistance.

Instructors – If you found that additional materials were necessary in order to teach this module effectively, please let us know so that we may include them in the Equipment/Materials list in the Instructor's Guide.

Write: Curriculum Revision and Development Department
National Center for Construction Education and Research
P.O. Box 141104, Gainesville, FL 32614-1104

Fax: 352-334-0932

E-mail: curriculum@nccer.org

Craft ______ Module Name ______

Copyright Date ______ Module Number ______ Page Number(s) ______

Description ______

______

______

______

(Optional) Correction ______

______

______

(Optional) Your Name and Address ______

______

______

Motor Controls

26311-02

MODULE OVERVIEW

This module introduces the Electrical trainee to the methods and procedures used in selecting and wiring motor controls.

PREREQUISITES

Please refer to the Course Map in the Trainee Module. Prior to training with this module, it is recommended that the trainee shall have successfully completed the following:

Core Curriculum; Electrical Level One; Electrical Level Two; Electrical Level Three, Modules 26301 through 26310

OBJECTIVES

Upon completion of this module, the trainee will be able to:

1. Identify contactors and relays both physically and schematically and describe their operating principles.
2. Identify pilot devices both physically and schematically and describe their operating principles.
3. Interpret motor control wiring, connection, and ladder diagrams.
4. Select and size contactors and relays for use in specific electrical motor control systems.
5. Select and size pilot devices for use in specific electrical motor control systems.
6. Connect motor controllers for specific applications according to NEC requirements.

PERFORMANCE TASKS

Under supervision of the instructor, the trainee should be able to:

1. Make all connections for a magnetic motor controller, controlled by two pushbutton stations, including the connections for holding the circuit interlock.

NCCER STANDARDIZED CRAFT TRAINING PROGRAMS

The National Center for Construction Education and Research (NCCER) provides a standardized national program of accredited craft training. Key features of the program include instructor certification, competency-based training, and performance testing. The program provides trainees, instructors, and companies with a standard form of recognition through a National Craft Training Registry. The program is described in full in the *Guidelines for Accreditation*, published by the NCCER. For more information on standardized craft training, contact the NCCER by writing us at P.O. Box 141104, Gainesville, FL 32614-1104, calling 352-334-0911, or e-mailing info@nccer.org. More information may be found at our Web site, www.nccer.org.

HOW TO USE THIS ANNOTATED INSTRUCTOR'S GUIDE

Each page presents two sections of information. The larger section displays each page exactly as it appears in the Trainee Module. The narrow column ties suggested trainee and instructor actions to each page and provides icons to call your attention to material, safety, audiovisual, or testing requirements. The bottom of each page includes space for your notes.

If you see the Teaching Tip icon, that means there is a teaching tip associated with this section. Also refer to any suggested teaching tips at the end of the module.

SAFETY CONSIDERATIONS

Ensure that the trainees are equipped with appropriate personal protective equipment.

PREPARATION

Before teaching this module, you should review the Module Outline, Objectives, Performance Tasks, and the Materials and Equipment List. Be sure to allow ample time to prepare your own training or lesson plan and gather all required equipment and materials.

MATERIALS AND EQUIPMENT LIST

Materials:
Transparencies
Markers/chalk
Copy of the latest edition of the *National Electrical Code®*
Assorted wire and connectors necessary for making control circuit wiring connections
Assorted manufacturer's motor control device catalogs/data sheets
Examples of wiring diagrams
Examples of circuit schedules/wire lists
Examples of control ladder diagrams
Examples of logic diagrams
Module Examinations*
Performance Profile Sheets*

Equipment:
Overhead projector and screen
Whiteboard/chalkboard
Appropriate personal protective equipment
Open-frame electromechanical power relays
Miniature electromechanical plug-in relays
Assorted NEMA and IEC magnetic and manual contactors and motor starters
Melting-alloy thermal overload relays
Bimetallic overload relays
Magnetic overload relays
Contactor/motor starter accessories including:
- Power-pole adder kit
- Timer attachment
- Fuse kit
- Transient suppression module
- Internal auxiliary contacts

Control transformers
Pushbutton switches
Push-pull pushbutton switches
Selector switches
Pilot lights
Assorted pushbutton stations
Temperature switches
Pressure switches
Mechanical limit switches
Flow switches
Float switches
Foot switches
Jogging and plugging switches
Inductive and capacitive proximity sensors
Photoelectric switches/sensors
Drum switches
Assorted NEMA enclosures
240V motor
Tools necessary for making wiring connections

*Located in the Test Booklet packaged with this Annotated Instructor's Guide.

ADDITIONAL RESOURCES

This module is intended to present thorough resources for task training. The following reference works are suggested for both instructors and motivated trainees interested in further study. These are optional materials for continued education rather than for task training.

Electrical Motor Controls, Gary Rockis and Glen Mazur. Homewood, IL: American Technical Publishers, Inc., 1997.

National Electrical Code Handbook. Quincy, MA: National Fire Protection Association, 2002.

NFPA 70B Recommended Practice for Electrical Equipment Maintenance. Quincy, MA: National Fire Protection Association, 1998.

TEACHING TIME FOR THIS MODULE

An outline for use in developing your lesson plan is presented below. Note that each Roman numeral in the outline equates to one session of instruction. Each session has a suggested time period of 2½ hours. This includes 10 minutes at the beginning of each session for administrative tasks and one 10-minute break during the session. Approximately 20 hours are suggested to cover *Motor Controls*. You will need to adjust the time required for hands-on activity and testing based on your class size and resources.

Topic	Planned Time
Session I. Introduction; Electromechanical Relays	
A. Introduction	____________
B. Electromechanical Relays	____________
1. General-Purpose Relay Configurations and Designations	____________
2. Typical Operation	____________
3. Relay Selection Criteria	____________
a. Coil Voltage Characteristics	____________
b. Contact Ratings	____________
c. Contact Materials	____________
4. Contact Arc Suppression	____________
Session II. Magnetic Contactors; Overload Protection	
A. Magnetic Contactors	____________
1. Magnetic Contactor Construction and Operation	____________
2. Contactor Marking Conventions	____________
a. UL Markings	____________
b. NEMA Nameplate Ratings	____________
c. IEC Nameplate Ratings	____________
d. NEMA and IEC Terminal Marking/Symbol Conventions	____________
3. Two-Wire and Three-Wire Contactor Control	____________
B. Overload Protection	____________
1. Melting-Alloy Thermal Overload Relays	____________
2. Bimetallic Overload Relays	____________
3. Magnetic Overload Relays	____________
Session III. Magnetic and Manual Motor Starters	
A. Magnetic and Manual Motor Starters	____________
1. Nonreversing and Reversing Magnetic Motor Starters	____________
2. NEMA Magnetic Contactor/Motor Starters	____________
a. Ratings	____________
b. Selection	____________
3. IEC Magnetic Contactors/Motor Starters	____________
a. Ratings	____________
b. Selection	____________
4. Manual Motor Starters	____________
5. Contactor/Motor Starter Accessories	____________

Session IV. Control Transformers and Pilot Devices

A. Control Transformers and Pilot Devices

1. Control Transformers
2. Pushbutton Switches, Selector Switches, and Pilot Lights
   a. Pushbutton Switches
   b. Push-Pull Pushbutton Switches
   c. Selector Switches
   d. Pilot Lights
   e. Pushbutton Stations
   f. Using Multiple Pushbutton Stations
3. Temperature Switches
4. Pressure Switches
5. Mechanical Limit Switches
6. Flow Switches
7. Float Switches
8. Foot Switches
9. Jogging and Plugging Switches
10. Proximity Switches/Sensors
11. Inductive Proximity Sensors
12. Capacitive Proximity Sensors
13. Load-Powered Proximity Sensors
14. Line-Powered Proximity Sensors
15. Proximity Sensor Installation Guidelines
16. Photoelectric Switches/Sensors

Session V. Drum Switches; Enclosures; Diagrams

A. Drum Switches

B. Enclosures

C. Diagrams

1. Wiring Diagrams
2. Wire Lists
3. Control Ladder Diagrams
   a. Sections of a Ladder Diagram
   b. Conventions Used on Ladder Diagrams
4. Logic Diagrams
5. Relating Diagrams to Equipment Wiring and Operation
   a. Relating Ladder Diagrams to System Operation
   b. Relating Logic Diagrams to System Operation

Session VI. NEC Regulations for the Installation of Motor Control Circuits; Connecting Motor Controllers for Specific Applications

A. NEC Regulations for the Installation of Motor Control Circuits

B. Connecting Motor Controllers for Specific Applications

1. Controlling an Air Compressor Motor
2. Controlling One-Pump Motors
3. Controlling Two-Pump Motors
4. Controlling a Motor from Multiple Locations
5. Controlling the Reversal of a Three-Phase Motor
6. Controlling Conveyor System Motors

Session VII. Laboratory

A. Laboratory

Under your supervision, have the trainees make all connections for a magnetic motor controller, controlled by two pushbutton stations, including the connections for holding the circuit interlock. Note the proficiency of each trainee.

Session VIII. Review; Module Examination; Performance Testing

A. Review

B. Module Examination

1. Trainees must score 70% or higher to receive recognition from the NCCER.
2. Record the testing results on Craft Training Report Form 200 and submit the results to the Training Program Sponsor.

C. Performance Testing

1. Trainees must perform each task to the satisfaction of the instructor to receive recognition from the NCCER.
2. Record the testing results on Craft Training Report Form 200 and submit the results to the Training Program Sponsor.

Module 26311-02

Motor Controls

Course Map

This course map shows all of the modules in the third level of the Electrical curriculum. The suggested training order begins at the bottom and proceeds up. Skill levels increase as you advance on the course map. The local Training Program Sponsor may adjust the training order.

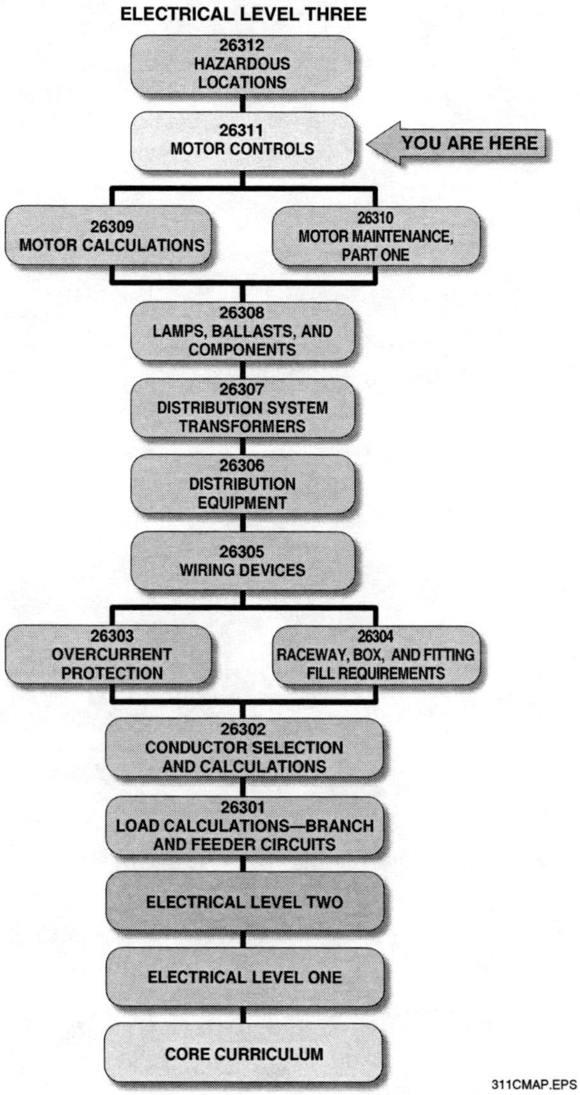

311CMAP.EPS

Homework

Assign reading of Module 26311.

MODULE 26311 CONTENTS

Instructor's Notes:

Figures

Instructor's Notes:

Tables

MODULE 26311

Motor Controls

Objectives

When you have completed this module, you will be able to do the following:

1. Identify contactors and relays both physically and schematically and describe their operating principles.
2. Identify pilot devices both physically and schematically and describe their operating principles.
3. Interpret motor control wiring, connection, and ladder diagrams.
4. Select and size contactors and relays for use in specific electrical motor control systems.
5. Select and size pilot devices for use in specific electrical motor control systems.
6. Connect motor controllers for specific applications according to NEC requirements.

Prerequisites

Before you begin this module, it is recommended that you successfully complete the following: Core Curriculum; Electrical Level One; Electrical Level Two; Electrical Level Three, Modules 26301 through 26310.

Required Trainee Materials

1. Pencil and paper
2. Appropriate personal protective equipment
3. Copy of the latest edition of the *National Electrical Code*

1.0.0 ◆ INTRODUCTION

This module is the second of three related modules that cover electrical motor controls. It expands upon the basic material presented in the Electrical Level Two module, *Contactors and Relays.* In addition to reviewing the principles of operation for contactors and relays, this module also provides information on selecting, sizing, and installing electromagnetic motor controllers and their control circuits. Included is coverage on common types of control circuit **pilot devices,** basic relay logic, and the different kinds of wiring diagrams used to describe motor control wiring schemes. Note that coverage of solid-state relays, controllers, and similar motor control devices is covered later in the Electrical Level Four module, *Advanced Controls.*

Relays and contactors/motor starters are used to control electrical power to various types of loads. Motor starters, also called *motor controllers,* are contactors with added motor overload protection devices sized for the motor load. Conventional contactors/motor starters and relays are electromagnetic, mechanical devices that operate very quickly and, if sized properly, minimize contact arcing caused by the opening (breaking) or closing (making) of an electrical circuit. Relays or contactors/motor starters are generally used in **motor control circuits** to amplify a signal. In this application, a low-current signal applied by a pilot device (switch or other control) to the relay or contactor/motor starter can be used to control the

Note: The designations "National Electrical Code," "NE Code," and "NEC," where used in this document, refer to the National Electrical Code®, which is a registered trademark of the National Fire Protection Association, Quincy, MA. *All National Electrical Code (NEC) references in this module refer to the 2002 edition of the NEC.*

Materials

Ensure that you have everything required to teach the course. Check the Materials and Equipment List at the front of this Instructor's Guide.

Classroom

Review the goals of the course.

Explain that terms shown in bold (blue) are defined in the Glossary at the back of this module.

Audiovisual

Show Transparency 1, Course Objectives.

Show Transparency 2, Performance Tasks.

Classroom

Discuss the functions provided by relays and contactors in motor circuits.

Classroom

List the types of overload protection used in motor circuits.

Explain that electromechanical relays (EMRs) are used in control circuits to operate low-voltage loads.

Discuss general-purpose EMRs. Refer trainees to Figures 1 and 2.

Discuss the common contact designations used with relays. Refer trainees to Table 1.

Audiovisual

Show Transparencies 3 and 4 (Figures 3 and 4).

application of higher-current power to a load. A large variety of relays or contactors/motor starters are used for lighting, motor, and HVAC control circuits. **Electromechanical relays (EMRs)** and contactors/motor starters are still common, although solid-state versions with no moving parts or contacts are rapidly finding their way into all types of applications. Special solid-state, thermal, or magnetic relay devices are used for motor protection. Depending on the device, some or all of the following types of protection can be provided:

- Current overload (time delay and/or instantaneous)
- Overvoltage and undervoltage
- Phase loss or unbalance
- Directional overcurrent
- Percentage of voltage or current differential

Contactors/motor starters are designed to handle higher current loads than most relays using relatively low-current control signals. Without overload protection, contactors are used to handle high-current, noninductive loads such as lighting or other resistance loads. Contactors/motor starters and their enclosures are available with National Electrical Manufacturers' Association (NEMA) or **International Electrotechnical Commission (IEC)** ratings. For identical load ratings, IEC-rated contactors are less expensive and of lighter-duty construction than the NEMA-rated contactors.

2.0.0 ◆ ELECTROMECHANICAL RELAYS

Electromechanical relays (EMRs) are normally used in control circuits to operate low-current loads. They range in size from subminiature versions rated for milliamp loads to power relays or mercury-displacement relays with average load ratings of 30A maximum at 277VAC or 15A maximum at 600VAC. Most relays, except for subminiature, reed, power, industrial, or mercury relays, are configured as plug-in devices. Subminiature relays are very small, sealed, electromechanical or solid-state relays that are typically used on printed circuit boards. Industrial relays are similar to the contactors covered later in this module. Reed relays are sensitive electromechanical relays with low-current contacts sealed in a glass envelope.

2.1.0 General-Purpose Relay Configurations and Designations

General-purpose EMRs are mechanical switches operated by a magnetic solenoid coil. They are available in various AC and DC voltage designs and current ratings. Coils can be specified in DC voltage ranges of 5V to 24V or in AC ranges of 12V to 240V, single phase. DC coils can be activated with as little as 4mA at 5VDC, making them compatible with certain integrated circuit logic gates. Relays can have up to 12 poles (separate, insulated, switching circuits) per device with various combinations of normally open and normally closed contacts. A normally closed (NC) contact or normally open (NO) contact is defined as the contact state that occurs when the coil is de-energized. *Figure 1* is an example of a single phase, 25A, 240V, open-frame AC power relay with two poles (double pole or DP). Each pole has an NC and NO contact with a common single break (SB) moving contact. When a pole is equipped with both NC and NO contacts, it is designated as a double-throw (DT) pole. The power relay shown is thus defined as a DPDT-SB relay. *Figure 2* is a miniature, single phase, 5A, 240VAC plug-in relay with four poles (4P). Each pole has an NC and NO contact (DT) with a common SB moving contact. This relay is defined as a 4PDT-SB relay.

As just explained, relays are designated by their number of poles, throws, and breaks. Relay pole and contact designations are shown in *Table 1*, and various relay contact configurations are shown in *Figure 3*. Some of these pole and contact designations are also used for contactors.

Manufacturers use a common code to simplify the contact sequence identification for relays. The code uses form letters to indicate the type of contact sequencing applicable to each pole of a relay, as shown in *Figure 4*.

Table 1 Relay Contact and Pole Designations

Designation	Meaning
ST	Single Throw
DT	Double Throw
NO	Normally Open
NC	Normally Closed
SB	Single Break
DB	Double Break (industrial relays)
SP	Single Pole
DP	Double Pole
3P	Three Pole
4P	Four Pole
5P	Five Pole
6P	Six Pole
(N)P	N = numeric number of poles

Instructor's Notes:

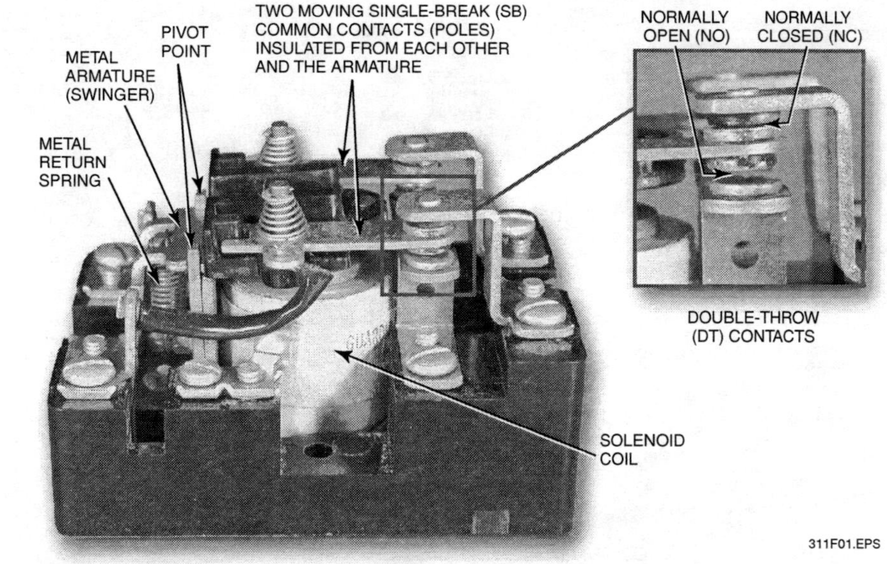

Figure 1 ◆ Typical double-pole, double-throw, single-break (DPDT-SB), open-frame power relay.

Figure 2 ◆ Miniature four-pole, double-throw, single-break (4PDT-SB), plug-in relay.

2.2.0 Typical Operation

In the single-pole, single-break version of an electromechanical power relay shown in *Figure 5*, operation occurs when a control signal voltage is applied to the relay solenoid coil. The magnetic field created through the core of the solenoid coil instantly attracts the metal armature and draws it into contact with the core. The moving contact, attached to the armature by the insulating block and contact pressure spring (see *Figures 5* and *6*), breaks from the NC contact and makes with the NO contact. The input power at the common (COM) terminal is routed through the moving contact to the circuits connected to the NO terminal. When the control signal is removed from the solenoid coil, a return spring pulls the armature back to a de-energized position, thus causing the moving contact to break with the NO contact and make with the NC contact. Input power from the COM terminal is then switched back from the NO contact circuits to the NC contact circuits.

When the solenoid coil is specified for use with an AC control signal, a shading coil (aluminum or copper) at the top of the core (see *Figures 5* and *6*) is used to create a weak, out of phase, auxiliary magnetic field. As the main field collapses when the AC current periodically drops to zero, the weak field generated by the shading coil is strong enough to keep the armature in contact with the core and prevent the relay from chattering at a 120Hz rate. If the shading coil is loose or missing, the relay will produce excessive noise and be subject to abnormal wear and coil heat buildup. In a power relay of this size, the moving contact arm must be made of relatively thick metal to accommodate the current rating of the relay. Lighter-duty relays use a flexible, thin-gauge copper spring stock for the moving contact arm. Because the armature is designed to swing over a slightly greater distance than necessary in both directions

Classroom

Describe how an EMR operates.

Audiovisual

Show Transparency 5 (Figure 5).

Demonstration

Disassemble a contractor and describe the function of each component. Refer trainees to Figure 6.

Teaching Tip

Draw various types of relay and contact configurations on the whiteboard/chalkboard and have the trainees identify them.

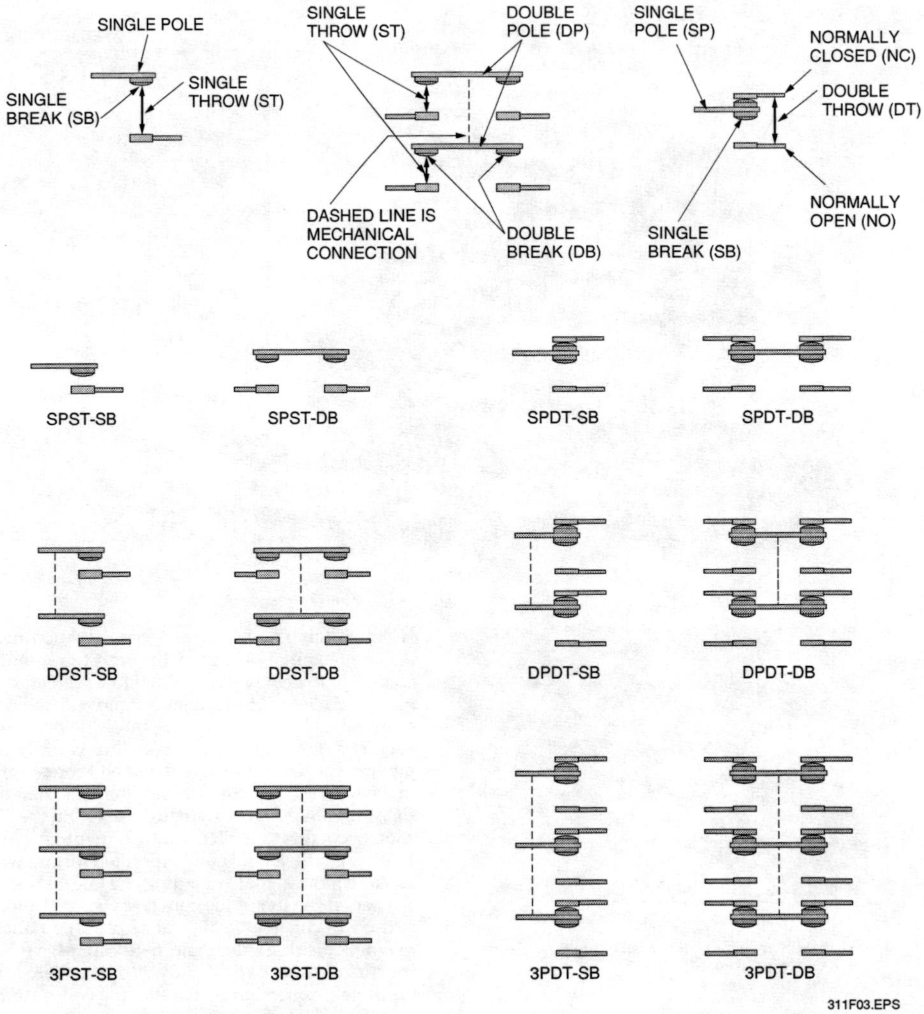

Figure 3 ◆ Examples of relay contact configurations.

Instructor's Notes:

(armature over travel), the flexible moving arm bends slightly in the normally open or closed positions. This eliminates any contact-pressure mating or bounce problems. In relays (or contactors) that use an inflexible moving contact arm, a contact-pressure spring is used to allow flexible positioning of the moving contact arm so that contact-pressure mating and bounce problems are eliminated. Like light-duty relays, this is accomplished by armature overtravel.

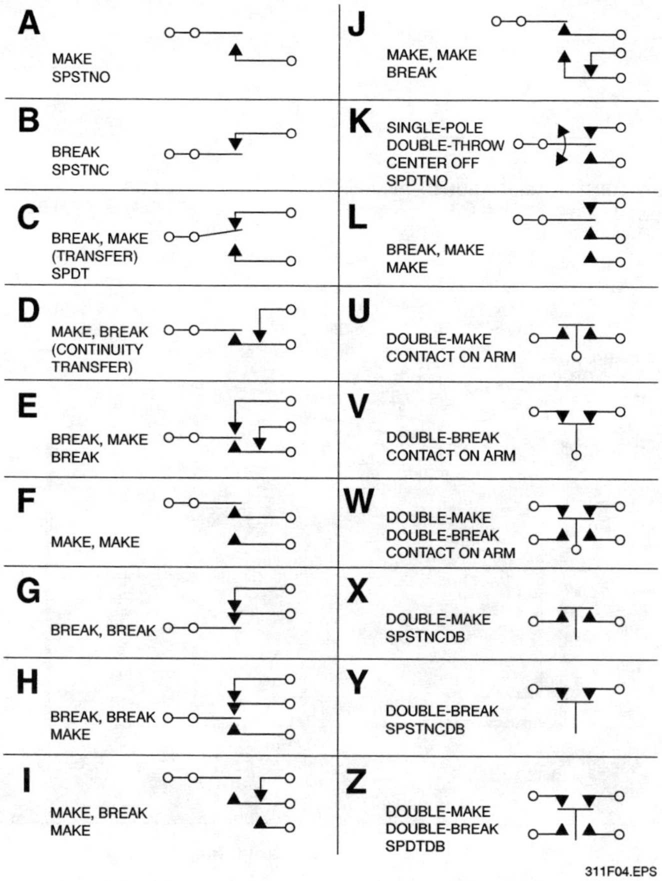

Figure 4 ◆ Relay contact form identification.

Contact Wiping

INSIDE TRACK

A degree of contact cleaning is accomplished by armature overtravel. As the contacts mate, the overtravel causes the contact surfaces to wipe slightly. This action removes oxides from the mating surfaces of the contacts and helps keep contact resistance low.

Classroom

Explain how the action of the armature helps to prevent the buildup of oxidation on the contact mating surfaces.

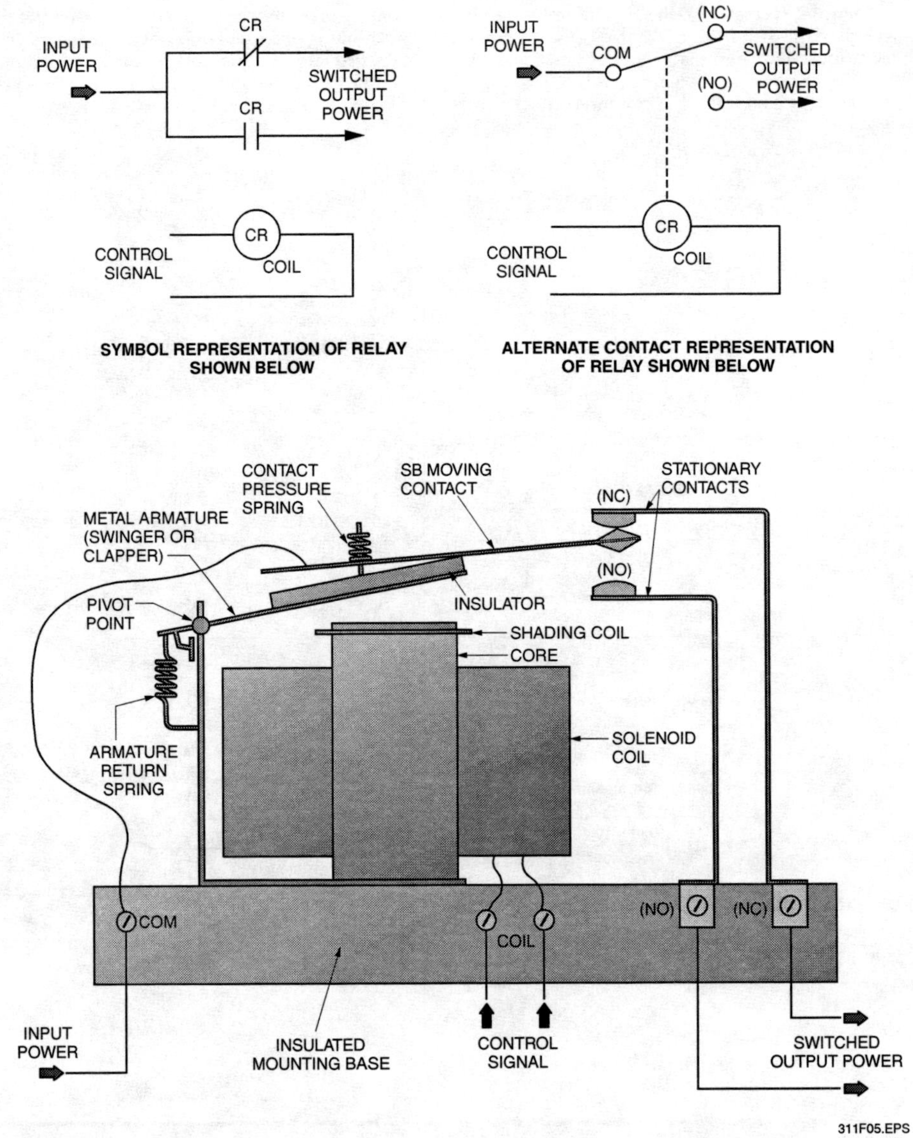

Figure 5 ◆ Typical SPDT-SB power relay and symbol representation.

Instructor's Notes:

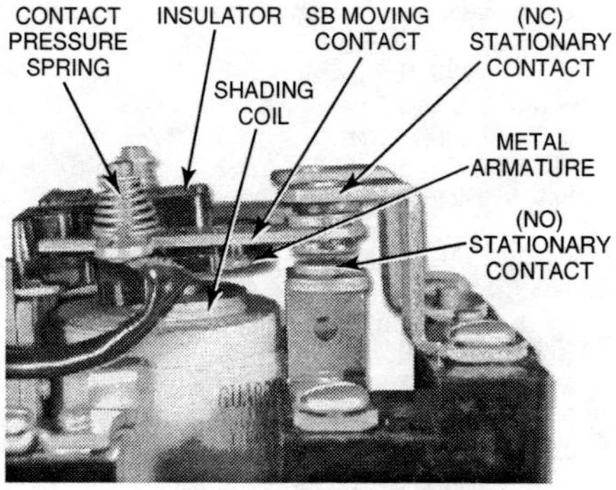

Figure 6 ◆ Detail view of a power relay.

2.3.0 Relay Selection Criteria

This section covers relay selection criteria, including coil voltage characteristics, contact ratings, and contact materials.

2.3.1 Coil Voltage Characteristics

The **pickup voltage** is the minimum allowable coil control voltage that will cause an electromechanical device to energize. Once energized, the **seal-in voltage** is the minimum allowable coil control voltage that will keep the device energized. It is usually less than the pickup voltage. The **dropout voltage** is defined as the coil voltage that is reached when the armature return spring overpowers the magnetic field of the coil and the contacts of the device change position. The dropout voltage is less than the seal-in voltage. Most coils on electromechanical devices such as relays and contactors must be designed so that the dropout does not occur until the voltage is reduced to 85% of the nominal coil voltage. They must also be designed so that they pick up when the voltage rises to more than 85% of the nominal coil voltage. These voltage levels are set by NEMA and are conservative. Most electromechanical devices manufactured today drop out and pick up at lower voltages.

2.3.2 Contact Ratings

The most important consideration in the selection of a relay (or contactor) for a particular application is the current rating of the contacts. However, the rating of contacts can be confusing. Sometimes only one rating is given, and it is important to know the definition of the rating. Many relays and other devices have three published ratings:

- Inrush current (make contact) capacity
- Normal or continuous-carrying capacity
- Current break (opening) capacity

For example, a typical industrial relay may have the following contact ratings at a particular AC voltage:

- 15A noninductive (resistive) continuous load
- 8A inductive continuous load
- 75A inrush, 50A break (inductive or resistive)

Two examples of resistive loads are heating elements and incandescent lights. Inductive loads are coils such as those used in solenoids, relay coils, and motor starter coils. The type of load and its inrush current, along with switching frequency, can cause contact welding. Therefore, for loads with inrush current, the steady-state and inrush currents should be measured to determine the selection of the proper contactor or relay contact rating. Some typical loads and their approximate inrush currents are summarized in *Table 2.*

2.3.3 Contact Materials

The contacts used in relays or contactors are available in a number of materials. These materials have certain advantages and disadvantages for various applications, as shown in *Table 3.*

2.4.0 Contact Arc Suppression

The typical contact life of a relay ranges between 100,000 and 500,000 operations. The contact rating

> **INSIDE TRACK**
>
> ***Hermetically Sealed Coils***
>
> Many coils used today are hermetically sealed (encapsulated) in plastic to protect against corrosive environments and can operate continuously at 110% of the nominal rated voltage without damage to the coil. However, to ensure proper and reliable relay operation, the specified nominal rated coil voltage should be maintained at the coil terminals when the coil is energized.

Classroom

Define the three voltages associated with coil control.

Discuss the three capacity ratings associated with contacts. Refer trainees to Table 2.

Discuss the types of materials used in the construction of contacts. Refer trainees to Table 3.

Explain that the contact life of a relay can support up to half a million operations.

Classroom

Point out that relays operated at excess voltages have reduced contact life due to arcing. Explain that arc suppression circuits provide an alternate path for the voltage.

Audiovisual

Show Transparency 6 (Figure 7).

Classroom

Describe how arc suppression devices operate.

of a relay is based on the contact's full-rated power. When contacts switch loads that are less than their full-rated power, contact life is increased. If loads exceed the contact rating, or if arcing occurs, the life of the contact is shortened due to burning and overheating.

In some applications involving control relays, arc protection circuits can be added to the relay-controlled circuit to reduce the arcing caused by inductive loads. The circuit provides a nondestructive path for the voltage generated by the collapsing field of the inductive load when the relay contacts open. *Figure 7* shows three different protection circuits that can be used for arc suppression.

In the DC diode protection circuit, a diode can be placed in parallel with the load so that it will oppose any current flow when the load is energized by the relay contacts. When the relay contacts open, the reverse polarity voltage generated by the collapsing field is shorted out by conduction through the diode. By shorting out the generated voltage, the resulting current is dissipated across the load instead of the relay contacts. The diode on-time and resulting dissipation time can be controlled by the use of a zener diode. The cutoff voltage characteristic of the zener can be used to halt the dissipation across the load very shortly after the contacts have opened.

In an AC circuit, arc suppression is less of a problem because the arc is extinguished wherever the current passes through zero. This is why contacts of relays rated for AC use can tolerate a much smaller separation at break than DC contacts. Arcing can still occur, however, and a resistor/capacitor (RC) circuit or a varistor can be used for contact protection to extend contact life. In the RC protection circuit, the resistor/capacitor combination is selected so that the circuit is effectively resistive and its time constant is approximately

Table 2 Typical Loads and Their Inrush Currents

Type of Load	Approximate Inrush Current
Resistive heating	Steady-state current
Sodium vapor lamps	1 to 3 times steady-state current
Mercury lamps	About 3 times steady-state current
Motors	5 to 10 times steady-state current
Transformers	5 to 15 times steady-state current
Incandescent lamps	10 to 15 times steady-state current
Solenoids	10 to 20 times steady-state current
Capacitive loads	20 to 40 times steady-state current

Table 3 Contact Material Characteristics

Material	Advantages	Disadvantages
Silver	This metal has the highest conductivity and thermal properties of any contact material.	This metal is subject to welding and sticking under arcing conditions. Rapid sulfidation (tarnishing) in many applications creates a film that increases contact resistance. Normal contact wiping usually removes the film. It is not used in intermittent or arcing applications.
Silver-cadmium alloy or silver-cadmium oxide alloy	These alloys have good conductivity and thermal properties. Cadmium oxide alloy conducts even when oxide forms on surface of contacts.	These alloys resist arcing damage but are subject to some sulfidation. Normal contact wiping usually removes the film. When used in circuits drawing several amperes at more than 12V, any sulfidation is burned off.
Gold-flashed silver	This metal has the same advantages as silver. Gold flashing protects against sulfidation. It is used in intermittent applications and is good for switching current of 1A or less.	This metal is not used in applications where arcing occurs because gold burns off quickly.
Tungsten or tungsten-carbide alloy	These alloys experience minimal damage from arcing due to their high melting temperature. They are good for high-voltage and repetitive switching applications.	These alloys offer higher contact resistance than other materials.
Silver tungsten-carbide alloy	This alloy has the same advantages as tungsten or tungsten-carbide but lower contact resistance.	This alloy is subject to minor sulfidation; however, wiping action or any arcing removes the film.

Instructor's Notes:

> **INSIDE TRACK — *Transient Suppression***
>
> Some manufacturers offer transient (arc) suppression devices as accessories for use across the coils of their relays or contactors. These devices are primarily designed for noise suppression caused by arcing.

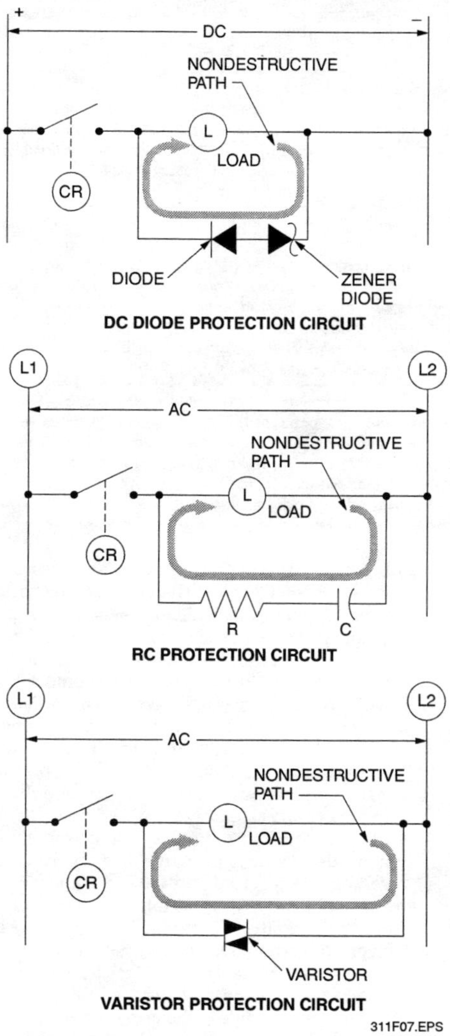

Figure 7 ◆ Contact protection circuits.

equal to the time constant of the load. This provides a path through the load to dissipate the generated voltage and current when the relay contact is opened. In the varistor protection circuit, the varistor is a solid-state device whose resistance is inversely proportional to the voltage applied across it. When the relay contacts are closed and voltage appears across the load and the varistor, the varistor resistance is high. When the relay contacts open, a low-resistance path is provided through the varistor to dissipate the generated voltage and current across the load. The disadvantages of these circuits are that they slow down coil reaction time and they cannot be used in electronic timer-controlled circuits.

Other circuits that are placed across the control relay contacts can be used for arc suppression, but they are not fail-safe. If a component shorts in a suppression circuit across the relay contacts, the load may remain energized even though the relay contact opens.

3.0.0 ◆ MAGNETIC CONTACTORS

Magnetic contactors, like relays, are control devices that energize or de-energize loads by the use of a control signal. They are used instead of relays to repeatedly establish and interrupt higher current loads. They are usually available from most manufacturers in four categories for the applications summarized in *Table 4*.

3.1.0 Magnetic Contactor Construction and Operation

Like a relay, a magnetic contactor is actuated by a solenoid. *Figure 8* shows three types of mechanical actions used for solenoid closure of contactors. In all cases, energizing the solenoid coil assembly, made up of a frame, coil, and moving armature, causes a moving contact assembly with spring-loaded bridging contacts (one contact for each pole) to mate with corresponding pairs of stationary contacts to close a circuit. Contactors typically have between one and six poles. The bridging contacts and the materials used for the contacts themselves allow the higher current ratings for

Classroom

Discuss the basic categories of magnetic contactors. Refer trainees to Table 4.

Describe the three types of mechanical actions used to operate magnetic contactors.

Audiovisual

Show Transparency 7 (Figure 8).

Classroom

Discuss the function and operation of typical mechanical contactors. Provide examples for the trainees to examine. Refer trainees to Figures 9 through 12.

Table 4 Contactor Categories and Applications

Category	Application
Lighting contactors	These contactors are current-rated only for resistive loads and do not have horsepower ratings for motor use. The current rating of these contactors is for the maximum continuous current required by a resistive load. At the maximum current rating, the contactors are designed to withstand the large initial inrush currents of tungsten and ballast lamp loads, as well as nonmotor (resistive) loads, without contact welding. They are generally available in versions that are locally or remotely controlled via AC or DC control circuits and are magnetically held, mechanically held, or magnetically latched. The mechanically held or latched contactors are quiet (no AC hum) and remain closed during power interruptions so that their loads will come back on when power is restored.
Definite-purpose contactors and motor starters	Low-duty cycle, definite-purpose contactors and motor starters are intended for use in applications where the control requirements are well defined. These contactors carry dual ratings (ampere and horsepower) for either resistive or inductive loads found in applications such as refrigeration, air conditioning, resistance heating, and other Standard Industrial Classification (SIC) applications. The motor starters are usually equipped with bimetallic motor overcurrent protective devices (overload relays).
NEMA-rated contactors and motor starters	These are contactors for general use and for use as motor starters when equipped with overcurrent protective devices (overload relays). They are identified in eleven overlapping ranges (sizes) and rated in horsepower (for motor starters) and/or continuous current capacity. They are primarily designed for use with inductive motor loads and to withstand the interrupt current of a locked rotor. They are designed with reserve capacity to perform over a broad range of applications without the need for an assessment of life requirements. The motor starters are either manual or magnetically actuated and are equipped with melting-alloy, bimetallic, or solid-state overload relays. Most of these contactors have replaceable contacts and encapsulated (sealed) coils.
IEC-rated contactors and motor starters	These contactors and motor starters perform the same functions as NEMA-rated devices but are smaller than NEMA devices for the same horsepower or current ratings. As a result, they are very application sensitive and may require rating and life assessment matches to the load. U.S. manufacturers normally provide size tables similar to NEMA tables to facilitate selection. The motor starters are usually equipped only with bimetallic overload relays. Most of these contactors do not have sealed coils and, except for the larger sizes, have nonreplaceable contacts.

contactors. The angled contacts allow a slight, random wiping (cleaning) action as the contacts open and close. Coil voltage characteristics and contact rating information is the same as for relays.

Figure 9 is a typical NEMA-rated magnetic contactor. It is a three-pole (3Ø) power device that is closed by a mechanical bell crank when the solenoid coil assembly is energized. The energized indicator shows if the contactor is de-energized (flush) or energized (depressed). It can also be manually depressed to check if the contactor mechanism is free to move and the contacts can be closed. Except for certain IEC or small contactors, most contactors can be disassembled and individual components replaced. For the contactor shown, the coil and contacts can be individually replaced. Other types of contactors, including definite-purpose and lighting contactors, are similarly constructed.

For coil replacement, the two cover screws (see *Figure 9*) can be loosened or removed and the cover removed (*Figure 10*). This exposes the solenoid coil assembly and its positioning clips that are released when the cover is removed. For this contactor, lugs on the cover are forced down between the clips and the contact mechanism housing to clamp the solenoid assembly in place horizontally when the cover is replaced. After the wires to the solenoid coil assembly are removed, the assembly can be lifted out of the contact mechanism housing (*Figure 11*). When assembled, the bell-crank lever hook of the armature fits over the spring-returned bell-crank lever. Energizing the coil causes the armature to pull in and move the bell crank (against a return spring) so that the moving contact assembly is pushed down to mate with the stationary contacts under the contact mechanism housing. When the solenoid coil assembly is energized and the armature is pulled in, a slight air gap is maintained between the frame and armature for two reasons. The first is to prevent the laminated steel armature from slamming against the laminated steel frame, and the second is to prevent any residual magnetism in the frame from holding the armature in when the coil is de-energized. *Figure 12* shows the laminated frame and armature separated from the encapsulated coil. Like a relay, the legs of the coil

Instructor's Notes:

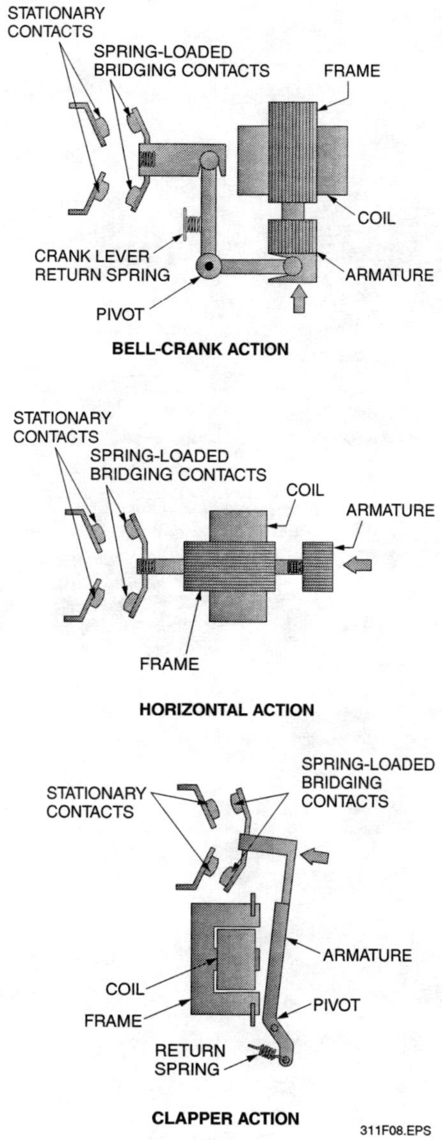

Figure 8 ◆ Mechanical actions used for magnetic contactors.

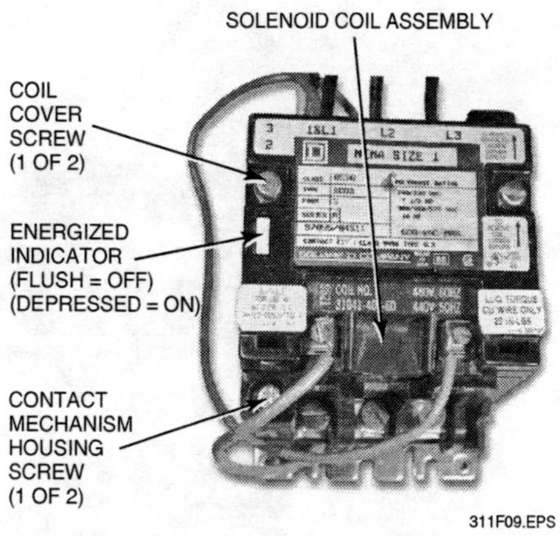

Figure 9 ◆ Typical bell-crank actuated three-pole magnetic contactor.

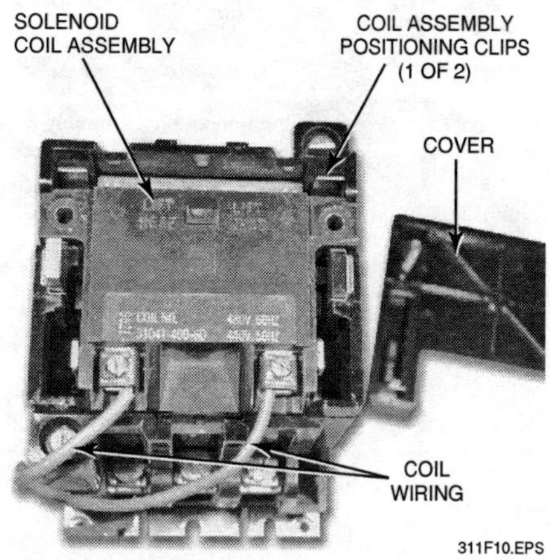

Figure 10 ◆ Contactor with cover removed.

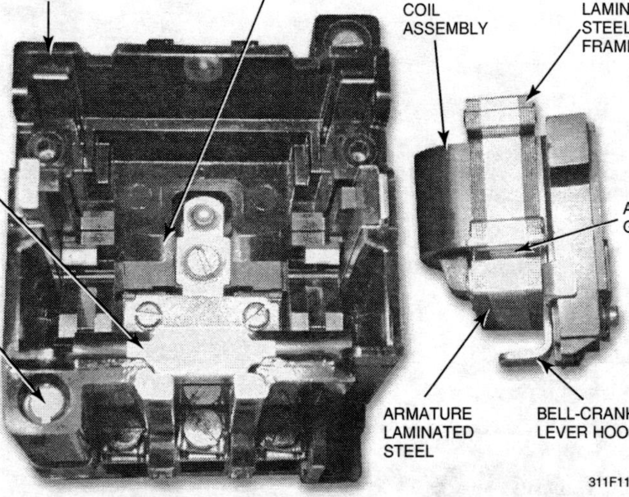

Figure 11 ◆ Contactor with solenoid coil assembly removed.

Figure 12 ◆ Disassembled solenoid coil assembly.

Instructor's Notes:

INSIDE TRACK

Shading Coils

Missing or loose shading coils can cause excessive contactor noise or wear, along with coil damage due to overheating.

Replacing Contacts

If any contacts must be replaced, all moving contacts and stationary contacts for all poles must be replaced to maintain equal contact resistance and pressure.

frame used with an AC coil are each fitted with shading coils to prevent chattering of the armature when the coil is energized.

To expose the contacts, the contact mechanism housing mounting screws can be loosened, and the contact mechanism housing removed from the contactor base (*Figure 13*). Even though shown disassembled from the housing, the solenoid coil assembly does not have to be removed to gain access to the contacts. A close-up of the moving contact assembly with replaceable spring-loaded bridging contacts and corresponding replaceable stationary contacts in the contactor base is shown in *Figure 14*. Each bridging contact is removed by sliding it out from under its spring-loaded clip in the moving contact assembly. Each pair of stationary contacts can be removed by removing the screws securing them to the contactor base.

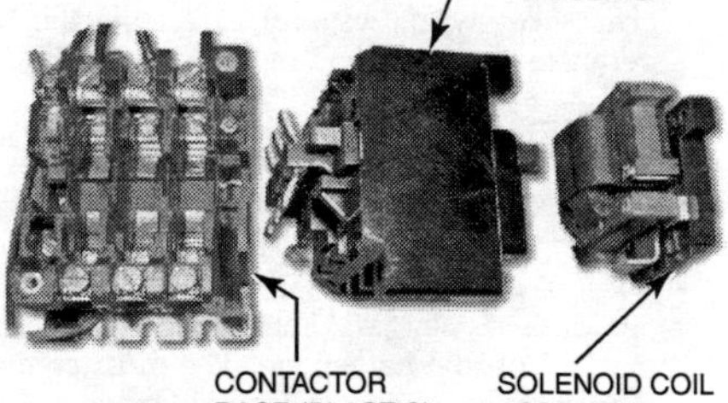

Figure 13 ◆ Disassembled contactor.

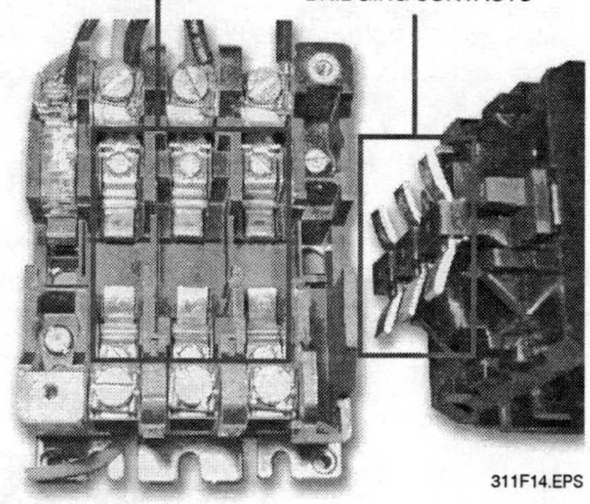

Figure 14 ◆ Close-up view of stationary and moving contacts.

3.2.0 Contactor Marking Conventions

Most contactors reflect various labeling/marking conventions that are in general use by manufacturers. Many of these conventions are because of Underwriters Laboratories, Inc. (UL) requirements or NEMA and IEC standards. Because many NEMA devices conform to IEC standards and vice versa, marking conventions used by manufacturers may include elements of both on the devices.

3.2.1 UL Markings

There are three categories of UL marking applied to contactors and motor starters:

- *UL Listed*—Products that are UL Listed must have a UL Listing mark (UL enclosed in a circle) on their nameplates adjacent to the ratings to which the marking applies (*Figure 15*). Next to the mark, the words "Listed Industrial Control Equipment" or "Listed Ind. Cont. Eq." must appear. Equipment that is UL Listed is suitable for installation with general-use tools and for application at the ratings to which the listing mark is related. A UL Listed product complies with a governing UL Standard and qualifies for installation under specific provisions of the NEC. Other ratings and certifications such as IEC horsepower or kW ratings or Canadian Certifications (CE or CSA) may also appear on the nameplate, but they are segregated from the ratings associated with the UL Listing mark.

Demonstration

Disassemble a contactor and describe the function of each component. Refer trainees to Figures 13 and 14.

Classroom

Discuss the three categories of UL markings.

Classroom

Discuss the ratings found on NEMA-rated contactors and motor starters. Refer trainees to Figure 15.

Define the ratings found on IEC-rated contactors and motor starters.

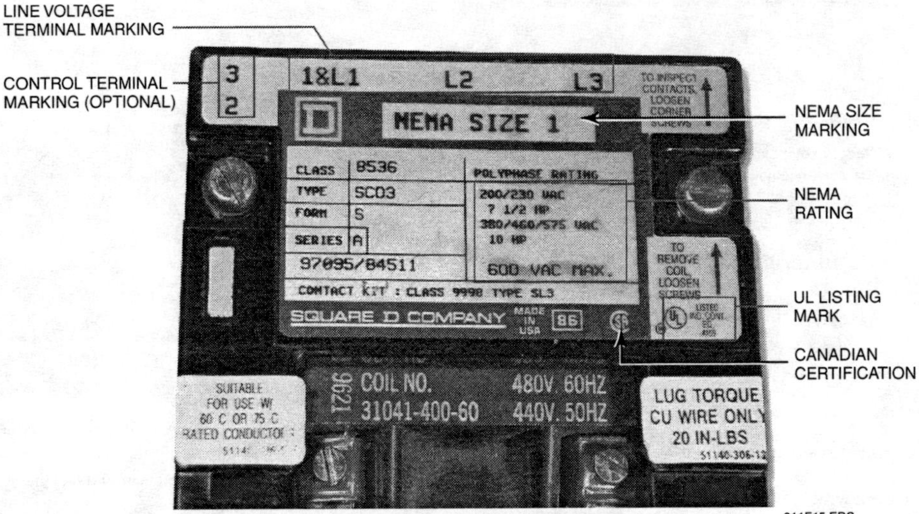

Figure 15 ◆ Typical NEMA contactor nameplate showing size, rating, and UL Listing mark.

- *UL Recognized*—A product that is UL Recognized carries the UL Recognized Component Mark (called a backward UR), a printed, boldfaced, back-slanted letter U and R, joined and reversed as they would appear in a mirror (ЯU). This mark must appear adjacent to the manufacturer's identification and catalog number. UL Recognized equipment is not suitable for general use and must be combined with other items, under stated condition of acceptability, into another product that can be UL Listed. The conditions of acceptability are published in a required manufacturer's UL Component Recognition Report.
- *UL Classified*—A product that is UL Classified carries the circled UL mark and the word classified along with a notation that Underwriters Laboratories has evaluated the product for compliance with a specific characteristic, standard, or part of a standard. UL Classification of a product for these items has no bearing on the product's ability to comply with the NEC.

It is important to note that not all functions or ratings shown on a nameplate of a UL Listed, Recognized, or Classified product are qualified for use under all articles of the NEC or meet other UL standards.

3.2.2 NEMA Nameplate Ratings

The following ratings are normally marked on NEMA-rated contactor or motor-starter nameplates:

- *NEMA size*—The NEMA size designation is a standardized rating system for contactors and motor-starters (see *Figure 15*). As defined by NEMA standards, horsepower, voltage, frequency, and/or current ratings are assigned for each size.
- *Horsepower and voltage*—The maximum horsepower at various voltages that correspond with values assigned for the designated NEMA size.
- *Continuous current*—In addition to or instead of horsepower marking, some contactors are marked with the maximum current that an enclosed starter or contactor may carry continuously or switch without exceeding the temperature rise permitted by NEMA standards.

3.2.3 IEC Nameplate Ratings

The following ratings are normally marked on IEC contactor or motor-starter nameplates:

- *HP and kW*—The maximum rating for each operational voltage (U_e) and utilization category marked on the nameplate. The most common ratings given are for Utilization Category AC-3.
- *Utilization category*—Describes the types of service for which the controller is rated, such as AC-1, AC-2, AC-3.

Instructor's Notes:

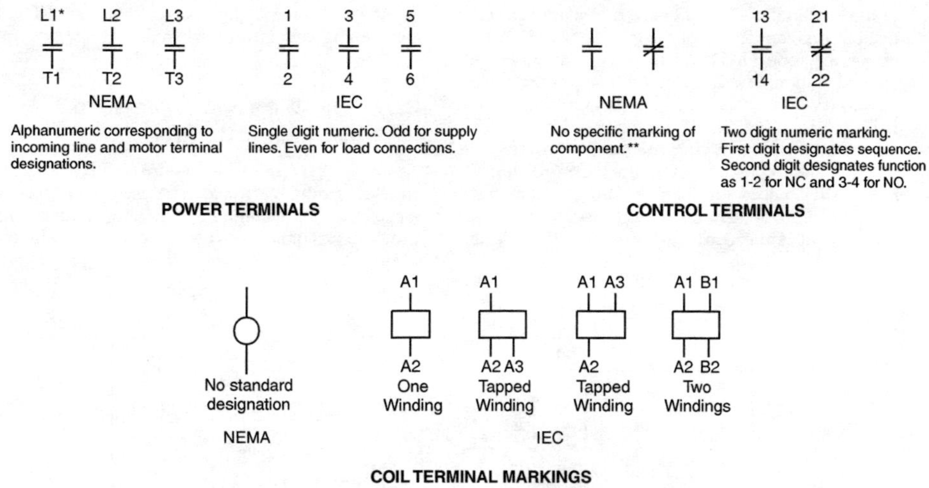

* Some manufacturers add a 1 before the L1 to designate a control-circuit voltage source connection (See *Figure 15*).
** Some manufacturers label a factory-installed control circuit (NO) with terminal numbers 2 and 3 (See *Figures 15* and *17*).

311F16.EPS

Figure 16 ◆ Conventional terminal markings.

- *Thermal current (I_{th})*—The maximum current a contactor or starter, outside an enclosure, can carry continuously without exceeding the temperature rise allowed by the IEC standard. This is not a load-switching rating.
- *Rated operational current (I_e)*—The maximum full-load current (FLC) at which a motor starter or contactor may be used for a given combination of voltage, frequency, and utilization category. A device may have more than one operational current.
- *Rated insulation voltage (U_i)*—A parameter sometimes shown that defines the insulation properties of the controller. This parameter is not usually used for selection or application purposes.
- *Rated operational voltage (U_e)*—The voltage required for each listed hp or kW rating.
- *Standard designation*—The specific IEC standard to which the product has been tested.

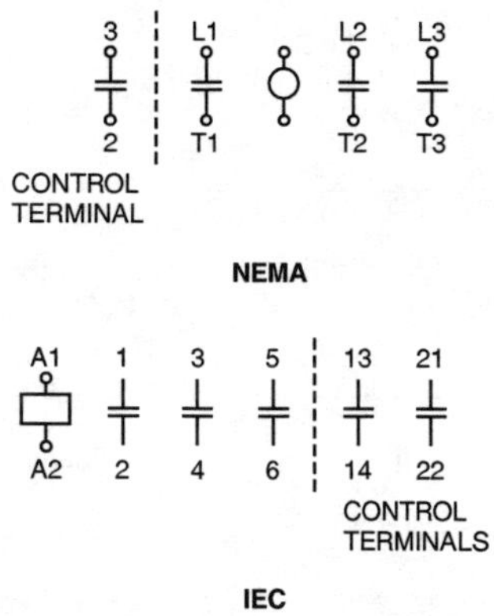

311F17.EPS

Figure 17 ◆ Typical contactor/motor starter marking and symbols.

3.2.4 NEMA and IEC Terminal Marking/Symbol Conventions

The conventions shown in *Figures 16* and *17* for NEMA are generally used for all North American and Canadian products including lighting, general-purpose, and IEC contactors sold in these countries. Many IEC contactors retain the IEC marking as additional information. However, imported equipment may only reflect the IEC marking.

The most commonly accepted wiring practice for NEMA-marked products is to connect all terminals together that have the same terminal marking. For example, in power circuits, the terminal marked T1 on a motor starter is connected

Classroom

Compare the NEMA and IEC conventions for terminal markings.

Audiovisual

Show Transparencies 8 and 9 (Figures 16 and 17).

Classroom

Explain the operation of two-wire and three-wire contactor controls using basic schematic diagrams.

Audiovisual

Show Transparency 10 (Figure 18).

to the terminal marked T1 on a motor, and in control circuits, terminals marked 1, 2, and 3 on the starter are connected to terminals 1, 2, and 3 on the control station. The IEC devices use power terminals marked 1, 3, 5, and 2, 4, 6 that correspond to L1, L2, L3 and T1, T2, T3 on NEMA devices. IEC device terminals carrying the same numbers are not normally connected together. Control terminals on IEC contactors have two-digit markings; the first digit designates location sequence, and the second is the control-device function. Most NEMA coils are single voltage; however, IEC coils can be single or multi-voltage.

3.3.0 Two-Wire and Three-Wire Contactor Control

As shown in *Figure 18,* two-wire or three-wire control using pilot devices can be used to energize an electrically held magnetic contactor. Pilot devices are switches or sensors used to control a contactor.

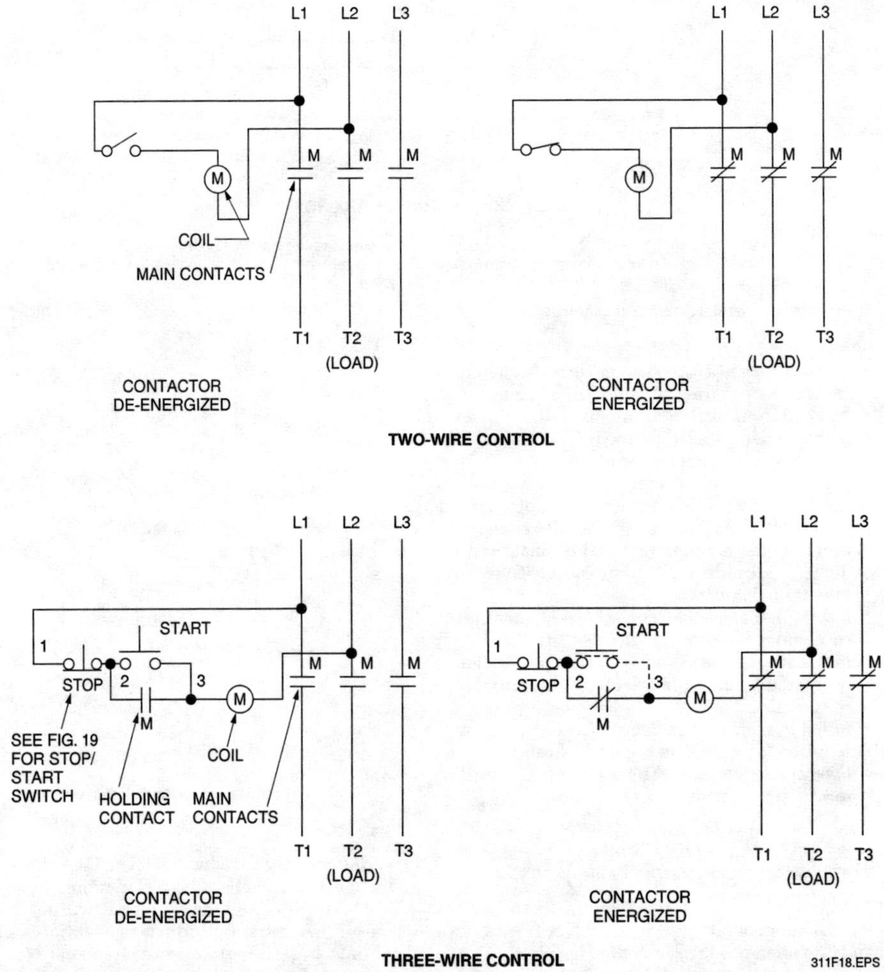

Figure 18 ◆ Two-wire and three-wire control schematics.

Instructor's Notes:

Three-wire control is accomplished with a pair of momentary contact pushbutton switches (*Figure 19*) and is used when loss-of-voltage protection and/or personnel safety is involved. In the simplest configuration (see *Figures 18* and *19*), a three-phase contactor is initially energized when the START pushbutton switch is momentarily depressed to connect L1 (via wire 1) through the closed STOP switch and the contactor coil to L2. To keep the contactor energized, a holding (seal-in) **auxiliary contact** is paralleled across the START switch via wires 2 and 3. The holding contact is normally located on the left side of NEMA contactors and is mechanically closed by a plunger on the side of the moving contact assembly when the contactor is initially energized. If factory installed, the holding contact on NEMA contactors is sometimes marked as terminals 2 and 3.

When closed by the moving contactor assembly, the holding contact bypasses the START pushbutton switch contacts so that after the pushbutton is released, control voltage will remain applied to the coil. If the source of the control voltage is momentarily interrupted or the STOP pushbutton is momentarily depressed, the contactor de-energizes, the auxiliary contact opens, and the contactor remains de-energized until the START pushbutton is depressed again. In the pictured contactor, the auxiliary contact is a momentarily actuated snap-action switch assembly that can be replaced by lifting it out of its mounting slot in the contactor base.

Two-wire control of an electrically held magnetic contactor is accomplished with a simple on/off toggle switch, latching start/stop switch, or other toggling pilot device without the use of a holding contact. In this case, loss of voltage causes the contactor to de-energize, but when power is restored, the contactor will automatically reenergize if the pilot device is still closed.

4.0.0 ◆ OVERLOAD PROTECTION

Overload protection for an electric motor is necessary to prevent burnout and ensure maximum operating life. If permitted, an electric motor will operate at an output of more than its rated capacity. Conditions of motor overload may be caused by an overload from driven machinery, a low line voltage, or an open line in a polyphase system that results in single-phase operation. Under any condition of overload, a motor draws excessive current that causes overheating. Since motor winding insulation deteriorates when subjected to overheating, there are established limits on motor operating temperatures. To protect a motor from overheating, overload relays are employed on motor controls to limit the amount of current drawn. Overload protection is also known as running protection.

The ideal overload protection for a motor is an element with current-sensing properties that are very similar to the heating curve of the motor (*Figure 20*), which would act to open the motor circuit when the full-load current is exceeded. The operation of the protective device should allow the motor to carry short overloads but should also disconnect the motor from the line when an overload has persisted for too long.

Single-element, nontime-delay fuses are not designed to provide overload protection. Their basic function is to protect against short circuits (overcurrent protection). Motors draw a high inrush current when starting, and conventional

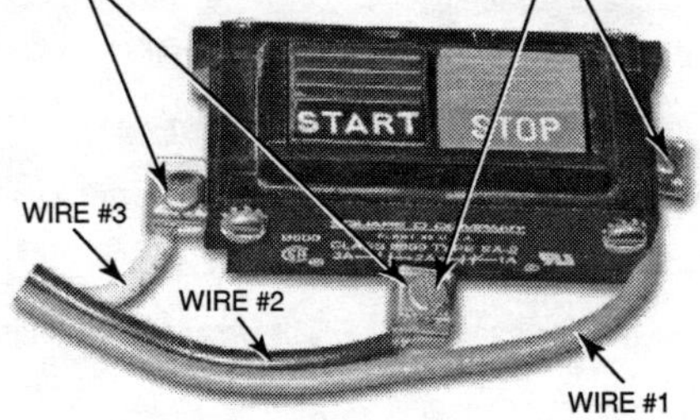

Figure 19 ◆ Momentary START/STOP switch assembly.

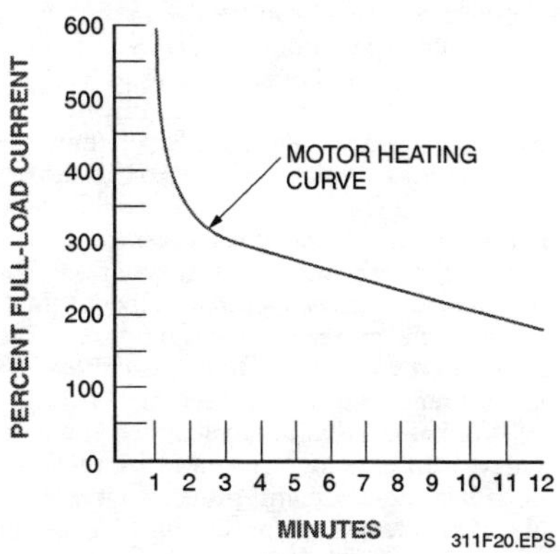

Figure 20 ◆ Typical motor heating curve.

Classroom

Describe the operation of a momentary START/STOP switch assembly. Refer trainees to Figure 19.

Explain that overload relays are essential to preventing overheating and burnout. Discuss the heating curve for a typical motor.

Audiovisual

Show Transparency 11 (Figure 20).

Classroom

Discuss the operating characteristics of overload relays.

List the basic categories of overload relays. Provide examples for the trainees to examine.

Discuss the construction of a melting-alloy overload relay. Refer trainees to Figure 21.

single-element fuses have no way of distinguishing between this temporary and harmless inrush current and a damaging overload. Such fuses, if chosen based on motor full-load current, will blow every time the motor is started. On the other hand, if a fuse is chosen large enough to pass the starting or inrush current, it will not protect the motor against smaller long-term, harmful overloads that might occur later. Dual-element, time-delay fuses can provide motor overload protection but suffer the disadvantages of being nonrenewable and must be replaced.

The overload relay is the heart of motor protection. As stated previously, it normally has inverse trip-time characteristics, permitting short high-current draw during the motor accelerating period (when inrush current is drawn), yet providing protection on overloads above the full-load current when the motor is running. Unlike dual-element fuses, overload relays are renewable and can withstand repeated trip and reset cycles without requiring replacement. However, overload relays, unless also equipped with an instantaneous overcurrent trip device, cannot take the place of overcurrent protective equipment.

The overload relay consists of a current-sensing unit connected in line with the motor, plus a mechanism that is actuated by the sensing unit and serves to directly or indirectly break the circuit. In a manual starter, an overload trips a mechanical latch and causes the starter contacts to open and disconnect the motor from the line. In a magnetic starter, an overload opens a set of contacts within the overload relay itself. These contacts are wired in series with the starter coil in the control circuit of the magnetic starter. Breaking the coil circuit causes the starter main contacts to open, disconnecting the motor from the line.

Overload relays can be thermal, magnetic, or solid-state. Solid-state overload relays are covered in the Electrical Level Four module, *Advanced Motor Controls*. As the name implies, thermal overload relays rely on rising temperatures caused by an overload current to trip the overload mechanism. They are also known as inverse-time overload relays because their response time varies inversely to the amount of current flow. The ideal temperature curve of an overload relay corresponds to a motor overheating curve. Magnetic overload relays react only to current excesses and are not reflective of motor temperature.

Thermal overload relays can be further subdivided into melting-alloy and bimetallic types. Both melting-alloy and bimetallic types are usually available in three or four NEMA-rated classes of trip time. Class 10 devices must trip within 10 seconds, Class 15 within 15 seconds, Class 20 within 20 seconds, and Class 30 within 30 seconds. The three most common classes available for NEMA devices are Class 10, Class 20 (Standard), and Class 30. IEC starters usually use equivalent type Class 10 devices.

Thermal units (heaters) for both melting-alloy and bimetallic overload relays should always be selected based on the full-load current (FLC) rating given on the motor nameplate. The FLC is also known as the full-load ampere (FLA) rating of the motor. In the event that the plate is missing, the full-load current can be measured when the motor is operating under its maximum load conditions. As a last resort, manufacturers have charts that give approximate FLC (FLA) values for motors with various horsepower ratings operating at different voltages. However, any heaters that are chosen based on tables or measurements may have to be changed if nuisance tripping occurs.

4.1.0 Melting-Alloy Thermal Overload Relays

Figure 21 is a typical melting-alloy thermal overload relay used with a three-pole contactor. It comprises a housing with three changeable melting-alloy heater assemblies, an internal spring-loaded trip mechanism that operates an NO switch (overload contact), and a reset button. The alloy, called a eutectic alloy, is a mixture of low-melting point metals that always melts at a fixed temperature and rapidly changes from a

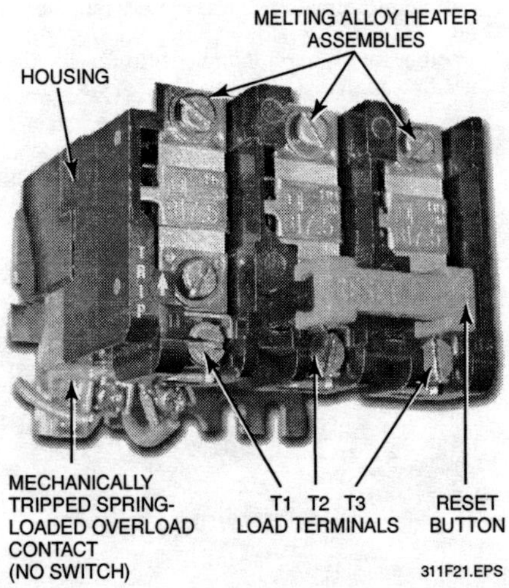

Figure 21 ◆ Typical melting-alloy overload relay.

Instructor's Notes:

solid to a liquid. The alloy is also unaffected by repeated melting and re-solidifying cycles. *Figure 22* shows a highly simplified version of a melting-alloy overload relay along with its symbol representation and a typical application. When the alloy in the solder pot is solid, the shaft of the ratchet wheel is locked in place and the trip lever holds the overload contacts closed. The closed overload contact is in series with the coil of a contactor and allows the contactor to be energized. Motor current flows from the closed contactor contacts through the overload heaters and out to the motor load connected to T1, T2, and T3. Any excessive overload motor current, passing through any one of the heater elements, eventually raises the temperature enough to cause the alloy in the integral solder pot to melt. The ratchet wheel shaft then turns in the molten alloy because of the upward pressure of the spring-loaded trip lever. By mechanical linkage, this action causes the overload contact to open, which, in turn, de-energizes the contactor, stopping the motor.

Figure 23 shows one of the melting-alloy heater assemblies removed from the overload relay shown in *Figure 21* and turned over to show the ratchet wheel and solder pot containing the alloy. Also visible is one of the three trip levers that, through a common mechanism, opens the spring-loaded overload contact. A cooling-off period is required to allow the alloy in the solder pot to harden before the overload relay assembly can be manually reset and motor service restored.

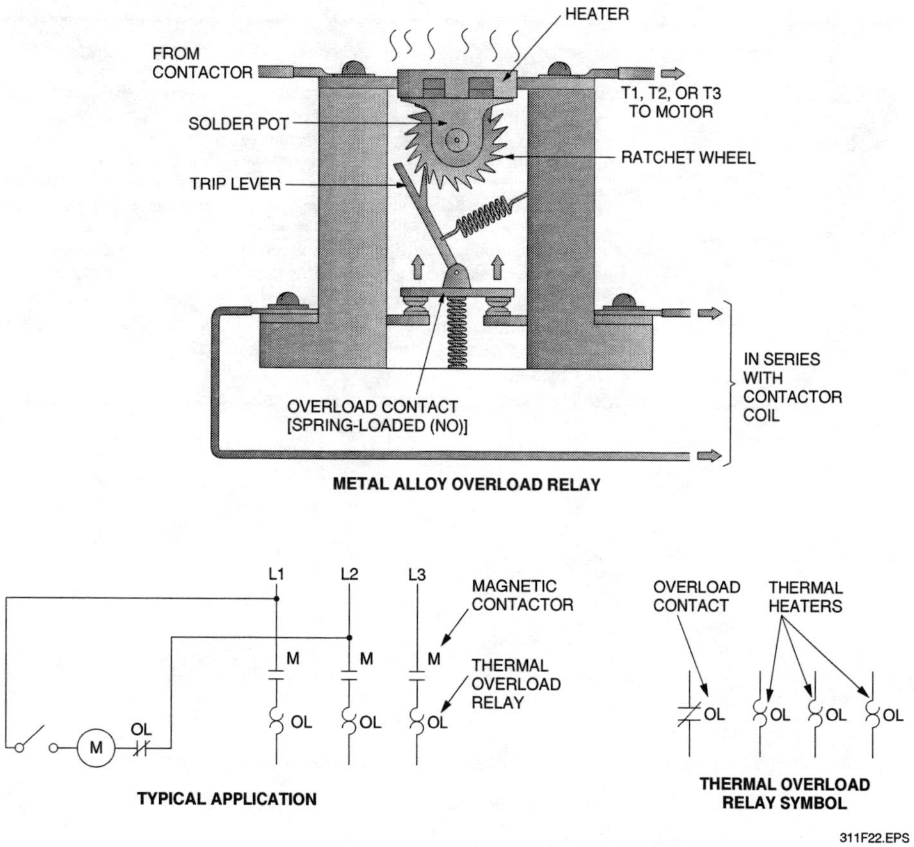

Figure 22 ◆ Mechanical and symbol representation of a thermal overload relay.

Classroom

Describe the operation of a melting-alloy overload relay.

Audiovisual

Show Transparency 12 (Figure 22).

Demonstration

Disassemble a melting-alloy overload relay and describe the function of each component. Refer the trainees to Figure 23.

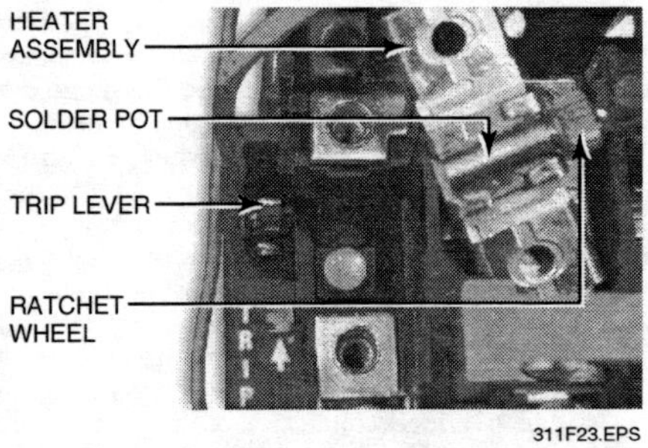

Figure 23 ◆ View of melting-alloy heater assembly and trip lever.

Melting-alloy heater assemblies are interchangeable and of one-piece construction, which ensures a constant relationship between the heater element and solder pot and allows factory calibration, making them virtually tamperproof in the field. These important features are not possible with any other type of overload relay construction. A wide selection of interchangeable heater assemblies is available to provide precise motor overload protection for any full-load current. This includes special heater assemblies used for partial correction of ambient temperature deviations at the overload relay location that are substantially different than at the motor location.

INSIDE TRACK

Overload Relays

Although seldom identified, overload relays are also categorized by NEMA for their ability to consider the cumulative heating effect of motor operation or overload. This characteristic is called operating memory and may be either volatile or nonvolatile, as shown below. All thermal overloads have a volatile operating memory (Category B) in that they have to cool off before they can be reset. This mimics the heat retention of the motor to some extent. Solid-state overload relays may have volatile and/or nonvolatile memory (Category B and/or A), depending on whether the memory is stored by charging a capacitor or as a value in a nonvolatile microprocessor memory. Magnetic overload relays normally have no memory (Category C) and can be instantly reset even though the motor may still be hot.

- *Category A*—Nonvolatile
- *Category B*—Volatile
- *Category C*—No memory

Instructor's Notes:

4.2.0 Bimetallic Overload Relays

Today a great many overload relays are of the bimetallic variety (*Figure 24*). These include the ones usually used in most IEC motor starters. Except for the IEC devices, which have non-changeable fixed heaters, bimetallic overload relays use interchangeable heaters to accommodate the FLC of various sizes of motors. If a different current range is required for an IEC fixed heater overload relay or its heater must be replaced, the entire overload relay must be changed.

A bimetallic overload relay has three advantages over a melting-alloy overload relay:

- The automatic reset feature is an advantage when the device is mounted in a location that is not easily accessible for manual operation or requires an automatic reset, such as in refrigeration units. Some bimetallic overload relays have a provision for selection of either automatic reset or manual reset.
- These relays usually have a motor service factor adjustment that is used to adjust to the trip range within 85% to 115% of the nominal trip rating of the heater unit. This feature is useful when the recommended heater size might result in unnecessary tripping, while the next larger size would not provide adequate protection.
- Most bimetallic overload relays employ automatic ambient temperature compensation so that the overload relay can be located in a different ambient temperature than the motor. The ambient compensation is accomplished by a second bimetal element that changes the trip point of the overload contact. If no automatic compensation is included, special heaters may sometimes be available to partially correct the difference.

Figure 25 is a single-pole, auto-reset, bimetallic overload relay with the side cover removed, showing the location of the contact, changeable heater, and heater bimetal element. As motor overcurrent increases, the increasing heat generated by the heater eventually causes the U-shaped heater bimetal element to straighten and press on the overload contact plunger. This action forces the snap-action opening of the ambient compensated overload contact. The contact is held open until the heater and bimetallic strip cool, allowing the snap action contact to toggle back to a closed position.

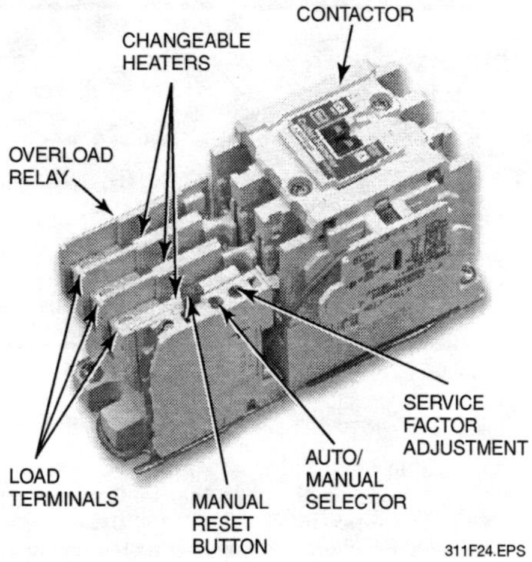

Figure 24 ◆ Typical three-pole bimetallic overload relay.

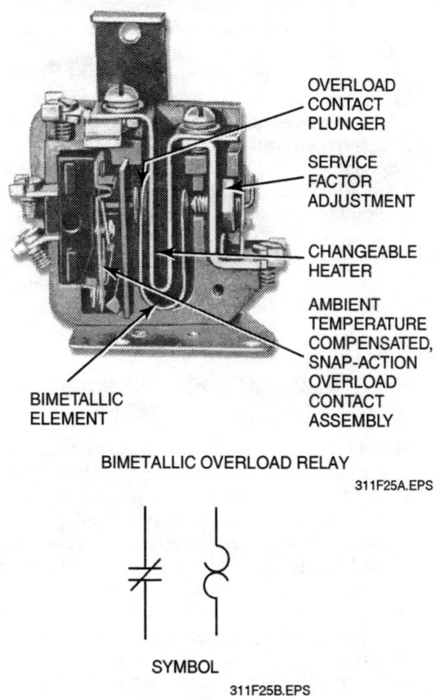

Figure 25 ◆ A single-pole, auto-reset, bimetallic overload relay and symbol.

Classroom

Discuss the construction and operation of a bimetallic overload relay.

List the advantages of bimetallic relays.

Demonstration

Disassemble a bimetallic overload relay and describe the function of each component. Refer the trainees to Figures 24 and 25.

Classroom

Discuss the operation and applications of magnetic overload relays.

Discuss magnetic and manual motor starters. Provide examples for the trainees to examine. Point out that they must withstand both the inrush and the running current.

Audiovisual

Show Transparency 13 (Figure 26).

Classroom

Compare IEC and NEMA nonreversing and reversing motor starters. Refer trainees to Figure 27 and Table 5.

INSIDE TRACK

Automatic Reset

NEC Section 430.43 prohibits the use of automatic motor overload reset devices in applications where automatic restart can endanger personnel.

Instantaneous Overload Relays

Instantaneous overload relays, including magnetic or solid-state versions, are sometimes referred to as jam relays. In some applications, a magnetic overload relay may be combined with a thermal overload relay to provide both loadjamming and running overload protection.

4.3.0 Magnetic Overload Relays

Magnetic overload relays are used for special applications in heavy industrial environments where the specific motor current draw must be detected or where prolonged inrush currents may be encountered when starting high inertia loads. Under either of these conditions, the time/current curves of thermal overload relays would be unsuitable. Magnetic overload relays may offer settings for either instantaneous or inverse-time trips along with adjustable time or current settings.

Magnetic overload relays operate by passing load current through a heavy coil that functions as a solenoid in each pole of the relay. The position of a moving core within the solenoid coil is adjustable so that the core is sensitive to current flow. When equipped or set for inverse-time trip purposes, a fluid-filled adjustable **dashpot** is attached to the moving core so that the time delay of the relay can be adjusted over a wide range for various currents. When activated, the instantaneous or time-delayed total travel of the core trips an overload contact that must be manually reset.

5.0.0 ◆ MAGNETIC AND MANUAL MOTOR STARTERS

NEMA-rated motor starters consist of NEMA magnetic contactors and overload relays assembled into a controller consisting of one or more contactors with one or more overload relays on a common base plate. They have complete control circuit wiring (except for connections to remote equipment) and complete power circuits from the line to load terminals. They are designed so that the factory wiring is not disturbed by the connections for line, load, or remote control circuit wiring. Typically, IEC-rated magnetic motor starters are not factory wired and must be assembled in the field after delivery. However, a number of North American manufacturers also offer IEC magnetic motor starters as factory-assembled, internally wired units, the same as NEMA units. Field assembly, when required, includes assembling and mounting the contactor(s) and overload relay(s), as well as furnishing and connecting all internal and external wiring. Both NEMA and IEC motor starters use contacts that are designed to withstand high-inrush current (*Figure 26*) encountered when motors are started. *Table 5* summarizes the basic differences between IEC and NEMA devices.

5.1.0 Nonreversing and Reversing Magnetic Motor Starters

The most commonly encountered NEMA or IEC magnetic motor starters are nonreversing and reversing types for wye, delta, and reconfigurable wye/delta motors. *Figure 27* shows typical nonreversing and reversing NEMA motor starters along with their symbol representations. IEC motor starters are similar except for the terminal marking that has been described previously.

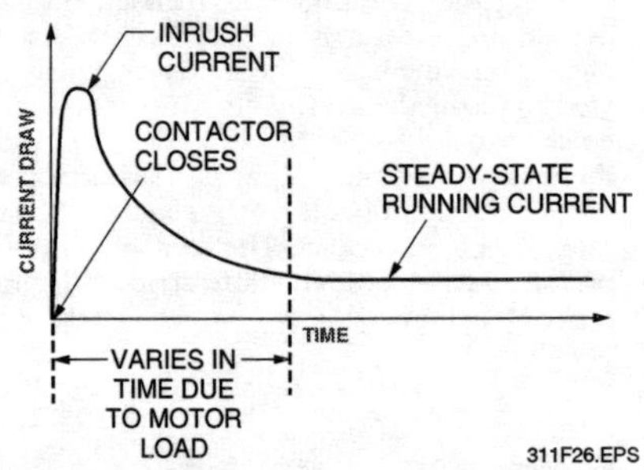

Figure 26 ◆ Typical motor inrush and running current.

Instructor's Notes:

Table 5 IEC and NEMA Product Comparison Summary

Subject	IEC	NEMA
Starter size	Physically smaller	Physically larger
Contactor performance	Electrical life = 1 million for AC-3 category operations with 30,000 for AC-4 category operations	Electrical life typically 2.5 to 4 times more than equivalently rated IEC device
Contactor application	Application sensitive—more knowledge and care in selection required	Application selection easier and less critical with fewer parameters to consider
Overload relay trip reset characteristics	Class 10 (fast) typical	Class 20 (Standard) typical
Overload relay adjustability	Fixed, noninterchangeable heaters; adjustable to suit different motors at the same horsepower	Field changeable heaters allow adjustment for motors of different horsepower
Overload relay reset mechanism	Manual/Auto typical; some use RESET/STOP dual function mechanism	Manual/Auto or Manual Only typical
Short circuit current rating	Typically designed for use with fast-acting, current-limiting fuses	Designed for use with common domestic fuses and circuit breakers

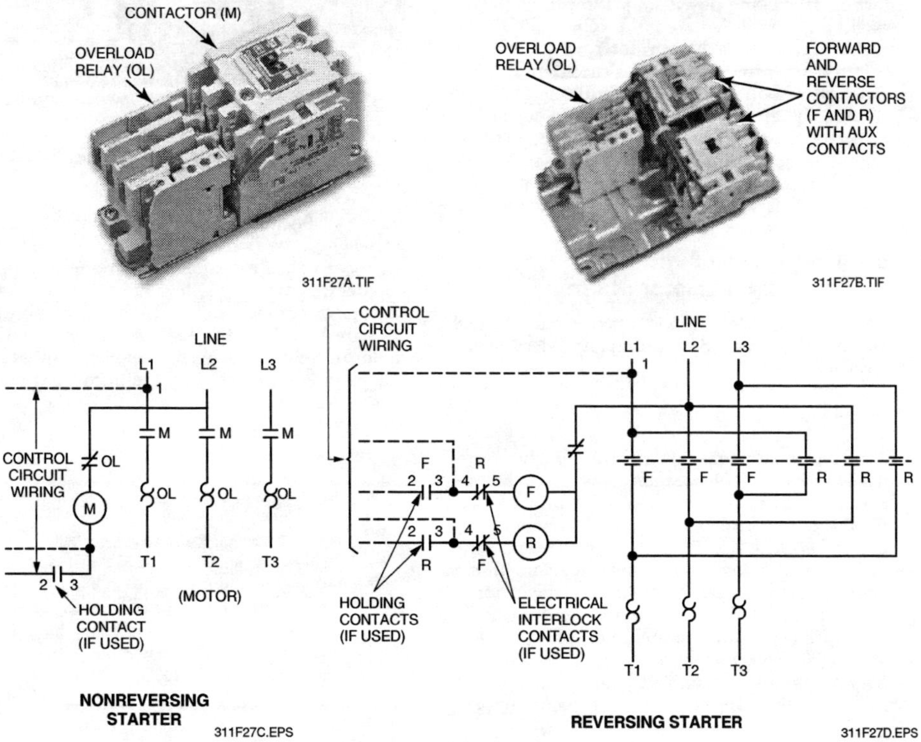

Figure 27 ◆ Typical NEMA nonreversing and reversing magnetic motor starters and symbol representations.

Classroom

Discuss the wiring conventions used with nonreversing and reversing motor starters.

Discuss NEMA-rated magnetic contactors and motor starters. Refer trainees to Figure 28 and Table 6.

Point out that the NEC has specific marking requirements for contactors/starters used with Design E motors.

INSIDE TRACK

Contactors Used for Plugging or Jogging Applications

Contactors/starters used regularly for **plugging** and **jogging (inching)** are derated. Plugging is the momentary stopping and/or reversing of a running motor. Jogging is the momentary starting of a stopped motor. Both plugging and jogging are used in industrial machine operations for drive-train positioning purposes that can subject contactors/starters to high inrush currents numerous times in a short period.

As shown in the symbol representations of the starters, normal convention for a nonreversing starter is to wire L1 to T1, L2 to T2, and L3 to T3 for forward motor rotation. In the case of the reversing motor starter, the same wiring convention is observed for the forward contactor (F). The wiring convention for the reverse contactor is to interchange L1 and L3 so that L1 is tied to T3, L2 is tied to T2, and L3 is tied to L1. Note that a mechanical interlock that is part of the contactor assembly is sometimes used to prevent both contactors from being closed at the same time. If pilot devices are used that are not mechanically interlocked, an electrical backup-interlock system can be used for protection against accidental energization of both contactors simultaneously. As shown, the electrical interlock uses a set of NC auxiliary contacts (F and R) that are wired in series with their opposite contactor coils. The NO auxiliary contacts can be used as holding contacts for start switches or other pilot devices if required.

5.2.0 NEMA Magnetic Contactor/Motor Starters

This section covers NEMA magnetic contactors and motor starters. Both rating and selection criteria are covered.

5.2.1 Ratings

A NEMA contactor/motor starter (*Figure 28*) is designed to meet the size rating specified in NEMA Standards. The standards are used to provide electrical interchangeability among manufacturers for a given NEMA size. *Table 6* (see pages 11.26 and 11.27) lists the NEMA sizes for contactors/motor starters and gives the maximum allowable horsepower ratings at different line voltages for various applications of devices, including plugging and jogging (inching) operations. The table also lists the continuous current, kW, kVA, and kVAR ratings for other applications of the contactors/starters. Because NEMA contactors/motor starters must be able to safely interrupt the locked rotor current of a motor under nonplugging or nonjogging conditions, the maximum allowable horsepower ratings shown for a particular NEMA size are based on the locker rotor current for motors rated at the listed horsepower for that size. For plugging and jogging, the maximum allowable horsepower within any NEMA size is derated as reflected in *Table 6*.

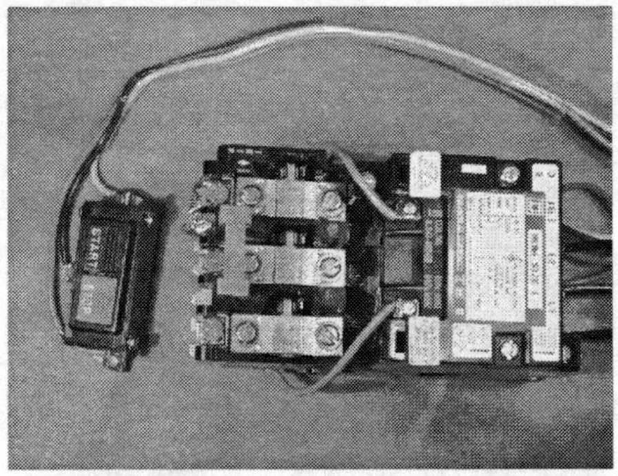

311F28.EPS

Figure 28 ◆ Typical NEMA Size 1 motor starter wired with a three-wire STOP/START switch.

CAUTION

Per ***NEC Section 430.83(A)(1)***, contactors/starters used for NEMA Design E motors of more than 2hp shall be marked for use with a Design E motor or shall have a horsepower rating of no less than 1.4 times the rating of a motor rated 3hp through 100hp, or no less than 1.3 times the rating of a motor over 100hp.

Instructor's Notes:

5.2.2 Selection

The information found on a motor nameplate is essential in the selection of a contactor/motor starter and must be used instead of actual current measurements or manufacturers' tables if possible. On the motor nameplate shown in *Figure 29*, the following parameters must be noted and used as indicated in the selection process:

- *Voltage (volts) and frequency (Hz)*—230V at 60Hz AC
- *FLC or FLA (amps)*—13.0A
- *Phase*—Three-phase
- *NEMA Design Letter*—B (For Design Letter E motors, refer to "CAUTION" on page 11.24.)
- *Horsepower (hp)*—5hp
- *Service factor (SF)*—1.0
- *Power factor*—85.7 (Necessary only if capacitor correction is connected on load side of overload relays. If so, overload relay manufacturer must be consulted to determine current elements for the relays.)

Assuming the motor is not used for plugging or jogging, *Table 6* shows that a 230V, polyphase, 60Hz, 5hp motor can be controlled with a NEMA Size 1 starter. The overload relay would be selected and equipped for a FLC (FLA) of 13A in accordance with the manufacturer's motor-starter overload relay selection tables and adjusted (if allowed) for a service factor (SF) of 1.0. The service factor rating is the amount of extra power demand that can be placed on the motor for a short period without damage to the motor. Common motor service factors range from 1.00 to 1.25, indicating that the motor can intermittently be required to produce 0% to 25% extra power over its normal rating. In the event that the SF is unknown, a value of 1.00 should be assumed.

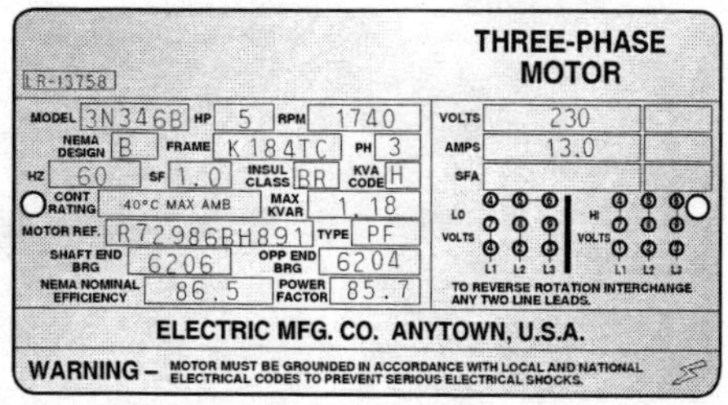

311F29.EPS

Figure 29 ◆ Typical motor nameplate.

The total excessive current required for intermittent power demands can be approximated by multiplying the motor FLC (FLA) by the motor SF. For the motor SF value of 1.0 shown in *Figure 29*, the allowed extra current is 0%. This means that any service factor adjustment would be equal to the value of the FLC (13A). If the SF had been 1.25, then the adjustment would be equal to 16.25A (13A × 1.25 = 16.25A). Any SF adjustment determined that is greater than the FLC permits the overload relay to allow any excessive intermittent power demands without nuisance tripping. For overload relays that are not ambient temperature compensated and where the starter is located in a different temperature environment than the motor, an appropriate overload relay/heating element must be selected as specified in special tables available from the motor-starter manufacturer.

Classroom

List the types of information found on a motor nameplate. Explain how to apply the nameplate information to the application information found in Table 6.

Audiovisual

Show Transparency 14 (Figure 29).

Teaching Tip

Provide the trainees with motor nameplate information and have them find the appropriate size motor starter.

Table 6 NEMA Size Ratings for Reversing and Nonreversing Contactors and Motor Starters

NEMA Size	Volts	Maximum Horsepower Rating—Nonplugging and Nonjogging Duty[1]		Maximum Horsepower Rating—Plugging and Jogging Duty[1]		Continuous Current Rating, Amperes 600V Max[2]
		Single Phase	Poly-Phase	Single Phase	Poly-Phase	
	115	⅓	—	—	—	9
	200	—	1½	—	—	9
	230	1	1½	—	—	9
00	380 (50Hz)	—	1½	—	—	9
	460	—	2	—	—	9
	575	—	2	—	—	9
	115	1	—	½	—	18
	200	—	3	—	1½	18
	230	2	3	1	1½	18
0	380 (50Hz)	—	5	—	1½	18
	460	—	5	—	2	18
	575	—	5	—	2	18
	115	2	—		—	27
	200	—	7½	—	3	27
	230	3	7½		3	27
1	380 (50Hz)	—	10	—	5	27
	460	—	10	—	5	27
	575	—	10	—	5	27
	115	3	—	2	—	45
	200	—	10	—	7½	45
	230	7½	15	5	10	45
2	380 (50Hz)	—	25	—	15	45
	460	—	25	—	15	45
	575	—	25	—	15	45
	115	7½	—	—	—	90
	200	—	25	—	15	90
	230	15	30	—	20	90
3	380 (50Hz)	—	50	—	30	90
	460	—	50	—	30	90
	575	—	50	—	30	90
	200	—	40	—	25	135
	230	—	50	—	30	135
4	380 (50Hz)	—	75	—	50	135
	460	—	100	—	60	135
	575	—	100	—	60	135
	200	—	75	—	60	270
	230	—	100	—	75	270
5	380 (50Hz)	—	150	—	125	270
	460	—	200	—	150	270
	575	—	200	—	150	270
	200	—	150	—	125	540
	230	—	200	—	150	540
6	380 (50Hz)	—	300	—	250	540
	460	—	400	—	300	540
	575	—	400	—	300	540
	230	—	300	—	—	810
7	460	—	600	—	—	810
	575	—	600	—	—	810
	230	—	450	—	—	1215
8	460	—	900	—	—	1215
	575	—	900	—	—	1215
	230	—	800	—	—	2250
9	460	—	1600	—	—	2250
	575	—	1600	—	—	2250

NOTES: 1. These horsepower ratings are based on the locked-rotor current ratings given in NEMA Standard ICS 2. For motors having higher locked-rotor currents, a controller should be used so that its locked-rotor current rating is not exceeded. 2. The continuous current ratings represent the maximum rms current, in amperes, which the controller shall be permitted to carry

Instructor's Notes:

Service-Limit Current Rating, Amperes[3]	Tungsten and Infrared Lamp Load, Amperes 250V Max[2]	Resistance Heating Loads, kW other than Infrared Lamp Loads		KVA Rating for Switching Transformer Primaries at 50 or 60 Cycles		3-Phase Rating for Switching Capacitors Kvar
		Single Phase	Poly-Phase	Single Phase	Poly-Phase	
11	5	—	—	—	—	—
11	5	—	—	—	—	—
11	5	—	—	—	—	—
11	—	—	—	—	—	—
11	—	—	—	—	—	—
11	—	—	—	—	—	—
21	10	—	—	0.9	1.2	—
21	10	—	—	—	1.4	—
21	10	—	—	1.4	1.7	—
21	—	—	—	—	2.0	—
21	—	—	—	1.9	2.5	—
21	—	—	—	1.9	2.5	—
32	15	3	5	1.4	1.7	—
32	15	—	9.1	—	3.5	—
32	15	6	10	1.9	4.1	—
32	—	—	16.5	—	4.3	—
32	—	12	20	3	5.3	—
32	—	15	25	3	5.3	—
52	30	5	8.5	1.0	4.1	—
52	30	—	15.4	—	6.6	11.3
52	30	10	17	4.6	7.6	13
52	—	—	28	—	9.9	21
52	—	20	34	5.7	12	26
52	—	25	43	5.7	12	33
104	60	—	—	0.9	7.6	—
104	60	—	—	—	13	23.4
104	60	—	—	1.4	15	27
104	—	—	—	—	19	43.7
104	—	—	—	1.9	23	53
104	—	—	—	1.9	23	67
156	120	—	45	—	20	34
156	120	30	52	11	23	40
156	—	—	86.7	—	38	66
156	—	60	105	22	46	80
156	—	75	130	22	46	100
311	240	—	91	—	40	69
311	240	60	105	28	46	80
311	—	—	173	—	75	132
311	—	120	210	40	91	160
311	—	150	260	40	91	200
621	480	—	182	—	79	139
621	480	120	210	57	91	160
621	—	—	342	—	148	264
621	—	240	415	86	180	320
621	—	240	515	86	180	400
932	720	180	315	—	—	240
932	—	360	625	—	—	480
932	—	450	775	—	—	600
1400	1080	—	—	—	—	360
1400	—	—	—	—	—	720
1400	—	—	—	—	—	900
8590	—	—	—	—	—	—
2590	—	—	—	—	—	—
2590	—	—	—	—	—	—

continuously without exceeding the temperature rises permitted by NEMA Standard ICS 1. 3. The service limit current ratings represent the maximum rms current, in amperes, which the controller shall be permitted to carry for protracted periods in normal service. At service-limit current, temperature rises may exceed those obtained by testing the controller at its continuous current rating.

Classroom

List the IEC utilization categories for lighting and motor control. Refer trainees to Tables 7 and 8.

Discuss the construction and operation of an IEC motor starter. Refer trainees to Figure 30.

5.3.0 IEC Magnetic Contactors/ Motor Starters

This section covers IEC magnetic contactors and motor starters. Both rating and selection criteria are covered.

5.3.1 Ratings

IEC Standard 60947 does not define any standard contactor or motor-starter sizes like NEMA. An IEC-rated contactor or motor starter (*Figure 30*) indicates that the contactor or motor starter has been evaluated by the manufacturer or a laboratory to meet the requirements of a number of defined applications called *utilization categories.* There are a number of categories (AC-1 to -8, -11 to -15, and -20 to -23) covering 600V or less AC current switchgear and control gear. The same is true for similar DC equipment (DC-1, -3, -5, -6, -11 to -14, and -20 to -23). Of these, the categories used for motor control or lighting control purposes are summarized in *Table 7*. *Table 8* lists and expands the definition of the most common categories used for motor control. Category A1 is included because most Category AC-3 and -4 devices have Category A-1 included on their nameplate. Category AC-2 is omitted because these motors are uncommon.

5.3.2 Selection

Convention has led to the assignment of several ratings to any particular contactor or motor starter for different categories and voltages. A designer can choose a preferred device for an application based on the device's ability to meet or exceed the required horsepower voltage rating and other factors, including performance. This technical data is available in the manufacturer's specifications. IEC contactors used in the U.S. are usually

Table 7 IEC Categories for Lighting and Motor Control

Category	Typical Application
AC-1	Noninductive or slightly inductive loads
AC-2	Slip-ring motor starting and switching off
AC-3	Squirrel-cage motor starting and switching off
AC-4	Squirrel-cage motor starting, plugging, and jogging
AC-5a	Switching electric discharge lighting loads
AC-5b	Switching large incandescent lighting loads
AC-6a	Switching transformer loads
AC-6b	Switching capacitor loads
AC-11	Control of AC electromagnetic circuits (auxiliary contacts)

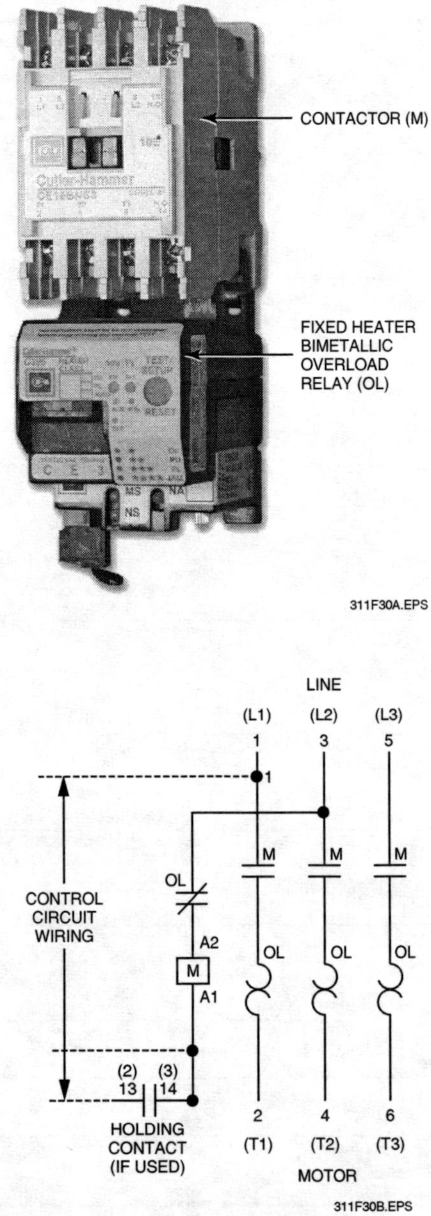

Figure 30 ◆ Typical IEC-rated, nonreversing motor starter and IEC symbol representation.

Instructor's Notes:

Table 8 Common IEC Categories for Motor Control Devices

Category	Application
AC-1	These devices are used with noninductive or slightly inductive loads, such as resistive furnaces, fluorescent lights, and incandescent lights.
AC-3	These devices are used in squirrel-cage motors for starting and switching off while running at rated speed. They include contact-make capability for locked-rotor current and break at full-load current. They are occasionally used for jogging and plugging for limited times such as machine setup. During such periods, the number of operations should not exceed five per minute nor more than ten in a ten-minute period.
AC-4	These devices are used in squirrel-cage motors for starting and switching off while running at less than rated speed. They provide jogging, plugging stop, and plugging reverse, and include contact make-and-break capability for locked-rotor current. Very few applications in the industry are totally Category AC-4.

marked with voltage and horsepower ratings for maximum AC-3 rated operational current. Most U.S. manufacturers provide tables that classify IEC contactors/motor starters in some sort of size ratings for horsepower, similar to NEMA, based on the maximum rated operational current in a range of standard line voltages. IEC contactor/motor starter selection is based on the percent that jogging and plugging (AC-4) is of the nonjogging and plugging (AC-3) condition in the duty cycle and the desired contact electrical life. In general, any time the duty cycle includes significant jogging or plugging, a larger size IEC contactor/motor starter is selected than would be needed for pure AC-3 applications, or an appropriate AC-4 rated device is selected. Like NEMA motor starters, the same motor nameplate data is required to select an IEC motor starter with an appropriate overload relay, as well as to set the adjustments on the overload relay.

5.4.0 Manual Motor Starters

Manual motor starters (*Figure 31*) are primarily used on small machine tools, fans, blowers, pumps, compressors, and conveyors. They are available as single-, double-, or three-pole devices. Instead of magnetically closed contacts, NEMA-rated and/or UL approved motor starters are operated by a mechanically linked and latched toggle handle or pushbutton. The operating mechanisms are quick make-and-break toggle type switches that cannot be teased into a partially open or closed position. They generally come equipped with melting-alloy or bimetallic overload relays.

311F31.EPS

Figure 31 ◆ Typical NEMA manual motor starter.

INSIDE TRACK

AC-3 Rated Contactors

U.S. manufacturers usually supply AC-3 derating tables for plugging and jogging duty.

Manual Starters with Melting-Alloy Overloads

In versions of manual starters that use melting-alloy overloads, the latched contacts are tripped mechanically by the overload relay and must be manually reset with a separate button or by setting the starter to off (stop) and then back to on (start). In bimetallic overload versions, the reset function may be selectable as automatic or manual, depending on the manufacturer.

Classroom

Discuss the operation and applications of manual motor starters. Refer trainees to Figure 31.

Classroom

Describe the types of accessories used with contactors and starters. Provide examples for the trainees to examine. Refer trainees to Figure 32.

5.5.0 Contactor/Motor Starter Accessories

Manufacturers offer a variety of accessories that can be added to various contactor/motor starters to customize them for a particular application. *Figure 32* shows a number of these accessories.

- *Power-pole adder kits*—These kits can be used to increase the number of high-current power contacts actuated by the contactor/motor starter. Depending on the manufacturer, some power-pole adder kits are mounted on the sides or the top of the contactor/motor starter. In some cases, the internal springs or coil must be changed to accommodate the mechanical loads imposed by the extra poles.
- *Timer attachment (not shown)*—Mechanical or solid-state control circuit timer devices with adjustable on-delay or off-delay are available for side mounting on the contactor/motor starter. Some devices have field-selectable on- or off-delay. They are equipped with one or more single- or double-throw snap-switch control-circuit contacts.
- *Fuse kit*—When control circuit power tapped from the line inputs of a contactor/motor-starter must be fused, side-mounted single- or double- fuse holder kits are available. Rated at 600V, they will usually accept $\frac{13}{32} \times 1\frac{1}{2}$ fuses up to 6A.
- *Transient suppression module*—Side-mounted transient suppression modules are available to reduce transient voltages and contact arcing in coil control circuits when the circuits are opened. This eliminates noise that interferes with operation of nearby electronic circuits. Most modules consist of an RC circuit that is designed to suppress coil voltage transients to approximately 200% of peak coil supply voltage.
- *Internal auxiliary contacts*—Some contactors/motor starters can have one or more internal NO or NC auxiliary contacts added for additional low-current circuits, including status feedback, or for electrical interlocking purposes. The devices are toggle snap switches with SB contacts.
- *External auxiliary contacts*—Side-mounted auxiliary contacts are available as single or double NO or NC contacts that are field convertible from NO to NC or vice versa. Some are available as DT contacts. The devices are toggle snap switches with SB contacts.

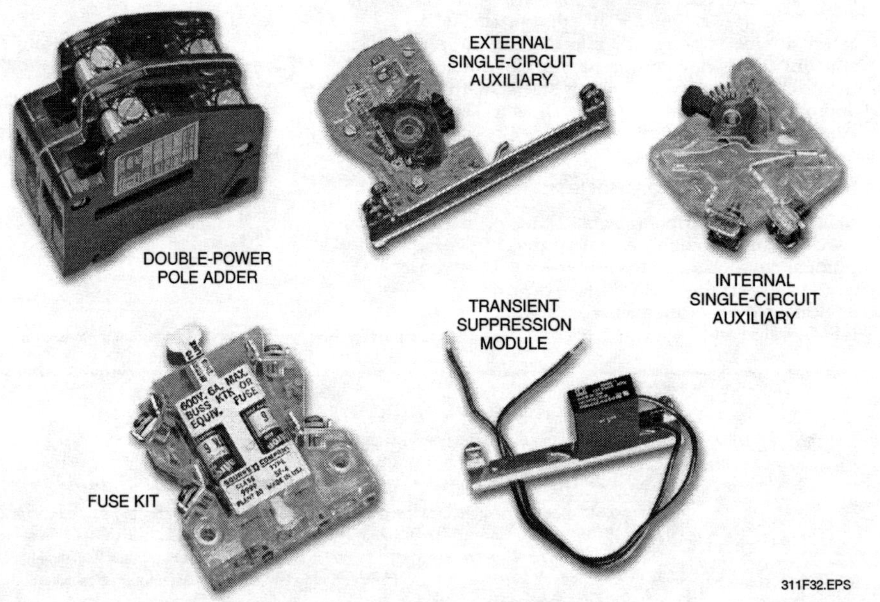

Figure 32 ◆ Typical contactor/motor starter accessories.

Instructor's Notes:

6.0.0 ◆ CONTROL TRANSFORMERS AND PILOT DEVICES

This section covers control transformers and several types of pilot devices widely used in motor control circuits. Pilot devices contain switch contacts. The opening, closing, or transfer of these contacts govern the operation of related relays or similarly controlled devices. Pilot devices are used to provide sequencing and automatic operation within certain parameters. Some commonly used types of pilot devices covered here include:

- Pushbutton switches
- Selector switches
- Pilot lights
- Temperature switches
- Pressure switches
- Limit switches
- Flow switches
- Float switches
- Foot switches
- Proximity switches/sensors
- Photoelectric switches/sensors

6.1.0 Control Transformers

Control transformers like the one shown in *Figure 33* provide the operating voltage for the motor control circuits and their components. Control transformers are step-down transformers that operate to reduce the line voltage applied to the equipment to the required operating voltage needed for the equipment's control circuits. They are designed to provide good transformer regulation where high inrush currents caused by contactors and relays are drawn. Secondary fuse protection kits are often used with control transformers for fusing the secondary winding in order to provide for control circuit protection. These kits consist of fuse blocks and either cartridge or glass fuses of the proper size. Stepping down a higher line voltage to a lower voltage for use in the control circuits has three advantages:

- Operator safety is increased at the control stations and other pilot devices.
- The reduced voltage decreases the chance of a fault occurring between lines of the control circuit wiring and ground.
- Use of components designed to operate at lower voltages lowers the cost for manufacturing the equipment.

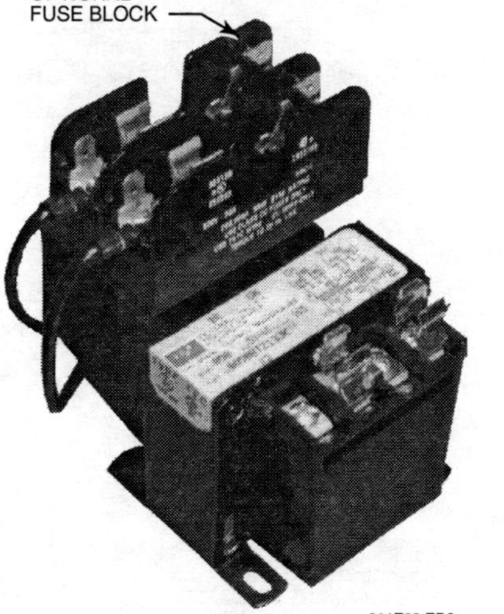

Figure 33 ◆ Industrial control transformer.

Control transformers are made to accommodate many different primary-voltage to secondary-voltage combinations. Control transformers used in most residential HVAC equipment and other equipment normally operate to step down an applied primary voltage of 120V or 240VAC to 24VAC. Control transformers used for industrial applications typically operate to produce a 24V or 120VAC secondary voltage. A typical industrial control transformer contains two primary windings and one or two secondary windings. *Figure 34* shows a control transformer that has two primary windings and one secondary winding, with each primary winding having a voltage rating of 240V and the secondary winding a rating of 24V. For the purpose of discussion, assume that there is a turns ratio of 10:1 between each of the primary windings and the secondary winding. Terminals of most control transformers are identified using an industry-standard method. As shown, the terminals of one of the primary windings are marked H1 and H2 and for the other primary winding, they are marked H3 and H4. The terminals for the secondary winding are marked X1 and X2. The primary windings of most control transformers have the H2 and H3 terminals crossed as shown in *Figure 34*. This is done to make the physical connection between the terminals of the primaries easier.

Classroom

Explain that pilot devices serve to control the operation of relays and other motor controls. List various types of pilot devices.

Discuss the function and operation of control transformers. Provide examples for the trainees to examine. Refer trainees to Figure 33.

Review the schematic diagram for a control transformer.

Audiovisual

Show Transparency 15 (Figure 34).

Classroom

Discuss the factors involved in selecting a control transformer. Point out that sizing and selection information is normally provided in the manufacturer's literature.

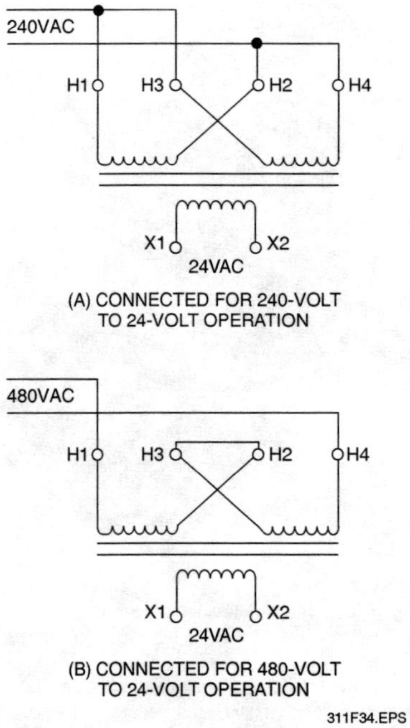

Figure 34 ◆ Control transformer schematic shown connected for 240V/480V primary and 24V secondary operation.

If this control transformer must be connected to step down a primary voltage of 240V to produce a secondary voltage of 24V, the two primary windings are connected in parallel, as shown in *Figure 34(A).* Since the two primary windings are connected in parallel, each will receive the same voltage. This will produce a turns ratio of 10:1 between the primary and secondary windings. When 240V is connected to the primary of a 10:1 ratio transformer, the secondary voltage produced is 24V.

If this same transformer is connected to step down a primary voltage of 480V to produce the secondary voltage of 24V, the primary windings are connected in series as shown in *Figure 34(B).* In this connection, terminal H2 of one primary winding is connected to H3 of the other primary winding. This series connection of the two primary windings produces a turns ratio of 20:1. When 480V is connected to the primary windings, 24V is produced by the secondary winding.

Selecting a control transformer for a specific application requires that you know the following factors about the transformer and its related control circuit:

- *Inrush VA*—The product of the load voltage (V) times the current (A) that is required during circuit startup. It is determined by adding the inrush VA requirements for all the circuit load devices such as contactors, timers, relays, or pilot lights that will be energized at the same time.
- *Sealed (steady-state) VA*—The product of the load voltage (V) times the current (A) that is required to operate the circuit after the initial startup or under normal operating conditions. It is determined by adding the sealed VA requirements of all electrical components of the circuit that will be energized at any given time. The sealed VA requirements for each component can be obtained from the component manufacturer data sheets.
- *Primary voltage*—The primary voltage applied to the primary of the transformer and its operational frequency.
- *Secondary voltage*—The operating voltage required for the control circuit connected to the transformer secondary.

Once these factors have been determined, selection of a proper transformer can be done by following the procedure normally given in the control transformer manufacturer's catalog or application bulletin. A typical procedure is described here.

INSIDE TRACK

Transformer Connections

The correct connections for achieving the desired voltage output from a specific control transformer are normally shown on the transformer's nameplate and in the manufacturer's data sheet provided with the transformer. Metal links are normally used to make the jumper connections between the terminals of the primary and/or secondary windings to configure the transformer for the desired voltage(s).

Instructor's Notes:

Step 1 Determine the application inrush VA for the control transformer using the formula:

$$\text{Application inrush VA} = \sqrt{(\text{INRUSH VA})^2 + (\text{SEALED VA})^2}$$

For example, assume that we have an inrush VA of 1,000VA and a sealed VA of 100VA. Using the formula, the application inrush VA is 1,005VA, calculated as shown:

$$\text{Application inrush VA} = \sqrt{(\text{INRUSH VA})^2 + (\text{SEALED VA})^2}$$

$$\text{Application inrush VA} = \sqrt{(1{,}000)^2 + (100)^2}$$

$$\text{Application inrush VA} = \sqrt{1{,}010{,}000}$$

$$\text{Application inrush VA} = 1{,}004.987\text{VA} = 1{,}005\text{VA (rounded off)}$$

Step 2 Using *Table 9,* select a proper secondary voltage column to use. If the primary voltage is stable and does not vary more than 5% from nominal, use the 90% secondary voltage column of the table. If the primary voltage varies between 5% and 10% of nominal, use the 95% secondary voltage column. Note that to comply with NEMA standards, which require all magnetic devices to operate properly at 85% of rated voltage, the 90% secondary column is most often used in selecting a transformer.

Step 3 After determining the proper secondary voltage column to use, read down the column in *Table 9* until you find a value equal to or greater than the application inrush VA calculated in Step 1. For our example, an application inrush VA of 1,150 is closest to the 1,005VA. In no case should a value less than the application inrush VA be used. Then read left to the Transformer VA Rating column to find the proper transformer VA for your application. For this example, your transformer should have a VA rating equal to or greater than 200VA. As a final check, make sure that the transformer VA rating is equal to or greater than the total sealed requirements. If not, select a transformer with a VA rating equal to or greater than the total sealed VA.

Step 4 Using the transformer VA rating found in *Table 9*, refer to the applicable transformer manufacturer's product data catalog or bulletin to identify the model and part number for the transformer. Your selection is based on the required transformer VA rating and the primary and secondary voltage requirements.

6.2.0 Pushbutton Switches, Selector Switches, and Pilot Lights

Manually operated pushbutton switches and selector switches (*Figure 35*) are two widely used types of pilot devices. They are made in standard-duty and heavy-duty versions. Heavy-duty switches are able to carry higher continuous and make-break currents. Different types of pushbutton and lever switches are needed to serve the wide variety of industrial motor applications. For this reason, switch manufacturers designed many of their industrial pushbutton and lever switches so that they can be assembled using different parts that can be interchanged according to the customer's specifications to meet the requirements of

Table 9 Regulation Data Chart

Transformer VA Rating	Application Inrush VA at 20% Power Factor		
	95% Secondary Voltage	90% Secondary Voltage	85% Secondary Voltage
25	100	130	150
50	170	200	240
75	310	410	540
100	370	540	730
150	780	930	1,150
200	810	1,150	1,450
250	1,400	1,900	2,300
300	1,900	2,700	3,850
350	3,100	3,650	4,800
500	4,000	5,300	7,000
750	8,300	11,000	14,000

Classroom

Review the procedure for sizing and selecting a control transformer. Refer trainees to Table 9.

Discuss the applications of manually operated pushbutton and selector switches. Provide examples for the trainees to examine. Refer trainees to Figure 35.

Demonstration

Disassemble a pushbutton switch and describe the function of each component. Refer the trainees to Figure 36.

Classroom

Discuss the schematic symbols used to represent pushbutton switches.

Audiovisual

Show Transparency 16 (Figure 37).

Classroom

Point out that pushbutton switches are often color-coded red and green for easy identification. Refer trainees to Figure 38.

a specific application. This modular approach to building a switch also lowers the manufacturer's costs by reducing the number and type of switches and switch components that must be manufactured and maintained in inventory.

6.2.1 Pushbutton Switches

The parts used to assemble a typical pushbutton switch include the **operator** (*Figure 36*), legend plate, one or more contact blocks, and related mounting adapters and hardware. The operator is

Figure 35 ◆ Typical pushbutton and selector switches.

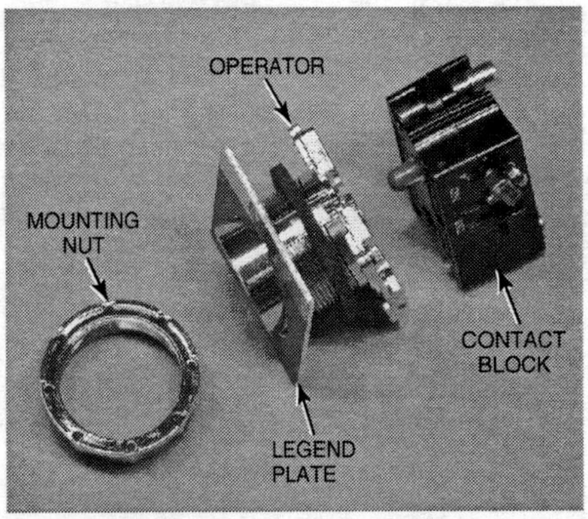

Figure 36 ◆ Typical parts of a pushbutton switch.

the part of a switch that is pressed or pulled by the person operating the switch in order to activate the switch contacts. Operators are made in many different shapes, sizes, and colors with red used to indicate a stop or off function. Operators can also be nonilluminated or illuminated to indicate when active. The legend is a plate that is placarded (marked) to show the function of the pushbutton, such as ON, OFF, and INCH. The contact block is the part of the switch containing the contacts that are activated when the operator is pressed.

As the name implies, the operator of a pushbutton switch is pushed to activate the electrical contacts. The switch contact block can have normally open (NO), normally closed (NC), or both NO and NC sets of contacts. Multiple contact blocks are often assembled (stacked) together to form a switch that has three, four, or more NO or NC sets of contacts. Normally open (NO) contacts usually are used to close the circuit when the operator is pressed in order to initiate a START or ON circuit function. Normally closed (NC) contacts are usually used to open the circuit when the operator is pressed in order to initiate STOP or OFF circuit functions. There are two types of pushbuttons: momentary and maintained. A normally open momentary pushbutton closes as long as the button is held down. A normally closed momentary pushbutton opens as long as the button is held down. A maintained pushbutton latches in place when the pushbutton is pressed. *Figure 37* shows the schematic symbols for pushbutton switches. *Figure 38* shows a simple line diagram of start and stop pushbutton switches used in a basic contactor control circuit.

6.2.2 Push-Pull Pushbutton Switches

Pushbutton switches can also be of the push-pull type. A push-pull switch is typically used to replace two separate pushbuttons, such as START and STOP pushbuttons. There are three types of push-pull switches:

- *Maintained*—This is a two-position switch that remains in the pulled or pushed position until manually actuated to the opposite position.
- *Momentary*—This is a three-position switch. A spring returns the switch to an intermediate position when pulled or pushed and released.
- *Momentary pull, maintained push*—This is a three-position switch. A spring returns the switch to an intermediate position when pulled. In the push position, it maintains its position until manually returned to the intermediate position.

Instructor's Notes:

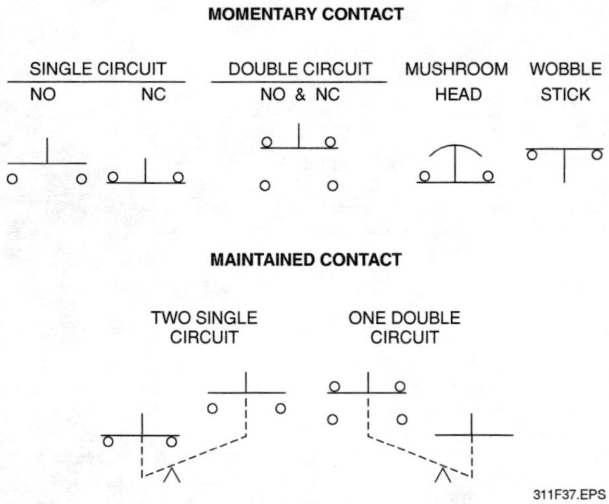

Figure 37 ◆ Schematic symbols for pushbutton switches.

Figure 38 ◆ Stop and start pushbutton switches used in a basic control circuit.

6.2.3 Selector Switches

Selector switches are used to select one of several different positions. Standard-duty selector switches have either two or three positions; heavy-duty selector switches have two or more positions. Some have as many as 12 positions. The operator of a selector switch is rotated, instead of pushed, to activate the contacts for each switch position. Selector switches are assembled the same way as pushbutton switches, with the customer selecting the required operator, legend, and contact block assemblies suitable for the application. *Figure 39* shows the schematic symbols for a two-position and a three-position selector switch. Schematic diagrams and switch manufacturers' product catalogs commonly show the contact positions for each position of a selector switch using a truth table placed near the switch. As shown, an X is placed in the truth table if a contact is closed in any position.

Figure 39 also shows a simple line diagram of a three-position selector switch being used to control a contactor. In the OFF position, all the contacts of the selector switch are open, preventing the contactor coil from being energized. This is a safety position. In the HAND position, the upper contacts of the three-way switch are closed, enabling the contactor coil to be energized or de-energized by pressing the START or STOP pushbuttons, respectively. In the AUTO position, the lower contacts are closed, enabling the contactor coil to be energized whenever the contacts of the

Classroom

Discuss the operation and applications of selector switches. Provide examples for the trainees to examine.

Audiovisual

Show Transparency 17 (Figure 39).

Classroom

Note that some selector switches are operated using a joystick. Refer trainees to Figure 40.

Explain that a pilot light provides a visual cue as to the condition of a circuit. Provide examples for the trainees to examine.

Discuss the operation and schematic representation of pilot lights. Refer trainees to Figure 41.

Discuss the operation and applications of pushbutton stations. Provide examples for the trainees to examine. Refer trainees to Figure 42.

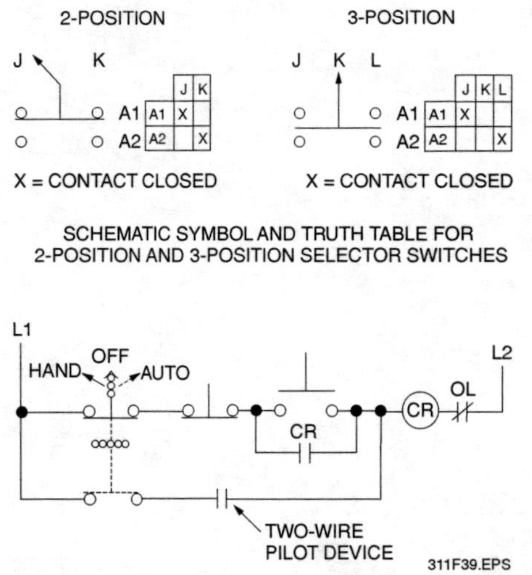

Figure 39 ◆ Selector switch schematic symbols and typical control circuit.

two-wire pilot device are closed. This device can be a temperature switch, pressure switch, limit switch, or similar device.

Some types of selector switches have a joystick operator (*Figure 40*). Joystick selector switches can have two to eight positions. This type of switch is normally used in applications where only one circuit is to be energized at one time, such as when operating a hoist or crane. In this application, movement of the operator for a five-position switch (four positions and center off) closes one circuit each in the up, down, left, and right positions, with all circuits open when the joystick is in the center position. Depending on design, the joystick operator may be a momentary contact with spring return of the handle to the center position, or it can be maintained in one of the four positions.

6.2.4 Pilot Lights

A pilot light is a small electric light used to visually show a specific condition of a circuit. Pilot lights can be separate assemblies, or they can be built into a pushbutton switch assembly. Pilot light assemblies, like the ones shown in *Figure 41*, typically have a polycarbonate or glass lens (cap). Colored lenses are made in red, green, amber, blue, clear, white, or yellow. Typically, pilot lights use a bayonet base incandescent lamp that operates at full voltage or at a reduced voltage via a transformer or resistor. Depending on the control

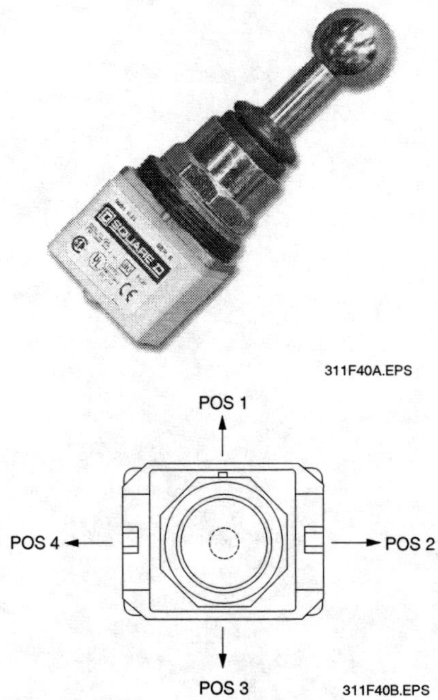

Figure 40 ◆ Joystick operated selector switch.

circuit arrangement, pilot lamps operate at a voltage ranging from 6VAC or DC to 120VAC or DC. Some pilot lights use a light-emitting diode (LED) instead of an incandescent lamp. Some are available with a local push-to-test capability, others with a remote test capability used to test the operation of the indicator lamp.

6.2.5 Pushbutton Stations

Pushbutton stations (*Figure 42*) are enclosures used to house one or more pushbuttons, selector switches, and/or pilot lights to protect them from dust, dirt, water, and corrosive fluids. They can be bought unassembled or with the pushbutton and/or lever switches installed. Cast metal, polyester, and stainless steel pushbutton stations are available in various NEMA enclosures and in various sizes. Always use the correct switch components and enclosure for the environment where they will be used.

Selecting the components for a typical pushbutton or lever switch and related switch station involves using the switch manufacturer's catalog or data sheets to identify the proper components

Instructor's Notes:

needed for each switch and switch station. You must make sure that the electrical characteristics of the selected switching device and its contacts are compatible with the requirements for your application. Switch manufacturers describe the electrical characteristics of their pushbutton switches, lever switches, and contacts using codes defined in *NEMA Standard ICS 5* for categories of utilization and AC and DC contact switching ratings. For example, category entries such as AC-14 or DC-13 define typical applications of the device. Contact rating designations such as A600 or B600 define the continuous current-carrying thermal capability of the device or contact. The codes and related operating characteristics are given in *Tables 10* and *11*.

A general procedure for selecting a pushbutton/lever switch is given here.

Step 1 Select the style of switch operator:
- Standard-duty or heavy-duty
- Chrome, plastic, or other material
- Oil, water, dust resistant, corrosion resistant

Step 2 Select the type of switch operator:
- Pushbutton or lever
- Nonilluminated or illuminated
- Color of pushbutton

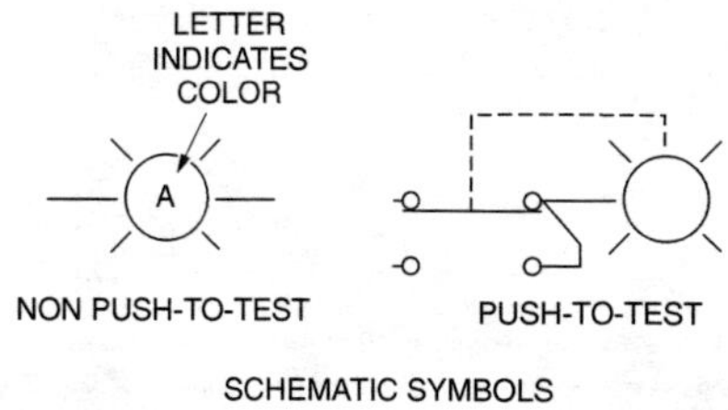

311F41A.EPS

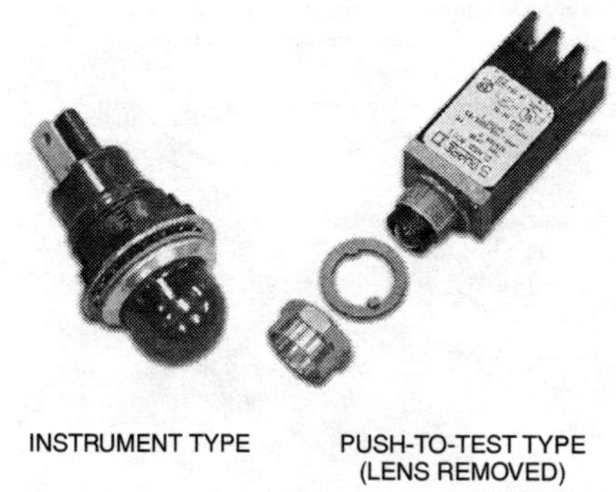

311F41B.EPS

Figure 41 ◆ Typical pilot light assemblies.

311F42.EPS

Figure 42 ◆ Example pushbutton station.

INSIDE TRACK — *Installation of Pushbutton Switches in a Pushbutton Station*

When installing pushbutton switches in a pushbutton station, follow these guidelines:

- *STOP button location*—On all single row pushbutton stations, the STOP button shall be located below or to the right of all other associated buttons, including lights and selector switches.
- *Pushbuttons used with multi-speed motors (mounted vertically)*—The lowest button shall be the STOP button. The lowest-speed button shall be above the STOP button, followed by those for consecutively higher speeds.
- *Pushbuttons used with multi-speed motors (mounted horizontally)*—The right-hand button shall be the STOP button. The lowest-speed button shall be to the left of the STOP button, followed by those for consecutively higher speeds.

Classroom

List the rating codes and utilization categories of pushbutton stations. Refer trainees to Tables 10 and 11.

Describe the procedure for selecting a pushbutton station.

List the installation requirements for pushbutton stations.

Table 10 Rating Codes for AC and DC Control Circuit Contacts

Rating Codes for AC Control Circuit Contacts at 50 and 60 Hertz(†)											
Contact Rating Code Designation*	Thermal Continuous Test Current (Amperes)	120 Volt		240 Volt		420 Volt		600 Volt		Maximum Volt-Amperes	
		Make	Break	Make	Break	Make	Break	Make	Break	Make	Break
A150	10	60	6.0	—	—	—	—	—	—	7200	720
A300	10	60	6.0	30	3.00	—	—	—	—	7200	720
A600	10	60	6.0	30	3.00	15	1.50	12	1.20	7200	720
B150	5	30	3.00	—	—	—	—	—	—	3600	360
B300	5	30	3.00	15	1.50	—	—	—	—	3600	360
B600	5	30	3.00	15	1.50	7.50	0.75	6	0.60	3600	360
C150	2.5	15	1.5	—	—	—	—	—	—	1800	180
C300	2.5	15	1.5	7.5	0.75	—	—	—	—	1800	180
C600	2.5	15	1.5	7.5	0.75	3.75	0.375	3.00	0.30	1800	180
D150	1.0	3.60	0.60	—	—	—	—	—	—	432	72
D300	1.0	3.60	0.60	1.80	0.30	—	—	—	—	432	72
E150	0.5	1.80	0.30	—	—	—	—	—	—	216	36

(*) The numerical suffix designates the maximum voltage design values, which are to be 600, 300, and 150 volts for suffixes 600, 300, and 150, respectively. The test voltage is to be 600, 240, or 120 volts. (†) For maximum ratings at voltages between the maximum design value and 120 volts, the maximum make and break ratings are to be obtained by dividing the volt-amperes rating by the application voltage. For voltages below 120 volts, the maximum make current is to be the same as for 120 volts, and the maximum break current is to be obtained by dividing the break volt-amperes by the application voltage, but these currents are not to exceed the thermal continuous test current.

Rating Codes for DC Control Circuit Contacts					
Contact Rating Code Designation*	Thermal Continuous Test Current (Amperes)	Maximum Make or Break Current, Amperes			Maximum Make or Break Volt-Amperes At 300 Volts or Less
		125 Volt	250 Volt	301 to 600 Volt	
N150	10.0	2.2	—	—	275
N300	10.0	2.2	1.1	—	275
N600	10.0	2.2	1.1	0.40	275
P150	5.0	1.1	—	—	138
P300	5.0	1.1	0.55	—	138
P600	5.0	1.1	0.55	0.20	138
Q150	2.5	0.55	—	—	69
Q300	2.5	0.55	0.27	—	69
Q600	2.5	0.55	0.27	0.10	69
R150	1.0	0.22	—	—	28
R300	1.0	0.22	0.11	—	28

(*)The numerical suffix designates the maximum voltage design values, which are to be 600, 300, and 150 volts for suffixes 600, 300, and 150, respectively. Test voltage shall be 600, 240, or 120 volts. (†) For maximum ratings at 300 volts or less, the maximum break ratings are to be obtained by dividing the volt-ampere rating by the application voltage, but the current values are not to exceed the thermal continuous test current.

Instructor's Notes:

Table 11 Utilization Categories for Control Circuit Switching Elements

Utilization Categories for Switching Elements		
Kind of Current	**Category**	**Typical Applications**
Alternating Current	AC-12	Control of resistive loads and solid-state loads with optical isolation
	AC-13	Control of solid-state loads with transformer isolation
	AC-14	Control of small electromagnetic loads (max. 72VA closed)
	AC-15	Control of electromagnetic loads (greater than 72VA closed)
Direct Current	DC-12	Control of resistive loads and solid-state loads with optical isolation
	DC-13	Control of electromagnets
	DC-14	Control of electromagnet loads having economy resistor in circuit

Step 3 Select the contact block(s) needed:
- Normally open or normally closed
- Standard or hazardous location
- Standard-duty or heavy-duty

Step 4 Select appropriate operator identification or legend nameplate and color.

Step 5 Select the proper station enclosure for the type and number of operators involved.

6.2.6 Using Multiple Pushbutton Stations

If motors are required to be started from more than one location, additional pushbutton stations must be installed in the circuit. In doing so, the START buttons in these stations must be connected in parallel with the original START button, and the STOP buttons must be connected in series with the original STOP button as shown in *Figure 43*. The auxiliary contactor must also be connected in parallel with the START buttons. For three or more control stations, all the START buttons must be connected in parallel with the auxiliary contactor and all the STOP buttons connected in series.

6.3.0 Temperature Switches

Temperature switches (*Figure 44*) are used to monitor temperatures in the control circuits for applications such as heating/cooling systems, damper systems, fire alarm systems, and process control systems. Temperature switch contacts open or close in response to a rise or fall in temperature detected by their sensing element. The temperature where the opening or closing of the contacts occurs is normally determined by operator-adjustable **setpoints.** The temperature-sensing elements used in mechanical temperature switches are usually either a bimetallic element or a bulb and capillary element. There are also electronic temperature-sensing elements, but these tend to be used more with temperature controller units rather than switches.

A bimetallic element is made of two different metals bonded together; one is usually copper or

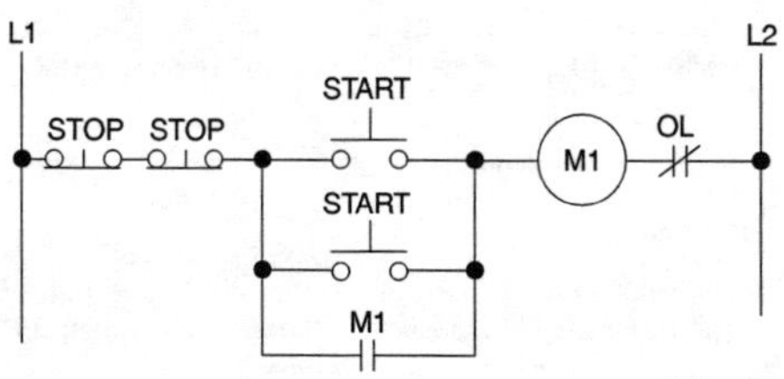

Figure 43 ◆ Multiple STOP and START pushbuttons.

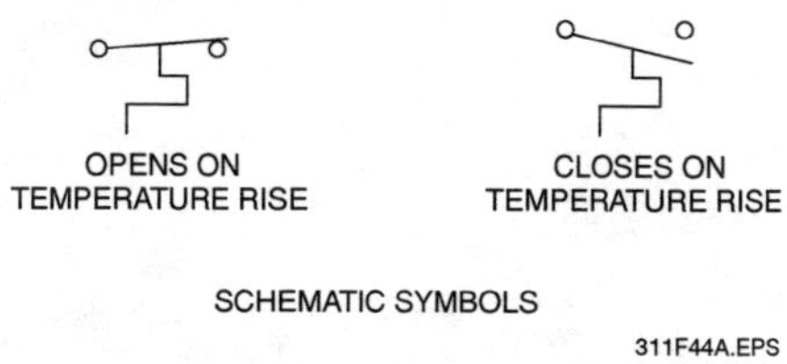

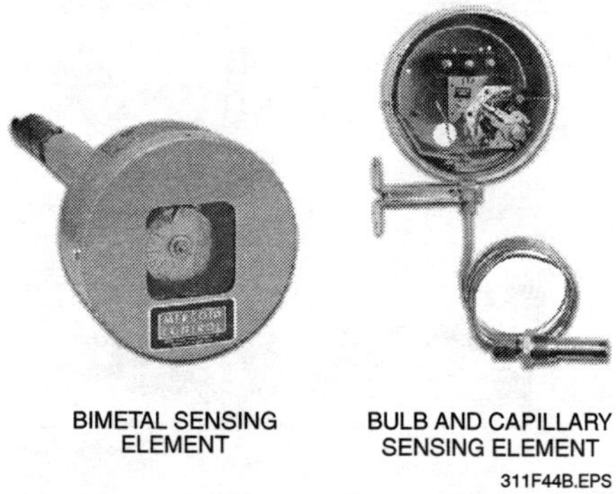

Figure 44 ◆ Typical temperature switches.

Classroom

Discuss the wiring requirements for multiple pushbutton stations.

Audiovisual

Show Transparency 18 (Figure 43).

Classroom

Discuss the operation and applications of temperature switches. Provide examples for the trainees to examine. Refer trainees to Figure 44.

Classroom

Explain how a bimetal strip operates.

Audiovisual

Show Transparency 19 (Figure 45).

Classroom

Describe the operation of Bourdon tube and bellows capillary switches.

Audiovisual

Show Transparency 20 (Figure 46).

brass (*Figure 45*). The other, a special metal called *Invar* steel, contains 36% nickel. When heated, the copper or brass has a more rapid expansion rate than the Invar steel and changes the shape of the element. The movement that occurs when the bimetal changes shape is used to open or close the switch contacts. Although bimetal elements are constructed in various shapes, the coil-wound element is the most commonly used. In one type of widely used bimetal-operated switch, the coiled bimetal strip is attached to a sealed glass tube containing a small pool of mercury (a conductor). When the bimetal strip causes the tube to be tipped in one direction, the mercury moves across the switch contacts, closing the contacts. When the tube is tipped in the opposite direction, the mercury moves to the other end of the tube, causing the switch contacts to be open. In some other types of bimetal-operated switches, the bimetal strip is connected directly to one of the switch contacts. This moving contact makes or breaks contact with a stationary switch contact in response to the expansion and contraction of the bimetal strip.

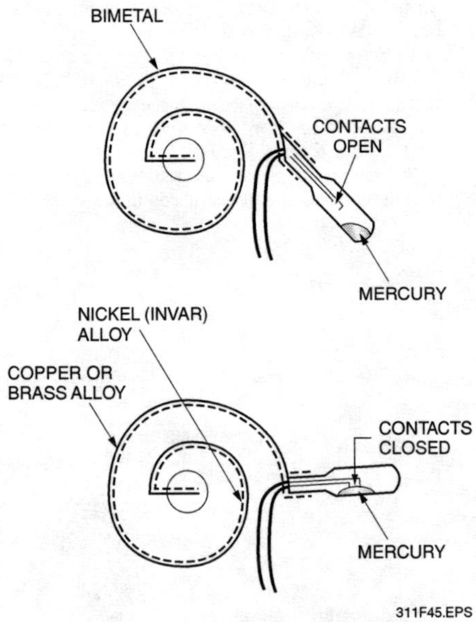

Figure 45 ◆ Bimetal strips used in temperature switches.

Temperature switches with bulb and capillary sensing elements consist of a tube (capillary tube) that connects the main body of the switch to a remotely located bulb that is partially filled with a liquid, gas, or vapor. This sensing bulb can be clamped to a pipe or duct, inserted in a cooling/heating coil, or placed in a tank well for sensing product temperature. The switch operates on the principle that when the temperature increases at the bulb, expansion of the bulb media (liquid, gas, or vapor) takes place. This causes a force to be transmitted through the capillary tube and exerted on the mechanism that operates the switch contacts. Two types of switching mechanisms are commonly used: Bourdon tube and bellows. *Figure 46(A)* shows a C-type Bourdon tube. It is a flattened metal tube that is open at one end and closed at the other. The tube straightens out when a rise in pressure is applied to the open end of the tube and curls inward with a decrease in pressure. This movement is transmitted by a mechanical linkage to the switch contacts, causing them to open or close. Spiral and helical Bourdon tubes are also used that work on the same principle as the C-type. However, both of these tubes produce more tip travel and higher torque capacity at the free end than the C-type.

In a bellows temperature switch, the pressure change in the bulb media is transmitted through the capillary tube for application to a bellows inside the temperature switch, shown in *Figure 46(B)*. One end of the bellows is closed, and the other end is connected to the bulb pressure source via the capillary tube. The bellows is a round device with several folds, much like an accordion, that expand or contract with the applied pressure. The expanding bellows moves against a calibrated spring switch mechanism that opens or closes the switch contacts. Usually, the spring tension is adjustable so that the switch can be adjusted for different temperature set points.

INSIDE TRACK

Environmental Hazard

Mercury is an environmental and health hazard. Never handle mercury. Do not dispose of thermostats containing a mercury tube in regular trash. Contact your local waste management or environmental authority for disposal/recycling instructions.

Instructor's Notes:

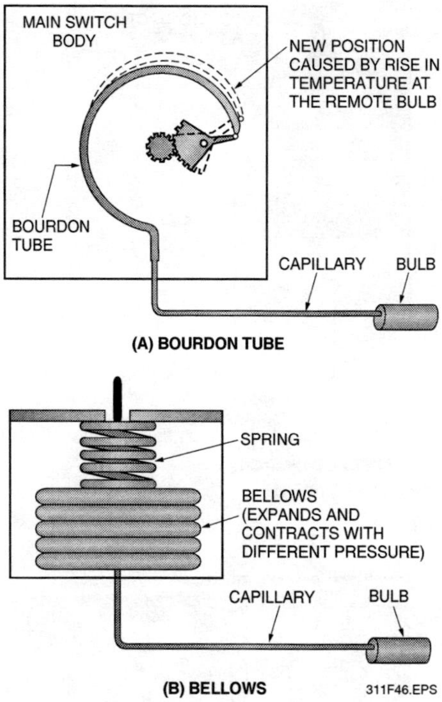

Figure 46 ◆ Bourdon tube and bellows temperature switch mechanisms.

To properly select temperature switches, you need to know and understand several terms used in switch manufacturer's catalogs:

- *Allowable temperature limits*—The maximum and minimum temperatures to which a temperature switch may be exposed without altering its performance characteristics. It is important that a temperature switch not be exposed to or used at temperatures beyond the manufacturer's allowable limits.
- *Cut-in temperature*—The temperature of the sensed medium at which a temperature switch is actuated to energize the load. The cut-in temperature may be the trip or reset point.
- *Cut-out temperature*—The temperature of the sensed medium at which a temperature switch is actuated to de-energize the load. The cut-out temperature may be the trip or reset point.
- *Operating temperature differential*—The difference between the cut-in and cut-out temperatures. The operating temperature differential is also called the **deadband.**
- *Operating temperature range*—The range between the maximum and minimum temperature settings at which a temperature switch will continue to operate within the manufacturer's specifications.
- *Remote temperature sensing*—The construction and installation of a temperature switch in which the sensing unit and the switch mechanism are thermally isolated so that the temperature of the sensed medium does not affect the performance of the switch mechanism.

6.4.0 Pressure Switches

Pressure switches (*Figure 47*) use mechanical motion in response to pressure changes to open or close contacts when a predetermined pressure level is reached. Depending on the switch design and application, NO and NC contacts may be activated on positive, negative (vacuum), or differential pressures. Single-stage switches are used in a wide variety of machine and process applications to protect the equipment from low or high pressures or to monitor the system pressure. Dual-stage pressure switches are made that contain two

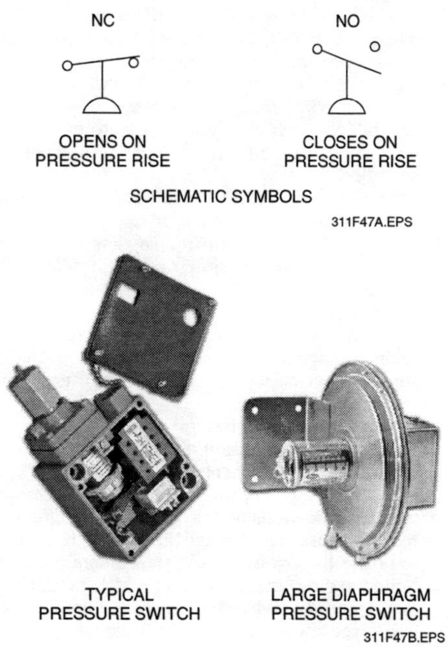

Figure 47 ◆ Typical pressure switches.

Classroom

Define the terms associated with temperature switches.

Discuss the operation and applications of pressure switches. Provide examples for the trainees to examine. Refer trainees to Figure 47.

Classroom

Compare the operation of sealed piston and diaphragm pressure switches.

Audiovisual

Show Transparency 21 (Figure 48).

Classroom

Define the terms associated with pressure switches.

INSIDE TRACK

Bourdon Tube Pressure Switches

Care must be taken not to apply pressures beyond the rating of a Bourdon pressure switch. Exposure to excessive pressures can bend the Bourdon tube beyond its ability to return to its original shape.

pressure switches. They are used in applications where it is practical to monitor two separate pressures using a single device. For example, in a refrigeration system, a dual pressure switch is commonly used to monitor the compressor suction and head pressures for a low-pressure and high-pressure condition, respectively. Differential pressure switches can signal that a predetermined pressure difference has been reached by an increasing difference between two pressures or a decreasing difference between two pressures.

Both adjustable and nonadjustable types of pressure switches are made. Use of an adjustable type allows a particular model switch to be used in more than one application. Different kinds of pressure-responsive elements are used in pressure switches. They can be a Bourdon tube, bellows, sealed piston, or diaphragm. Bourdon tube and bellows switch sensors operate in the same way as described earlier for use in temperature switches.

In a sealed piston pressure switch, the switch contacts are actuated by a piston that moves up and down in its cylinder in response to the applied pressure, as shown in *Figure 48(A)*. The top of the piston moves against a calibrated spring, the tension of which actuates the switch contacts when the source pressure increases. Normally, the spring tension is adjustable so that the switch can be adjusted to operate at different pressure setpoints.

In a diaphragm pressure switch, the edge of a disk-shaped diaphragm is firmly attached to the case of the switch, as shown in *Figure 48(B)*. The pressure being monitored is applied against the full underside area of this diaphragm. In response to this pressure, the center of the diaphragm opposite the pressure side moves against a calibrated spring, the tension of which actuates the switch contacts when the pressure increases. Normally, the spring tension is adjustable so that the switch can be adjusted to operate at different pressure set points.

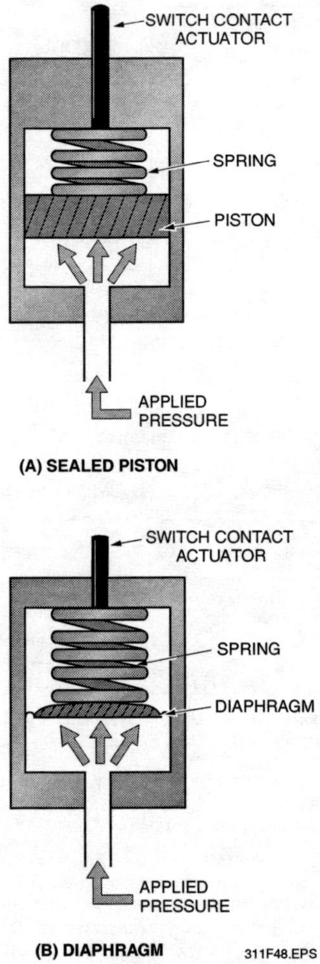

Figure 48 ◆ Sealed piston and diaphragm pressure switch mechanisms.

To properly select pressure switches, you need to know and understand several terms used in switch manufacturers' catalogs:

- *Allowable pressure limits*—The maximum and minimum pressures, stated by the manufacturer, to which a pressure switch may be exposed for brief or extended periods without altering the performance characteristics of the switch. Maximum allowable pressure is also referred to as *overrange* and includes surge pressures.

Instructor's Notes:

> ### *Pressure Switches*
>
> INSIDE TRACK
>
> A pressure switch should not be used where its pressure-containing envelope is subjected to a surge that is greater than its maximum allowable pressure. A pressure switch used in a system pressurized by a reciprocating pump should be protected from surges by a pulsation snubber.

- *Cut-in pressure*—The pressure at which the switch is actuated to energize the load. The cut-in pressure may be the trip or reset point.
- *Cut-out pressure*—The pressure at which the switch is actuated to de-energize the load. The cut-out pressure may be the trip or reset point.
- *Deadband*—The pressure difference between the setpoint at which a switch activates when the pressure increases and the point at which the switch resets when the pressure drops.
- *Drift*—An inherent change in operating value for a given setting over a specified number of operations and specified environmental conditions.
- *Maximum static pressure*—The continuous pressure that the pressure-containing envelope of a pressure switch sustains without rupture. The maximum static pressure is sometimes called the *rated static pressure.*
- *Operating differential pressure*—The difference between cut-in pressure and cut-out pressure. The operating differential pressure is also called the *deadband.*
- *Operating pressure range*—The range between the maximum and minimum pressure settings at which a pressure switch will operate and continue to operate within the manufacturer's specifications.
- *Proof pressure*—The nondestructive static test pressure, in excess of the maximum allowable pressure, which causes no permanent deformation or malfunction.
- *Pulsation snubber*—A device used with pressure switches to reduce the effect of pressure surges within the pressure system on the pressure responsive element of the switch.

6.5.0 Mechanical Limit Switches

Mechanical limit switches are switches used to detect the position of an object by having the object make direct physical contact with the switch **actuator.** They commonly are used in applications in which it is desired to limit the travel of machine tools, detect moving items on a conveyor belt, monitor an object's position, or detect that machinery safety guards are properly in place. Limit switches come in many designs and sizes with single-pole, double-throw (SPDT) and double-pole, double-throw (DPDT) versions being the most common. There are two types of limit switches: rotary lever-actuated (*Figure 49*) and plunger-actuated (*Figure 50*). Both consist of an actuator, switch body, and terminals. The switch actuator is the part of the switch which, when moved by physical

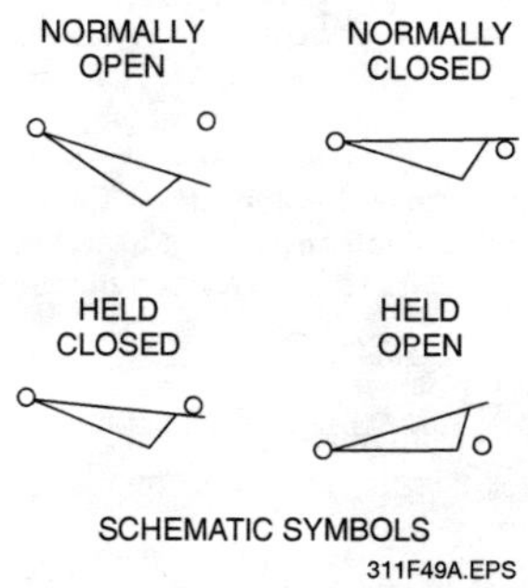

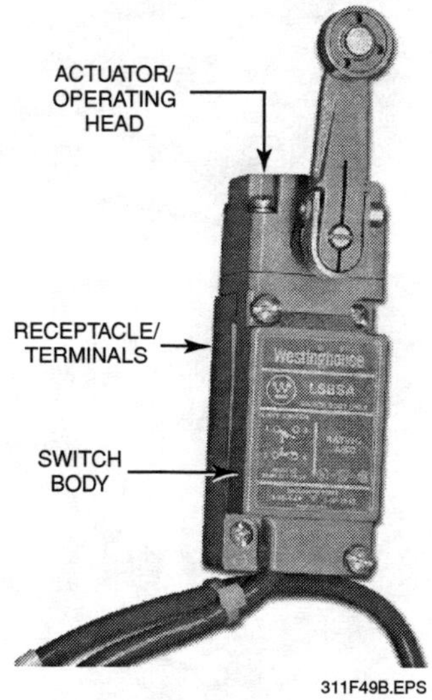

Figure 49 ◆ Rotary lever-actuated limit switch.

Classroom

Discuss the operation and applications of limit switches. Provide examples for the trainees to examine. Refer trainees to Figure 49.

Audiovisual

Show Transparency 22 (Figure 50).

Classroom

Discuss the operation of plunger and rotary-type limit switch actuators.

Audiovisual

Show Transparencies 23 and 24 (Figures 51 and 52).

Classroom

Define the terms associated with limit switches.

contact with an object, operates the switch contacts. The switch body houses the electrical contacts, and the terminals are the point of switch connection for the circuit wiring.

A rotary lever-actuated limit switch works on the principle that a **cam** or plate hits the end of the lever arm, which rotates a shaft that operates the switch contacts. In some rotary lever-actuated limit switches, the actuator attached to the shaft can be interchanged with a variety of different kinds of actuators (*Figure 51*). This allows the same switch to be used in many different applications. A plunger-activated limit switch works on the principle that a cam or plate hits the end of the plunger, which is pressed in to operate the contacts of the switch.

Limit switch operating heads can be of two types: **momentary contact switch** or **maintained contact switch.** Momentary contacts, sometimes called spring return contacts, return from the actuated position to their normal (nonactuated) position when the actuating force is removed. Maintained contacts remain in the actuated position until actuated to another position even after the actuator is released. They are reset only by further mechanical action of the operating head. For example, the contacts may be reset by shaft rotation in the opposite direction.

Manufacturers of limit switches describe the position or motion of the switch actuator in specific terms in their product literature. These terms are shown in *Figure 52* and are described in the following list:

- *Movement differential*—The distance or angle from the operating position to the releasing position.

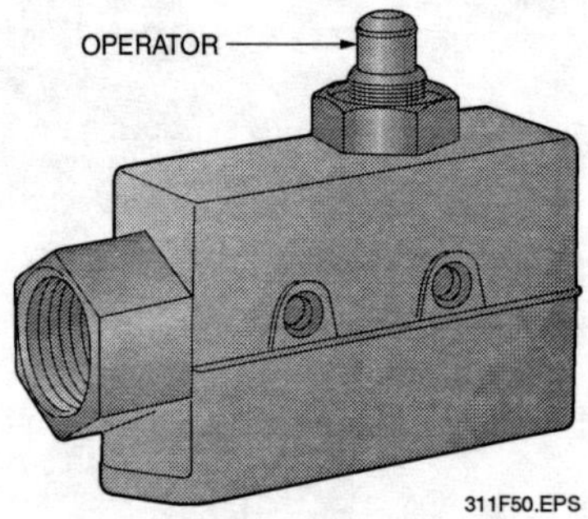

Figure 50 ◆ Plunger-actuated limit switch.

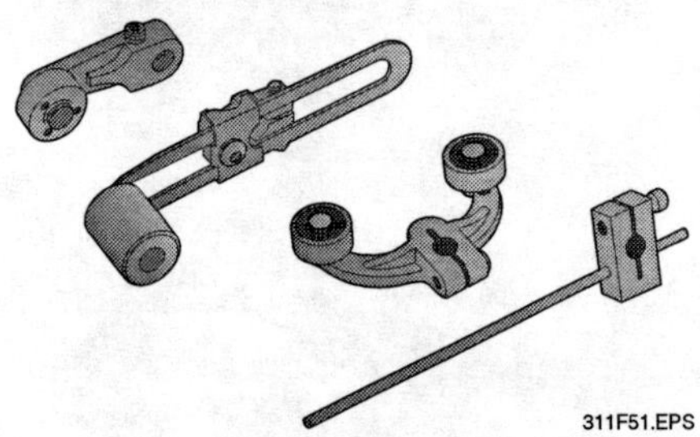

Figure 51 ◆ Rotary lever-actuated limit switch actuators.

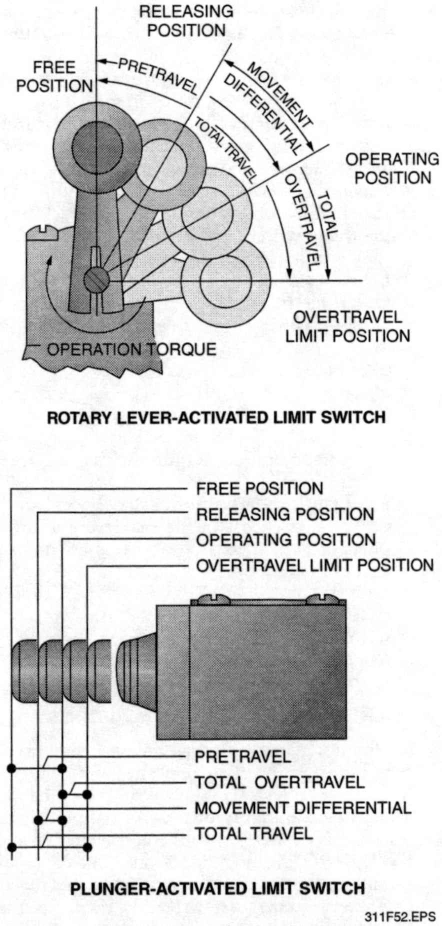

Figure 52 ◆ Positions of limit switch actuators.

Instructor's Notes:

- *Free (initial) position*—The switch actuator is positioned such that the limit switch contacts are in their normal (untriggered) position.
- *Operating position or point*—The position of the actuator at which the contacts change from their untriggered position to the operated (triggered) position.
- *Operating torque*—The minimum force that must be applied to the actuator to cause the contacts to change from the untriggered to triggered state.
- *Overtravel*—In the case of the rotary lever-actuated limit switch, it is the distance or angle through which the actuator moves when traveling past the triggered position. In a plunger-actuated limit switch, the overtravel distance is the safety margin for the sensor to avoid breakage.
- *Pre-travel*—The distance or angle through which the actuator moves from the free position to the position just before the contacts change state. The contacts are still at their normal (untriggered) position.
- *Release position or point*—The position of the actuator at which the contacts change from the triggered position and return to the normal untriggered position.
- *Release torque*—The value to which the torque on the actuator must be reduced to allow the contacts to change from the triggered position to the normal untriggered position.

Depending on the limit switch design, the normally open and normally closed contacts may or may not be conducting current simultaneously during operation. SPDT limit switches can have make-before-break, break-before-make, and simultaneous make-and-break contacts. With make-before-break contacts, the normally open contact closes before the normally closed contact opens. With break-before-make contacts, the normally closed contact opens before the normally open contact closes. With simultaneous make-and-break contacts, the normally closed contact opens at the same time the normally open contact closes. Manufacturers of limit switches often use contact function diagrams like the one shown in *Figure 53* to define the operation of their switches and show any overlap or nonoverlap of contact operation.

When selecting and installing limit switches, the following guidelines should be taken into consideration:

- A rotary lever-actuated limit switch is the best choice for most applications. It can be used in any application in which the cam moves perpendicular to the lever's rotational shaft.
- A plunger-actuated limit switch is the best choice to monitor short, controlled machine movements, or where space or mounting restrictions will not permit the use of a lever-actuated switch.
- The switch contacts must be selected according to the proper voltage and current size for the load. If the load current exceeds the switch contact rating, a relay, contactor, or motor starter must be used to interface the limit switch with the load.
- The switch contacts must be connected to the same polarity. Otherwise, arcing or welding of the contacts may occur. Do not connect voltages from different sources to the contacts of the same switch unless the switch is designed for such service.
- Mount the switch so that there is no possibility of unintentional actuation by the movement of an operator or moving parts of the machine or equipment. Mount it firmly in an easily accessible location with suitable clearances to allow for service and replacement. Do not place the switch in a location where machining chips or other materials can accumulate under normal operating conditions or where the temperature and atmospheric conditions are beyond those for which the switch has been designed.
- Do not expose the switch to oils, coolants, or other liquids unless designed for such service. If liquid entry is possible, the switch should be mounted face down to prevent seepage through the seals on the operating head. All conduit connections must also be tightly sealed.

INSIDE TRACK

Limit Switches

Most newer industrial limit switches are of a plug-in design, which means that the operating head, switch body, and receptacle are separate components. Should the switch body become damaged or fail, it can easily be replaced in the field without rewiring the switch. Simply remove the switch body, and the wiring remains intact in the receptacle.

Classroom

Review the selection and installation guidelines for limit switches.

Audiovisual

Show Transparency 25 (Figure 53).

Classroom

Discuss the operation and applications of flow and float switches. Provide examples for the trainees to examine. Refer trainees to Figures 54 and 55.

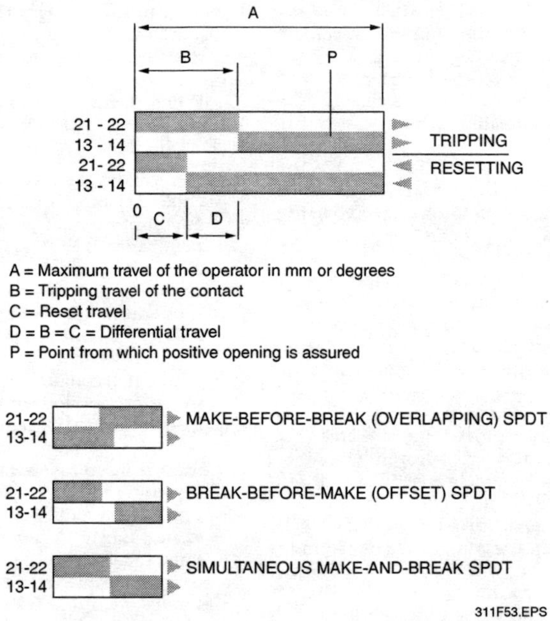

Figure 53 ◆ Example of a limit switch contact function diagram.

- For rotary lever-actuated limit switches, it is important that the correct lever actuator be selected. The manufacturer's recommendations in selecting and applying limit switches and actuators should always be followed. It is important that the selected actuator be securely fastened to the switch body shaft to prevent it from slipping.
- Where relatively fast motions are involved with a rotary lever-actuated limit switch, the cam arrangement should be such that the actuator does not receive a severe or sharp impact. The cam should be tapered to extend the time it takes to engage the electrical contacts. This prevents wear on the switch and allows the contacts a longer closing time, giving relays, valves, and other related devices enough time to operate.

6.6.0 Flow Switches

Flow switches are typically used to detect (prove) the presence of a liquid flowing through a pipe or airflow in a duct. Mechanical flow switches (*Figure 54*) have a vane that extends into the pipe or duct. When the force of the liquid flowing in the pipe or air flowing in the duct is sufficient to overcome the spring tension of the vane, the vane moves and actuates the switch contacts. When there is a loss or reduction in flow below the predetermined vane actuation setpoint, the vane returns to the central position. For some flow switches, the spring tension on the vane is adjustable. This allows the switch vane to be adjusted so that it actuates the switch contacts at different flow rates. The NO or NC switch contacts may be used to start or stop pumps, open or close valves, actuate a warning lamp, or sound an alarm.

6.7.0 Float Switches

Float switches are used to maintain or monitor the level of a liquid in a storage tank or other vessel. For example, when the level of a liquid reaches a predetermined high (or low) point in a tank, the float switch operates to shut off or turn on a pump. Another example is that of a float switch used to activate an alarm, shut down a pump, or both as it monitors a safe level or a not-to-exceed level of a liquid in a tank. There is a wide variety of float switches including mechanical, electronic, ultrasonic, and optical designs. A basic mechanical type of float switch is shown in *Figure 55.* It

Instructor's Notes:

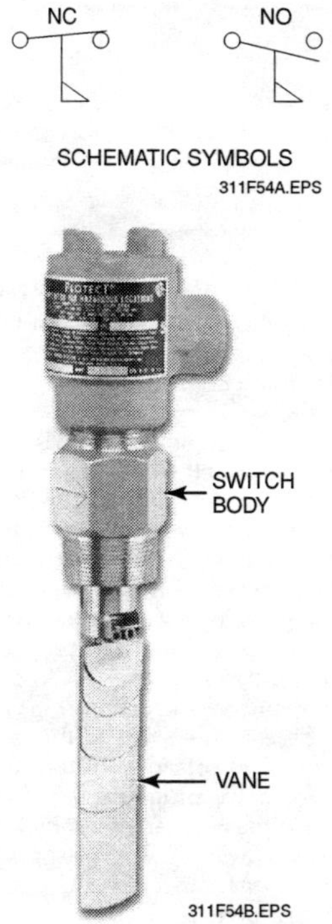

Figure 54 ◆ Vane liquid flow switch.

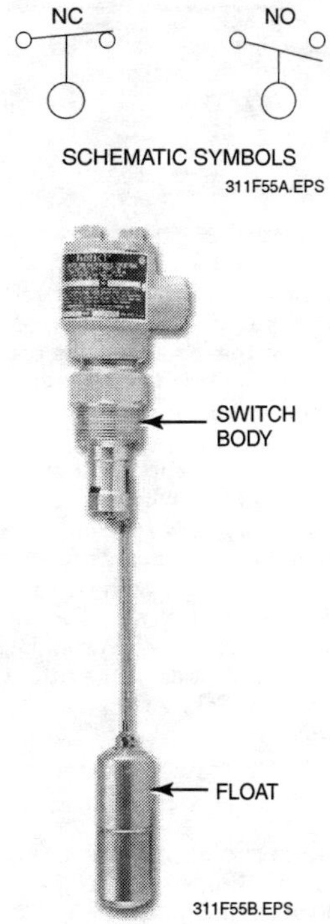

Figure 55 ◆ Liquid level float switch.

consists of a float and hermetically sealed, magnetically operated snap switch. As the float moves up and down with the level of the liquid in a tank, the NO or NC switch contacts are activated when the liquid level reaches a predetermined height. Two switches can be used to monitor and maintain minimum and maximum liquid levels.

6.8.0 Foot Switches

A foot switch (*Figure 56*) is a control device operated by a foot pedal. It is used where the process or machine requires that the operator have both hands free. Foot switches usually have momentary contacts but some are available with latches that enable them to be used as maintained-contact devices. Most foot switches typically have two positions, a toe-operated position and a spring-loaded off position. Some foot switches are available with three positions, two positions that allow for toe and heel control and a spring-loaded off position.

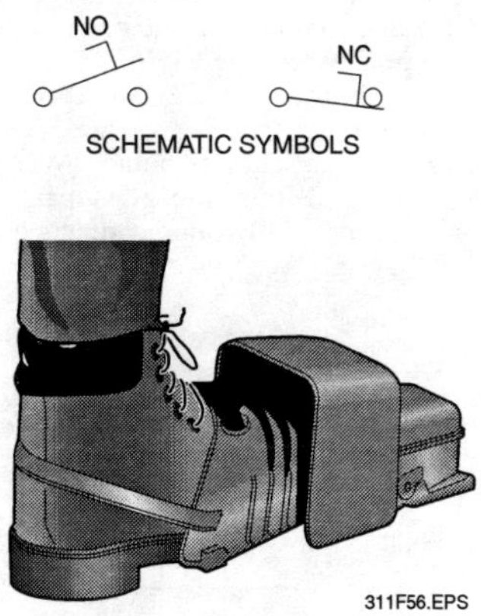

Figure 56 ◆ Foot-operated switch.

Classroom

Discuss the operation and applications of foot switches. Provide examples for the trainees to examine.

Audiovisual

Show Transparency 26 (Figure 56).

Classroom

Discuss the applications of jogging and plugging switches. Provide examples for the trainees to examine.

Describe the operation of a basic jogging circuit.

Audiovisual

Show Transparency 27 (Figure 57).

Classroom

Explain that plugging switches provide rapid speed reversal to ensure precise mechanical action.

Audiovisual

Show Transparencies 28 and 29 (Figures 58 and 59).

INSIDE TRACK

Use of Foot Switches

When using a foot switch in applications such as power presses, additional operator protection, such as point-of-operation guarding, must be provided. This is necessary since the operator's hands and other parts of the body are free to enter the pinch-point area and serious injury can occur.

In applications in which more than one foot switch is required, as when two or more persons are operating the machine, it is required that the foot switches be wired in series, making it necessary that each operator's foot switch be actuated before the machine can operate.

The use and selection of a foot switch for an application is based on the customer's knowledge of the conditions and factors present during the setup, operation, and maintenance of the particular machine or process with which the switch is to be used. When selecting a foot switch, always refer to the applicable ANSI standards and OSHA regulations. Additional information can be found in the National Safety Council's *Accident Prevention Manual.*

6.9.0 Jogging and Plugging Switches

Many motor control applications require the use of switches in the control circuit to initiate the jogging (inching) or plugging of a motor. Jogging a motor involves the frequent closure of a switch to stop and start the motor for short periods of time, which is required when it is necessary to precisely position a crane or machine tool. There are many ways that a pushbutton switch can be connected in a control circuit to facilitate jogging. *Figure 57* shows a line diagram of one basic control circuit where a pushbutton switch is used to jog a motor. In this circuit, the normally closed contacts of the JOG pushbutton switch are connected in series with the holding circuit contact (M) of the magnetic starter. When the JOG pushbutton is pressed, the normally open contacts energize the starter magnet. At the same time, the normally closed contacts disconnect the holding circuit. Therefore, when the JOG pushbutton is released, the starter immediately opens to disconnect the motor from the line.

Plugging is an operation in which a motor is brought to a rapid stop (braked) by reversing the phase sequence of the power applied to the motor. Plugging typically is done in machine tool applications when the tool must be stopped rapidly at some point in its cycle of operation in order to prevent inaccuracies in the work or damage to the machine. Plugging can only be done if the driven machine and its load will not be damaged by the reversal of the motor.

A plugging switch (*Figure 58*) is connected mechanically to the shaft of the motor or driven machinery. The rotating motion of the motor is transmitted to the plugging switch contacts either by a centrifugal mechanism or by magnetic induction. The contacts of a plugging switch are designed to open and close as the shaft speed of the plugging switch varies.

There are several ways that a plugging (speed) switch can be connected in a control circuit to facilitate plugging. *Figure 59* shows a line diagram of a basic control circuit that uses a plugging switch to plug a motor to a stop from one direction only. In this circuit, the forward rotation of the motor closes the normally open plugging switch

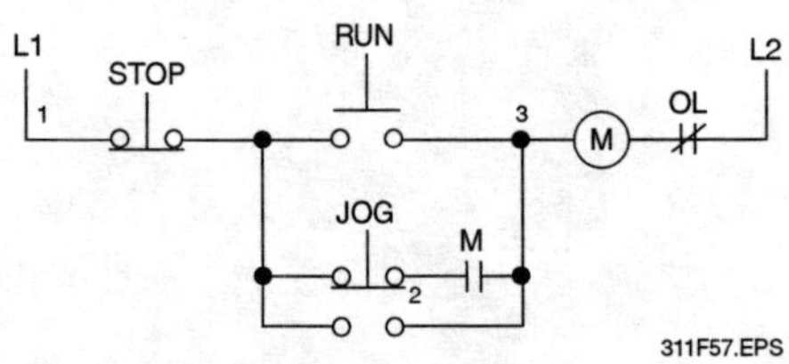

Figure 57 ◆ Basic motor jogging control circuit.

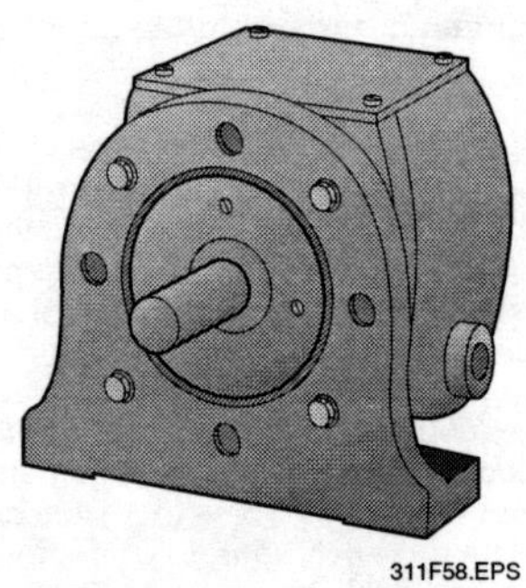

Figure 58 ◆ Plugging switch.

Instructor's Notes:

> **INSIDE TRACK**
> ***Plugging***
> Capacitor-start motors cannot be plugged to a stop because they cannot be reversed at full speed.

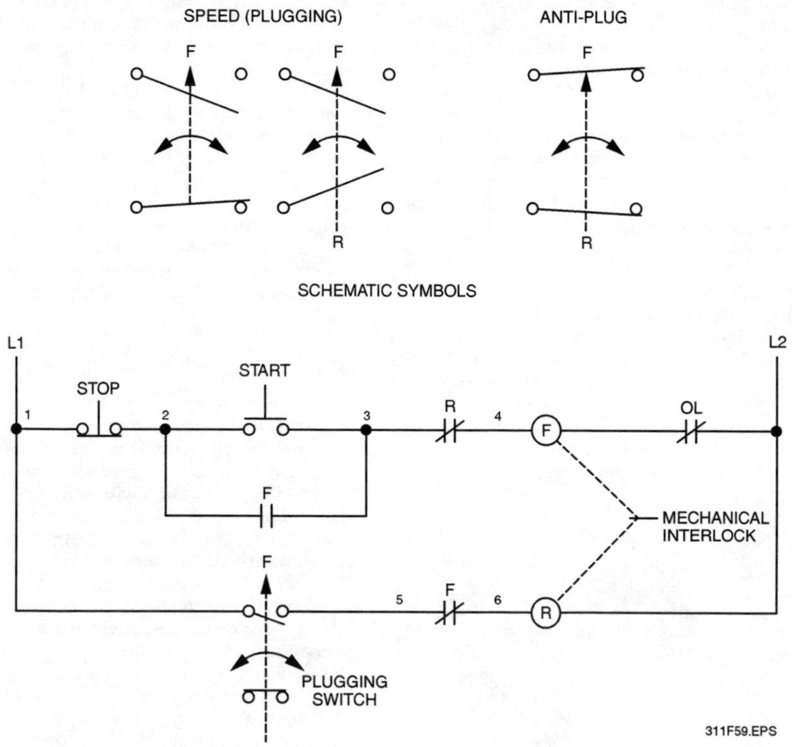

Figure 59 ◆ Basic motor plugging control circuit.

contact. When the stop button is pushed, the forward starter de-energizes. At the same time, the reverse starter is energized through the plugging switch and the normally closed forward interlock. This reverses the motor connections, and the motor is braked to a stop. When the motor is stopped, the plugging switch opens to disconnect the reverse contactor.

Some plugging switches have an anti-plugging protection capability. This means the switch design is such that it prevents the application of a counter torque until the motor speed is reduced to an acceptable value. With this switch, a contact on the switch opens the control circuit of the contactor used to reverse the rotation of the motor and is prevented from closing until the motor speed is reduced. Then the other contactors can be energized.

6.10.0 Proximity Switches/Sensors

Proximity switches, commonly called **proximity sensors,** are encapsulated solid-state devices used where it is required or desired to detect the presence of objects (targets) within a short sensing range without having direct physical contact with

Classroom

Discuss the operation and applications of proximity switches. Provide examples for the trainees to examine.

Classroom

Explain that inductive proximity sensors are used to detect metal targets. Refer trainees to Figure 60. Describe the operation of an inductive proximity sensor.

Audiovisual

Show Transparency 30 (Figure 61).

Classroom

Explain that capacitive proximity sensors can be used to detect both metallic and nonmetallic targets. Describe the operation of a capacitive proximity sensor.

Audiovisual

Show Transparency 31 (Figure 62).

the target(s). They can detect all sizes of objects, including very small ones, in applications such as presence, passage, flow of parts, end of travel, rotation, and counting. Proximity switches are commonly used in place of mechanical limit switches, float switches, or level switches for many applications. This is because they have faster switching speeds and no moving parts, making them more reliable and accurate. Being solid-state, they are compatible with electronic controllers and automated systems. Proximity switches are intended for use in circuits with rated voltages not exceeding 250 or 300VDC. Two widely used proximity switches are the inductive proximity sensor and the capacitive proximity sensor. Physically, both types of sensors look the same. *Figure 60* shows two circularly shaped proximity sensors. Block-shaped sensors are also available.

6.10.1 Inductive Proximity Sensors

An inductive proximity sensor is used to detect only metal targets within its sensing field, which is typically 0.5mm to 40mm (0.020" to 1.575"). Its sensitivity depends on the size of the target and the metallic material of which the target is made. The internal circuitry of a solid-state inductive proximity sensor consists of an oscillator circuit, output driver, and output switch (*Figure 61*). The oscillator circuit produces radio frequency (RF) oscillations. These oscillations are present at a coil (part of the oscillator circuit) that is located at the sensing face of the proximity sensor. This coil has two functions: it determines the tuning of the oscillator circuit, and it radiates an electromagnetic field created by the RF oscillations into the sensing field in front of the switch face.

When a target enters the sensing field, the electromagnetic field emitted by the proximity sensor causes **eddy currents** to be induced in the target. These eddy currents disrupt the electromagnetic field at the sensor face coil, causing a change in the tuning of the oscillator circuit. As a result, the oscillations cease. The output driver senses the loss of the oscillations and commands the output switch to operate, producing an ON or OFF output signal depending on the switch design. After the target passes out of the sensing field, the oscillator signal is again generated, causing the switch output signal to be returned to its normal inactive state.

6.10.2 Capacitive Proximity Sensors

The capacitive proximity sensor can be used to detect the absence or presence of both metallic and nonmetallic targets within its sensing field, which is typically 0.3 mm to 15 mm (0.012" to 0.59"). Target detection with this type of sensor is based on the principle that when a target moves into the sensing field, it changes the coupling capacitance of the sensor's sensing capacitor. Remember that a capacitor consists of two plates separated by an insulator (dielectric) and that any change in a capacitor's **dielectric constant** causes a change in the capacitor's capacitance value. The internal circuitry of a solid-state capacitive proximity sensor consists of an oscillator, output driver, and output switch circuit (*Figure 62*). The sensing capacitor is part of the oscillator's tuning circuit. It is formed by two small plates separated by a dielectric (air) and is located behind the front of the sensing face of the sensor. In the absence of a target, the tuning of the oscillator is such that the

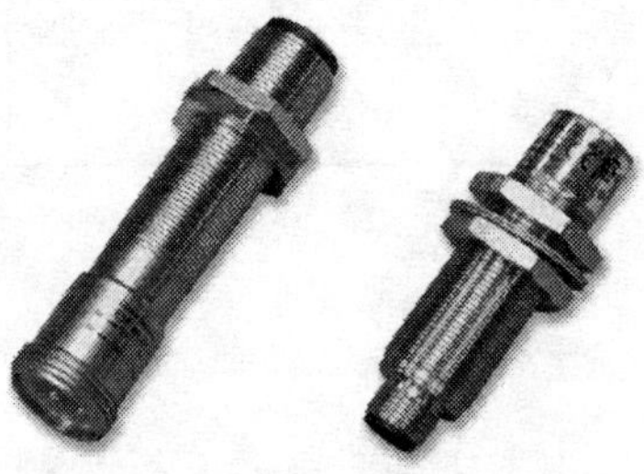
311F60.EPS

Figure 60 ◆ Examples of typical proximity sensors.

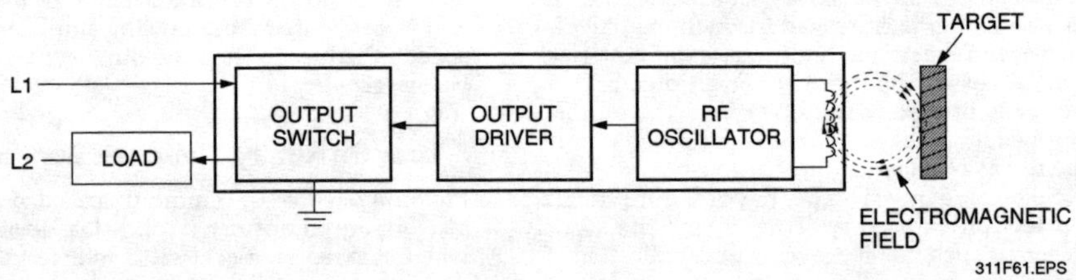

311F61.EPS

Figure 61 ◆ Block diagram of an inductive proximity sensor.

Instructor's Notes:

Capacitive Proximity Sensors

INSIDE TRACK

Because capacitive proximity sensors measure a dielectric gap, many have a sensitivity adjustment. This allows the sensor to be adjusted in order to compensate for target and application conditions. Do not handle a capacitive proximity sensor during adjustment. This is because your hands have a dielectric constant greater than air, which can cause the sensor to lose its calibration.

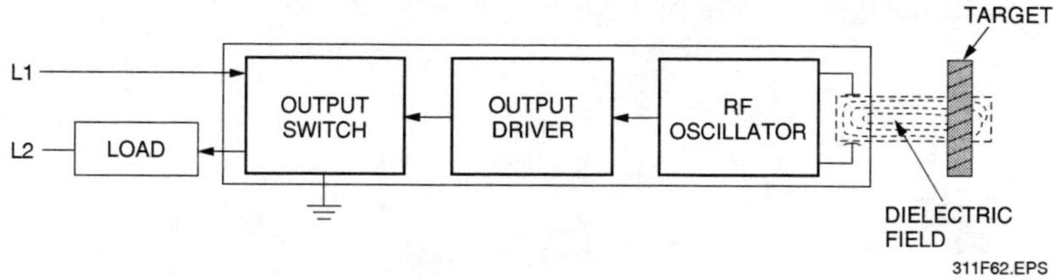

Figure 62 ◆ Block diagram of a capacitive proximity sensor.

oscillator is inoperative. Note that this is just the opposite of the inductive proximity sensor.

Capacitive proximity sensors can detect any target that has a dielectric constant slightly greater than that of air. Air has a dielectric constant of 1. When a target with a dielectric constant greater than 1 moves into the sensing field, its presence modifies the coupling capacitance. This change in capacitance tunes the oscillator circuit so that it begins to generate oscillations. The output driver senses these oscillations and commands the output switch to operate, producing an ON or OFF output signal depending on the switch design. After the target passes out of the sensing field, the oscillator signal again ceases, causing the switch output signal to return to its normal inactive state.

6.10.3 Load-Powered Proximity Sensors

There are two categories of proximity sensors: **load-powered sensors** and **line-powered sensors.** Load-powered sensors are two-wire sensors (excluding ground) that are connected in series with the controlled load (*Figure 63*). They draw their operating current, also commonly called leakage or residual current, through the load. When the sensor is in the open state (absence of a target), it must draw a minimum operating current, sometimes called the off-state leakage current, through the de-energized load device in order to power the proximity sensor's electronics. However, this operating current must be low enough so that it will not inadvertently energize the load.

When the sensor is in the closed state (target present), the current flowing in the control circuit is a combination of the proximity sensor operating current and the current drawn by the energized load. This means that the proximity sensor must have a current rating high enough to carry the load current. Another factor to consider is that the current draw of the load must be high enough to ensure that the sensor will operate when the load is energized. This current level is called the minimum holding current. If the current drawn by the load falls below the required minimum holding current, the sensor will be inoperative.

Two-wire proximity sensors can be connected in series or in parallel like other pilot devices. However, it is recommended that no more than three devices be connected in series or in parallel because they can affect the operation of the load. No more than three sensors should be connected in series because the sum of the individual voltage drops across each of the sensors can lower the voltage available to the load sufficiently enough to render it inoperative. Sensor voltage drops vary for different sensors. Typically, two-wire sensors have a voltage drop ranging anywhere between 3V and 8V. The lower the voltage drop, the more sensors that can be connected in series.

No more than three sensors should be connected in parallel because the sum of the operating currents flowing through the individual sensors, if excessive, can be enough to inadvertently energize the load. Also, when sensors are connected in parallel, the total amount of current

Classroom

Explain that proximity sensors can be either line-powered or load-powered. Describe how each type is wired into the circuit.

Audiovisual

Show Transparency 32 (Figure 63).

Audiovisual

Show Transparency 33 (Figure 64).

Classroom

Discuss the basic guidelines for selecting and installing proximity sensors.

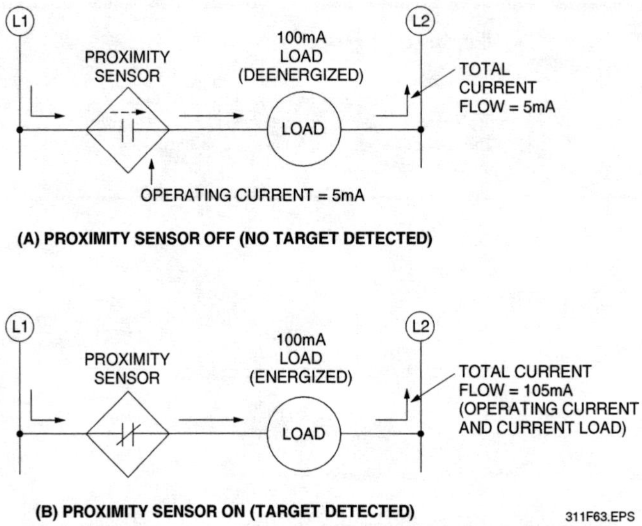

Figure 63 ◆ Load-powered proximity sensor circuit.

the load draws when operating must be less than the maximum current rating of the lowest rated sensor.

6.10.4 Line-Powered Proximity Sensors

Line-powered proximity sensors are three-wire sensors (excluding ground) that derive their power from the power lines, not the load (*Figure 64*). They have two wires for the power supply and one output wire. Some four-wire models are also available. These produce complementary NO and NC outputs. The operating current for a line-powered sensor drawn from the power lines is called the burden current.

Connecting line-powered sensors in series and in parallel can affect the operation of the load. Connecting these sensors in series can affect the load because the first sensor in the line must carry the load current plus the operating currents for each of the following (downstream) sensors. When connected in parallel, there is a danger that a nonconducting sensor may be damaged by reverse polarity. Note that this problem is usually eliminated by connecting a blocking diode in the output line of each sensor.

6.10.5 Proximity Sensor Installation Guidelines

Proximity sensors should always be installed in accordance with the manufacturer's recommendations. Some guidelines for selecting and installing inductive and capacitive proximity sensors are given here:

- The output switch in inductive and capacitive proximity sensors can use either a triac, SCR, or transistor as the switching device. Proximity sensors with triacs are used to switch AC loads. Those with SCRs and transistors are used to switch high-power DC loads and low-power DC loads, respectively. Many proximity sensors are designed for use in either AC or DC circuits. For DC and AC/DC versions, the switch can be connected to either positive (PNP) or negative (NPN) logic inputs.
- A shielded (flush mount) inductive sensor can be fully embedded in a metal mounting block without affecting the range. A nonshielded sensor needs a clearance around it called a metal-free zone that is determined by the sensing range. Otherwise, the sensor will sense the metal mounting and operate continuously.

Instructor's Notes:

- When mounting two or more sensors near each other, make sure to mount them in accordance with the manufacturer's instructions so that their radiated detection fields do not interfere with each other. A rule of thumb for flush-mounted sensors is to use a distance equal to or greater than twice the diameter of the sensors. Note that when sensors of different diameters are used, base the separation distance on the largest diameter. For nonflush mounted sensors, use a distance equal to or greater than three times the diameter of the sensors.
- When inductive and capacitive sensors are installed next to each other, their sensing fields can cause false readings on each other. For this reason, leave a space three times the diameter of the largest sensor between them. When installed opposite to each other, leave a distance of six times the rated sensing distance.
- Where metal chips or similar debris is created, the sensor should be mounted to prevent the chips from building up on the sensor face.
- Avoid installing inductive sensors in locations where the magnetic field from nearby electrical wiring can affect sensor operation.
- Avoid installing sensors in locations where electrical interference generated by nearby motors, solenoids, or relays can have an effect on sensor operation.

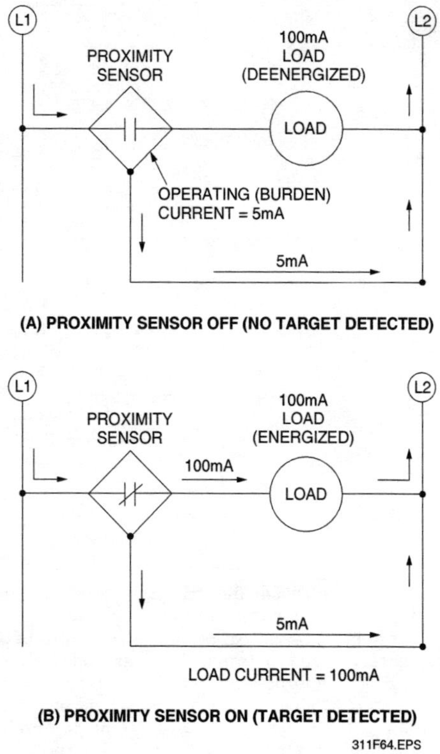

Figure 64 ◆ Line-powered proximity sensor circuit.

6.11.0 Photoelectric Switches/Sensors

Photoelectric sensors (*Figure 65*), like proximity sensors, are solid-state devices used to detect the absence or presence of objects (targets) within their sensing range. Generally, they can detect objects more quickly and at further distances than proximity switches. A photoelectric sensor sends out a beam of light, which is picked up by a photodetector. When an object (target) moves into the path of the light beam, the beam is interrupted, indicating that a target is detected.

Figure 66 shows a block diagram for a basic photoelectric switch. As shown, a photoelectric sensor consists of transmitter and receiver circuits and an output switch that is turned on or off by the receiver output. Depending on the application, these circuits can be housed together in the same encapsulated enclosure or be in separate enclosures. Basically, the transmitter emits a beam of light, which is picked up by the receiver. When an object (target) moves into the path of the light beam, the beam is interrupted, indicating that a target has been detected.

The transmitter circuit operates to generate **infrared (IR)** light that is emitted from an LED through a lens placed just behind the face of the

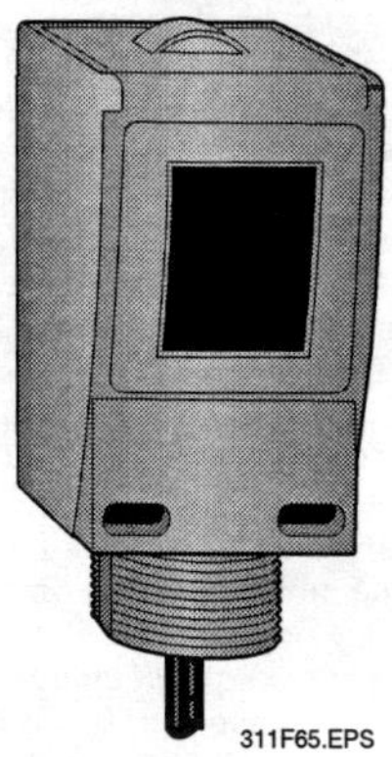

Figure 65 ◆ Photoelectric sensor.

Classroom

Discuss the operation and applications of photoelectric switches. Provide examples for the trainees to examine.

Audiovisual

Show Transparencies 34 and 35 (Figures 65 and 66).

Classroom

Discuss the basic guidelines for selecting and installing photoelectric switches.

Describe the various scanning arrangements used with photoelectric sensors.

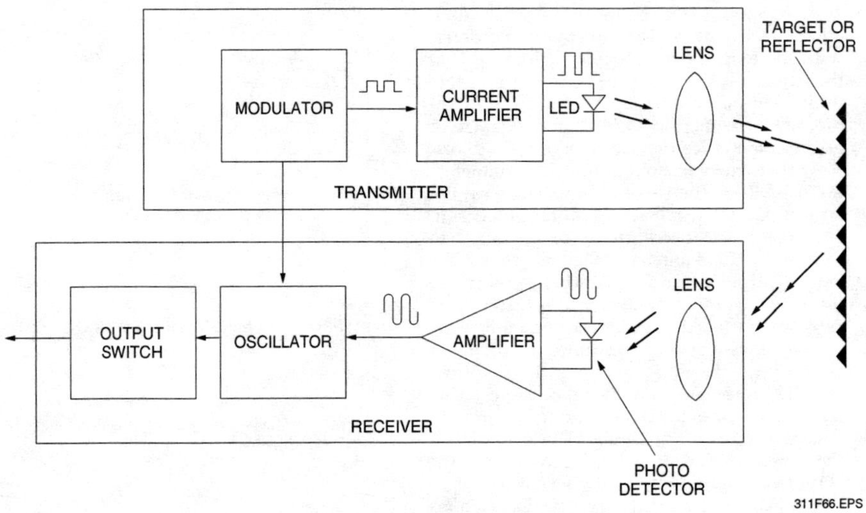

Figure 66 ◆ Block diagram of a modulated-beam photoelectric sensor.

sensor. By sensor design, the emitted infrared light beam is either an unmodulated beam or modulated beam. An unmodulated beam is one that is emitted continuously; a modulated beam is one that is turned on and off at a very high frequency. Modulated-beam sensors tend to be more popular since they do not respond to ambient light or other forms of light noise. Unmodulated-beam sensors are typically used where the scanning range is very short and where dirt, dust, and bright ambient light are not a problem.

A photodetector (photodiode or phototransistor) placed just behind a lens at the face of the sensor is the front end of the sensor's receiver circuit. Being light sensitive, the photodetector detects any received infrared light and, in response, produces an output signal. This signal is further processed in the receiver to amplify it to a useable level. In the receiver of a modulated-beam sensor, the amplified received signal is also applied to a demodulator circuit. There, the characteristics of the received signal are compared against those of the original modulated transmitted signal. This comparison helps to select the desired received signals and reject unwanted noise signals. The receiver output is sent to the output switch commanding it on or off, depending on the application.

Photoelectric sensors generate an output any time an object is detected. If this occurs when the photodetector sees light, the sensor is classified as working in the light operate mode. This means that the sensor's output switch is energized when the target is missing. If the sensor generates an output when the photodetector does not see light, the sensor is classified as working in the dark operate mode. This means the sensor's output switch is energized when a target is present (breaks the beam).

Photoelectric sensors can be used in several different scanning arrangements to meet the requirements of different applications. These include:

- *Direct scan (thru-beam)*—The transmitter and receiver are separate assemblies located opposite each other. The light beam from the transmitter is aimed directly at the receiver. The target to be detected passes between them, interrupting the light beam. This is the most basic approach.
- *Retroreflective scan*—The transmitter and receiver are housed in one assembly. The light beam transmitted from the transmitter is reflected to the receiver by means of a reflector located opposite the sensor. The target to be detected passes between the sensor and reflector, interrupting the light beam.
- *Polarized scan*—The transmitter and receiver are housed in one assembly. The light beam transmitted from the sensor is filtered by a special lens so that it projects in one plane only. The receiver responds only to depolarized light reflected from a corner cube reflector or polarized sensitive tape.

Instructor's Notes:

- *Specular scan*—The transmitter and receiver are separate assemblies placed at equal angles from a highly reflective surface, requiring the receiver to be positioned precisely to receive reflected light. The transmitter sends the signal to the receiver by reflecting the signal off the reflective surface to be detected.
- *Diffused scan*—The transmitter and receiver are housed in one assembly. The light beam transmitted from the transmitter is reflected back to the receiver by the target to be detected.
- *Convergent scan*—The transmitter and receiver are housed in one assembly. The light beam transmitted from the transmitter is simultaneously focused and converged to a fixed point in front of the sensor. The detection point is fixed such that targets before or beyond the focal point are not detected.

Photoelectric sensors should be selected and installed in accordance with the manufacturer's recommendations. General guidelines for connecting and installing load-powered and line-powered photoelectric sensors are the same as those described for proximity sensors.

7.0.0 ◆ DRUM SWITCHES

Drum switches are totally enclosed, multi-pole switches used in machine operation applications to start, stop, and reverse the direction of rotation of single-phase, three-phase, or DC motors. As shown in *Figure 67*, the switching elements are contained in a cylindrical housing that resembles a drum, from which it gets its name. Drum switches are rated by maximum horsepower and are made in several sizes and types of enclosures. They can have either maintained or momentary (spring return to center) contact operation. Drum switches do not contain protective overloads; therefore, circuit overload protection must be provided by installing a manual or magnetic starter in line before the drum switch.

When the motor being controlled by a drum switch is not running in forward or reverse, the handle is in the OFF (center) position. To reverse the direction of a running motor, the handle must first be moved to the OFF position until the motor stops, then moved to the reverse position. Motor reversing occurs in single-phase motors by the drum switch contacts causing the starting winding connections to be reversed. For a three-phase motor, it occurs because the switch contacts reverse two of the three motor leads. For a DC motor, it is done by the switch contacts reversing the direction of current through the motor fields without changing the direction of the current through the armature or vice versa.

311F67.EPS

Figure 67 ◆ Drum switch.

8.0.0 ◆ ENCLOSURES

The correct selection and installation of an enclosure for a particular application can contribute considerably to the length of life and trouble-free operation. To shield electrically live contactors/motor starters and pilot devices from accidental contact, some form of enclosure is always necessary. This function is usually filled by a general-purpose sheet-steel cabinet. However, as specified by the NEC, dust, moisture, or explosive gases make it necessary to employ a special enclosure to protect contactors/motor starters and pilot devices from corrosion or the surrounding equipment from explosion.

When selecting and installing any electrical equipment, it is always necessary to carefully consider the conditions under which the equipment must operate so that compliance with the NEC or local codes can be accomplished. A general-purpose enclosure does not afford the required protection in many applications.

Underwriters Laboratories has defined the requirements for protective enclosures according to various hazardous conditions, and NEMA has standardized enclosures from these requirements.

> INSIDE TRACK
>
> ***Enclosures***
>
> Most manufacturers provide a selection of engineered combination contactor/motor starter assemblies that include a NEMA enclosure as well as the fused-disconnect device required by the NEC.

Classroom

Discuss the operation and applications of drum switches. Provide examples for the trainees to examine. Refer trainees to Figure 67.

Emphasize the importance of selecting the appropriate enclosure for the equipment being used.

Classroom

Review the NEMA enclosure classifications. Provide examples for the trainees to examine. Refer trainees to Figure 68. Note that the IEC equivalents are listed in Table 12.

Explain that wiring diagrams show the actual connections for the device and its control circuits.

NEMA enclosure types (*Figure 68*) and the IEC IP enclosure types that they conform to include the following:

- *General-purpose (NEMA Type 1–IP40)*—A general-purpose enclosure is intended primarily to prevent accidental contact with the enclosed apparatus. It is suitable for general-purpose applications indoors where it is not exposed to unusual service conditions. A NEMA Type 1 enclosure serves as protection against dust, light, and indirect splashing but is not dust-tight.
- *Dust-tight, rain-tight (NEMA Type 3–IP52)*—This enclosure is intended to provide suitable protection against rain and dust. It is suitable for application outdoors, such as for construction work.
- *Rainproof, sleet-resistant (NEMA Type 3R–IP52)*—This enclosure protects against interference in operation of the contained equipment due to rain and resists damage from exposure to sleet. It is designed with conduit hubs and external mounting, as well as drainage provisions.
- *Watertight (NEMA Type 4–IP65)*—This enclosure is designed to withstand high-pressure hose-directed water and windblown dust or rain. It is constructed of steel with a gasketed cover. Pilot device holes are located on the enclosure cover and are provided with hole plugs.
- *Watertight, corrosion-resistant (NEMA Type 4X–IP65)*—These enclosures are similar to NEMA Type 4 enclosures except that they are made of a material that is highly resistant to corrosion. They are used in applications such as meat-packing and chemical plants, where contaminants would ordinarily destroy a steel enclosure over a period of time.
- *Hazardous locations, Class I (NEMA Type 7C or D)*—These enclosures are designed to meet the application requirements of NEC Class I hazardous locations in which flammable gases or vapors are or may be present in the air in quantities sufficient to produce explosive or ignitable mixtures. They must withstand the pressure generated by explosion of internally trapped gases and be able to contain the explosion so that gases in the surrounding atmosphere are not ignited. The letter(s) following the type number indicate(s) the particular group(s) of hazardous locations (as defined in the NEC) for which the enclosure is designed. The designation is incomplete without the suffix letter(s).
- *Hazardous locations, Class II (NEMA 7 Type 9E, F, or G)*—These enclosures are designed to meet the application requirements of NEC Class II locations for operation in the presence of combustible dust. The letter(s) following the type number indicate(s) the particular group(s) of hazardous locations (as defined in the NEC) for which the enclosure is designed. The designation is incomplete without the suffix letter(s).
- *Industrial use (NEMA Type 12–IP62)*—This type of enclosure is designed to exclude dust, lint, fibers and other flying materials, and oil or coolant seepage. There are no conduit openings or knockouts in the enclosure, and mounting is by means of flanges or mounting feet. Pilot device holes are located on the enclosure cover and are provided with hole plugs.
- *Oil-tight, dust-tight (NEMA Type 13–IP62)*—These enclosures are generally made of cast iron and are used in the same areas as NEMA Type 12 enclosures. The main difference is that, due to its cast housing, a conduit entry is provided as an integral part of the NEMA Type 13 enclosure, and mounting is by means of blind holes rather than mounting brackets. Pilot device holes are located on the enclosure cover and are provided with hole plugs.

IEC compliant enclosures are identified by a two-digit IP protection rating system that is defined in *Table 12*. Additional information for selecting enclosures for use in specific locations other than hazardous locations can be found in ***NEC Section 430.91*** and ***NEC Table 430.91.***

9.0.0 ◆ DIAGRAMS

The wiring scheme for motor control circuits and their components can be shown in several forms. These include wiring diagrams, circuit schedules, control ladder diagrams, and logic diagrams.

9.1.0 Wiring Diagrams

Wiring diagrams, commonly called connection diagrams, show the actual point-to-point wiring connections for a motor-driven device and its control circuits. This type of diagram is normally found in the service literature for a specific piece of factory-wired equipment. Connection diagrams are useful when it is necessary to trace and locate a specific wire connected between two components in a circuit, such as might be required when it is necessary to find and replace an open or grounded wire.

Instructor's Notes:

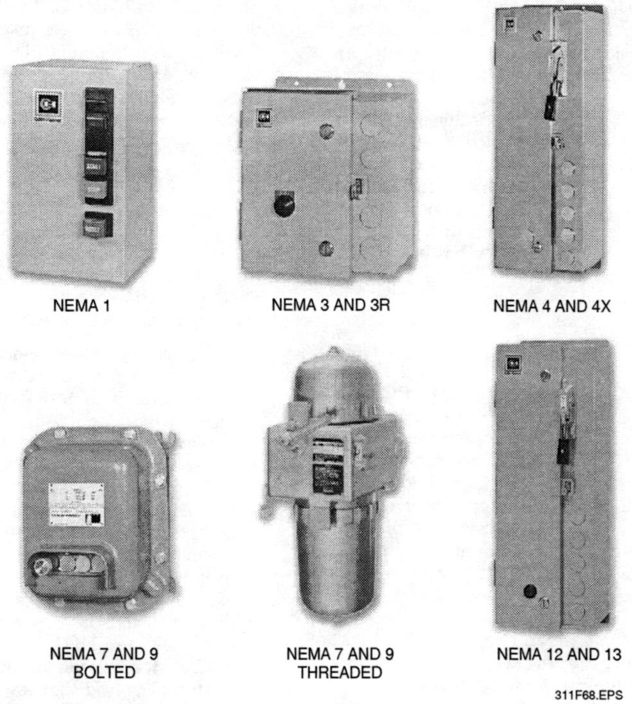

Figure 68 ◆ Typical NEMA enclosures.

Table 12 IEC Enclosure IP Two-Digit Protection Code Definitions

First Digit	Description	Second Digit	Description
0	No protection	0	No protection
1	Protection against solid objects greater than 50mm	1	Protection against vertically falling drops of water
2	Protection against solid objects greater than 12mm	2	Protection against dripping water when tilted up to 15°
3	Protection against solid objects greater than 2.5mm	3	Protection against spraying water
4	Protection against solid objects greater than 1mm	4	Protection against splashing water
5	Total protection against dust; limited ingress (dust protected)	5	Protection against water jets
6	Total protection against dust (dust tight)	6	Protection against heavy seas
		7	Protection against the effects of immersion
		8	Protection against submersion

Audiovisual

Show Transparency 36 (Figure 69).

Figure 69 shows an example of a typical connection diagram. The format of connection diagrams varies depending on the type of equipment and manufacturer. Some connection diagrams show power-circuit wiring as heavy-weight lines and control circuit wiring as lighter-weight lines. Some also show the actual color of each wire in the point-to-point wiring scheme. Where applicable, the component terminal identification number or letter to which a specific wire is to be connected is shown. To help the technician interpret the connection diagram, it normally has a legend that describes any nonstandard symbols and abbreviations that are used on the diagram.

Classroom

Explain that circuit schedules show wiring connections in a tabular format. Refer trainees to Table 13.

Discuss the types of information found on ladder diagrams. Point out that ladder diagrams can be divided into a signal section, a decision (logic) section, and an action section.

Review typical ladder and logic diagrams.

9.2.0 Circuit Schedules

Circuit schedules, sometimes called wire lists, show actual wire or cable point-to-point connections for a motor and its control circuits. They are used mainly when installing a new system or modifying an existing one. Connection information in a circuit schedule is typically presented in tabular form; each individual wire or cable is given a sequence or circuit identification number. The information given normally includes the connection points for the wire and the type and size of the wire. Some circuit schedules also include the color of the wire and type(s) of termination used at each end of the wire. *Table 13* shows an example of a typical circuit schedule.

9.3.0 Control Ladder Diagrams

Control circuits that use various discrete components such as pilot devices and relay coils/contacts and/or those that use a microprocessor to energize contactor/motor starter coils are generally presented as control ladder (line) diagrams (*Figure 70*). Without identifying them as such, ladder diagrams have been used in many illustrations of control circuits presented earlier in this module. A ladder diagram is a diagram that shows, with single lines and symbols, the operational sequence of an electrical circuit.

9.3.1 Sections of a Ladder Diagram

All ladder diagrams are usually presented in three sections between the two wires from the source of the control voltage, as shown in *Figure 71*. The three sections consist of a signal source section, a decision section (or combined signal/decision section), and an action section.

- *Signal section*—Pilot devices are the source of the control signal. These devices can be manual (pushbuttons or footswitches), mechanical (limit switch), or automatic reset devices (flow, pressure, or temperature switches).
- *Decision section*—The decision section of the circuit adds, sorts, selects, and redirects the signals from the pilot devices to the load. The function of this area must be logical. In many cases, the pilot devices are part of the decision section. The way pilot device contacts are connected in the circuit provides the circuit logic. There are three basic logic functions or circuits: AND, OR, and NOT. The electrical relay logic symbols for these logic functions, their related truth tables, and logic symbols are shown in *Figure 72* where contacts A and B represent pilot device contacts. The status of each set of contacts (open or closed) is represented in the truth table by a zero (0) when the contacts are open and by a one (1) when they are closed. The status of the output, on (1) or off (0), called *Y*, is shown for all the possible combinations of contact A and B positions.

Audiovisual

Show Transparencies 37 through 39 (Figures 70 through 72).

Table 13 Example of a Circuit Schedule

Circuit/Cable ID	Cable Type	Length	From	To
P108	3/c #10 w/grd	140'	MCC 10 Section 4A	M108
C108	9/c #14	120'	MCC 10 Section 4A	PLC10-TB1
C108A	12/c #14	160'	MCC 10 Section 4A	JB108
C108B	5/c #14	40'	JB108	HS108
C108C	3/c #14	30'	JB108	HSS108A
C108D	3/c #14	10'	HSS108A	HSS108B
C108E	3/c #14	30'	JB108	AA108A
C108F	3/c #14	80'	JB108	AA108B
C108G	5/c #14	70'	JB108	SSL108
C108H	3/c #14	10'	JB108	ZS108A
C108J	3/c #14	10'	ZS108A	ZS108B
C108K	3/c #14	60'	JB108	ZS108C
C108L	3/c #14	10'	ZS108C	ZS108D

Instructor's Notes:

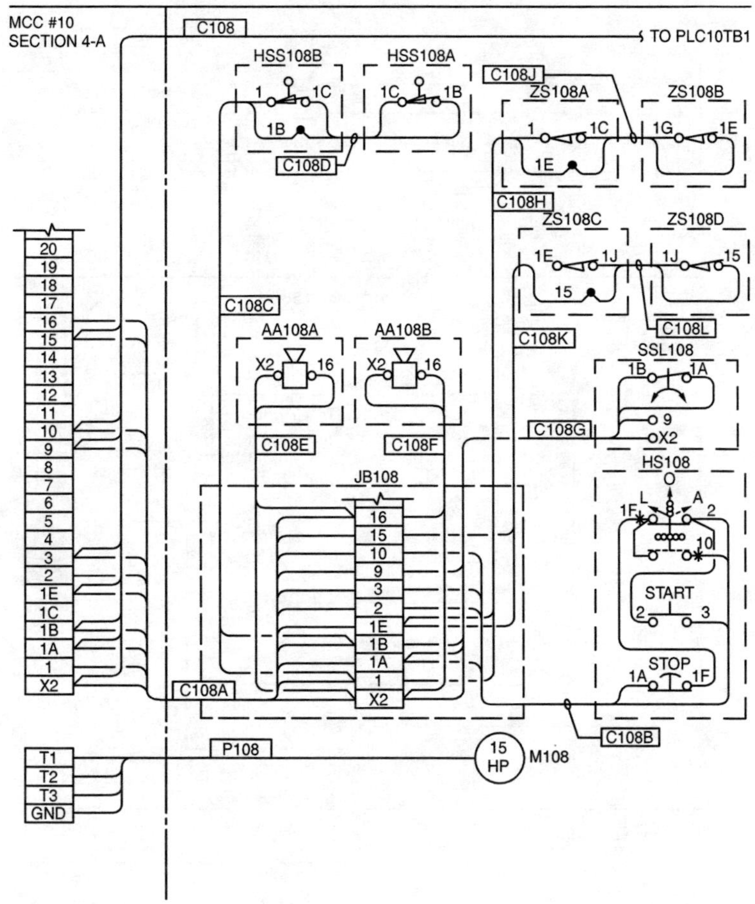

LEGEND

AA108A-D	Conveyor start warning horns
HS108	Local control station with stop mushroom maintained pushbutton, start pushbutton, and Local/Off/Auto selector switch
HSS108A-B	Emergency pullcord safety switches
PLC	Programmable Logic Controller
SSL108	Conveyor zero-speed switch
ZS108A-D	Belt runoff switches

311F69.EPS

Figure 69 ◆ Example of a wiring (connection) diagram.

Teaching Tip

Have the trainees identify various components on the diagram.

Teaching Tip

Have the trainees identify various components on the diagram.

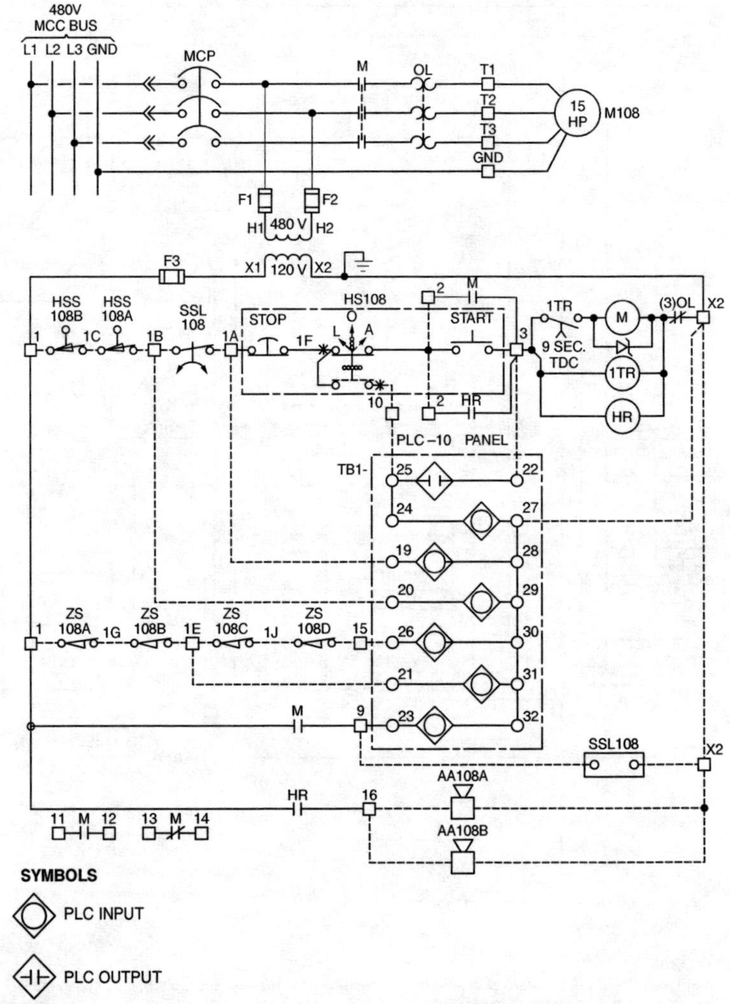

LEGEND

AA108A-D	Conveyor start warning horns
HS108	Local control station with stop mushroom maintained pushbutton, start pushbutton, and Local/Off/Auto selector switch
HSS108A-B	Emergency pullcord safety switches
PLC	Programmable logic controller
SSL108	Conveyor zero-speed switch
ZS108A-D	Belt runoff switches

311F70.EPS

Figure 70 ◆ Example of a ladder diagram.

Instructor's Notes:

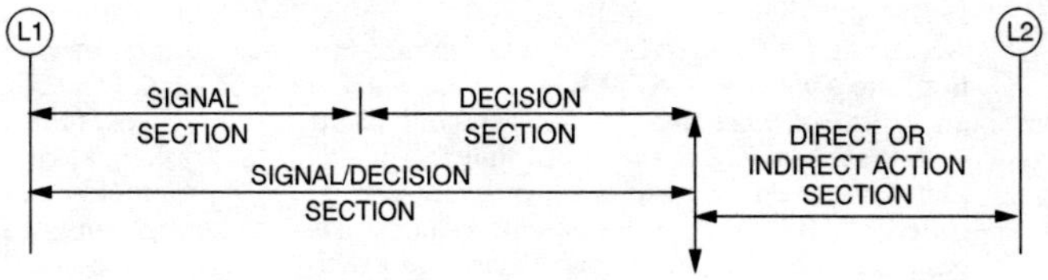

Figure 71 ◆ Typical ladder diagram sections.

311F71.EPS

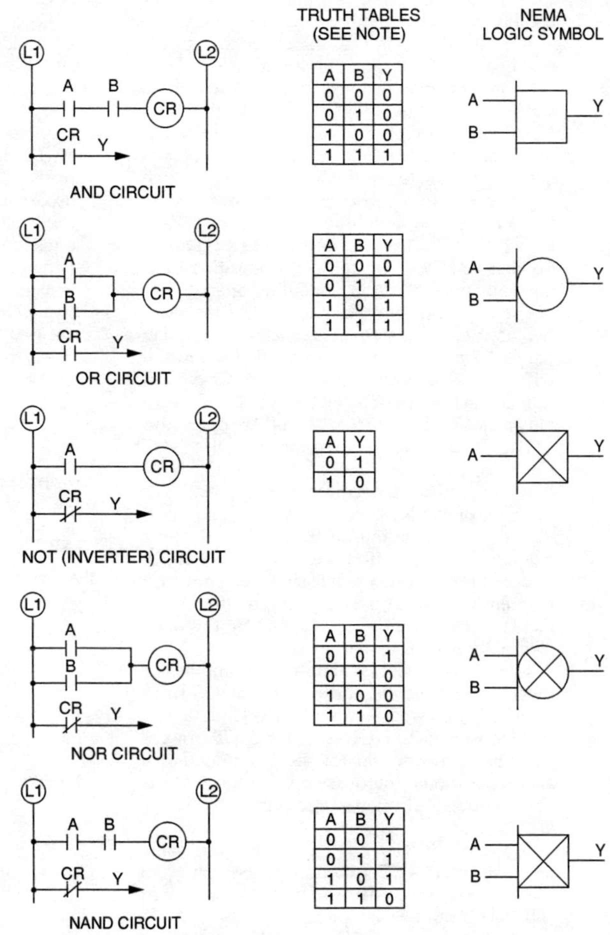

NOTE:
When contacts A and/or B are shown as zero (0) in the truth table, it means that the contacts are open. If shown as one (1), it means that the contacts are closed.

Figure 72 ◆ Basic logic functions.

311F72.EPS

Classroom

Explain the operation of various logic functions.

Audiovisual

Show Transparency 40 (Figure 73).

The logic symbols used in *Figure 72* for the AND, OR, NOT, NAND, and NOR logic functions are a version of NEMA logic symbols that are typical of those used for a motor control circuit. It's important to point out that the particular logic symbols used on a diagram can be different depending on whether the diagram is related to electrical, electronic, hydraulic, pneumatic, or some other type of equipment. Even for similar types of equipment, the symbols used by different manufacturers may vary.

As shown in *Figure 72,* the electrical equivalent of an AND function is two normally open sets of contacts (or switches) connected in series. As indicated by its truth table, the AND function has an output that is on (1) only when both contacts A and B are closed (1). This is because the related coil CR is energized, causing its normally open contacts to close. For all the other possible contact position combinations, the Y output remains off (0). This is because coil CR is de-energized, resulting in its normally open contacts remaining open.

The electrical equivalent of an OR function is two normally open sets of contacts connected in parallel. An OR function is so called because it will produce a 1 output (on) whenever contact A *or* contact B is closed (1). Output Y will be a 0 only when both contacts A and B are open (0).

The third basic logic function is the NOT function, commonly called an inverter. Note that it has only one normally open contact. The function of the NOT circuit is simple. It operates to produce an output that is always opposite the contact position. Thus, when contact A is closed (1), output Y is 0 (off): when contact A is open (0), output Y is 1 (on). For the circuit shown, the inversion is achieved by the open (normally closed) contacts of coil CR.

A NOR (NOT-OR) function is the same as an inverted OR function. It consists of an OR function combined with an inverter (NOT) function. Comparison of the NOR truth table with the OR truth table shows that for the same contact A and B positions, the output Y for the NOR is always opposite (inverted) from that of the basic OR function.

A NAND (NOT-AND) function is the same as an inverted AND function. It consists of an AND function combined with an inverter (NOT) function. Comparison of the NAND truth table with the AND truth table shows that for the same contact A and B positions, the output Y for the NAND is always opposite from that of the basic AND function.

Except for the NOT function, any of the functions just described can have more than two contacts added in series or in parallel with those shown in the figure, and the truth table can be expanded to include them. For example, if you have three series connected contacts in an AND function, all three contacts must be closed to get a Y output of 1. Otherwise, the output is 0. Similarly, three parallel connected contacts in an OR function would mean that the Y output is 1 if any one of the three contacts is closed.

Up to this point, relay coils and contacts have been used to describe the operation of the basic AND, OR, NOT, NOR, and NAND logic functions. These logic functions are commonly incorporated electronically into integrated chips, PLCs, and similar devices used in modern solid-state motor control equipment.

- *Action section*—The action section is usually the energization of a relay or contactor coil. As shown in the truth tables, when Y is 1, power is applied to the coil, and when it is 0, the coil is de-energized.

As a simple example, examination of *Figure 73* reveals an AND function overlapped by an OR function. The AND function is the STOP/START switch and MOL contact. Both the STOP contacts and the START contacts and the MOL contacts must be closed to energize the contactor coil M. Once energized, an OR function consisting of the START switch *or* the M holding contacts keeps the coil energized.

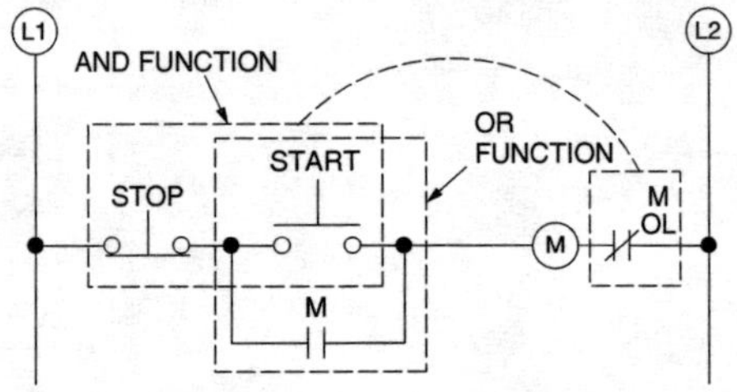

Figure 73 ◆ Example AND function and OR function in a motor-starter control circuit.

Instructor's Notes:

9.3.2 Conventions Used on Ladder Diagrams

Any ladder diagram normally references the type of machine or process to which the diagram applies. Careful study of the applicable ladder diagram can help the technician determine or confirm the correct sequence of operation for a particular piece of motor control equipment. As previously described, ladder diagrams are divided into sections horizontally and read as follows:

- *Left to right*—Control devices are connected between L1 and the operating coil (or load) from left to right in accordance with the following rules:
  - —All control functions shall occur on the left side (L1 side) of the coil.
  - —Only overload contacts are connected on the right side (L2 side) of the coil.
  - —Stop and shutdown circuits should occur before start and run circuits (closer to L1).
  - —Where practical, external connections such as buttons and switches should occur before internal connections such as relay connections.
- *Top to bottom*—Ladder diagrams are usually reference-line numbered from top to bottom with the first function starting on line one and subsequent or dependent functions proceeding down the ladder until the last function occurs.

Figure 74 shows the conventions that should be used on a typical ladder diagram. However, diagrams furnished by manufacturers or designers vary widely and not all the conventions pointed out on the figure may appear in all diagrams for similar functions. The conventions that should be used on a diagram are explained as follows:

- *Component identification*—The diagram should identify by reference designator (PB1, S1, CR1) and/or by function (low-pressure switch, on-delay timer) the various control circuit devices and loads. If a control device is remotely located from the main equipment enclosure, the diagram should also identify the location of a device.
- *Line fuse*—Many codes require that line-voltage control circuits must be fused if they exist outside the enclosure for the applicable relays or contactors, or if a step-down transformer is not power limited or is located outside the enclosure.
- *Step-down transformer and fuse*—In many cases, a low-voltage step-down transformer along with a fuse may be required for control circuits. If not, these devices are omitted and line voltage is used as the control voltage.
- *Component terminal numbers*—If a component has terminals that are marked with numbers or letters, they should be shown on the diagram.
- *Ladder line numbers*—The horizontal lines on the ladder should be numbered sequentially from top to bottom for contact cross-reference purposes. This aids in locating all contacts of a relay, contactor, or other device.
- *Mechanical linkage symbols*—If the items are located on adjacent lines, a dashed line may be used to indicate the linkage. An alternate approach is the arrowed box containing the ladder line reference to the location of the other linked device. Appropriately numbered arrowed boxes are used at both locations of the linked item.
- *Wire numbers*—Unique wire numbers are assigned to each wire or group of wires that are electrically common. Labeling the wires in the actual equipment with these numbers eases the installation and future troubleshooting.
- *Line references for contacts*—Relays and most contactors/motor starters have one or more sets of control circuit contacts. At the right-hand side of the line containing the coil for the device, the location of any control contacts not shown on the same line are referenced by ladder line numbers, separated by commas. Underlined line numbers are for any NC contacts, and nonunderlined numbers are for any NO contacts.

INSIDE TRACK

Fail-Safe Designs

Motor controls must be fail-safe. This means that in the event of a power loss, the equipment should be designed so that it cannot be re-energized upon the restoration of power without operator intervention. This is necessary in order to protect personnel from injury and/or the equipment from damage should the power be inadvertently reapplied, such as often occurs during power failures associated with electrical storms. Relays, contactors, and similar control devices should be wired so that if they fail, the operation of the equipment will cease, or an appropriate alarm will be generated to alert personnel to the failure. Otherwise, the failure of a device that provides for personnel and/or equipment protection can go undetected.

Classroom

Discuss the conventions used on ladder diagrams.

Audiovisual

Show Transparency 41 (Figure 74).

Teaching Tip

Have the trainees identify various components on the diagram.

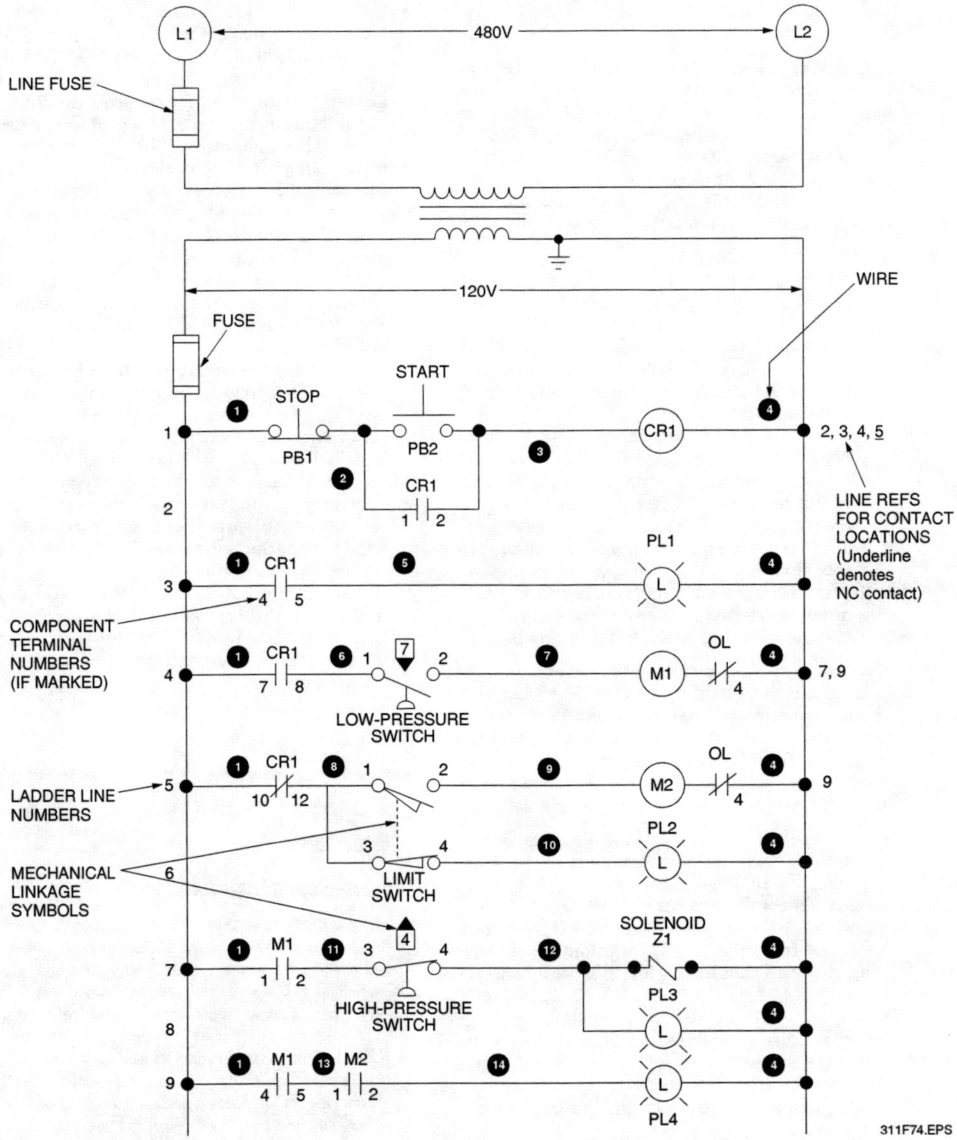

Figure 74 ◆ Typical ladder diagram conventions.

Instructor's Notes:

9.4.0 Logic Diagrams

Logic diagrams show a pictorial representation of circuit elements in the form of logic symbols in order to convey the logic behind the operation for a particular system, machine, or network. Typically they are used to communicate logic requirements to a control systems supplier or programmer. They are also used collectively by engineers, designers, technicians, and operators to agree on the rules of system operation before programming and circuit design of a system begin. Logic diagrams are useful for technicians maintaining a piece of equipment because they show the decision-making processes that are made by the equipment with regard to system operation and sequencing. These decision processes often are not obvious by the examination of a related ladder or schematic diagram because many are accomplished solely by the system software.

The logic symbols used on logic diagrams do not represent the actual types of electronic components used, but represent only their logical functions (AND, OR, NOR, NAND, NOR, INVERTER). As mentioned earlier, there are many styles of logic symbols used on logic diagrams. The specific set of symbols used are different, depending on whether the diagram is related to electrical, electronic, hydraulic, pneumatic, or some other type of equipment. Even for similar types of equipment, the symbols used by different manufacturers may vary. *Figure 75* shows a typical logic diagram of a motor control circuit. The symbols used on this diagram are the same NEMA logic symbols described earlier in this module.

The presentation of information on a logic diagram typically flows from the top of the page to the bottom. Logic functions associated with permissive signals normally are shown first; logic functions associated with the generation of running signals are shown second; and logic functions associated with the generation of miscellaneous signals, such as alarm signals, are shown last. Permissive signals typically are ones generated by safety switches, stop switches, mode switches, or related software that function to enable or inhibit machine or system operation, depending on their status. Running signals are those signals that cause the machine to operate and control its operation.

The logic flow path shown on the diagram is from left to right. Sometimes the output line from the last logic element on the right-hand side of the drawing may loop back to the left side of the drawing if this is necessary to continue showing the remainder of a logic function. Labels shown on the left side of the diagram normally identify a device name or signal name input to the logic element; those on the right side show the output or action from a logic element. Note that the inputs and outputs shown on a logic diagram can be applied from or sent to hardware devices, while others may be applied from or sent to the system-controlling software. Understanding the information shown on a logic diagram will help the electrician to:

- Understand how a programmable logic controller (PLC) controls the operation of a system or device without the need for sorting through the specific PLC programming details.
- Understand the details of system operation that are not obvious by using a ladder diagram.
- Generate a ladder diagram.
- Determine the status of field devices shown on the drawings.
- Determine how and where to implement wiring modifications because of changes in operating requirements.
- Explain to others how the equipment works.

9.5.0 Relating Diagrams to Equipment Wiring and Operation

Up to this point, you have been introduced to the different types of diagrams that can be used to show the wiring and operation for a system or piece of equipment. The diagrams and circuit schedules shown earlier in this module all are related to a mill conveyor system, designated as 120-CNV-05. It has a 15hp head-mounted drive motor that serves to transfer screened oversize materials to a ball mill, designated as 120-BM-01. Operation of this conveyor system is described in the following section. Based on this description of system operation, you should examine the different diagrams and relate what they show to the actual equipment and its operation. The connection diagram and circuit schedule (*Figure 69* and *Table 13*) show the wiring connections between the different components of the conveyor system. The ladder and logic diagrams (*Figures 70* and *75*) show the control sequence and design logic, respectively, pertaining to the equipment operation.

9.5.1 Relating Ladder Diagrams to System Operation

Refer to the ladder diagram shown in *Figure 70* while following the description of conveyor system operation given here. Placing the LOCAL/OFF/AUTO switch (part of HS108) in the AUTO position selects the normal operation mode for the conveyor. In this mode, the conveyor is controlled by the programmable logic controller (PLC10)

Classroom

Point out that logic diagrams can be used to understand how a piece of equipment operates.

Audiovisual

Show Transparency 42 (Figure 75).

Classroom

Trace the operation of a motor circuit on a ladder diagram.

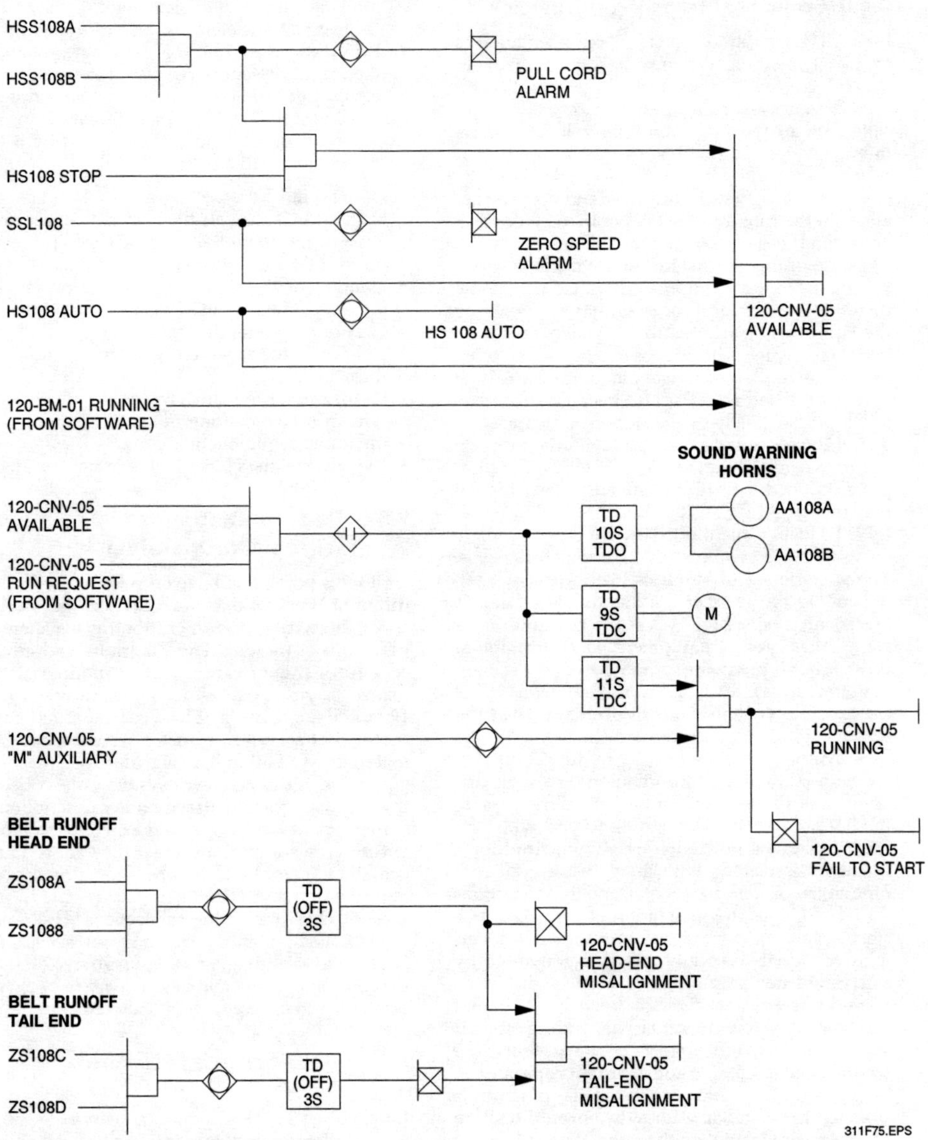

Figure 75 ◆ Example of a logic diagram.

Instructor's Notes:

that functions to provide sequential interlocking for the 120-BM-01 ball mill.

Emergency pull-cord safety switches (HSS108A and HSS108B) prevent the conveyor motor from starting if one or both are activated (contacts open). If activated when the conveyor motor is running, interruption of control power will cause the conveyor motor to stop immediately. It will also cause the PLC to generate an alarm.

A conveyor zero-speed switch (SSL108) is mounted on the tail end of the belt pulley. It provides protection against belt freeze-up or slippage. This switch is energized when the conveyor motor is started. It contains an internal time-on delay circuit whose contacts allow sufficient time (typically one to five seconds depending on belt length) for the conveyor belt to accelerate to normal speed after the conveyor is turned on. If the conveyor belt is rotating at the correct speed at the end of the delay period, the contacts remain closed, allowing the conveyor motor to continue operation. Should belt rotation slow down below the correct speed at any time or if belt rotation does not begin within the delay period, the switch contacts open, opening the control circuit and causing the motor to turn off and an alarm to be generated via the PLC.

Belt runoff switches (ZS108A through ZS108D) serve to detect belt misalignment (side travel) at the head end (ZS108A and ZS108B) and tail end (ZS108C and ZS108D) of the conveyor. Should a misalignment be detected by one or more of the switches, the switch signals cause the PLC to generate an alarm. They do not cause the conveyor motor to be turned off.

Conveyor start warning horns (AA108A and AA108B) provide a local area alarm that sounds for ten seconds as controlled by horn relay HR when a motor start command is initiated. This warns people that the conveyor is about to be started. Starting of the conveyor motor is delayed for nine seconds by time delay relay 1TR after a start command is issued.

Placing the LOCAL/OFF/AUTO switch (part of HS108) in the LOCAL position allows the conveyor to be operated for maintenance purposes independent of PLC control and its sequential interlocking. The safety features provided by the emergency pull-cord switches, zero-speed switch, and conveyor start warning horns remain active when in the local mode. The functions of belt runoff switches are inactive during the local mode of operation.

9.5.2 Relating Logic Diagrams to System Operation

Remember that the purpose of a logic diagram is to convey the design logic behind the operation for a particular system, machine, or network. It is important to point out that many of the functions shown on a logic diagram may be accomplished by the system operating software, rather than by hardware. For this reason, there is not always a direct correlation between what is shown on the logic diagram and what is shown on the related ladder diagram. In order to get the entire picture of how a particular system or machine operates, you should refer to both types of diagrams.

Refer to the logic diagram shown in *Figure 75* for the conveyor system. The top part of the logic diagram shows the logic elements that develop the inputs to a four-input AND. As shown, the field inputs to this part of the logic diagram are applied from the system emergency pull cord safety switches (HSS108A and HSS108B), the STOP button (part of HS108), the conveyor zero-speed switch (SSL108), and the AUTO switch (part of HS108). Also applied from the system software is an input (interlock input) signaling that the 120-BM-01 ball mill is running. Each of these permissive signal inputs is applied from a contact that must be closed (supplying a logical one) in order for the conveyor to run.

Tracing the emergency pull cord safety switch and the STOP button inputs through their respective two-input AND logic gates shows that when all three input signals are a logical one, the related input to the four-input AND is a logical one. The remaining three inputs applied to the four-input AND from the conveyor zero-speed switch, AUTO switch, and the system software are all a logical one. This causes the output signal from the four-input AND to be a logical one. This signal (*120-CVN-05 available*) is applied as an input to the next group of decision logic. Should any one of the input signals be a logical zero, the output from the four-input AND will be a logical zero, indicating that the conveyor system is not available for use.

The middle of the diagram shows the development of the system running control signals. The first logic decision made here is whether the conveyor system is ready to run and whether the software or operator wants it to run. As shown, this decision is made by a two-input AND gate. The decision depends on the status of its input signals, *120-CVN-05 available* and *120-CVN-05 run request*

Classroom

Trace the operation of a motor circuit on a logic diagram.

Classroom

Continue tracing the operation of a motor circuit on a logic diagram. Refer trainees to Table 14.

applied from the software. If both inputs are a logical one, the output from the AND is a logical one. It causes a PLC contact closure that begins the conveyor starting sequence by enabling a ten-second, timed open, timing function that causes the conveyor start warning horns to sound for ten seconds. A look at the ladder diagram shows that this function is accomplished in the conveyor system hardware by the ten-second horn relay (HR) and the conveyor start warning horns (AA108A and AA108B). Simultaneously, a nine-second, timed closed, timing function is also enabled that causes the conveyor motor contactor (M) to energize after a nine-second delay, causing the conveyor motor to turn on. In the hardware, this function is performed by nine-second timing relay 1TR and contactor M.

The remainder of the decision logic shown in this section consists of an 11-second, time closed, time delay (TD) function; a PLC input; a two-input AND gate; and an inverter. The purpose of this logic is to verify that when the conveyor is commanded to run, it is indeed running. If the input to the 11-second TD function is a logical one for 11 seconds after conveyor turn on, the TD function outputs a logical one to the two-input AND. The purpose for the 11-second delay is to allow time for contactor M to pull in and the internal time delay of the conveyor zero-speed switch (SSL108) to expire. The second input to the two-input AND (*input 120-CNV-05 M aux*) via the PLC is the status of the contactor M auxiliary contact signal. With both inputs to the two-input AND at a logical one, it outputs a logical one signal (*120-CNV-05 running*) to the software that verifies the conveyor is running. If either the TD function output or the *120-CVN-05 M aux* signal input is a logical zero, the AND gate output (*120-CNV-05 running*) will be a logical zero, indicating that the conveyor is not running. The logical zero output from the AND gate is applied through an inverter to generate a logical one *120-CVN-05 fail to start signal* for application to the software.

The bottom of the diagram shows the decision logic used to determine if a belt runoff misalignment condition exists, and if it does, whether the problem is caused by a head-end misalignment or a tail-end misalignment. The individual status of belt runoff switches ZS108A and B (head end) and Z108C and D (tail end) are applied as inputs to two-input AND gates. The outputs from the head-end and tail-end AND gates are applied via PLC inputs to three-second, timed off, time-delay (TD) functions. The purpose for using the three-second, timed off TD functions is to eliminate the generation of nuisance alarms.

When no conveyor belt misalignment exists, the inputs applied to both three-second TD functions are a logical one, causing the outputs from both three-second TD functions to be a logical one. Should a head-end belt misalignment exist (or both a head-end and tail-end belt misalignment exist) for more than three seconds, the outputs from both three-second TD functions are a logical zero. Both are a logical zero in this case because the tail-end run-off switches (Z108C and D) are in series with the head-end run-off switches (ZS108A and B), as shown on the ladder diagram. Therefore, if the head-end input is open, the tail-end input will also be open. When only a tail-end belt misalignment occurs, the output of the head-end three-second TD function will be a logical one, and the output of the tail-end three-second TD function will be a logical zero. The statuses of the three-second TD function outputs for the various belt conditions are summarized in *Table 14.*

Table 14 Three-Second TD Function Outputs

Belt Alignment Condition	Logical Signal Output from Head-End Three-Second Timed Off Function	Logical Signal Output from Tail-End Three-Second Timed Off Function	120-CNV-05 Head-End Misalignment Logical Signal Output	120-CNV-05 Tail-End Misalignment Logical Signal Output
No belt misalignment	1	1	0 (no alarm)	0 (no alarm)
Head-end belt misalignment or both head-end and tail-end belt misalignment	0	0	1 (alarm)	0 (no alarm)
Tail-end belt misalignment	1	0	0	1 (alarm)

Instructor's Notes:

As shown, the output from the head-end three-second TD function is applied as an input to an inverter and to one input of a two-input AND gate. The output from the tail-end three-second TD function is applied through an inverter as the second input to the two-input AND gate. The resultant logic status of the inverter output signal (*120-CNV-05 head-end misalignment*) and the AND gate output signal (*120-CNV-05 tail-end misalignment*) for the different belt conditions is shown in the truth table given above. These signals are used by the PLC software to determine the condition of belt alignment.

10.0.0 ◆ NEC REGULATIONS FOR THE INSTALLATION OF MOTOR CONTROL CIRCUITS

NEC Article 430 governs the installation of motors, motor circuit, and controllers. Where equipment incorporates a hermetic refrigerant compressor, such as in refrigeration or air conditioning equipment, *NEC Article 440* also applies. Other articles that pertain to motor circuits involving special equipment and hazardous locations are referenced in *NEC Section 430.5.* You have studied the NEC requirements for motor circuits in your earlier training on motors. *Figure 76* shows a summary of the *NEC Article 430* requirements that apply to the installation of motors, motor circuits, and their controllers. Detailed information can be found in the NEC under the articles or sections referenced in the diagram.

Some *NEC Article 430* requirements pertaining to motor controllers and related control circuit devices are overviewed here:

- A magnetic starter cannot serve as a disconnecting means.
- Controllers must be able to start and stop the motors they control and be able to interrupt the locked rotor current of the motor.
- A controller must have a horsepower rating not lower than the rating of the motor it controls. Controllers used with high-efficiency Design E motors rated up to 100hp shall be marked by the manufacturer as rated for use with Design E motors or have a horsepower rating not less than 1.4 times the rating of the motor. For Design E motors over 100hp, the rating shall be 1.3 times the motor rating.
- Branch circuit protective devices (circuit breaker, fuse) can be used as the controller for small stationary motors (⅛hp or less) and portable motors (⅓hp or less). A general-use switch rated at not less than twice the full loaded motor current may be used with stationary motors up to 2hp, rated at 300V or less.
- Where a controller does not also serve as both the controller and disconnect switch, and a fused switch provides the required disconnect as well as the circuit short circuit and overload protection, the controller need not open all conductors to the motor. It needs to open only enough conductors to stop and start the motor.
- The rating of time-delay fuses used in a combination fuseholder and switch as the motor-running protective devices must not exceed 125% or, in some cases, 115% of the full-load motor current.
- *NEC Table 430.91* provides the selection criteria for selecting the type of enclosure to use for controllers in nonhazardous (classified) locations.
- All conductors of a remote motor control circuit outside of the controller, such as to remote pushbutton stations, must be installed in a raceway or otherwise protected.
- Where one side of the motor control circuit is grounded, the circuit must be wired so that an accidental ground in the control device or related wiring will not start the motor or bypass manually operated shutdown devices or automatic safety shutdown devices.

Classroom

Discuss the NEC requirements for motor control circuits. Refer the trainees to *NEC Articles 430* and *440.*

Audiovisual

Show Transparency 43 (Figure 76).

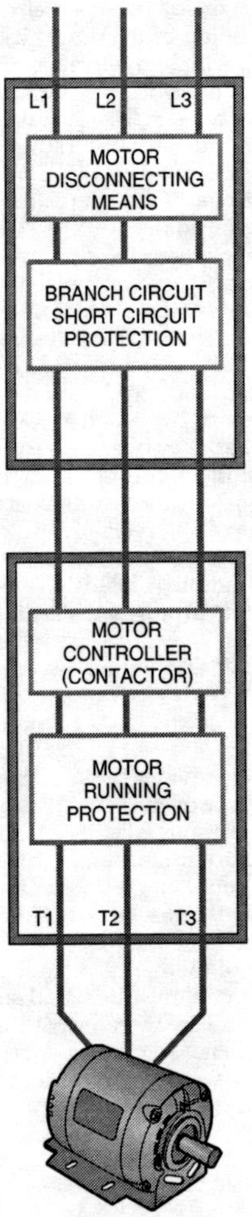

NEC Article 430 Part IX Sections 430.101 through 430.113

Disconnects motor and controllers from circuit.

1. Continuous rating of 115% or more of motor FLC. Also see ***NEC Article 430 Part II.***
2. Disconnecting means shall be as listed in ***NEC Section 430.109.***
3. Must be located in sight of motor location and driven machinery. The controller disconnecting means can serve as the disconnecting means if the controller disconnect is located in sight of the motor location and driven machinery.

NEC Article 430 Part IV Sections 430.51 through 430.58

Protects branch circuit from short circuits or grounds.

1. Must carry starting current of motor.
2. Rating must not exceed values in ***NEC Table 430.52*** unless not sufficient to carry starting current of motor.
3. Values of branch circuit protective devices shall in no case exceed exceptions listed in ***NEC Section 430.52.***

NEC Article 430 Part VII Sections 430.81 through 430.91

Used to start and stop motors.

1. Must have current rating of 100% or more of motor FLC.
2. Must be able to interrupt LRC.
3. Must be rated as specified in ***NEC Section 430.83.***

NEC Article 430 Part III Sections 430.31 through 430.44

Protects motor and controller against excessive heat due to motor overload.

1. Must trip at following percent or less of motor FLC for continuous motors rated more than one horsepower.
   a) 125% FLC for motors with a marked service factor of not less than 1.15 or a marked temperature rise of not over 40°C.
   b) 115% FLC for all others. (See NEC for other types of protection.)
2. Three thermal units required for any three-phase AC motor.
3. Must allow motor to start.
4. Select size from FLC on motor nameplate.

NEC Article 430 Part II Sections 430.21 through 430.29

Specifies the sizes of conductors capable of carrying the motor current without overheating.

1. To determine the ampacity of conductors, switches, branch circuit overcurrent devices, etc., the full-load current values given in ***NEC Tables 430.147 thru 430.150*** shall be used instead of the actual current rating marked on the motor nameplate. ***(See NEC Section 430.6.)***
2. According to ***NEC Section 430.22,*** branch circuit conductors supplying a single motor shall have an ampacity of not less than 125% of motor FLC, as determined by ***NEC Section 430.6(A)(1).***

311F76.EPS

Figure 76 ◆ Summary of requirements for motors, motor circuits, and controllers.

Instructor's Notes:

11.0.0 ◆ CONNECTING MOTOR CONTROLLERS FOR SPECIFIC APPLICATIONS

Motor controllers and their related control circuits can be connected in many ways to satisfy the requirements needed for specific applications. This section gives a few examples of basic control circuit wiring of controllers for some common applications including:

- Controlling an air compressor motor
- Controlling a pump motor
- Controlling two pump motors
- Controlling a motor from multiple locations
- Controlling the reversing of a three-phase motor
- Controlling conveyor system motors

11.1.0 Controlling an Air Compressor Motor

Controlling an air compressor motor is done in a similar way as controlling an AC motor with a contactor. However, an NC pressure switch (high pressure switch) is added in the control circuit to cause the compressor motor to turn on and off, depending on the amount of air pressure in the system. *Figure 77* shows a line diagram for this circuit. In this circuit, setting the ON/OFF switch to ON allows current to pass from L1 through the NC pressure switch contacts, the contactor coil C, NC overloads, and on to L2. This energizes the contactor coil, closing its NO power contacts (not shown). This turns on the air compressor motor.

The air compressor motor will continue to run until the air pressure in the system builds up to and exceeds the set point of the pressure switch. When this occurs, the NC contacts of the pressure switch open, causing the contactor coil to de-energize and its related NO power contacts to open. This turns off the air compressor motor. It will remain turned off until the pressure in the system falls below the pressure switch reset point, causing the NC contacts of the pressure switch to close again. This energizes the contactor coil again and turns on the air compressor motor. This automatic sequence of turning on and off the air compressor motor in response to the system pressure level continues until the ON/OFF switch is set to OFF.

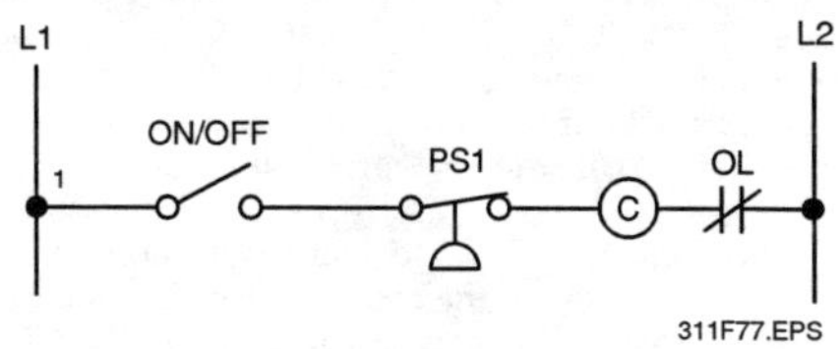

Figure 77 ◆ Control circuit for an air compressor.

11.2.0 Controlling One-Pump Motors

A control circuit used for controlling a pump that maintains the level of water (or any liquid) in a tank or other vessel can use one or more float switches. *Figure 78* shows the line diagram for a liquid pump being controlled by a HAND/OFF/AUTO switch and a single float switch. When the HAND/OFF/AUTO switch in this circuit is set to OFF, no current can flow through the contactor coil to L2; therefore, the pump is turned off. When the switch is placed in the HAND position, current from L1 can pass through the NO switch contacts, the contactor coil C, NC overloads, and on to L2. This energizes the contactor coil, closing its NO power contacts (not shown), thus turning on the pump motor. The pump motor will continue running until the HAND/OFF/AUTO switch is set to OFF. This provides a manual mode of operation.

When the HAND/OFF/AUTO switch is set to AUTO, current can flow from L1 through the closed switch contacts, the NC float switch contacts, the contactor coil, NC overloads, and on to L2. This energizes the contactor coil, closing its NO power contacts, causing the pump motor to turn on. The pump will continue running until either the HAND/OFF/AUTO switch is set to OFF or the water in the tank reaches a level high enough to actuate the float switch, causing its NC contacts to open. When this occurs, the contactor coil is de-energized and its NO power contacts open. This turns off the pump motor. The pump motor will remain turned off until the water level in the tank falls below the float switch reset point, causing the NC contacts of the float switch to close again. In turn, the contactor coil is energized again and turns on the pump motor. This automatic sequence of turning on and off the pump motor in response to the water level in the tank continues until the HAND/OFF/AUTO switch is set to OFF.

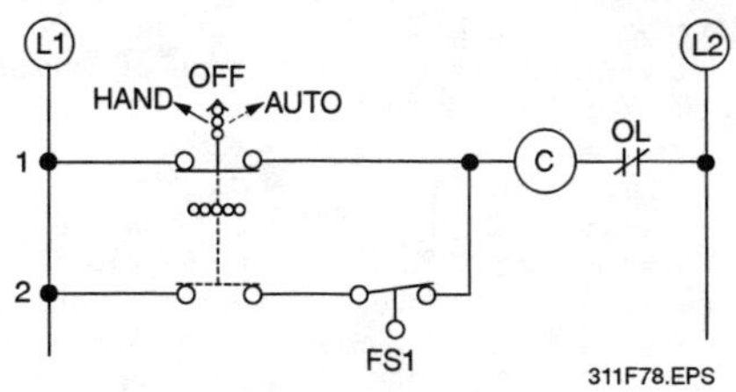

Figure 78 ◆ Control circuit for a pump motor.

Classroom

List some of the specific applications for motor controllers.

Describe the operation of an air compressor motor control circuit.

Audiovisual

Show Transparency 44 (Figure 77).

Classroom

Describe the operation of a single-pump motor control circuit.

Audiovisual

Show Transparency 45 (Figure 78).

Describe the operation of a two-pump motor control circuit.

Show Transparency 46 (Figure 79).

Explain that some controls allow operation of the motor from multiple locations.

Show Transparency 47 (Figure 80).

Describe the operation of a magnetic reversing starter.

Show Transparency 48 (Figure 81).

11.3.0 Controlling Two-Pump Motors

For the purpose of an example, assume that there is a requirement that involves controlling two-pump motors so that they alternately operate to automatically fill and drain a tank, based on the level of the liquid in the tank. *Figure 79* shows one way of connecting such a circuit. Pressing the START pushbutton in this circuit allows current to flow from L1 through the STOP switch NC contacts, the closed START switch contacts, the control relay coil CR, and on to L2. This energizes the control relay, causing its three sets of NO contacts to close. Assuming the tank is empty or only partially full, coil M2 is de-energized. This allows current to flow from L1 through the closed control relay contacts, closed M2 auxiliary contacts, the NC float switch contacts, coil M1, NC overloads, and on to L2. This energizes coil M1, closing its NO power contacts (not shown) to turn on the fill pump motor. It also causes its NC auxiliary contacts to open. This provides an electrical interlock in the control circuit for coil M2 that prevents coil M2 from being energized at the same time as coil M1. The fill pump motor will continue to operate until the water in the tank reaches a level high enough to actuate the float switch, causing its NC contacts to open and its NO contacts to close. This causes coil M1 to de-energize and its related NO power contacts to open, turning off the fill pump motor.

Activation of the float switch also allows current to flow from L1 through the closed control relay contacts, closed auxiliary M1 contacts, the closed NO float switch contacts, coil M2, NC overloads, and on to L2. This energizes coil M2, closing its NO power contacts (not shown) to turn on the drain pump motor. It also causes its NC auxiliary contacts to open. This provides an electrical interlock in the control circuit for coil M1 that prevents coil M1 from being energized at the same time as coil M2. The drain pump motor will continue to operate until the water in the tank is drained to a level that is low enough to actuate the float switch, causing its NC contacts and NO contacts to close and open again, respectively. This causes coil M2 to de-energize and its related NO power contacts to open, turning off the drain pump motor. It also allows the fill pump motor to be started again, as described earlier. This cycle of turning on and off the fill and drain pumps will continue until the STOP pushbutton is pressed. Pressing the STOP pushbutton opens its NC contacts, removing power from the control relay coil. This causes all its NO contacts to open, preventing both coils M1 and M2 from being energized.

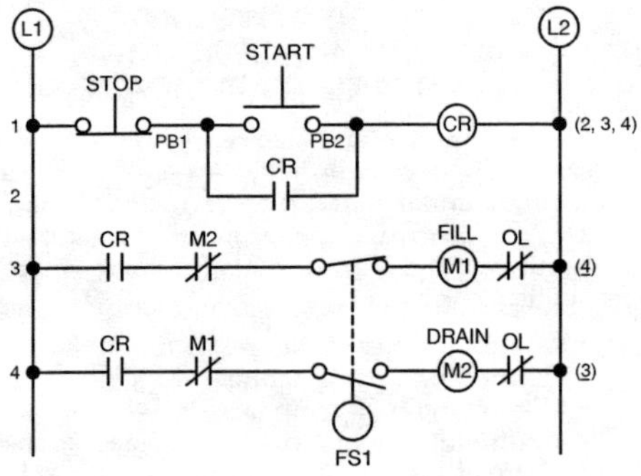

311F79.EPS

Figure 79 ◆ Control circuit for two pumps.

11.4.0 Controlling a Motor from Multiple Locations

It is often required to start and stop a motor from more than one location. *Figure 80* shows a line diagram for a control circuit where a magnetic starter may be energized to start and stop a motor from any one of three locations. In this circuit, pressing any one of the START pushbuttons causes coil M1 to energize. This causes the auxiliary contacts M1 to close, adding memory to the circuit until coil M1 is de-energized by pressing any one of the stop buttons. The important thing to remember is that no matter how many stations, the START pushbuttons must all be connected in parallel and the STOP pushbuttons in series.

11.5.0 Controlling the Reversal of a Three-Phase Motor

Reversing of three-phase motors is a common requirement. It is usually done by reversing the L1 and L3 power leads to the motor via the operation of a magnetic reversing starter and related control circuit (*Figure 81*). Circuit operation is started by pressing either the FORWARD or REVERSE pushbutton. Pressing the FORWARD pushbutton allows current to flow from L1 through the NC contacts of the STOP switch, NC contacts of the REVERSE switch, the closed contacts of the FORWARD switch, NC auxiliary contacts of the reverse coil, the forward coil F, the NC closed overload contacts, and on to L2. This energizes coil F, closing its NO forward-power contacts (not shown), causing the motor to turn on and run in the forward direction. It also activates the NO and

Instructor's Notes:

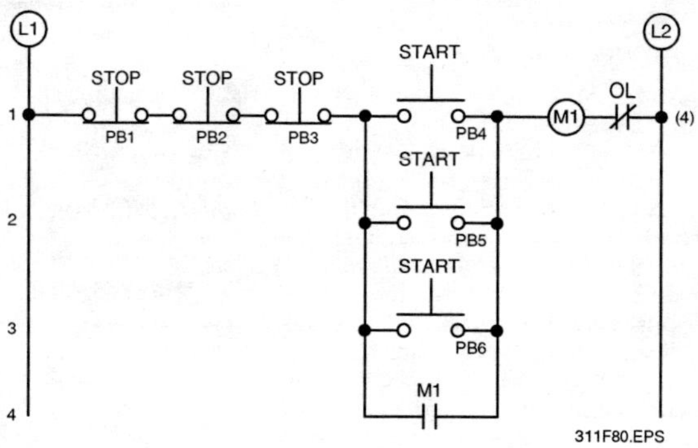

Figure 80 ◆ Control circuit for starting and stopping a single motor from three locations.

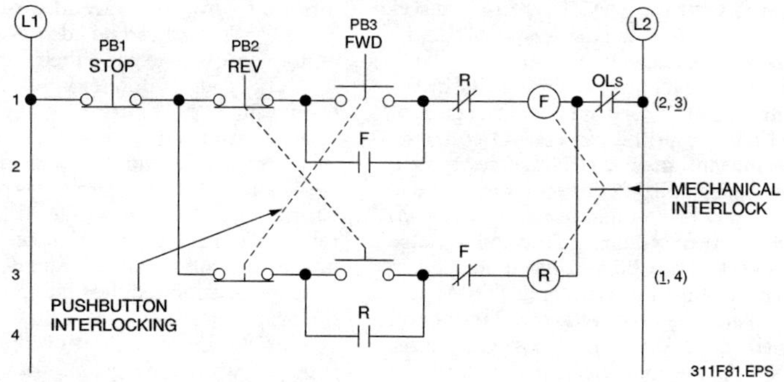

Figure 81 ◆ Control circuit for reversing a three-phase motor.

NC auxiliary contacts of coil F. The NO auxiliary contacts close, providing memory for the FORWARD switch. The NC auxiliary contacts (located in the coil R control circuit) open, providing an electrical interlock that prevents coil R from being energized simultaneously with coil F.

Pressing the REVERSE pushbutton allows current to flow from L1 through the NC contacts of the STOP switch, NC contacts of the FORWARD switch, the closed contacts of the REVERSE switch, NC auxiliary contacts of the forward coil, the reverse coil R, the NC closed overload contacts, and on to L2. This energizes coil R, closing its NO reverse-power contacts (not shown), causing the motor to turn on and run in the reverse direction. It also activates the NO and NC auxiliary contacts of coil R. The NO auxiliary contacts close, providing memory for the REVERSE switch. The NC auxiliary contacts (located in the coil F control circuit) open, providing an electrical interlock that prevents coil F from being energized simultaneously with coil R. Pressing the STOP pushbutton de-energizes, as appropriate, coil F or coil R, stopping motor operation.

This circuit uses three methods of interlocking. One method is a built-in mechanical interlock protection incorporated in the magnetic reversing starter. Another method is the pushbutton interlocking of the FORWARD and REVERSE pushbuttons. Pushbutton interlocking uses both NO and NC contacts mechanically connected on each pushbutton. This provides a safety backup to ensure that both the forward and reverse coils cannot be energized simultaneously. As has already been pointed out, the third method involves the NC auxiliary contacts for the forward and reverse coils being used as safety interlocks to prevent the forward and reverse coils from being energized simultaneously.

Classroom

Explain that some controls allow several motors to energize in sequence.

Audiovisual

Show Transparency 49 (Figure 82).

Teaching Tip

Discuss the "Think About It."

Demonstration

Demonstrate how to make the connections for a magnetic motor controller, controlled by two pushbutton stations, including the connections for holding the circuit interlock.

Laboratory

Under your supervision, have the trainees make all connections for a magnetic motor controller, controlled by two pushbutton stations, including the connections for holding the circuit interlock. Note the proficiency of each trainee.

11.6.0 Controlling Conveyor System Motors

In manufacturing or other controlled processes, it is often required to sequence the operation of multiple conveyor belt systems so that one conveyor system is turned on before a second one can be started, a second one must be turned on before a third one can be started, and so on. *Figure 82* shows one basic method for wiring a control circuit to sequentially turn on three conveyor systems. Pressing the START pushbutton energizes the conveyor 1 starter coil M1 via current flow through the closed contacts of the STOP and START switches and NC overload contacts. When coil M1 energizes, it closes its NO power contacts (not shown), causing the conveyer 1 motor to turn on. Energizing coil M1 also closes its three sets of NO auxiliary contacts. One set of closed NO auxiliary contacts provides memory for the START switch. The second and third sets are located in the control circuits for coils M2 and M3, respectively.

When the set of NO auxiliary contacts in the coil M2 control circuit closes, it provides a current path from L1 through the closed NO contacts, through coil M2, and the NC overload contacts to L2. This energizes coil M2. Energizing coil M2 closes its NO power contacts, causing the conveyer 2 motor to turn on. Energizing coil M2 also closes its NO set of auxiliary contacts located in the coil M3 control circuit. With both the M1 and M2 sets of NO auxiliary contacts closed in the coil M3 control circuit, coil M3 energizes. This closes the coil 3 NO power contacts, causing conveyor motor 3 to turn on. Since there are no additional conveyors to energize, the NO auxiliary contacts for coil M3 are not used. Pressing the STOP pushbutton de-energizes, as applicable, coils M1, M2, and M3, stopping all conveyor motor operation.

THINK ABOUT IT

Overload Contacts

Why are all three overload contacts shown in *Figure 82* connected in series with the conveyor 1 starter coil?

Summary

Magnetic contactors and motor starters use a small control current to energize or de-energize the load connected to it. Motor starters are contactors that contain overload protection. Various auxiliary devices can be added to contactors and motor starters to expand their capability.

Many types of pilot (control) devices, ranging from simple pushbutton switches to proximity sensors, are used in motor control circuits. The device(s) selected for use depends on the specific application. Manufacturers' specifications sheets describe required amperage, voltage, and sizing information for control devices. For safety and proper operation in the intended environment, their installation requires that they be installed in the proper position and location.

Knowledge of the definitions, symbols, and types of wiring diagrams that have been presented throughout this module must be understood. This is necessary to properly select, size, and install electromagnetic relays, contactors, motor starters, and the pilot devices in their related control circuits.

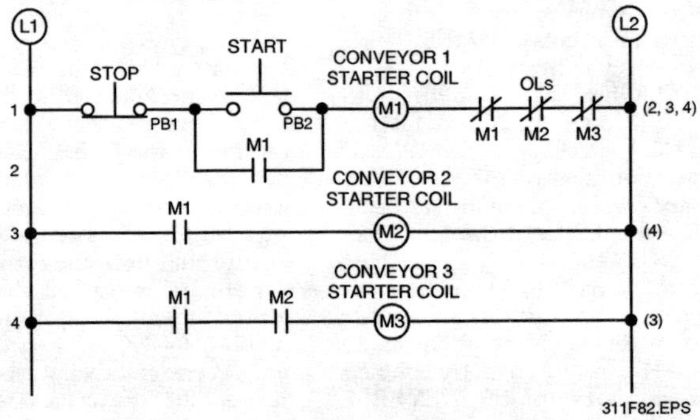

Figure 82 ◆ Control circuit for turning on three conveyor systems in sequence.

Instructor's Notes:

Review Questions

1. Contactors without overload protection may be used to control ____.
   a. fractional horsepower single-phase motors
   b. single-phase motors
   c. three-phase motors
   d. noninductive loads

2. Relay contacts that are defined as being *normally open (NO)* have contacts that are ____.
   a. open when the relay coil is energized
   b. not being used in a particular application
   c. open when the relay coil is de-energized
   d. closed when the relay coil is de-energized

3. The voltage level at which the armature spring of a power relay overcomes the magnetic field of the coil causing the contacts to change state is called the ____ voltage.
   a. dropout
   b. seal-in
   c. pickup
   d. transition

4. IEC-rated contactors and motor starters perform the same functions as NEMA-rated contactors and motor starters but are usually ____ than the NEMA devices for the same horsepower or current ratings.
   a. physically larger
   b. more application sensitive
   c. more expensive
   d. better designed

5. When replacing the contacts of a magnetic contactor, all moving and stationary contacts for all poles must be replaced.
   a. true
   b. false

6. Overload protection for motors is provided by ____.
   a. circuit breakers
   b. general-purpose, current-limiting fuses
   c. an overload relay
   d. backup current-limiting fuses

7. All of the following are NEMA-rated classes for the trip time of an overload relay *except* ____.
   a. Class 10
   b. Class 20
   c. Class 30
   d. Class 40

8. The thermal units for overload relays should always be selected based on the motor ____.
   a. locked rotor amps
   b. full-load currents from *NEC Tables 430.147 through 430.150*
   c. nameplate full-load current
   d. horsepower

9. The part of a melting-alloy overload that responds to an overload current is known as the ____.
   a. bimetal element
   b. eutectic alloy
   c. ratchet wheel
   d. solder pot

10. A NEMA-rated motor starter consists of ____ assembled and wired in the same enclosure.
    a. a disconnect switch and magnetic contactor
    b. a magnetic contactor and stop and start pushbuttons
    c. one or more magnetic contactors with associated overload relays
    d. a circuit breaker and magnetic contactor

11. All of the following are considerations when selecting overload relays *except* the ____.
    a. type of motor
    b. full-load current rating of the motor
    c. maximum horsepower rating for nonplugging and nonjogging duty
    d. manufacturer of the motor

12. The motor starter that must be used with a 230V, single-phase, 60Hz, 10hp motor not used for plugging or jogging applications is the ____.
    a. NEMA Size 2
    b. NEMA Size 3
    c. NEMA Size 4
    d. NEMA Size 5

13. When selecting a control transformer for a specific application, you have calculated its application inrush to be 1,850VA. If the primary input voltage is expected to vary between 5% and 10% from nominal, a ____ transformer should be used.
    a. 150VA
    b. 250VA
    c. 300VA
    d. 350VA

Classroom

Have the trainees complete the Review Questions and go over the answers prior to administering the Module Examination.

Examination

Administer the Module Examination. Record the results on Craft Training Report Form 200 and submit the results to the Training Program Sponsor.

Performance Testing

Administer the Performance Test and fill out Performance Profile Sheets for each trainee. If desired, trainee proficiency noted during laboratory sessions may be used to complete the Performance Test. Be sure to record the results on Craft Training Report Form 200 and submit the results to the Training Program Sponsor.

14. Pushbutton switches with normally open (NO) contacts are usually used to initiate a STOP or OFF circuit function.
a. true
b. false

15. A manufacturer states that the contacts of a particular switch are rated as A600. In a 120VAC application, the maximum break current for these contacts is ____.
a. 1A
b. 1.5A
c. 3A
d. 6A

16. A manufacturer lists the utilization category for a particular switching device as AC-15. The intended application for this device is control of ____.
a. electromagnetic loads greater than 72VA closed
b. solid-state loads with transformer isolation
c. small electromagnetic loads up to 72VA closed
d. resistive loads and solid-state loads with optical isolation

17. A line-powered proximity sensor draws its operating current through the load.
a. true
b. false

18. The scanning arrangement for a photoelectric sensor where the light beam transmitted from the transmitter is reflected back to the receiver by the target is called ____.
a. direct scan
b. retroreflective scan
c. diffused scan
d. convergent scan

19. Ladder diagrams include all of the following *except* a(n) ____.
a. signal section
b. decision section
c. action section
d. termination section

20. ***NEC Article 430,*** ____ covers motor and controller circuit overload protection.
a. Part III
b. Part IV
c. Part VII
d. Part X

Instructor's Notes:

GLOSSARY

Trade Terms Introduced in This Module

Actuator: The mechanism of a switch or switch enclosure that operates the contacts.

Auxiliary contact: A contact device that is mechanically connected to a contactor. It can be used to give an electrical signal related to the contactor's position.

Cam: A machine part or component that applies a force to the actuator of a limit switch, causing the actuator to move as intended.

Dashpot: A device that provides a time delay by controlling how fast air or liquid is allowed to pass into or out of a container through an orifice (opening) that is either fixed or variable in diameter.

Deadband: The temperature or pressure difference between the setpoint at which a switch activates when the temperature or pressure increases and the point at which the switch resets when the temperature or pressure decreases.

Dielectric constant: The ratio of the capacitance of a capacitor with a given dielectric to the capacitance of an identical capacitor having air for its dielectric.

Dropout voltage: The coil voltage at which the armature return spring of an electromechanical device overpowers the magnetic field of the coil and the contacts of the device change state.

Eddy current: Currents induced in the body of a metallic material by an oscillating magnetic field.

Electromechanical relay (EMR): A switching device with a set of contacts that are closed by a magnetic effect.

Infrared (IR): A portion of the light spectrum that is not visible (wavelengths beyond 690 nanometers). IR radiation is emitted by certain light-emitting diodes (LEDs).

International Electrotechnical Commission (IEC): An international organization that produces recommended performance and safety standards for electrical products.

Jogging (inching): Repeated closure of the circuit used to start and stop a motor. Inching is used to accomplish small movements in the driven machine.

Line-powered sensor: A three-wire sensor that draws its operating current (burden current) directly from the line. Its operating current does not flow through the load.

Load-powered sensor: A two-wire sensor in series with the load that draws its operating current (residual current) through the load.

Maintained contact switch: A switch with contacts that remain in the operating position when the actuating force is removed. The switch has provisions for resetting.

Momentary contact switch: A switch with contacts that return from the operating position to the release position when the actuating force is removed.

Motor control circuit: A circuit that carries the electric signals that direct the performance of a controller or other device, but does not carry the main power current.

Operator: A switch component that is pressed, pulled, or rotated by the person operating the circuit in order to activate the switch contacts.

Pickup voltage: The minimum allowable coil control voltage that will cause an electromechanical device to energize.

Pilot devices: Devices such as electrical pushbuttons and switches that contain contacts which govern the operation of related relays or similarly controlled devices.

Plugging: Braking a motor by reversing the phase sequence of the power to the motor. It is used to achieve both rapid stop and quick reversal.

Proximity sensors (switches): Sensing switches that detect the presence or absence of an object using an electronic noncontact sensor.

Seal-in voltage: The minimum allowable coil control voltage that will keep an electromechanical device energized. It is usually less than the pickup voltage.

Setpoint: The desired value to which a switching device is set or adjusted.

Instructor's Notes:

Additional Resources

This module is intended to present thorough resources for task training. The following reference works are suggested for further study. These are optional materials for continued education rather than for task training.

Electrical Motor Controls. Gary Rockis and Glen Mazur. Homewood, IL: American Technical Publishers, Inc., 1997.

National Electrical Code Handbook. Quincy, MA: National Fire Protection Association, 2002.

NFPA 70B Recommended Practice for Electrical Equipment Maintenance. Quincy, MA: National Fire Protection Association, 1998.

ANSWER KEY

Answers to Review Questions

Answer	Section
1. d	1.0.0
2. c	2.1.0
3. a	2.3.1
4. b	3.0.0; Table 4
5. a	3.1.0
6. c	4.0.0
7. d	4.0.0
8. c	4.0.0
9. b	4.1.0
10. c	5.0.0
11. d	5.2.1; 5.2.2; Table 6
12. b	5.2.2; Table 6
13. c	6.1.0; Table 9
14. b	6.2.1
15. d	6.2.5; Table 10
16. a	6.2.5; Table 11
17. b	6.10.4
18. c	6.11.0
19. d	9.3.1
20. a	10.0.0; Figure 76

Section 11.6.0 *Think About It—Overload Contacts*

Placing the three overloads in series with the conveyor 1 starter coil protects the conveyor system by shutting down the entire system should an overload occur in any one of the three conveyors.

CONTREN™ LEARNING SERIES — USER UPDATES

The NCCER makes every effort to keep these textbooks up-to-date and free of technical errors. We appreciate your help in this process. If you have an idea for improving this textbook, or if you find an error, a typographical mistake, or an inaccuracy in NCCER's Contren™ textbooks, please write us, using this form or a photocopy. Be sure to include the exact module number, page number, a detailed description, and the correction, if applicable. Your input will be brought to the attention of the Technical Review Committee. Thank you for your assistance.

Instructors – If you found that additional materials were necessary in order to teach this module effectively, please let us know so that we may include them in the Equipment/Materials list in the Instructor's Guide.

Write: Curriculum Revision and Development Department
National Center for Construction Education and Research
P.O. Box 141104, Gainesville, FL 32614-1104

Fax: 352-334-0932

E-mail: curriculum@nccer.org

Craft ______ Module Name ______

Copyright Date ______ Module Number ______ Page Number(s) ______

Description ______

______

______

______

(Optional) Correction ______

______

______

(Optional) Your Name and Address ______

______

______

Hazardous Locations

26312-02

MODULE OVERVIEW

This module introduces the Electrical trainee to the NEC requirements and installation procedures related to electrical equipment installed in hazardous locations.

PREREQUISITES

Please refer to the Course Map in the Trainee Module. Prior to training with this module, it is recommended that the trainee shall have successfully completed the following:

Core Curriculum; Electrical Level One; Electrical Level Two; Electrical Level Three, Modules 26301 through 26311.

OBJECTIVES

Upon completion of this module, the trainee will be able to:

1. Define the various classifications of hazardous locations.
2. Describe the wiring methods permitted for branch circuits and feeders in specific hazardous locations.
3. Select seals and drains for specific hazardous locations.
4. Select wiring methods for Class I, Class II, and Class III hazardous locations.
5. Follow NEC requirements for installing explosionproof fittings in specific hazardous locations.

PERFORMANCE TASKS

Under supervision of the instructor, the trainee should be able to:

1. Using two rigid metal conduit nipples, a sealing fitting, three pieces of No. 12 THHN conductors, and a packing fiber/sealing kit, perform the following operations:
   - Secure one conduit nipple in each end of the seal.
   - Make sure the required amount of threads are engaged.
   - Pull the three THHN conductors through the nipples and seal so that about 6" is protruding from each nipple.
   - Pack the fiber as per the instructions furnished with the sealing kit.
   - Mix the sealing compound.
   - Position the unit in the required location and pour in the sealing compound.

NCCER STANDARDIZED CRAFT TRAINING PROGRAMS

The National Center for Construction Education and Research (NCCER) provides a standardized national program of accredited craft training. Key features of the program include instructor certification, competency-based training, and performance testing. The program provides trainees, instructors, and companies with a standard form of recognition through a National Craft Training Registry. The program is described in full in the *Guidelines for Accreditation*, published by the NCCER. For more information on standardized craft training, contact the NCCER by writing us at P.O. Box 141104, Gainesville, FL 32614-1104, calling 352-334-0911, or e-mailing info@nccer.org. More information may be found at our Web site, www.nccer.org.

HOW TO USE THIS ANNOTATED INSTRUCTOR'S GUIDE

Each page presents two sections of information. The larger section displays each page exactly as it appears in the Trainee Module. The narrow column ties suggested trainee and instructor actions to each page and provides icons to call your attention to material, safety, audiovisual, or testing requirements. The bottom of each page includes space for your notes.

If you see the Teaching Tip icon, that means there is a teaching tip associated with this section. Also refer to any suggested teaching tips at the end of the module.

SAFETY CONSIDERATIONS

Ensure that the trainees are equipped with appropriate personal protective equipment.

PREPARATION

Before teaching this module, you should review the Module Outline, Objectives, Performance Tasks, and the Materials and Equipment List. Be sure to allow ample time to prepare your own training or lesson plan and gather all required equipment and materials.

MATERIALS AND EQUIPMENT LIST

Materials:
Transparencies
Markers/chalk
Copy of the latest edition of the *National Electrical Code®*
Sealoff fittings, packing fiber, and sealing compound
Short conduit nipples
No. 12 THHN conductors
Module Examinations*
Performance Profile Sheets*

Equipment:
Overhead projector and screen
Whiteboard/chalkboard
Appropriate personal protective equipment
Various types of explosionproof fittings
Various types of sealing fittings used in hazardous locations, including those with drains

*Located in the Test Booklet packaged with this Annotated Instructor's Guide.

ADDITIONAL RESOURCES

This module is intended to present thorough resources for task training. The following reference works are suggested for both instructors and motivated trainees interested in further study. These are optional materials for continued education rather than for task training.

American Electrician's Handbook, Terrell Croft and Wilfred I. Summers. New York, NY: McGraw-Hill, 1996.

Code Digest. Syracuse, NY: Cooper Crouse-Hinds, 2002.

National Electrical Code Handbook. Quincy, MA: National Fire Protection Association, 2002.

TEACHING TIME FOR THIS MODULE

An outline for use in developing your lesson plan is presented below. Note that each Roman numeral in the outline equates to one session of instruction. Each session has a suggested time period of 2½ hours. This includes 10 minutes at the beginning of each session for administrative tasks and one 10-minute break during the session. Approximately 15 hours are suggested to cover *Hazardous Locations.* You will need to adjust the time required for hands-on activity and testing based on your class size and resources.

Topic	**Planned Time**
Session I. Introduction	
A. Introduction	________
1. Class I Locations	________
2. Class II Locations	________
3. Class III Locations	________
4. Applications	________

Session II. Prevention of External Ignition/Explosion

A. Prevention of External Ignition/Explosion
   1. Sources of Ignition
   2. Combustion Principles

Session III. Explosionproof Equipment

A. Explosionproof Equipment
   1. Intrinsically Safe Equipment
   2. Explosionproof Conduit and Fittings
   3. Seals and Drains
      a. Seals
      b. Drains
      c. Selection and Installation of Seals and Drains
      d. Sealing Compounds and Dams

Session IV. Laboratory

A. Laboratory

   Under instructor supervision, have the trainees practice installing sealoff fittings and pouring seals. Note the proficiency of each trainee.

Session V. Garages and Similar Locations; Airport Hangars; Hospitals; Petrochemical Hazardous Locations; Manufacturers' Data

A. Garages and Similar Locations

B. Airport Hangars

C. Hospitals

D. Petrochemical Hazardous Locations

E. Manufacturers' Data

Session VI. Review; Module Examination; Performance Testing

A. Review

B. Module Examination
   1. Trainees must score 70% or higher to receive recognition from the NCCER.
   2. Record the testing results on Craft Training Report Form 200 and submit the results to the Training Program Sponsor.

C. Performance Testing
   1. Trainees must perform each task to the satisfaction of the instructor to receive recognition from the NCCER.
   2. Record the testing results on Craft Training Report Form 200 and submit the results to the Training Program Sponsor.

Module 26312-02

Hazardous Locations

Course Map

This course map shows all of the modules in the third level of the Electrical curriculum. The suggested training order begins at the bottom and proceeds up. Skill levels increase as you advance on the course map. The local Training Program Sponsor may adjust the training order.

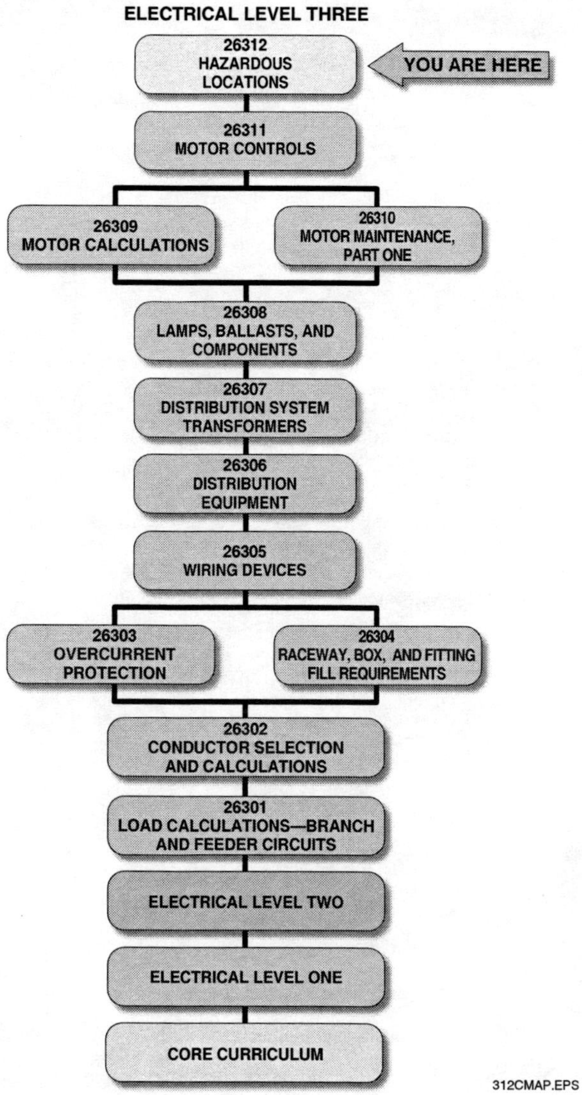

312CMAP.EPS

Homework

Assign reading of Module 26312.

MODULE 26312 CONTENTS

Figures

Tables

Instructor's Notes:

MODULE 26312

Hazardous Locations

Objectives

When you have completed this module, you will be able to do the following:

1. Define the various classifications of hazardous locations.
2. Describe the wiring methods permitted for branch circuits and feeders in specific hazardous locations.
3. Select seals and drains for specific hazardous locations.
4. Select wiring methods for Class I, Class II, and Class III hazardous locations.
5. Follow NEC requirements for installing explosionproof fittings in specific hazardous locations.

Prerequisites

Before you begin this module, it is recommended that you successfully complete the following: Core Curriculum; Electrical Level One; Electrical Level Two; Electrical Level Three, Modules 26301 through 26311.

Required Trainee Materials

1. Pencil and paper
2. Appropriate personal protective equipment
3. Copy of the latest edition of the *National Electrical Code*

1.0.0 ◆ INTRODUCTION

NEC Articles 500 through 504 cover the requirements of electrical **equipment** and wiring for all voltages in locations where fire or explosion hazards may exist due to flammable gases or vapor, flammable liquids, combustible dust, or ignitable fibers or other flying materials. Locations are classified depending on the properties of the flammable vapors, liquids, gases, or combustible dusts or fibers that may be present, as well as the likelihood that a flammable or combustible concentration or quantity is present.

Any area in which the atmosphere or a material in the area is such that the arcing of operating electrical contacts, components, and equipment may cause an explosion or fire is considered a **hazardous (classified) location.** In all such cases, **explosionproof apparatus,** raceways, and fittings are used to provide an **explosionproof** wiring system.

The NEC divides hazardous materials into three classes (Class I, Class II, and Class III), with two divisions for each class (Division 1 and Division 2). Of these, Class I, Division 1 represents the most hazardous location. These classes have been established on the basis of the explosive character of the atmosphere for the testing and approval of equipment for use in each class. However, it must be understood that considerable skill and judgment must be applied when deciding to what degree an area contains hazardous concentrations of vapors, combustible dusts, or easily ignitable fibers and flying materials. Furthermore, many factors, such as temperature, barometric pressure, quantity of release, humidity, ventilation, and distance from the vapor source, must be considered. When information on all factors concerned is properly evaluated, a consistent classification for the selection and location of electrical equipment can be developed.

Note: The designations "National Electrical Code," "NE Code," and "NEC," where used in this document, refer to the National Electrical Code®, which is a registered trademark of the National Fire Protection Association, Quincy, MA. *All National Electrical Code (NEC) references in this module refer to the 2002 edition of the NEC.*

Materials

Ensure you have everything required to teach the course. Check the Materials and Equipment List at the front of this Instructor's Guide.

Classroom

Review the goals of the course.

Explain that terms shown in bold (blue) are defined in the Glossary at the back of this module.

Audiovisual

Show Transparency 1, Course Objectives.

Show Transparency 2, Performance Tasks.

Classroom

Define *hazardous location.* Explain that *NEC Articles 500* through *504* cover the basic requirements for installations in hazardous locations.

Explain that hazardous locations are grouped into three classes, with two divisions within each class. Point out that the most hazardous location is Class I, Division 1.

Classroom

Discuss Class I locations. Explain that Class I divisions are subdivided into groups, with Group A being the most hazardous. List examples of various Class I locations.

Audiovisual

Show Transparency 3 (Figure 1).

Classroom

Discuss Class II locations. Explain that Class II divisions are subdivided into groups, with Group E being the most hazardous. List examples of various Class II locations.

Audiovisual

Show Transparency 4 (Figure 2).

> INSIDE TRACK
>
> ### *Zone Classifications*
>
> The zone system covered in ***NEC Article 505*** is an alternative to the division classification covered in ***NEC Article 500.*** All electrical components and devices must be rated for the specific zone or division in which they are installed, regardless of the level of protection provided by the device. See ***NEC Section 505.6(B).***

NEC Article 505 allows classification and application to international standards, but it will be applied only under engineering supervision. In addition, few American-made products are listed for international application. As a result, this module will not cover *NEC Article 505* in any detail.

1.1.0 Class I Locations

Class I atmospheric hazards are divided into Divisions 1 and 2, and also into four groups (A, B, C, and D). Group A represents the most hazardous location.

Those locations in which flammable gases or vapors may be present in the air in quantities sufficient to produce explosive or ignitable mixtures are identified as Class I locations. If these gases or vapors are present during normal operation, frequent repair or maintenance operations, or where breakdown or faulty operation of process equipment might also cause simultaneous failure of electrical equipment, the area is designated as Class I, Division 1. Examples of such locations are interiors of paint spray booths where volatile, flammable solvents are used, inadequately ventilated pump rooms where flammable gas is pumped, anesthetizing locations of hospitals (to a height of 5' above floor level), and drying rooms for the evaporation of flammable solvents (see *Figure 1*).

Class I, Division 2 covers locations in which volatile flammable gases, vapors, or liquids are handled either in a closed system or confined within suitable enclosures, or where hazardous concentrations are normally prevented by positive mechanical ventilation. Areas adjacent to Division 1 locations, into which gases might occasionally flow, also belong in Division 2.

1.2.0 Class II Locations

Class II locations are those that are hazardous because of the presence of combustible dust. Class II, Division 1 locations are areas in which combustible dust may be present in the air under normal operating conditions in quantities sufficient to produce explosive or ignitable mixtures; examples are working areas of grain-handling and storage plants and rooms containing grinders or pulverizers (*Figure 2*). Class II, Division 2 locations are areas in which dangerous concentrations of suspended dust are not likely, but where dust might accumulate.

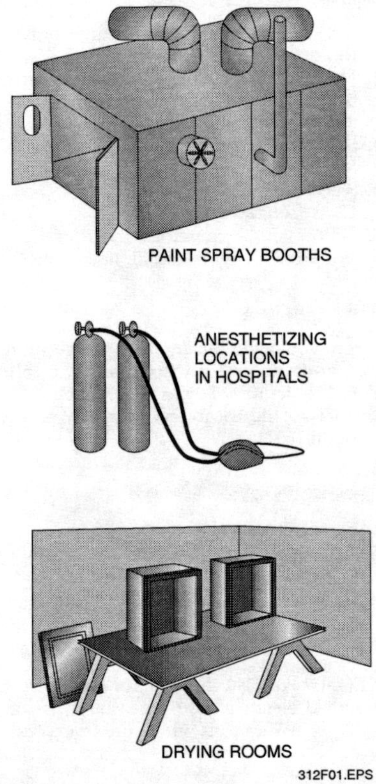

Figure 1 ◆ Typical NEC Class I locations.

Instructor's Notes:

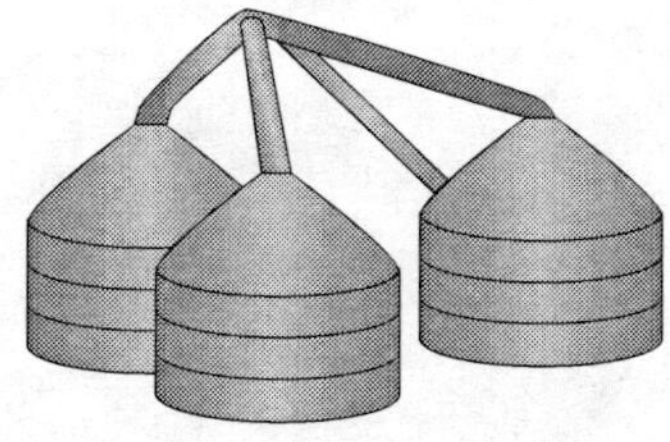

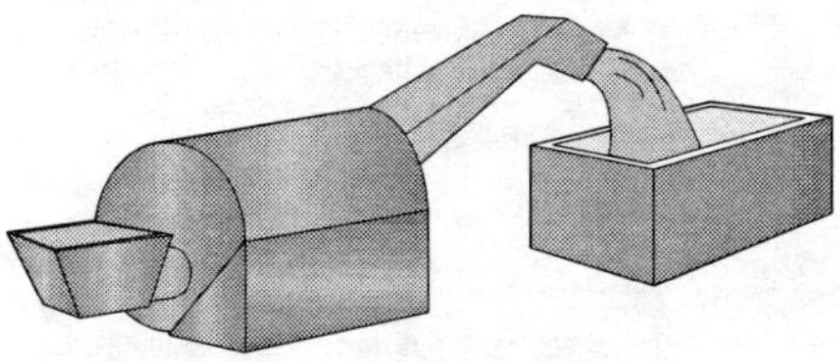

Figure 2 ◆ Typical NEC Class II locations.

Besides the two divisions, Class II atmospheric hazards also cover three groups of combustible dusts (E, F, and G). The groupings are based on the resistivity of the dust. Group E is always Division 1. Groups F and G may be either Division 1 or 2, depending on their resistivity. Because the NEC is considered the definitive classification tool and contains explanatory data about hazardous atmospheres, refer to *NEC Section 500.5* for exact definitions of Class II, Divisions 1 and 2.

1.3.0 Class III Locations

Class III locations are those areas that are hazardous because of the presence of easily ignitable fibers or other flying materials, but such materials are not likely to be in suspension in the air in quantities sufficient to produce ignitable mixtures. Such locations usually include certain areas of rayon, cotton, and textile mills, clothing manufacturing plants, and woodworking plants (*Figure 3*).

1.4.0 Applications

Hazardous atmospheres are summarized in *Table 1*. For a more complete listing of flammable liquids, gases, and solids, see *Recommended Practice for the Classification of Flammable Liquids, Gases, or Vapors and of Hazardous (Classified) Locations for*

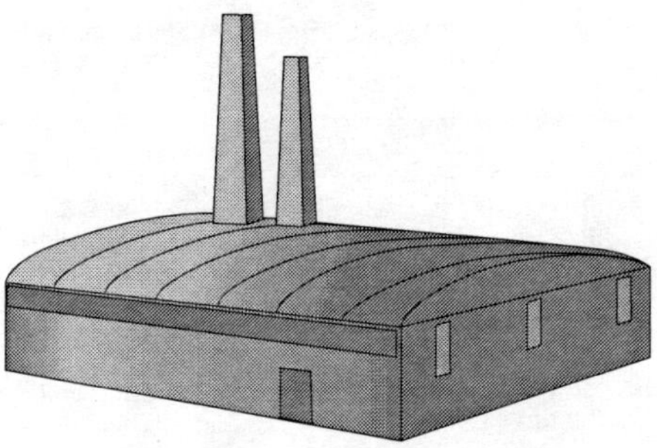

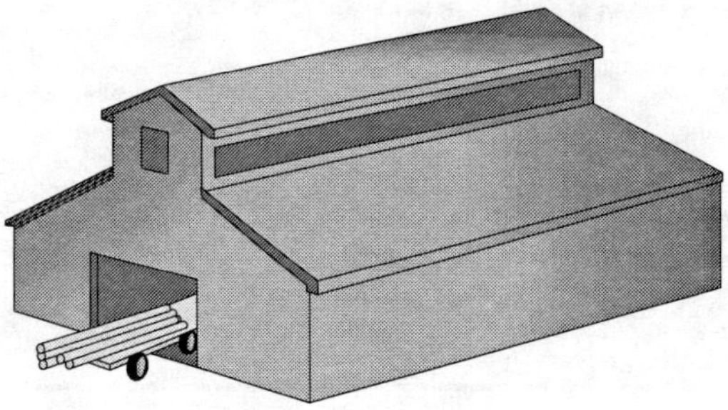

Figure 3 ◆ Typical NEC Class III locations.

Classroom

Discuss Class III locations. Explain that Class III is not divided into groups. List examples of various Class III locations.

Audiovisual

Show Transparency 5 (Figure 3).

Classroom

Point out that hazardous locations are summarized in Table 1 of the Trainee Module.

Review the table and explain that the zone system is covered in *NEC Article 505* and is the international equivalent to the division classifications defined in *NEC Article 500.*

Table 1 Summary of Hazardous Atmospheres

Hazardous Area	Class Subdivisions	Groups
	Class I Divisions	**Class I, Division Groups**
Class I: Material present is a flammable gas or vapor	Division 1: Locations in which hazardous concentrations of flammable gases or vapors are present normally or frequently	Group A: Atmospheres containing acetylene
	Division 2: Locations in which hazardous concentrations of flammable gases or vapors are present as a result of infrequent failure of equipment or containers	Group B: Atmospheres containing hydrogen, manufactured gases containing more than 30% hydrogen by volume, or gases or vapors of equivalent hazard Group C: Atmospheres containing ethylene, cyclopropane, or gases or vapors of equivalent hazard Group D: Atmospheres containing propane, gasoline, or gases or vapors of equivalent hazard
	Class I Zones	**Class I, Zone Groups**
	Zone 0: Locations in which combustible material is present continuously or for long periods	Group IIC: Atmospheres containing acetylene or hydrogen or other gases or vapors meeting Group IIC criteria
	Zone 1: Locations in which combustible material is likely to be present normally or frequently because of repair or maintenance operations or leakage	Group IIB: Atmospheres containing acetaldehyde, ethylene, or other gases or vapors meeting Group IIB criteria
	Zone 2: Locations in which combustible material is not likely to occur in a normal operation and, if it does occur, will exist only for a short period	Group IIA: Atmospheres containing propane, gasoline, or other gases or vapors meeting Group IIA criteria
		Class II, Division Groups
Class II: Material present is a combustible dust	Division 1: Locations in which hazardous concentrations of combustible dust are present normally or may exist because of equipment breakdown or where electrically conductive combustible dusts are present in hazardous quantities	Group E: Atmospheres containing combustible metal dusts including aluminum, magnesium, and other metals of similar hazards
	Division 2: Locations in which hazardous concentrations of combustible dust are not normally suspended in the air but may occur as a result of infrequent malfunction of equipment or where dust accumulation may interfere with safe dissipation of heat or may be ignitable by abnormal operation of electrical equipment	Group F: Atmospheres containing combustible carbonaceous dusts, including carbon black, charcoal, coals, or dusts that have been sensitized by other materials so that they present an explosion hazard
		Group G: Atmospheres containing combustible nonconductive dusts not included in Group E or F, including flour, grain, wood, and plastic
Class III: Material present is an ignitable fiber or flying	Division 1: Locations in which easily ignitable fibers or materials producing combustible flyings are handled, manufactured, or used	No Groups
	Division 2: Locations in which easily ignitable fibers are stored or handled, except in the manufacturing process	

Instructor's Notes:

Electrical Installations in Chemical Process Areas, National Fire Protection Association Publication No. 497.

Once the class of an area is determined, the conditions under which the hazardous material may be present determine the division. In Class I and Class II, Division 1 locations, the hazardous gas or dust may be present in the air under normal operating conditions in dangerous concentrations. In Division 2 locations, the hazardous material is not normally in the air, but it might be released if there is an accident or if there is faulty operation of equipment.

Tables 2 through *6* provide a summary of the various classes of hazardous locations as defined by the NEC.

Table 2 Application Rules for Class I, Division 1

Components	Characteristics	NEC Reference
Boxes, fittings	Explosionproof and threaded for connection to conduit	*NEC Section 501.4(A)*
Sealoffs	Approved for purpose	*NEC Section 501.5(A)*
Wiring methods	Rigid metal conduit, steel intermediate metal conduit, Type MI cable, and, under certain conditions, ITC and MC cable	*NEC Section 501.4(A)(1)*
Receptacles	Approved for the location	*NEC Section 501.12*
Lighting fixtures	Approved for Class I, Division 1	*NEC Section 501.9(A)*
Panelboards	Class I enclosure	*NEC Section 501.6(A)*
Circuit breakers	Class I enclosure	*NEC Section 501.6(A)*
Fuses	Class I enclosure	*NEC Section 501.6(A)*
Switches	Class I enclosure	*NEC Section 501.6(A)*
Motors	Class I, Division 1, totally-enclosed or submerged	*NEC Section 501.8(A)*
Liquid-filled transformers	Installed in an approved vault	*NEC Section 501.2(A)*
Dry-type transformers	Class I, Division 1 enclosure	*NEC Section 501.7(A)*
Utilization equipment	Class I, Division 1	*NEC Section 501.10(A)*
Flexible connections	Class I, explosionproof	*NEC Section 501.4(A)(2)*
Portable lamps	Class I, Division 1, approved as a portable assembly	*NEC Section 501.9(A)*
Generators	Class I, Division 1, totally enclosed or submerged	*NEC Section 501.8(A)*
Alarm systems	Class I, Division 1	*NEC Section 501.14(A)*

Table 3 Application Rules for Class I, Division 2

Components	Characteristics	NEC Reference
Boxes, fittings	Do not have to be explosionproof unless current interrupting contacts are exposed	*NEC Section 501.4(B)*
Sealoffs	Approved for purpose	*NEC Section 501.5(C)*
Wiring methods	Rigid metal conduit, steel intermediate metal conduit, Types MI, MC, MV, TC, ITC, or PLTC cables, or enclosed gasketed busways or wireways	*NEC Section 501.4(B)*
Receptacles	Approved for the location	*NEC Section 501.12*
Lighting fixtures	Protected from physical damage	*NEC Section 501.9(B)(2)*
Panelboards	General purpose with exceptions	*NEC Section 501.6(B)*
Circuit breakers	Class I enclosure	*NEC Section 501.6(B)(1)*
Fuses	Class I enclosure	*NEC Section 501.6(B)(3)*
Switches	Class I enclosure	*NEC Section 501.6(B)(2)*
Motors	General purpose unless motor has sliding contacts, switching contacts, or integral resistance devices; if so, use Class I, Division 1	*NEC Section 501.8(B)*
Motor controls	Class I, Division 1	*NEC Section 501.6(A)*
Liquid-filled transformers	General purpose	*NEC Section 501.2(B)*
Dry-type transformers	Class I, general purpose except switching mechanism Division 1 enclosures	*NEC Section 501.7(B)*
Utilization equipment	Class I, Division 1	*NEC Section 501.10(B)*
Flexible connections	Class I, explosionproof	*NEC Section 501.4(B)*
Portable lamps	Explosionproof	*NEC Section 501.9(B)(1)*
Generators	Class I, totally enclosed or submerged	*NEC Section 501.8(A)*
Alarm systems	Class I, Division 1	*NEC Section 501.14(B)*

Classroom

Explain how to locate the requirements for specific hazardous environments in the NEC.

Refer the trainees to Tables 2 through 6 in the Trainee Module.

Teaching Tip

Provide the trainees with a list of hazardous locations and have them cite the applicable sections of the NEC.

Table 4 Application Rules for Class II, Division 1

Components	Characteristics	NEC Reference
Boxes, fittings	Class II boxes required when using taps, joints, or other connections; otherwise, use dust-tight boxes with no openings	***NEC Section 502.4(A)(1)***
Wiring methods	Rigid metal conduit, steel intermediate metal conduit, or Types MI and, under certain conditions, MC cables listed for use in Class II, Division 1 locations	***NEC Section 502.4(A)***
Receptacles	Class II	***NEC Section 502.13(A)***
Lighting fixtures	Class II	***NEC Section 502.11(A)***
Panelboards	Dust/ignitionproof	***NEC Section 502.6(A)(1)***
Circuit breakers	Dust/ignitionproof enclosure	***NEC Section 502.6(A)***
Fuses	Dust/ignitionproof enclosure	***NEC Section 502.6(A)***
Switches	Dust/ignitionproof enclosure	***NEC Section 502.6(A)***
Motors	Class II, Division 1 or totally enclosed	***NEC Section 502.8(A)***
Motor controls	Dust/ignitionproof	***NEC Section 502.6(A)***
Liquid-filled transformers	Install in an approved vault	***NEC Section 502.2(A)***
Dry-type transformers	Class II, vault	***NEC Section 502.2(A)***
Utilization equipment	Class II	***NEC Section 502.10(A)***
Flexible connections	Extra-hard usage cord, liquid-tight, and others	***NEC Section 502.4(A)(2)***
Portable lamps	Class II	***NEC Section 502.11(A)***
Generators	Class II, Division 1 or totally enclosed	***NEC Section 502.8(A)***

Table 5 Application Rules for Class II, Division 2

Components	Characteristics	NEC Reference
Boxes, fittings	Use tight covers to minimize entrance of dust	***NEC Section 502.4(B)(1)***
Wiring methods	Rigid metal conduit, steel intermediate metal conduit, electrical metallic tubing (EMT), Types MI, MC, TC, ITC, or PLTC cables, or enclosed dust-tight busways or wireways	***NEC Section 502.4(B)***
Receptacles	Exposed live parts are not allowed	***NEC Section 502.13(B)***
Lighting fixtures	Class II	***NEC Section 502.11(B)***
Panelboards	Dust-tight enclosure	***NEC Section 502.6(B)***
Circuit breakers	Dust-tight enclosure	***NEC Section 502.6(B)***
Fuses	Dust-tight enclosure	***NEC Section 502.6(B)***
Switches	Dust-tight enclosure	***NEC Section 502.6(B)***
Motors	Class II, Division 1 or totally enclosed	***NEC Section 502.8(B)***
Motor controls	Dust-tight enclosure	***NEC Section 502.6(B)***
Liquid-filled transformers	Install in vault	***NEC Section 502.2(B)***
Dry-type transformers	Class II vault	***NEC Section 502.2(B)***
Utilization equipment	Class II	***NEC Section 502.10(B)***
Flexible connections	Extra-hard usage cord, liquid-tight, and others	***NEC Section 502.4(B)(2)***
Portable lamps	Class II	***NEC Section 502.11(B)(1)***
Generators	Class II, Division 1 or totally enclosed	***NEC Section 502.8(B)***

Instructor's Notes:

Table 6 Application Rules for Class III, Divisions 1 and 2

Components	Characteristics	NEC Reference
Boxes, fittings	Dust-tight	***NEC Section 503.3(A)(1)***
Wiring methods	Rigid metal conduit, steel intermediate metal conduit, EMT, Types MI and MC cables, or enclosed dust-tight busways or wireways	***NEC Section 503.3(A)***
Receptacles	Minimize accumulation of fibers or flyings	***NEC Section 503.11***
Lighting fixtures	Tight enclosure with no openings	***NEC Section 503.9***
Panelboards	Dust-tight enclosure	***NEC Section 503.4***
Circuit breakers	Dust-tight enclosure	***NEC Section 503.4***
Fuses	Dust-tight enclosure	***NEC Section 503.4***
Switches	Dust-tight enclosure	***NEC Section 503.4***
Motors	Totally enclosed	***NEC Section 503.6***
Motor controls	Dust-tight enclosure	***NEC Section 503.4***
Liquid-filled transformers	Install in an approved vault	***NEC Section 503.2***
Dry-type transformers	Dust-tight enclosure	***NEC Section 503.2***
Utilization equipment	Class III	***NEC Section 503.8***
Flexible connections	Extra hard usage cord and other flexible conduit/fittings	***NEC Section 503.3(A)(2)***
Portable lamps	Unswitched, guarded with tight enclosure for lamp	***NEC Section 503.9(D)***
Generators	Totally enclosed	***NEC Section 503.6***

INSIDE TRACK

Delayed Action Receptacles

The receptacle shown here is rated for Class I, Division 1 and 2 locations and features a delayed action rotating sleeve that prevents complete withdrawal of the plug in one continuous movement. This delay allows any arc-generated heat to be dissipated before the plug is released.

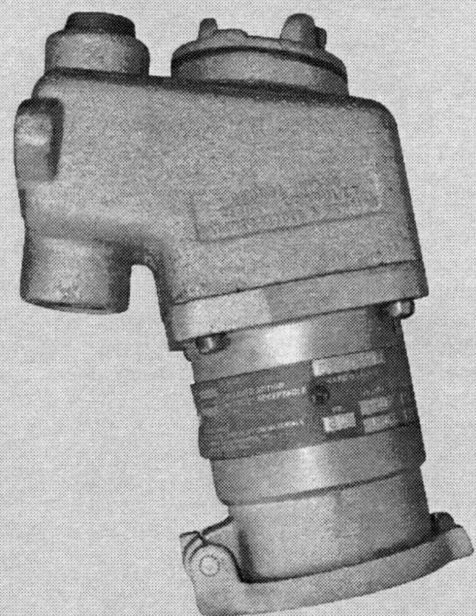

312P1202.EPS

Classroom

Discuss the operation of delayed-action receptacles.

Classroom

Point out the sources of energy that can cause an explosion in a hazardous environment. Stress the importance of electrical safety.

List the three conditions required for a fire or explosion to occur. Point out that the aim of fire safety is to eliminate one or more of these conditions.

2.0.0 ◆ PREVENTION OF EXTERNAL IGNITION/EXPLOSION

The main purpose of using explosionproof fittings and wiring methods in hazardous areas is to prevent ignition of flammable liquids or gases and to prevent an explosion.

2.1.0 Sources of Ignition

In certain atmospheric conditions when flammable gases or combustible dusts are mixed in the proper proportion with air, any source of energy is all that is needed to touch off an explosion.

One prime source of energy is electricity. Equipment such as switches, circuit breakers, motor starters, pushbutton stations, or plugs and receptacles can produce arcs or sparks in normal operation when contacts are opened and closed. This could easily cause ignition.

Other hazards are devices that produce heat, such as lighting fixtures and motors. In this case, the surface temperatures may exceed the safe limits of many flammable atmospheres.

Finally, many parts of the electrical system can become potential sources of ignition in the event of insulation failure. This group includes wiring (particularly splices in the wiring), transformers, impedance coils, solenoids, and other low-temperature devices without make-or-break contacts.

Non-electrical hazards such as sparking metal can also easily cause ignition. A hammer, file, or other tool that is dropped on masonry or on a ferrous surface can cause a hazard unless the tool is made of non-sparking material. For this reason, portable electrical equipment is usually made from aluminum or other material that will not produce sparks if the equipment is dropped.

Electrical safety is of crucial importance. The electrical installation must prevent accidental ignition of flammable liquids, vapors, and dusts released to the atmosphere. In addition, because much of this equipment is used outdoors or in corrosive atmospheres, the material and finish must be such that maintenance costs and shutdowns are minimized.

2.2.0 Combustion Principles

Three basic conditions must be satisfied for a fire or explosion to occur:

- A flammable liquid, vapor, or combustible dust must be present in sufficient quantity.
- The flammable liquid, vapor, or combustible dust must be mixed with air or oxygen in the proportions required to produce an explosive mixture.
- A source of ignition must be applied to the explosive mixture.

In applying these principles, the quantity of the flammable liquid or vapor that may be liberated and its physical characteristics must be recognized.

Vapors from flammable liquids also have a natural tendency to disperse into the atmosphere and rapidly become diluted to concentrations below the lower explosion limit, particularly when there is natural or mechanical ventilation.

WARNING!

The possibility that the gas concentration may be above the upper explosion limit does not afford any degree of safety, because the concentration must first pass through the explosive range to reach the upper explosion limit.

INSIDE TRACK

Explosionproof Equipment

Explosionproof equipment must be marked to show class and group. See ***NEC Section 500.8(B) plus Exceptions.*** Always make sure that the device selected is rated for the hazardous location in which it will be installed.

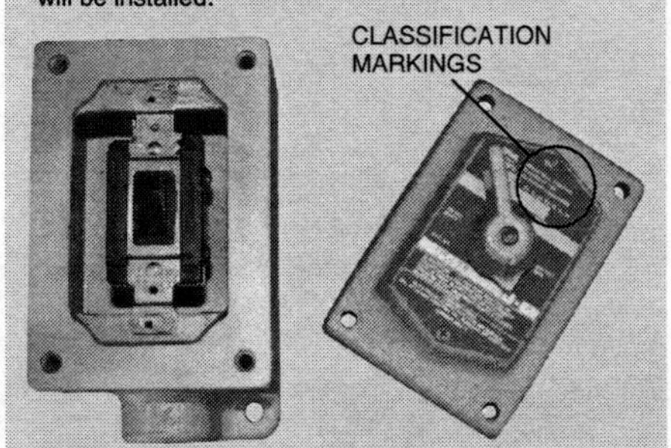

312P1201.EPS

Instructor's Notes:

3.0.0 ◆ EXPLOSIONPROOF EQUIPMENT

Each area that contains gases or dusts considered hazardous must be carefully evaluated to make certain the correct electrical equipment is selected. Many hazardous atmospheres are Class I, Group D or Class II, Group G. However, certain areas may involve other groups, particularly Class I, Groups B and C. Conformity with the NEC requires the use of fittings and enclosures **approved** for the specific hazardous gas or dust involved.

The wide assortment of explosionproof equipment now available makes it possible to provide adequate electrical installations under any of the various hazardous conditions. However, you must be thoroughly familiar with all NEC requirements and know what fittings are available, how to install them properly, and where and when to use the various fittings. For example, some electricians are under the false impression that a fitting rated for Class I, Division 1 can be used under any hazardous conditions. However, remember the groups. For example, a fitting rated for Class I, Division 1, Group C cannot be used in areas classified as Groups A or B. On the other hand, fittings rated for use in Group A may be used for any group beneath A; fittings rated for use in Class I, Division 1, Group B can be used in areas rated as Group B areas or below, and so on.

> **WARNING!**
> Never interchange fittings or covers between one hazardous area and another. Such items must be rated for the appropriate class, division, and group.

Explosionproof fittings are rated for both classification and groups. All parts of these fittings (including covers) are rated accordingly. Therefore, if a Class I, Division 1, Group A fitting is required, a Group B (or below) fitting cover must not be used. The cover itself must be rated for Group A locations. Consequently, when working on electrical systems in hazardous locations, always make certain that fittings and their related components match the condition at hand.

3.1.0 Intrinsically Safe Equipment

Intrinsically safe equipment is incapable of releasing sufficient electrical energy under normal or abnormal conditions to cause ignition of a specific hazardous atmospheric mixture in its most easily ignited concentration. The use of intrinsically safe equipment is primarily limited to process control instrumentation because these electrical systems lend themselves to the low energy requirements.

What's Wrong with This Picture?

312P1203.EPS

Installation rules for intrinsically safe equipment are covered in *NEC Article 504.* In general, intrinsically safe equipment and its associated wiring must be installed so that it is positively separated from the non-intrinsically safe circuits, because induced voltages could defeat the concept of intrinsically safe circuits. Underwriters Laboratories and Factory Mutual list several devices in this category.

3.2.0 Explosionproof Conduit and Fittings

A typical floor plan for a hazardous area is shown in *Figure 4*.

In hazardous locations where threaded metal **conduit** is required, the conduit must be threaded with a standard conduit cutting die (*Figure 5*) that provides ¾" taper per foot. The conduit should be made up wrench-tight to prevent sparking in the event fault current flows through the raceway system [*NEC Section 501.4(A)(1)(a)*]. All boxes, fittings, and joints shall be threaded for connection to the conduit system and shall be an approved, explosionproof type (*Figure 6*). Threaded joints must be made up with at least five threads fully engaged. Where it becomes necessary to employ flexible connectors at motor or fixture terminals (*Figure 7*), flexible fittings approved for the particular class location shall be used. Unions are provided to facilitate the installation and removal of equipment.

Classroom

Explain that explosionproof equipment is designed to separate the electrical system from the hazardous atmosphere.

Safety

Emphasize that all equipment and materials in a hazardous location must be rated for use in that environment.

Classroom

Define *intrinsically safe equipment.*

Teaching Tip

Discuss the "What's Wrong with This Picture?"

Classroom

Discuss the NEC requirements for installing explosionproof conduit and fittings.

Audiovisual

Show Transparency 6 (Figure 4). Also refer to Figures 5 through 7 in the Trainee Module.

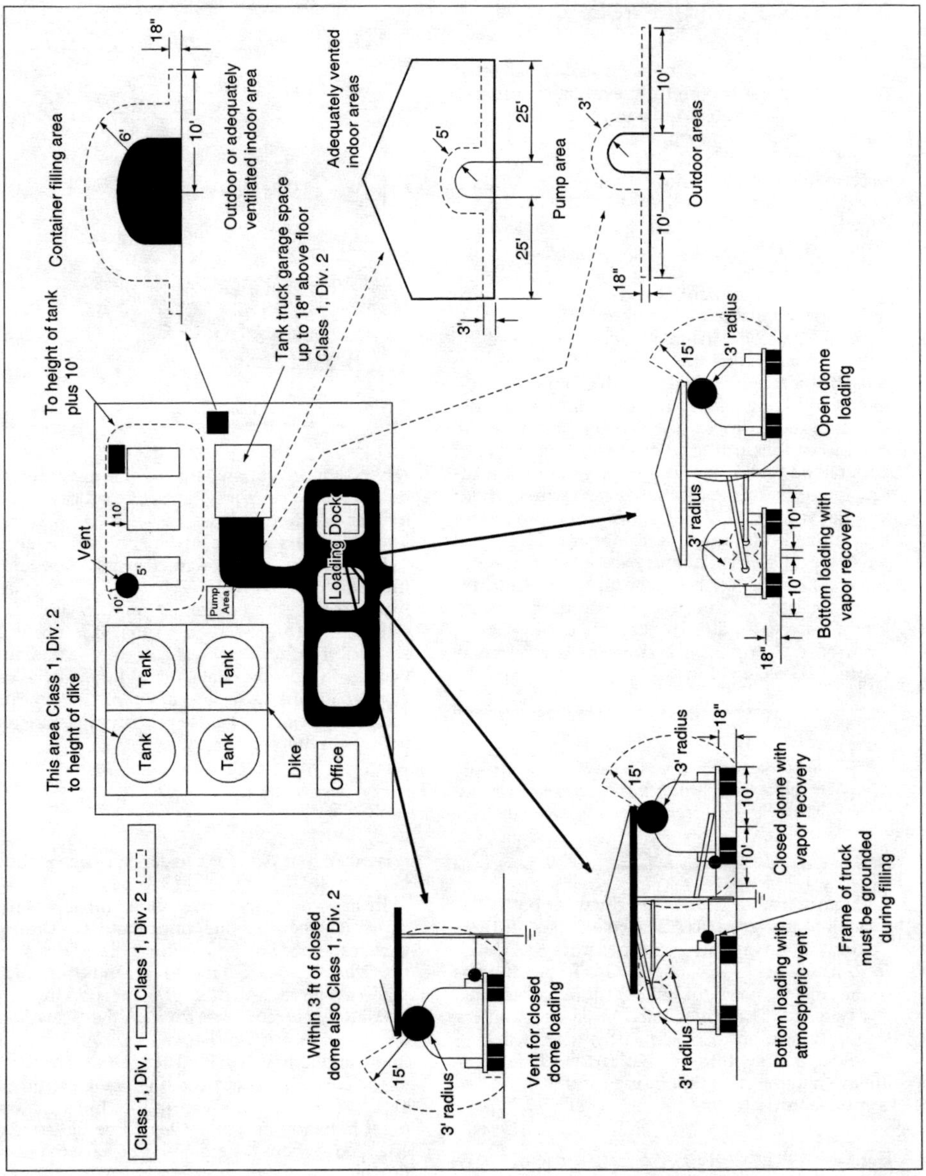

312F04.EPS

Figure 4 ◆ Floor plan of a hazardous location.

Instructor's Notes:

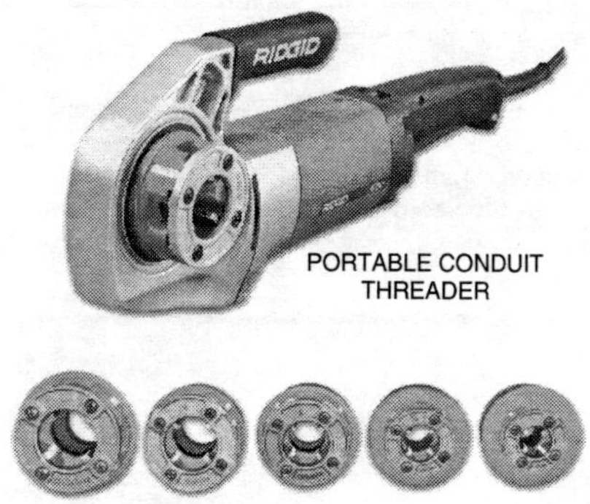

312F05.EPS

Figure 5 ◆ Portable conduit threader.

312F06.EPS

Figure 6 ◆ Typical fittings approved for hazardous areas.

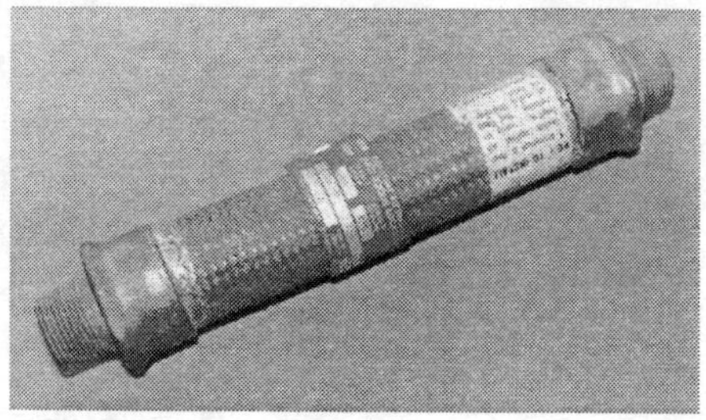

312F07.EPS

Figure 7 ◆ Explosionproof flexible connector.

3.3.0 Seals and Drains

Seals and drains are both used to protect conduit systems.

3.3.1 Seals

Sealoff fittings, also known as sealing fittings or seals *(Figure 8),* are required in conduit systems to prevent the passage of gases, vapors, or flames from one portion of the electrical installation to another at atmospheric pressure and normal ambient temperatures. Furthermore, sealoffs limit explosions to the enclosure and prevent precompression of pressure piling in conduit systems.

For Class I, Division 1 locations, ***NEC Section 501.5(A)(1)*** states that in each conduit run entering an enclosure for switches, circuit breakers, fuses, relays, resistors, or other apparatus that may produce arcs, sparks, or high temperatures, seals shall be installed within 18" from such enclosures. Explosionproof unions, couplings, reducers, elbows, capped elbows, and **conduit bodies** similar to L, T, and cross types shall be the only enclosures or fittings permitted between the sealing fitting and the enclosure. The conduit bodies shall not be larger than the largest trade size of the conduit.

However, one exception to this rule is that conduits are not required to be sealed if the current interrupting contacts are enclosed within a chamber hermetically sealed against the entrance of gases or vapors, immersed in oil in accordance with ***NEC Section 501.6(B),*** or enclosed within a factory-sealed explosionproof chamber within an enclosure approved for the location and marked *FACTORY SEALED* or equivalent.

Seals are also required in Class II locations where a raceway provides communication between an enclosure that is required to be dust/ignitionproof and one that is not (***NEC Section 502.5***).

Classroom

Discuss the NEC requirements for sealoff fittings.

Audiovisual

Show Transparency 7 (Figure 8).

Teaching Tip

Discuss the "Think About It."

Pressure Piling

INSIDE TRACK

Pressure piling occurs when an explosion in one section of a raceway creates expanding gases that cause the vapors to compress further down in the raceway that will then cause a secondary explosion of a greater magnitude. This explosion will result in additional expanding gases and compressed vapors, continuing to create additional explosions, which may potentially exceed the containment capabilities of the raceway system.

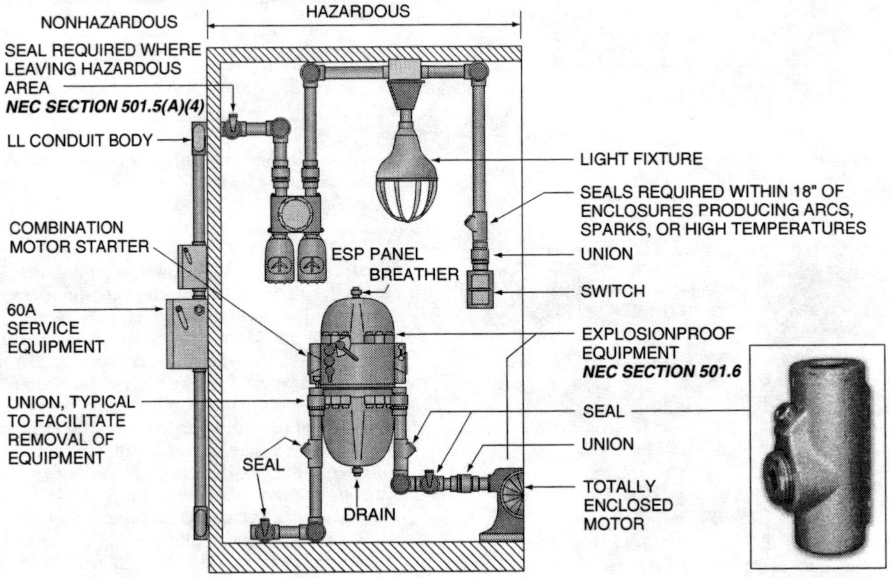

Figure 8 ◆ Installation of seals in Class I, Division 1 locations.

A permanent and effective seal is one method of preventing the entrance of dust into the dust/ignitionproof enclosure through the raceway. A horizontal raceway, not less than 10' long, is another approved method, as is a vertical raceway not less than 5' long and extending downward from the dust/ignitionproof enclosure.

Where a raceway provides communication between an enclosure that is required to be dust/ignitionproof and an enclosure in an unclassified location, seals are not required.

Where sealing fittings are used, all must be accessible.

Fill Requirements

THINK ABOUT IT

Why do the standards for listing sealoff fittings allow only a 25% fill for these fittings, while conduits are permitted to have a 40% fill?

Instructor's Notes:

While it is not an NEC requirement, many electrical designers sectionalize long conduit runs by inserting seals not more than 50' to 100' apart, depending on the conduit size. This is done in order to minimize the effects of pressure piling.

In general, seals are installed at the same time as the conduit system. However, the conductors are installed after the raceway system is complete and prior to packing and sealing the fittings.

3.3.2 Drains

In humid atmospheres or wet locations where it is likely that water will gain entrance to the raceway system, the raceways should be inclined so that water will not collect in enclosures or on seals, but will be led to low points where it may pass out through integral drains.

If the arrangement of raceway runs makes this impractical, special drain/seal fittings should be used, such as the type shown in *Figure 9*. These fittings prevent water from accumulating above the seal and meet the requirements of ***NEC Section 501.5(F)***.

Even if the location is not typically humid or wet, surprising amounts of water may still collect in conduit systems. This is because no conduit system is completely airtight. Alternate increases and decreases in temperature and/or barometric pressure due to weather changes or to the nature of the process carried on in the location where the conduit is installed will cause the introduction of outside air. If this air carries sufficient moisture, it will condense within the system when the temperature drops. Because the internal conditions are unfavorable to evaporation, the resultant water will remain and accumulate over time.

To avoid the accumulation of moisture, install drain/seal fittings with drain covers or fittings with inspection covers. This is a recommended practice even if prevailing conditions at the time of planning or installation do not indicate a moisture problem.

3.3.3 Selection and Installation of Seals and Drains

Always select the proper sealoff fitting for the hazardous location (such as Class 1, Groups A, B, C, or D) and for the proper use in respect to its mounting position. This is particularly critical when the conduit run crosses between hazardous and nonhazardous areas. The improper positioning of a seal may permit hazardous gases or vapors to enter the system beyond the seal and escape into another portion of the hazardous area or enter a nonhazardous area. Some seals are designed to be mounted in any position; others are restricted to horizontal or vertical mounting. *Figure 10* shows various types of seals.

Install the seals on the proper side of the partition or wall, as recommended by the manufacturer. The installation of seals should be made

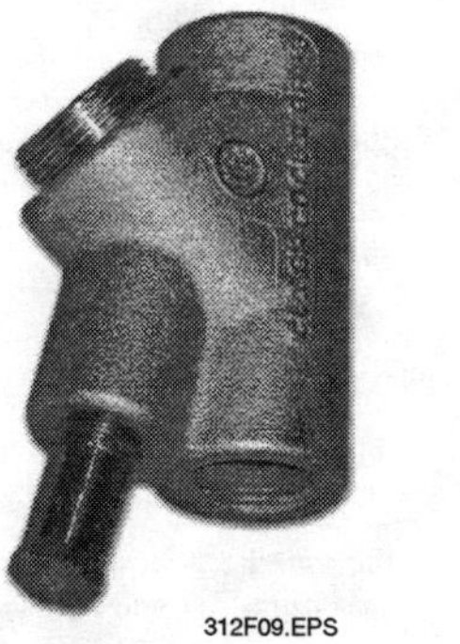

Figure 9 ◆ Typical drain seal.

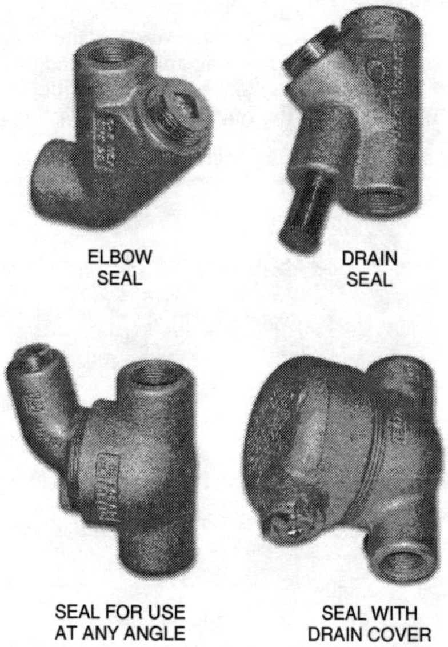

Figure 10 ◆ Various types of seals.

Classroom

Explain that many applications require the use of drains to remove accumulated moisture from the raceways. Refer to Figure 9 in the Trainee Module.

Note that seals are available in many configurations to suit the environment. Refer to Figure 10 in the Trainee Module. Emphasize the importance of selecting the appropriate seal for the application.

Provide examples of various seals for the trainees to examine.

Classroom

Explain that the proper installation of the sealing compound is essential to a safe and effective installation.

Safety

Emphasize that the compound used must be compatible with the packing, fitting, and type of conductors.

Classroom

Describe the procedure for pouring a seal.

only by trained personnel in strict compliance with the instruction sheets furnished with the seals and **sealing compound.** *NEC Section 501.5(C)(4)* prohibits splices or taps in sealoff fittings. Sealoff fittings are listed by UL for use in Class I hazardous locations with approved sealing compound only. This compound, when properly mixed and poured, hardens into a dense, strong mass that is insoluble in water, not attacked by chemicals, and not softened by heat. It is designed to withstand the pressure of the exploding trapped gases or vapors. Conductors sealed in the compound may be any approved thermoplastic or rubber insulated type. Conductors may or may not be lead-covered.

3.3.4 Sealing Compounds and Dams

Poured seals should be made only by qualified personnel following the manufacturer's instructions. Improperly poured seals serve no purpose. Sealing compound must be approved for the purpose, not be affected by the surrounding atmosphere or liquids, and not have a melting point of less than 200°F (93°C). The sealing compound and dams must also be approved for the type and manufacturer of the fitting. For example, Crouse-Hinds CHICO® A sealing compound is the only sealing compound approved for use with Crouse-Hinds ECM fittings.

To pack the sealoff, remove the threaded plug or plugs from the fitting and insert the fiber supplied with the packing kit. Tamp the fiber between the wires and the hub before pouring the sealing compound into the fitting, then pour in the sealing cement and reset the threaded plug tightly. The fiber packing prevents the sealing compound from entering the conduit lines in the liquid state.

Sealing compound is poured after the conduit system and seals are installed and the conductors and packing fiber have been installed. Most sealing compound kits contain a powder in a polyethylene bag within an outer container. Remove the bag of powder, fill the outside container, pour in the powder, and mix.

> **CAUTION**
> Always make certain that the sealing compound is compatible for use with the packing material, brand and type of fitting, and type of conductors used in the system.

In practical applications, dozens of seals may be required for a particular installation. Consequently, after the conductors are pulled, each seal in the system is first packed. To prevent the possibility of overlooking a seal, a certain color of paint is normally sprayed on the seal hub to indicate that the seal has been packed. When the sealing compound is poured, a different color paint is sprayed on the seal hub to indicate a finished job. This method permits the job supervisor to visually inspect the conduit run, and if a seal is not painted the appropriate color, he or she knows that the proper installation on this seal was not

INSIDE TRACK

Sealing Conductors

On a multi-conductor cable traveling between areas where sealoff fittings are required, strip the insulation from the cable in the middle of the fitting and slightly unravel and separate (birdcage) the conductors. This ensures that the sealing compound fully surrounds each conductor to prevent the passage of gas through the seal. See ***NEC Section 501.5(D)(2).*** This may require special consideration when selecting a sealoff fitting.

Pouring Vertical Seals

When pouring a vertical seal that penetrates the top of an enclosure, such as starters or panelboards, open the door while you are pouring the seal to observe whether or not the seal is leaking into the enclosure.

Anti-Seize Compound

When preparing sealing fittings, apply an approved anti-seize compound to the screw threads and interior surfaces of the caps and plugs. This will facilitate removal of the plugs for inspection purposes.

Instructor's Notes:

done; therefore, action can be taken to correct the situation immediately. The sealoff fittings in *Figure 11* are typical of those used. The type in *Figure 11(A)* is for vertical mounting and is provided with a threaded, plugged opening into which the sealing cement is poured. The sealoff in *Figure 11(B)* has an additional plugged opening in the lower hub to facilitate packing fiber around the conductors to form a dam for the sealing cement. *Figure 11(C)* shows a sealoff fitting along with fiber material and sealing compound.

The following guidelines should be observed when preparing sealing compound:

- Use a clean mix vessel for every batch. Particles of previous batches or dirt will spoil the seal.
- Recommended proportions are by volume—usually two parts powder to one part clean water. Slight deviations in these proportions will not affect the results.
- Do not mix more than can be poured in 15 minutes after water is added. Use cold water; warm water increases the setting speed. Stir immediately and thoroughly.
- If the batch starts to set, do not attempt to thin it by adding water or by stirring. Such a procedure will spoil the seal. Discard the partially set material and make a fresh batch. After pouring, close the opening immediately.
- Do not pour compound in sub-freezing temperatures or when these temperatures are likely to occur during curing.
- Ensure that the compound level is in accordance with the instruction sheet for the specific fitting.

Most other explosionproof fittings are provided with threaded hubs for securing the conduit as described previously. Typical fittings include switch and junction boxes, conduit bodies, unions and connectors, flexible couplings, explosionproof lighting fixtures, receptacles, and panelboard and motor starter enclosures. A practical representation of these and other fittings is shown in *Figures 12* through *14*.

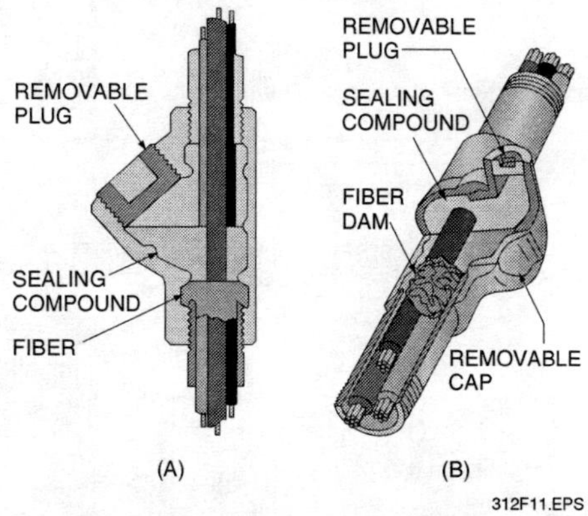

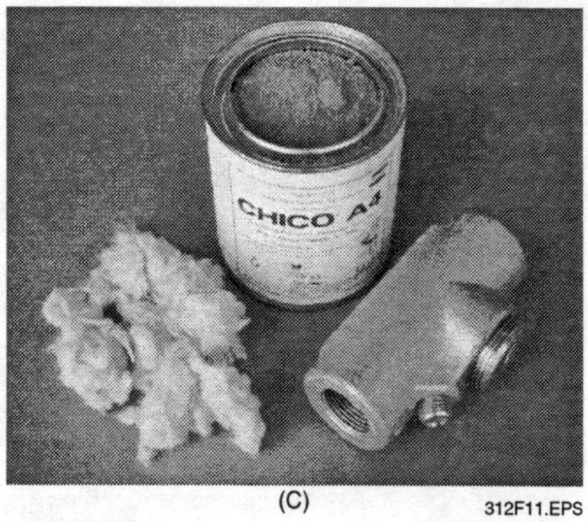

Figure 11 ◆ Seals made with fiber dams and sealing compound.

Classroom

Refer to Figure 11 in the Trainee Module. List the guidelines for pouring an effective seal.

Demonstration

Demonstrate how to install a sealoff fitting and pour the seal.

Laboratory

Have the trainees practice installing sealoff fittings and pouring seals. Note the proficiency of each trainee.

Classroom

Discuss the other types of fittings used in hazardous locations. Where available, provide examples for the trainees to examine.

Audiovisual

Show Transparencies 8 through 10 (Figures 12 through 14).

Teaching Tip

Take the trainees on field trips to one or more hazardous locations and point out the explosionproof equipment and fittings.

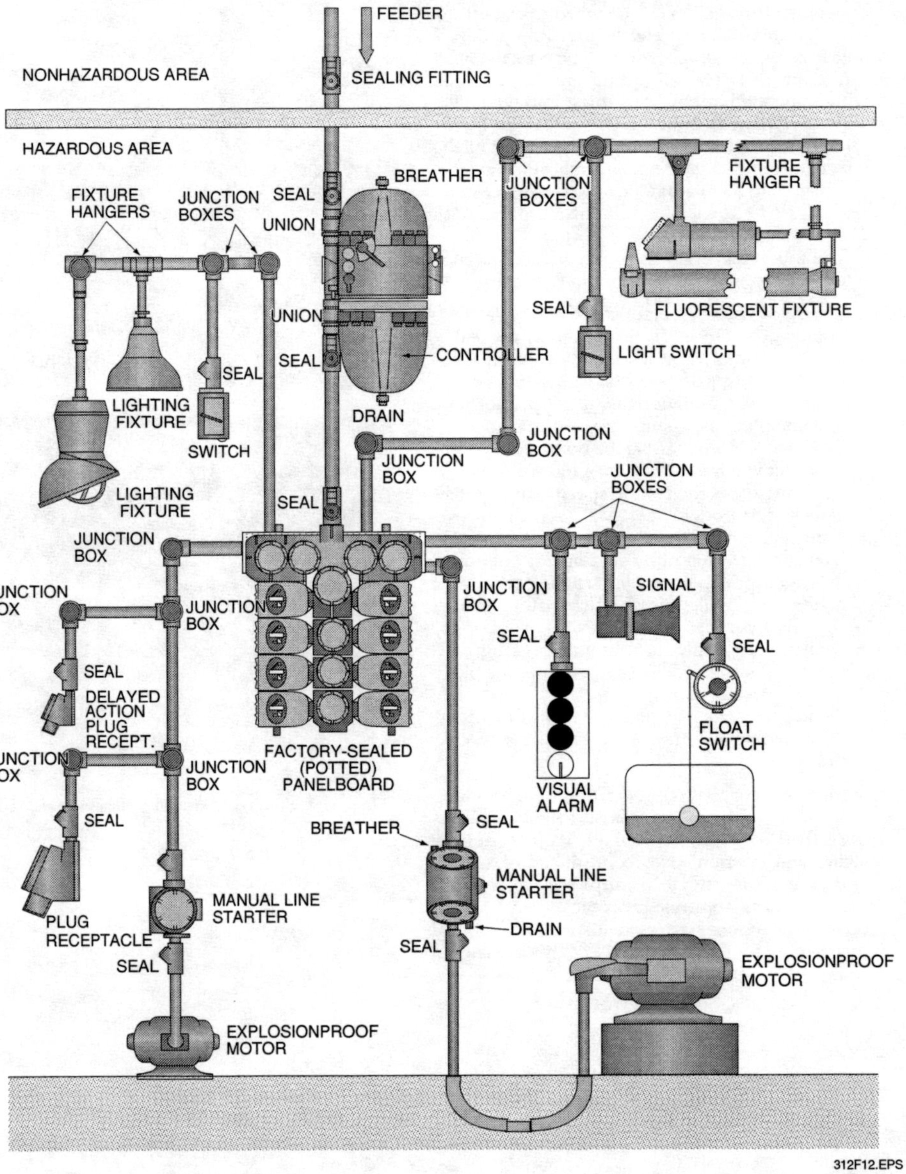

Figure 12 ◆ Class I, Division 1 electrical installation.

Instructor's Notes:

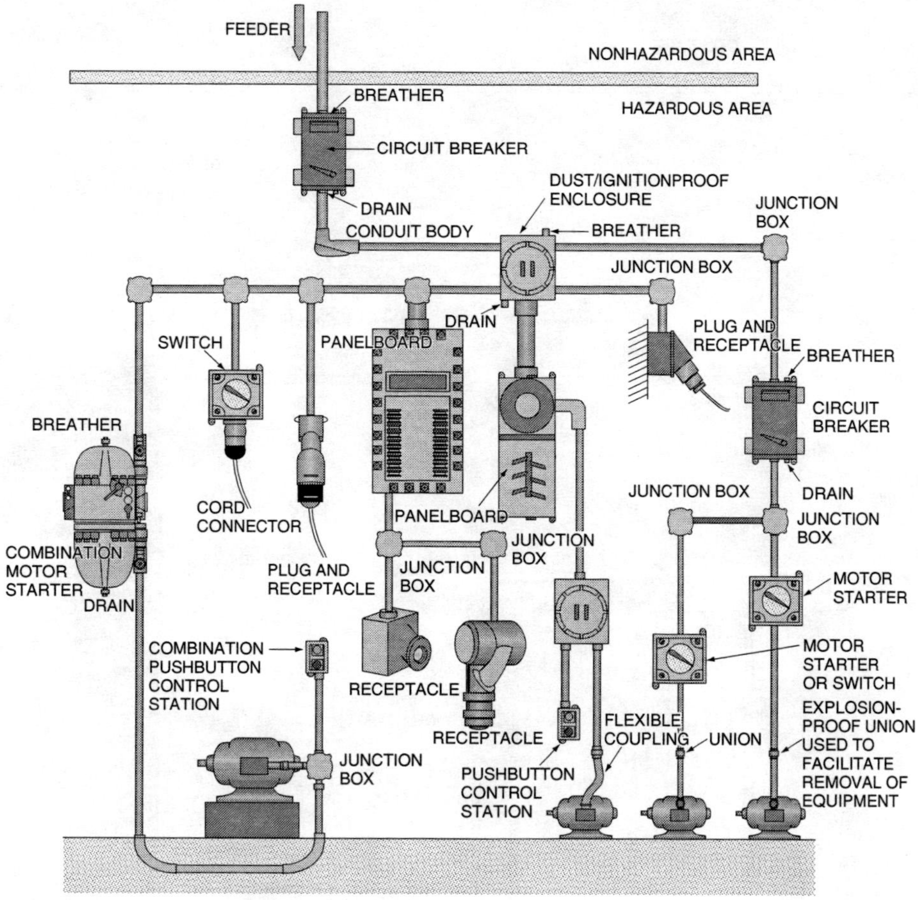

Figure 13 ◆ Class II, Division 1 electrical installation.

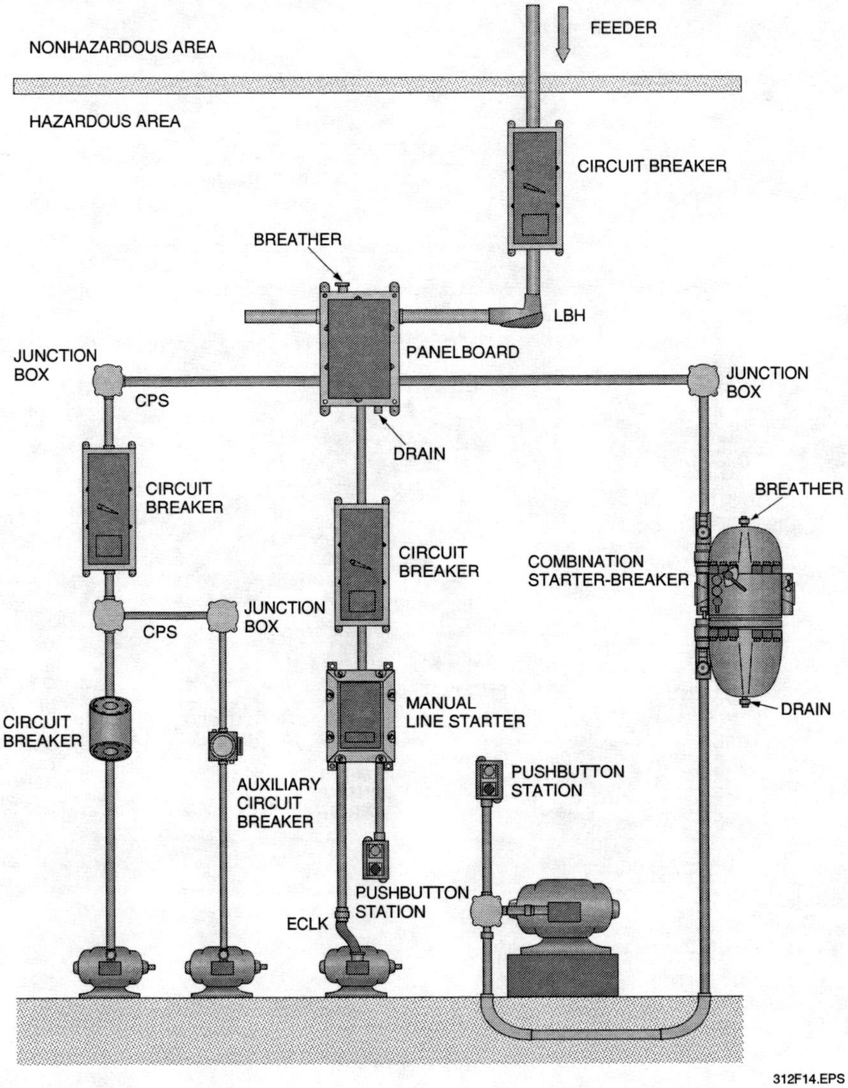

Figure 14 ◆ Class II power installation.

Instructor's Notes:

4.0.0 ◆ GARAGES AND SIMILAR LOCATIONS

Garages and similar locations where volatile or flammable liquids are handled or used as fuel in self-propelled vehicles (including automobiles, buses, trucks, and tractors) are not usually considered critically hazardous locations. However, the entire area up to a level 18" above the floor is considered a Class I, Division 2 location, and certain precautionary measures are required by the NEC. Likewise, any pit or depression below floor level shall be considered a Class I, Division 2 location, and the pit or depression may be judged as a Class I, Division 1 location if it is unvented.

Normal raceway (conduit) and wiring may be used for the wiring method above this hazardous level, except where conditions indicate that the area concerned is more hazardous than usual. In this case, the applicable type of explosionproof wiring may be required.

Approved sealoff fittings should be used on all conduit passing from hazardous areas to nonhazardous areas. The requirements set forth in *NEC Article 501* apply to horizontal as well as vertical boundaries of the defined hazardous areas. Raceways embedded in a masonry floor or buried beneath a floor are considered to be within the hazardous area above the floor if any connections or extensions lead into or through such an area. However, conduit systems terminating to an open raceway in an outdoor unclassified area shall not be required to be sealed between the point at which the conduit leaves the classified location and enters the open raceway.

Figure 15 (see page 12.20) shows a typical automotive service station with applicable NEC requirements. The space in the immediate vicinity of the gasoline dispensing island is denoted as Class I, Division 1. The surrounding area, within a radius of 20' of the island, falls under Class I, Division 2 to a height of 18" above grade. Bulk storage plants for gasoline are subject to comparable restrictions.

NEC Article 514 covers gasoline dispensing and service stations. *NEC Article 511* covers commercial garages.

A summary of NEC rules governing the installation of electrical wiring at and near gasoline dispensing pumps is shown in *Table 7*.

5.0.0 ◆ AIRPORT HANGARS

Buildings used for storing or servicing aircraft in which gasoline, jet fuels, or other volatile flammable liquids or gases are used fall under *NEC Article 513*. In general, any depression below the level of the hangar floor is considered to be a Class I, Division 1 location. The entire area of the hangar, including any adjacent and communicating area not suitably cut off from the hangar, is considered to be a Class I, Division 2 location up to a level of 18" above the floor. The area within 5' horizontally from aircraft power plants, fuel tanks, or structures containing fuel is considered to be a Class I, Division 2 hazardous location; this area extends upward from the floor to a level 5' above the upper surface of wings and engine enclosures.

Adjacent areas in which hazardous vapors are not likely to be released, such as stock rooms and

Table 7 NEC Application Rules for Service Stations

Application	NEC Regulation	NEC Reference
Equipment in hazardous locations	All wiring and components must conform to the rules for Class I locations.	*NEC Section 514.4*
Equipment above hazardous locations	All wiring must conform to the rules for such equipment in commercial garages.	*NEC Section 514.4*
Gasoline dispenser	A disconnecting means must be provided for each circuit leading to or through a dispensing pump to disconnect all voltage sources, including feedback, during periods of service and maintenance. An approved seal (sealoff) is required in each conduit entering or leaving a dispenser.	*NEC Sections 514.7 and 514.13*
Grounding	Metal portions of all noncurrent-carrying parts of dispensers must be effectively grounded.	*NEC Section 514.16*
Underground wiring	Underground wiring installed within 2' of ground level shall be in threaded rigid metal conduit or IMC. If underground wiring is buried 2' or more, rigid nonmetallic conduit may be used along with the types mentioned above; Type MI cable may also be used in some cases.	*NEC Section 514.8*

Classroom

Discuss the NEC requirements for garages and similar locations. Refer the trainees to Table 7 in the Trainee Module and *NEC Article 514*.

Audiovisual

Show Transparency 11 (Figure 15).

Classroom

Discuss the NEC requirements for airport hangars. Refer the trainees to *NEC Article 513*.

Audiovisual

Show Transparency 12 (Figure 16).

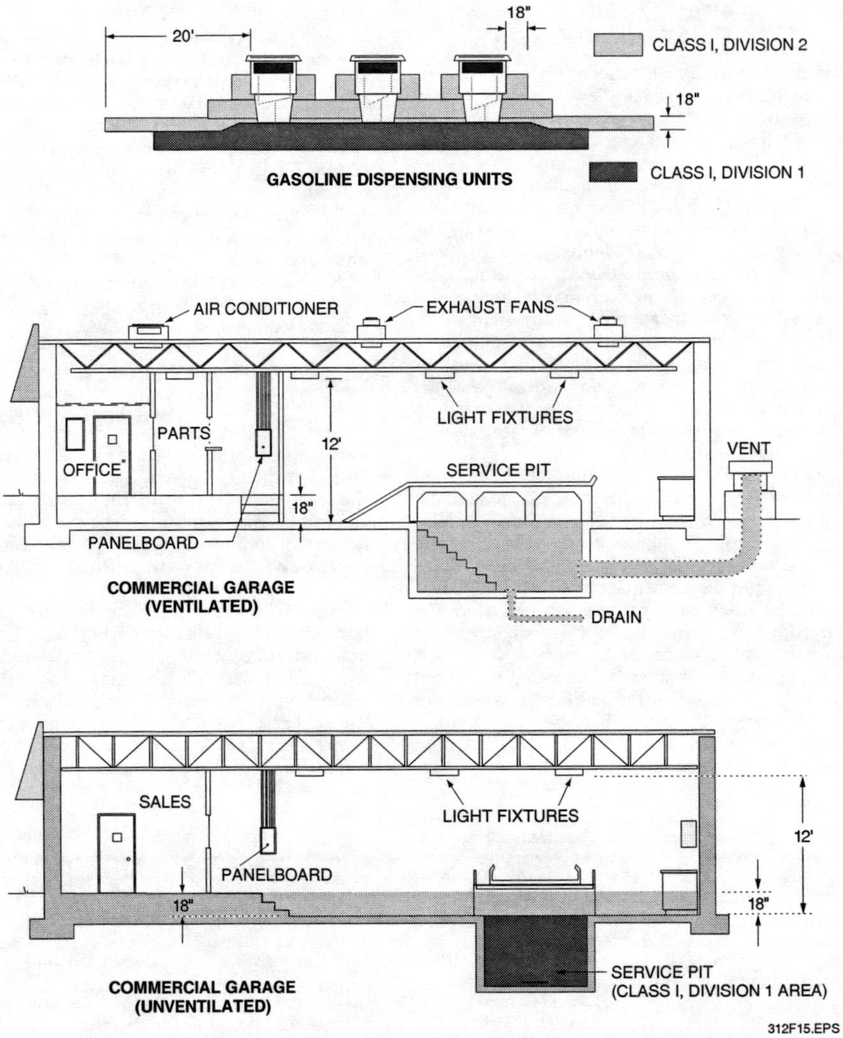

Figure 15 ◆ Commercial service station and garage classifications.

electrical control rooms, should not be classified as hazardous when they are adequately ventilated and effectively cut off from the hangar itself by walls or partitions. All fixed wiring in a hangar not within a hazardous area as defined in *NEC Section 513.3* must be installed in metallic raceways or shall be Type MI, TC, or MC cable; the only exception is wiring in nonhazardous locations as defined in *NEC Section 513.3(D)*, which may be of any type recognized in *NEC Chapter 3*. *Figure 16* summarizes the NEC requirements for airport hangars.

Instructor's Notes:

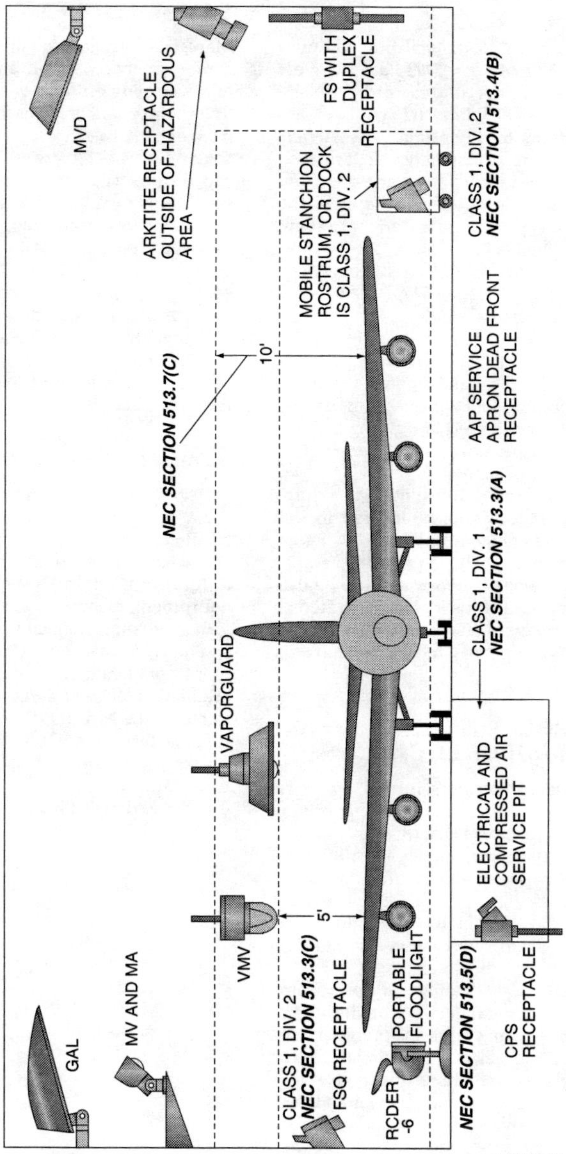

Figure 16 ◆ Sections of an airport hangar showing hazardous locations.

Classroom

Discuss the NEC requirements for hospitals. Refer the trainees to *NEC Article 517.*

Explain that the requirements for petrochemical plants are specific to the particular installation and will be listed in the drawings and specifications for the job.

Point out that equipment manufacturers are an excellent source of application and installation material.

6.0.0 ◆ HOSPITALS

Hospitals and other healthcare facilities fall under *NEC Article 517. NEC Article 517, Part II* covers the general wiring in patient areas of healthcare facilities. *NEC Article 517, Part III* covers essential electrical systems for hospitals. *NEC Article 517, Part IV* gives the performance criteria and wiring methods used in inhalation anesthetizing locations. *NEC Article 517, Part V* covers the requirements for electrical wiring and equipment in X-ray installations. *NEC Article 517, Part VI* covers communications, signaling systems, and fire alarm systems. *NEC Article 517, Part VII* covers isolated power systems.

Anesthetizing locations of hospitals are considered Class I, Division 1 to a height of 5' above the floor. Gas storage rooms are designated as Class I, Division 1 throughout. Most of the wiring in these areas, however, can be limited to lighting fixtures only by locating all switches and other devices outside of the hazardous area.

The NEC recommends that electrical equipment for hazardous locations be located in less hazardous areas wherever possible. It also suggests that by adequate, positive-pressure ventilation from a clean source of outside air, the hazards may be reduced or hazardous locations limited or eliminated. In many cases, the installation of dust-collecting systems can greatly reduce the hazards in a Class II area.

7.0.0 ◆ PETROCHEMICAL HAZARDOUS LOCATIONS

Most manufacturing facilities involving flammable liquids, vapors, or fibers must have their wiring installations conform strictly to the NEC as well as governmental, state, and local ordinances. Therefore, the majority of electrical installations for these facilities are carefully designed by experts in the field—either the plant in-house engineering staff or an independent consulting engineering firm.

Industrial installations dealing with petroleum or some types of chemicals are particularly susceptible to several restrictions involving many governmental agencies. Electrical installations for petrochemical plants will therefore have many pages of electrical drawings and specifications, which require approval from all the agencies involved. Once approved, these drawings and specifications must be followed exactly, because any change whatsoever must once again go through the various agencies for approval.

8.0.0 ◆ MANUFACTURERS' DATA

Manufacturers of explosionproof equipment and fittings expend a lot of time, energy, and expense in developing guidelines and brochures to ensure that their products are used correctly and in accordance with the latest NEC requirements. The many helpful charts, tables, and application guidelines available from manufacturers are invaluable to anyone working on projects involving hazardous locations. Therefore, it is recommended that you obtain as much of this data as possible. Once obtained, study this data thoroughly. Doing so will enhance your qualifications for working in hazardous locations of any type. Manufacturers' data is usually available to qualified personnel at little or no cost and can be obtained from local distributors or directly from the manufacturer.

Summary

Any area in which the atmosphere or a material in the area is such that the arcing of operating electrical contacts, components, and equipment may cause an explosion or fire is considered a hazardous location. In all such cases, explosionproof equipment, raceways, and fittings are used to provide an explosionproof wiring system.

The wide assortment of explosionproof equipment now available makes it possible to provide adequate electrical installations under any of the various hazardous conditions. However, you must be thoroughly familiar with all NEC requirements and know what fittings are available, how to install them properly, and where and when to use the various fittings.

Instructor's Notes:

Review Questions

1. The NEC lists ____ classification(s) of hazardous atmospheres.
   a. one
   b. two
   c. three
   d. four
2. There is/are ____ division(s) for each classification.
   a. one
   b. two
   c. three
   d. four
3. There is/are ____ group(s) listed under Class I, Division 1.
   a. one
   b. two
   c. three
   d. four
4. Which of the following are the groups listed under Class II, Division 1?
   a. A, B, C, and D
   b. E, F, G
   c. H, I, J
   d. L, M, N
5. When installing circuit breakers in Class II, Division 1 hazardous locations, a ____ must be used.
   a. dust/ignitionproof enclosure
   b. dust-tight enclosure
   c. tight metal enclosure with no openings
   d. standard enclosure with several openings
6. When rigid metal conduit is required in hazardous locations, the threads must be cut at ____ taper per foot.
   a. ½"
   b. ¾"
   c. 1"
   d. 1¼"
7. The main purpose of a union in conduit runs is to ____.
   a. facilitate the installation and removal of equipment
   b. ensure grounding continuity
   c. seal the system from flammable gases or vapors
   d. form tighter joints in the system
8. When installing switches or other arc-producing apparatus in Class I, Division 1 locations, within what distance of the switch (or other apparatus) must sealing fittings be installed?
   a. 6"
   b. 12"
   c. 18"
   d. 24"
9. The purpose of packing fiber in a sealing fitting is to ____.
   a. identify the type of seal and the sealing compound to use
   b. prevent any liquids, gases, or vapors from passing through the fitting
   c. prevent flammable vapors from mixing with the sealing compound
   d. provide a dam to contain the sealing compound until it hardens
10. Which of the following best describes the time to pour in the sealing compound in a sealing fitting?
    a. Within five minutes after it is installed in a raceway system
    b. After the conduit system and seals are installed and the conductors and packing fiber have been installed
    c. Prior to installing the packing fiber
    d. After the threaded plug is set tight

Classroom

Have the trainees complete the Review Questions and go over the answers prior to administering the Module Examination.

Examination

Administer the Module Examination. Record the results on Craft Training Report Form 200 and submit the results to the Training Program Sponsor.

Performance Testing

Administer the Performance Test and fill out Performance Profile Sheets for each trainee. If desired, trainee proficiency noted during laboratory sessions may be used to complete the Performance Test. Be sure to record the results on Craft Training Report Form 200 and submit the results to the Training Program Sponsor.

GLOSSARY

Trade Terms Introduced in This Module

Approved: Acceptable to the authority having jurisdiction.

Conduit: A tubular raceway such as electrical metallic tubing (EMT); rigid metal conduit, rigid nonmetallic conduit, etc.

Conduit body: A separate portion of a conduit or tubing system that provides access through removable covers to the interior of the system at a junction of two or more sections of the system or at a terminal point of the system.

Equipment: A general term including material, fittings, devices, appliances, fixtures, apparatus, and the like used as a part of (or in connection with) an electrical installation.

Explosionproof: Designed and constructed to withstand an internal explosion without creating an external explosion or fire.

Explosionproof apparatus: Apparatus enclosed in a case that is capable of withstanding an explosion of a specified gas or vapor that may occur within it; also capable of preventing the ignition of a specified gas or vapor surrounding the enclosure by sparks, flashes, or explosion of the gas or vapor within; which operates at such an external temperature that a surrounding flammable atmosphere will not be ignited thereby.

Hazardous (classified) location: A location in which ignitable vapors, dust, or fibers may cause a fire or explosion.

Sealing compound: The material poured into an electrical fitting to seal and minimize the passage of vapors.

Sealoff fittings: Fittings required in conduit systems to prevent the passage of gases, vapors, or flames from one portion of the electrical installation to another through the conduit. Also referred to as *sealing fittings* or *seals.*

Instructor's Notes:

Additional Resources

This module is intended to present thorough resources for task training. The following reference works are suggested for further study. These are optional materials for continued education rather than for task training.

American Electrician's Handbook. Terrell Croft and Wilfred I. Summers. New York, NY: McGraw-Hill, 1996.

Code Digest. Latest Edition. Syracuse, NY: Crouse-Hinds.

National Electrical Code Handbook. Quincy, MA: National Fire Protection Association, 2002.

ANSWER KEY

Answers to Review Questions

Answer	Section
1. c	1.0.0
2. b	1.0.0
3. d	1.1.0
4. b	1.2.0
5. a	1.4.0; Table 4
6. b	3.2.0
7. a	3.2.0
8. c	3.3.1
9. d	3.3.4
10. b	3.3.4

TEACHING TIPS—MODULE 26312-02

Section 3.0.0 *What's Wrong with This Picture?*

Note the unplugged openings in this explosionproof box.

Section 3.3.1 *Think About It—Fill Requirements*

This is to allow sufficient room to ensure proper application of the sealing compound, per *NEC Section 501.5(C)(6).*

CONTREN™ LEARNING SERIES — USER UPDATES

The NCCER makes every effort to keep these textbooks up-to-date and free of technical errors. We appreciate your help in this process. If you have an idea for improving this textbook, or if you find an error, a typographical mistake, or an inaccuracy in NCCER's Contren™ textbooks, please write us, using this form or a photocopy. Be sure to include the exact module number, page number, a detailed description, and the correction, if applicable. Your input will be brought to the attention of the Technical Review Committee. Thank you for your assistance.

Instructors – If you found that additional materials were necessary in order to teach this module effectively, please let us know so that we may include them in the Equipment/Materials list in the Instructor's Guide.

Write: Curriculum Revision and Development Department
National Center for Construction Education and Research
P.O. Box 141104, Gainesville, FL 32614-1104

Fax: 352-334-0932

E-mail: curriculum@nccer.org

Craft ____________ Module Name ____________

Copyright Date ____________ Module Number ____________ Page Number(s) ____________

Description

____

____

____

____

(Optional) Correction

____

____

____

(Optional) Your Name and Address

____

____

____

Photo Credits

Module 26301-02

Dan Lamphear, 301P0101, 301P0108
Chuck Rogers, 301P0102, 301P0103, 301P0104, 301P0105, 301P0106, 301P0107

Module 26302-02

Dan Lamphear, 302P0204
Mike Powers, 302P0206
Chuck Rogers, 302P0201, 302P0202, 302P0203
Veronica Westfall, 302P0205

Module 26303-02

Ed Cockrell, 303F07
Veronica Westfall, 303P0301

Module 26304-02

Tim Ely, 304P0402
Veronica Westfall, 304P0401, 304P0403, 304F11

Module 26305-02

Ed Cockrell, 305F18
Chuck Rogers, 305P0502, 305F19, 305P0504
Veronica Westfall, 305F06, 305P0501, 305F17, 305P0503

Module 26306-02

Boltswitch, Inc., 306F09
Tim Ely, 306P0605
Federal Pacific, 306F11
GE Industrial Systems, 306F38
Infrared Solutions, Inc., 306F05
Mike Powers, 306P0602, 306P0604
Chuck Rogers, 306F08, 306F39
Gerald Shannon, 306F02
Square D/Schneider Electric, 306F03, 306F04
Veronica Westfall, 306P0601, 306P0603, 306F10, 306F33

Module 26307-02

Dan Lamphear, 307P0701

Module 26308-02

GE Lighting Systems, Inc., 308F02, 308F03, 308F04
Hubbell Lighting, Inc., 308F16, 308F18, 308A01, 308A02, 308A03, 308A04, 308A05
Leviton Mfg. Co., Inc., 308F10
OSRAM Sylvania, 308F06, 308F07, 308P0802, 308F08, 308F09, 308P0803, 308F11, 308F12, 308P0805
Chuck Rogers, 308P0801, 308P0804
Gerald Shannon, 308F19

Module 26309-02

Baldor Electric Co., 309P0901
Walter Johnson, 309F01

Module 26310-02

Baldor Electric Co., 310F01
Electrical Apparatus Service Association, Inc., 310P1001, 310P1002

Module 26311-02

Cutler-Hammer, 311F68
Dwyer Instruments, Inc., 311F44, 311F47 (right), 311F54, 311F55
Federal Pacific, 311F33
Gerald Shannon, 311F01, 311F02, 311F06, 311F09, 311F10, 311F11, 311F12, 311F13, 311F14, 311F15, 311F19, 311F21, 311F23, 311F24, 311F27, 311F28, 311F30, 311F31, 311F32, 311F35, 311F36, 311F40, 311F41, 311F42, 311F47 (left), 311F49, 311F60, 311F67
Square D/Schneider Electric, 311F25
Veronica Westfall, 311F38

Module 26312-02

Dan Lamphear, 312P1203
Ridge Tool Co., 312F05 (top)
Gerald Shannon, 312F09, 312F10
Veronica Westfall, 312P1201, 312P1202, 312F05 (bottom), 312F06, 312F07, 312F08, 312F11

Index